MIDDLE SCHOOL
LIFE SCIENCE

Jefferson County Public Schools
Golden, Colorado

Judy Capra, Principal Author
Harold Pratt, Project Director
District Life Science Teachers

KENDALL/HUNT PUBLISHING COMPANY
2460 Kerper Boulevard P.O. Box 539 Dubuque, Iowa 52004-0539

Project Editor: Martha Hayes
Designer/Art Director: Bonnie Baumann
Illustrators: Scott Bodell, Lyle Miller, Mary Nelson, Seda Sookias-Maurer
Photo Director: Janet Long
Photo Researcher: Jenna Nelson

Cover Photos:

- *Young Girl:* Beermann/Superstock, Inc.
- *Hand Holding Plant:* Mike Mitchell/After-Image
- *Fetus and Placenta:* J. Baker/FPG International
- *Computer Enhanced Skeleton:* Howard J. Sochurek, Inc./The Stock Market
- *Heart:* Al Francekevich, Inc./The Stock Market

"Coyote Control" adapted from *You and the Environment: An Investigative Approach* by Educational Research Council of America. Copyright © 1976 by Educational Research Council of America. Published by Houghton Mifflin Company, Boston. Used by permission of Center for Learning, Rocky River, Ohio.

Opening passage of "A Survivor" from *Defiant Songdog of the West* by Francois Leydet. Copyright 1988. Published by University of Oklahoma Press.

"How Many Coyotes?" adapted from *Project WILD.* Copyright © 1983 by Project WILD. Published by Project WILD, Boulder, Colorado. Adapted with permission of the publisher.

This material is based upon works supported by the National Science Foundation under grant number MDR-8550202. Any opinions, findings, conclusions, or recommendations expressed in this publication are those of the author(s) and do not necessarily reflect the views of the Foundation.

Library of Congress Catalog Card Number: 90–61649

ISBN 0–8403–5098–8

Printed in the United States of America
10 9 8 7 6 5 4 3 2 1

Acknowledgments

STAFF WRITER Kathleen Ranwez, Moore Junior High, Jefferson County, Colorado

CONTRIBUTORS From Jefferson County, Colorado Schools

Carol Bollig	Sharon Close	Jeff Ginsberg	Bonnie Hayes	Connie McLaughlin
Mary Braukman	Bruno DeSimone	Merikay Haggerty	Mary Ann Larsen	Bette Schwartz

FIELD TEST TEACHERS From Jefferson County, Colorado

Ray Atkinson	Jim Ellis	Dennis Karsten	Tim Middle	Erle Swanson
Gary Ballengee	Jan Ensminger	David Kim	Ruth Mitchell	Margaret Suzukida
Linda Behm	Ken Evridge	Frank Krein	Vicki Mitchell	Jerry Trebilcock
Gary Blubaugh	Larry Franca	Bob LeCour	Jim Mundell	Marian True
Gary Buker	Joan Fretz	Walt Loeblein	Barb Nelson	Vicki Weaver
Betsy Cantrall	Dennis Giullian	Gayle LoPiccolo	John Nelson	Jim Whitmore
Bernard Ciarvella	Ralph Hancock	Julie McClellan	Bruce Nicholls	Chris Williams
Peg Coats	Rod Hayes	Sue McCool	Sabina Raab	Lindon Wood
Joe Creber	Bruce Hogue	Mike McGraw	Jim Rothrock	
Kelly Curran	Diderick Iversen	Cathy McIntosh	Wynne Simpson	

From other Colorado School Districts
Dale Brinker, *Douglas County Schools, Parker Junior High*
Lori Bryner, *Adams County School District #12, Westlake Village Junior High*
Donna Dale, *Adams County School District #12, Eastlake Campus*
Margo Kramer, *Boulder Valley Schools, Broomfield Heights Middle School*
Mike Mancini, *Adams County School District #12, Westlake Village Junior High*
Scott Miller, *Boulder Valley Schools, Broomfield Heights Middle School*
Linda Morris, *Sheridan School District #2, Sheridan Middle School*

ADVISORY COMMITTEE
Ronald Anderson, *Professor of Science Education, University of Colorado*
Mary Braukman, *Junior High Science Teacher, Jefferson County Schools*
Faye Hudson, *Middle School Science Teacher, Denver Public Schools*
Pat Patrick, *Senior High Science Teacher, Jefferson County Schools*
Gary Ranck, *Biologist, Regis College, Denver*
John Seiner, *Pediatric Allergist, private practice*
Eva Sujansky, *Pediatric Geneticist, University of Colorado Health Sciences Center*

CONTENT ADVISORS
David Armstrong, *Ecologist, University of Colorado, Boulder*
Sue Brandon, *Elementary Science Resource Specialist, Jefferson County*
Daniel Chiras, *Ecologist, University of Colorado, Denver*
Kurt Cunningham, *Environmental Educator, Colorado State University Extension Services*
Pat DeRosla, *Language Arts Resource Specialist, Jefferson County*
Cindy Fite, *Senior High Science Teacher, Jefferson County*
John Gapter, *Botanist, University of Northern Colorado*
Barbara Gentry, *Junior High Science Teacher, Jefferson County*
Karen Hollweg, *Environmental Educator, Denver Audubon Society*
Susan Loucks-Horsley, *Education Researcher, The NETWORK, Andover, Massachusetts*
Roger Pool, *Secondary Science Resource Specialist, Jefferson County*
Carse Pustmueller, *Ecologist, Colorado Natural Areas Program*
Rick Spitzer, *Naturalist, Rocky Mountain National Park*
Al Swanson, *Secondary Science Teacher, Jefferson County*
Robert Tully, *Wildlife Manager, Colorado Division of Wildlife*

CONTENT REVIEWERS
David Armstrong, *Ecologist, University of Colorado, Boulder*
Roberta Beech, *Pediatrician—Adolescent Medicine, Westside Clinic, Denver, Colorado*
Daniel Chiras, *Ecologist, University of Colorado, Denver*
Amy Cronister, *Genetic Associate, The Children's Hospital, Colorado*
Gordon Guist, *Bioengineer, FMC Corporation, Maine*
Ann Halbower, *Pediatrician, University of Colorado Health Sciences Center*
Joanne Johnson, *Dietician, The Allergy and Respiratory Institute of Colorado*
David Kessel, *Pediatrician—Sports Medicine, private practice*
Ted Ning, *Urologist, private practice*
Marylou Rottman, *Botanist, University of Colorado, Denver*
Sharon Schilling, *Health Educator, Denver Public Schools*
Reginald Washington, *Pediatric Cardiologist, The Children's Hospital, Colorado*
Jan Watson, *Obstetrician, University of Colorado Health Sciences Center*

CONTENTS

MIDDLE SCHOOL

LIFE SCIENCE

MESSAGE TO STUDENTS

Welcome to Life Science! You are about to start a year-long study of life. In this course, you will be focusing on your own body and how it works and you will be considering your place in the environment. All the information and skills you learn here will help you make wiser decisions. By reading this page before you start, you will better understand the design and purpose of this program.

Take time to flip through the pages of this book. You will discover ways that it does focus on your role in the environment, on your body and how it works, and on how to take care of yourself. Notice that you will learn by doing many activities, recording your findings, and analyzing your own data. The readings will help you understand the activities you have done. While working, you will be learning more than facts about science. You will be *doing* science. You will find out how science is used in your everyday life and you will be encouraged to use your newly acquired knowledge to make decisions. More than anything else, the purpose of this course is to help you learn to think. Keep in mind that if, after thinking, you still don't understand something, it is all right to *ask questions!* Asking questions is the basis of all science.

Have you ever thought about where your textbooks come from? Most people don't. Sometimes, however, it's helpful to know what went into developing a program. In the case of this text, no one person sat down and thought, "I'm going to write a Life Science book." Instead this was a group effort involving many people—scientists, teachers, parents, other educators, and, of course, students. Students suggested topics, gave us their papers to read, and turned in their questions. They also tested all the activities and experiments. Based on their trials, the activities were rewritten and rewritten until we got them right. Finally, after five years of revising, this book came together.

Anytime during the coming year, if you are struck by an idea about how to improve this program, share it with us. Write down your thoughts and send them on to the publisher. (The address is at the bottom of the title page.) The publisher will forward your letter to us. We would love to hear from you. Your input could make the next edition of this book even better.

Good luck! We hope that you will enjoy using this book as much as we have enjoyed writing it for you.

Judy Capra and Harold Pratt

UNIT 1

ECOSYSTEMS AND ECOLOGY

CHAPTER 1

INTRODUCTION TO ECOSYSTEMS

How would you describe this location? Is there a site like it near you?

1-1 YOU AND YOUR ENVIRONMENT

Where in North America do you live? Along the Mississippi River? Within sight of an ocean? If not by these, where? Your answer will determine what you study in this unit. If you live in the deserts of Nevada, for instance, you don't need to spend much time studying the characteristics of oaks or animals that live in cypress swamps. If you live in Florida, however, studying cypress swamps may be very appropriate. And anyone who lives in Massachusetts would probably want to know something about oaks.

Figure 1-1 How would you describe each of these environments? Top left: Miami, Florida. Bottom left: Laguna Pueblo, New Mexico. Above right: Ouray, Colorado.

Figure 1–2 The United States and southern Canada

Now look at Figure 1–2 and find where you live. Then locate this area in Figure 1–3 on page 9. What is used to illustrate your state in the map in Figure 1–3? Think about what is important about your state and what was chosen to illustrate it. If you had drawn this map, what would you have used to represent your state?

Every area has its own characteristics that make it unique. Some of these are a result of human effort—roads, buildings, bridges, and reservoirs, for example. Others, however, are part of the natural environment. A particular combination of climate, plants, animals, and land forms make each area unique.

Think about your environment for a minute—the one in which you live. Try to think of at least 10 characteristics that describe it but that are not likely to fit any other environment.

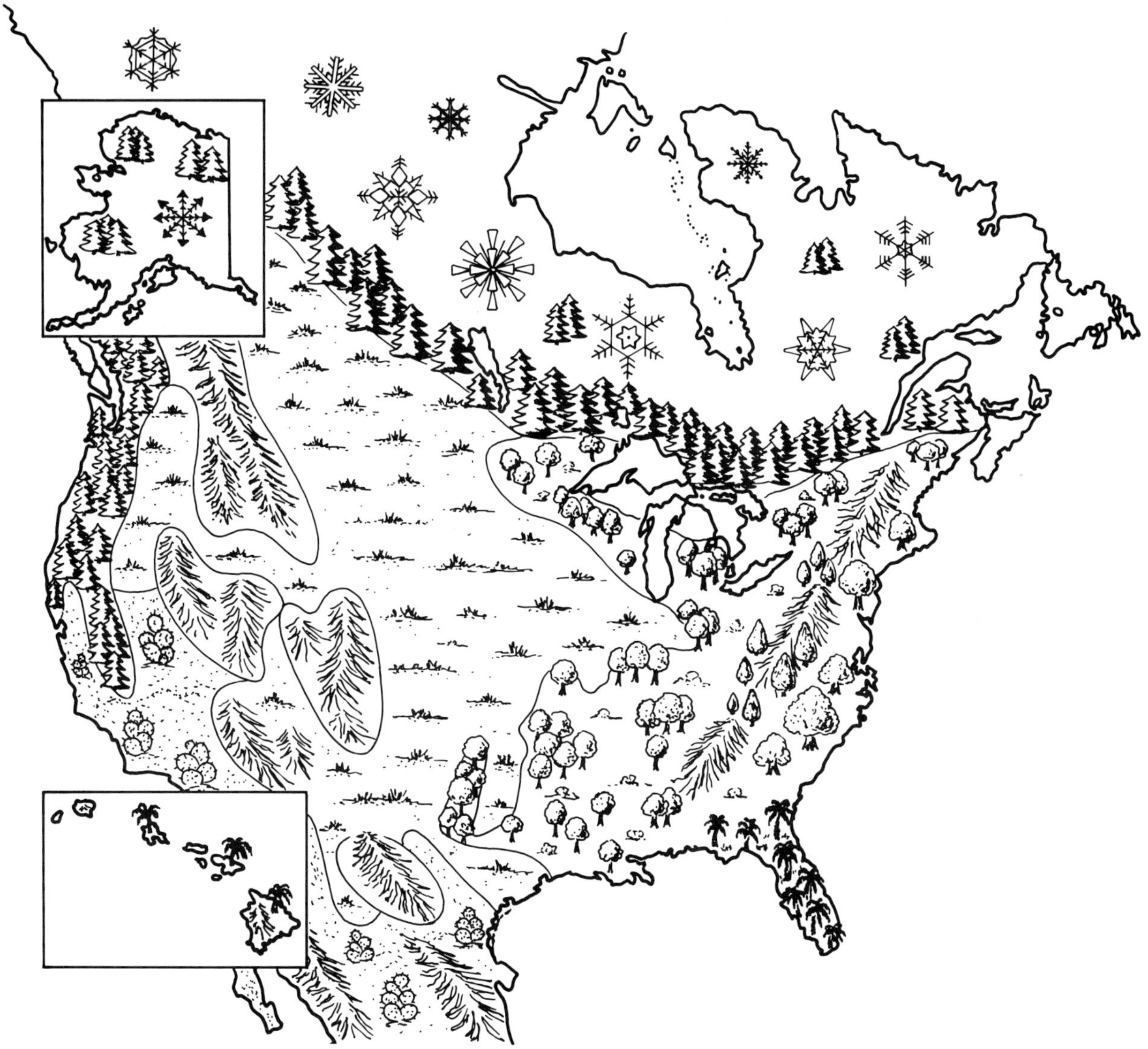

Figure 1–3 Different parts of the country have different environmental characteristics. What are the characteristics of your area?

How easy was it for you to list 10 characteristics? What kinds of things did you list—buildings, plants, weather patterns? You may have found this task somewhat difficult since we often don't notice what's around us. For example, think about your trip to school today. What did you see of your environment? How cloudy was it? How many birds flew overhead? Were any new buildings going up or old ones being torn down?

If you were aware of your surroundings, you are on your way to becoming an ecologist . . . or at least an aware citizen. The first step in learning about your environment is learning how to observe. Of course, it's not enough just to observe; you also need to think about what you see. Observing and thinking—these are two important skills that you will be working on throughout this unit.

1-2 A SITE STUDY

You may live in a large city, in a small town, or in the middle of "nowhere." However, no matter where you live, you are part of an environment. In order to understand this environment, you need to be a good observer. How observant are you? When you're outside, what kinds of things do you notice?

- How many new observations can you make about a specific study site?

You will be able to answer this question after you have done a brief study of a nearby site.

MATERIALS

map of study area
colored pencils

PROCEDURE

Work with a partner.

Part A. Getting Ready

1. With your partner, quickly look over this activity so you have a general idea of what you will be doing.
2. Read through the items listed in the chart on page 11. These are the things you will be locating and placing on a map of the study area. If there is something you don't understand, ask your teacher about it before you go outside.
3. Get your materials together. Also, take your textbook and a lead pencil outside with you.

Part B. Working Outside

You have only 20 minutes to complete this part of the activity so work quickly and carefully.

1. Imagine that you are seeing this area for the first time. Look around. How would you describe the neighborhood? Write your description along the bottom of the map.
2. Check your map for accuracy.
   a. Look around to make sure you see the items that are indicated on the map.
   b. If you observe other items in the area that you think would make the map more useful, go ahead and add them.
3. Now, on your map, locate and mark the items listed in the chart.
   a. Use your lead pencil so you can make changes if you need to.
   b. Plan to spend the last few minutes of your time adding color to the map as explained below.
4. Using your colored pencils, do the following:
   a. Put a *blue* star by the most attractive spot.
   b. Using *purple*, circle the most puzzling thing you noticed.
   c. With *yellow*, underline the one thing you noticed that you don't think anyone else will notice.
   d. Use *orange* to circle 3 things that you never noticed before.

Items to Locate on Your Map

the most attractive spot
the least attractive spot

the wettest spot
the driest spot

a spot that gets sun all day
a spot that is shaded all day

a plant smaller than a dime
the largest plant

the smallest animal
the largest animal

the most abundant animal
a one-of-a kind animal

the noisiest spot
the quietest spot

the most abundant plant
a one-of-a-kind plant

something you notice that no one else will notice

the most puzzling thing you notice

a 1-meter square area that has more than 10 kinds of plants
a 1-meter square area that has the fewest kinds of plants

a 1-meter square area that has more than 5 kinds of animals
a 1-meter square area that has the fewest kinds of animals

3 things you noticed today that you never saw before

evidence of pollution—air, water, and/or litter

5. Now imagine that you do not know what season of the year it is. On the back of your map, list at least 5 clues you could use to help you figure it out. Look around! There are lots of clues.

Part C. Thinking About Your Observations
Once you are back in the classroom, answer the analysis questions.

ANALYSIS QUESTIONS

1. Of all the things you found and put on your map, which one was hardest to locate? Why?
2. Think about the things you observed and write 3 sentences that begin, "I wonder why . . ." (For example, "I wonder why the grass outside the math room is so much greener that the rest of the lawn.")
3. Imagine that you are going to plant some cattails, a cactus, and some dandelions somewhere on the site you surveyed.
   a. Which one do you think would grow the best? Why?
   b. Which one(s) do you think would not survive? Why?
4. Imagine that the following animals are going to be released onto the site you surveyed.

a sparrow	a frog	a coyote
an earthworm	a bee	a mouse

   a. Which one(s) do you think could survive on the site? Why?

b. Which one(s) do you think would die? Why?

5. If you were on the planning committee to make your school grounds more attractive, what 3 things would you suggest?

CONCLUSION

How good do you think your observational skills were? Write a few sentences to answer this question and the focusing question on page 10.

EXTENSION ACTIVITIES

1. Do a temperature study. Use a thermometer to find the hottest and coolest spots on your study site. Plan your experiment and then do it. When you are finished, ask yourself why one spot is hotter than others and why one spot is colder than others.

2. Native American children were raised to be aware of nature. If you are interested in learning about their education, check your library for books such as *Indian Boyhood* by Charles Eastman or *Prettyshield, Medicine Woman of the Crows* by Frank B. Linderman. As you read, think about what aspects of their lives you would have enjoyed.

3. Read *INDIAN BOYHOOD* on page 13. While you do so, keep these questions in mind:

   How would you describe Hakadah's schooling?

   Would you have enjoyed this type of education?

   Think about what you've observed today. How many different kinds of birds did you see? Were there more clouds in the east or west? When did the ants start to work this morning?

INDIAN BOYHOOD—Learning to Observe

It seems to be a popular idea that all the characteristic skill of the Indian is instinctive and hereditary. This is a mistake. . . . My uncle, who educated me up to the age of fifteen years, was a strict disciplinarian and a good teacher. When I left the tepee in the morning, he would say: "Hakadah, look closely to everything you see"; and at evening, on my return, he used to quiz me for an hour or so.

"On which side of the trees is the lighter colored bark? On which side do they have the most regular branches?"

It was his custom to let me name all the new birds that I had seen during the day. I would name them according to the color or the shape of the bill or their song or the appearance and locality of the nest—in fact, anything about the bird that impressed me as characteristic. I made many ridiculous errors, I must admit. He then usually informed me of the correct name. . . .

He went much deeper into this science when I was a little older, that is, about the age of eight or nine years. He would say, for instance: "How do you know there are fish in yonder lake?"

"Because they jump out of the water for flies at mid-day."

He would smile at my prompt but superficial reply. "What do you think of the little pebbles grouped together under the shallow water? . . . Where do you find the fish-eating birds? Have the inlet and the outlet of a lake anything to do with the question?"

He did not expect a correct reply at once to all the questions that he put to me on these occasions, but he meant to make me observant and a good student of nature.

from *Indian Boyhood* by Charles Eastman (Hakadah). Originally published by McClure, Phillips & Company in 1902.

1-3 ECOSYSTEMS

When you describe your environment, you may comment about the local plants, the animals, and the land characteristics. By doing so, you are also describing your ecosystem. The term **ecosystem** refers to everything living and non-living in a given area. Your schoolyard could be thought of as an ecosystem; so could a city park, a person's backyard, a small town, or a large city. However, it is more usual to think of natural ecosystems. For instance, a freshwater pond is an ecosystem; so is a mountain forest, a marsh, a prairie grassland, or even a corn-field. As you might guess, there are hundreds of different ecosystems and each one has its own unique set of characteristics.

Even though ecosystems may be very different from each other, they all are made up of the same four components. These are the **producers,** the **consumers,** the **decomposers,** and the **abiotic factors.**

The Living Components of An Ecosystem

The producers, consumers, and decomposers are the living components of an ecosystem. Since they are living, they need food in order to survive. Food provides the nutrients necessary for growth and the energy for life processes. Where an organism gets its food determines in which group it belongs. For example, the **producers** do not eat; they produce their own food. They do this by a process known as *photosynthesis.* All green plants are producers.

Figure 1-4 Producers, consumers, and decomposers are found in every ecosystem.

However, many living organisms cannot produce their own food; they must eat in order to get the nutrients and energy they need for survival. These organisms are called **consumers.** All animals are consumers. No animal can live without eating.

The **decomposers** cannot produce their own food and yet they do not eat either. They get their energy from dead plant and animal materials. Decomposers live off such things as fallen leaves, dead insects, and mammal carcasses. The decomposers are the fungi,

Figure 1–5 What can you tell about the abiotics of this area by looking at this picture?

bacteria, and yeast. If it weren't for decomposers, we would be living on stacks of dead leaves, old trees, animal bodies, and insects shells.

The Nonliving Component of An Ecosystem

The nonliving part of an ecosystem determines which organisms can live there. These nonliving, or **abiotic,** factors can vary considerably from one ecosystem to another. For instance, rainfall is an example of an abiotic factor. An area that receives lots of rain may be a rainforest. An area that only receives a few inches of moisture is likely to be a desert. Other examples of abiotic factors include soil type, altitude, and day length. (See Figure 1–5.)

Producers. Consumers. Decomposers. Abiotic factors. Throughout the rest of this unit you will be studying these four components and seeing how they interact in your own ecosystem.

The Local Ecosystem

1. With a partner, use a sheet of paper to make a copy of Figure 1–6 on page 16. Remember to include the arrows and labels on your drawing.

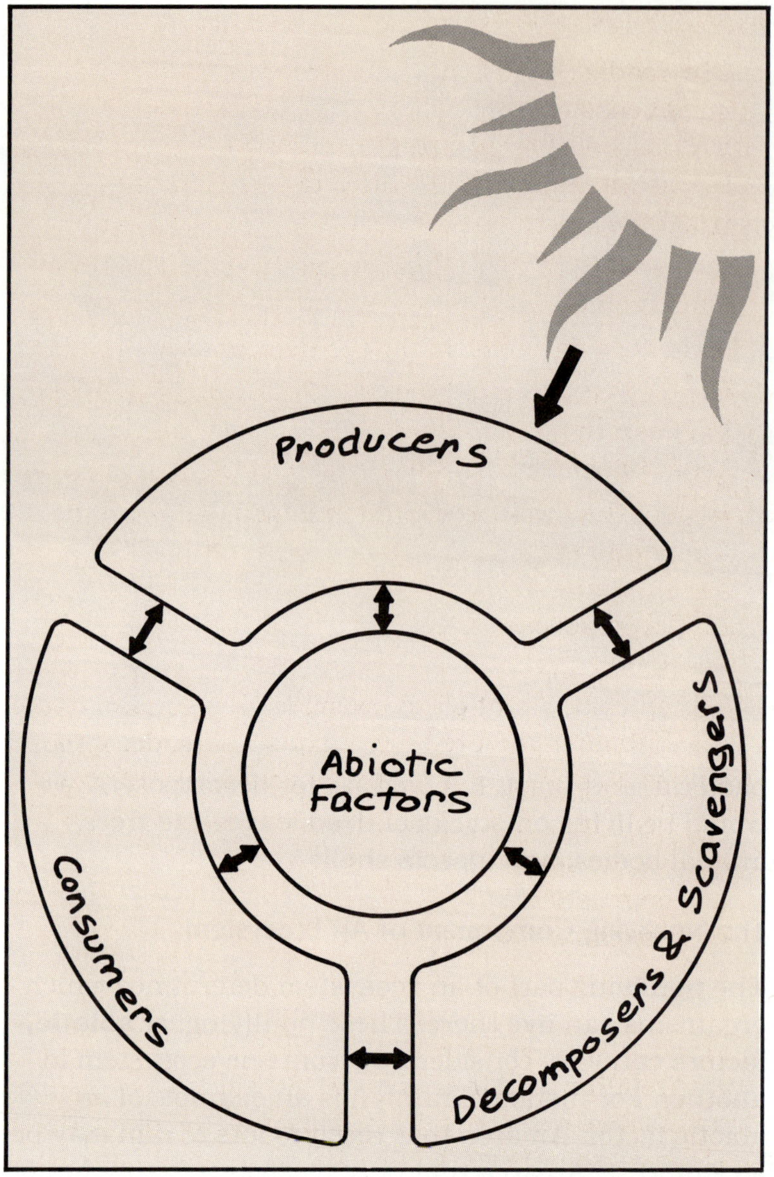

Figure 1–6

2. Now think about your own ecosystem and what you observed during your site study.
   a. Think of 8 consumers that live there. List them in the correct section.
   b. Think of 8 producers that live there. Add them to your drawing.
   c. Put examples of decomposers on the drawing.
   d. Use the center to describe the abiotic conditions of your ecosystem.
3. Color your drawing to make it more interesting and to help you remember the four parts of an ecosystem.

ANALYSIS QUESTIONS

1. The reading says that there are hundreds of different ecosystems.
   a. List 4 of the ones that are named in the reading.
   b. List at least 4 other kinds of ecosystems.
2. What are the 4 parts that make up every ecosystem? (Don't look them up yet! Try to list them from memory first.)
3. Why does an ecosystem need decomposers?
4. On your drawing of Figure 1–6 you made an arrow between the producer and decomposer sections. What do you think that arrow means?
5. If all of the producers suddenly died, what would happen to the
   a. consumers?
   b. decomposers?
   c. abiotics?
6. Which component of an ecosystem do you know the most about—the producers, consumers, decomposers, or abiotic factors? Why do you think this is so?
7. Over the last 300 years, which component of your ecosystem do you think has changed
   a. the most? Why?
   b. the least? Why?

EXTENSION ACTIVITY

Look through magazines and cut out several scenic pictures. List the examples of producers, consumers, decomposers, and abiotic factors that are shown in each picture. Where in the world do you think these places are located?

CHAPTER 2

PRODUCERS

Where could you stand in this place and see the greatest number of different kinds of plants?

2–1 FOCUS ON PLANTS

When you want to learn about an ecosystem, it makes sense to start with the producers. Unlike animals, plants don't get up and run away when you try to observe them and they "wait" while you make any necessary measurements. It's also convenient to be able to go back and find the same plant in the same spot in case you want to study it again.

The best place to start studying producers is in your own neighborhood. Even though you see these plants every day and may take them for granted, they can teach you a lot about your ecosystem.

- How many local plants can you recognize?
- What characteristics do you use when you identify plants?

Keep these questions in mind while you go outside on a "plant hunt."

MATERIALS

worksheet: ***Focus on Plants*** (Figure 2–1)
colored pencils

PROCEDURE

Work with a partner.

1. Read all of this procedure and then look over the worksheet so you know what you will be doing.
2. Plan your time. You will have 20–25 minutes to look at the plants outside and complete the worksheet.
3. Get your materials together. You may want to bring along a book or some other firm surface on which to write.
4. When you have finished the worksheet, remember to look back at the focusing questions and write your conclusion. Also, read the extension activity and decide if you would like to do it.

CONCLUSION

Write a list of at least 10 characteristics that you can use when identifying plants.

EXTENSION ACTIVITY

Look around your neighborhood for an area that has a variety of plants. Try to find someplace that has at least 20 kinds of plants in an area one meter square. When you find such a place, look around and ask yourself, "Why can so many different kinds of plants grow here?"

FOCUS ON PLANTS

NAMES ______________________

DATE ______________ PER. ______

1. Look around. Decide which plant is the most abundant.
   a. Sketch a leaf from this plant in the box at the right.
   b. How tall is the largest specimen of this plant that you can find? ______________
   c. Is there a flower on this plant? ________ If not, do you think this plant can produce flowers? ________ Explain your answer.
   d. Where on the site do you think this plant is growing best?
   e. Think of a way to estimate the number of these plants on this study site. Explain your method here.

      Using this method, how many of these plants do you think are on this study site? ________
   f. Would this plant be here if there were no people? ________ Explain your answer.
   g. Do you consider this plant to be a weed? ________ Explain.
2. Find one example of a plant, or a part of a plant, that has been eaten by an animal.
   a. In the box at the right, sketch the part of the plant that has been used for food.
   b. What do you think could have been eating this plant? (List at least 2 organisms.)
   c. Look at the plant next to the one you sketched. Has it been eaten, too? ________
   d. Find another plant like this second one. Does it look like it has been chewed on? ________

Figure 2–1 (Worksheet, page 1)

FOCUS ON PLANTS
(PAGE 2)

NAMES ______________________

3. Find a dead plant.
   a. Describe or sketch it.
   b. How do you know it is dead?
   c. List at least 3 reasons why it might have died.
4. Try to name at least 10 plants found on this study site.
   a. List them here.

      ____________ ____________ ____________
      ____________ ____________ ____________
      ____________ ____________ ____________
      ____________ ____________ ____________
   b. Put *blue stars* by the plants that you think are weeds.
   c. Put *red checks* by the ones that you think someone planted.
   d. Put *green circles* around the ones that you think might have been found here three hundred years ago.
5. Think of these plants from another point of view.
   a. Which would you like best if you were a

      woodpecker? ______________

      honeybee? ______________

      cow? ______________

      gardener? ______________
   b. Which one do you like best? ______________ Why?

Figure 2–1 (Worksheet, page 2)

Even in the City

Each city has its characteristic life forms. Whether you are walking on a city street or watching the city blocks pass in front of you on a television screen, you will be able to see plant life. Look closely. Around the edges of parking lots, in the cracks of sidewalks, and high on the walls of buildings, stubborn green leaves reach for the sun. Why do you think some plants can survive in the city while so many others cannot?

Figure 2–2 If you wanted to find wild plants in these cities, where would you look?

2-2 NATIVE PLANTS

Today few people can recognize plants that are native to their environment. However, knowing about such plants made it possible for the early Native Americans to survive. All Indians could learn about a new area by "reading the clues" they got from the plants. To some extent, the same is true today . The more you know about native plants, the better you will understand your environment.

- What are four examples of plants that are native to your area?
- What can these plants tell you about the environment?

Listen to a slide talk about your native plants to learn the answers to these questions.

PROCEDURE

1. Make a chart like the one at the bottom of the page.
2. Fill in your chart while you watch the slides. Take neat, complete notes so they will be helpful to you later.
3. When the slide show is over, look back over your notes. For each plant, ask yourself these questions:
   - Could I identify this plant if I saw it? What characteristics would help me recognize it?
   - What is the most interesting fact about this plant?
   - What growing conditions does this plant prefer?
4. Answer the analysis questions.

ANALYSIS QUESTIONS

1. Which plant that is native to your area do you think is most interesting? Why?
2. List 5 locations in your area where you think you are most likely to find native plants.
3. Which native plants do you think you could identify?

CONCLUSION

When you write a conclusion for this activity, make sure you answer the two focusing questions above. Also include a few sentences that tell how you felt about this activity.

Plant Name	Identifying Characteristics	Growing Conditions	Other Information

EXTENSION ACTIVITY

Find out more about one of your native plants. Look it up in several plant identification books. Call places where there might be someone who can give you more information about the plant. Try places such as a museum, plant club, botanic garden, or the biology department of a college. Use what you learn to prepare a poster for your classroom. Hang the poster so others will be able to learn more about this plant, too.

Not a Native

You may think of the hardy dandelion as being native to your area. Actually, this plant did not originate in America; it was brought in from Europe. How did dandelions get here? No one knows. They may have been planted intentionally since some people consider dandelions useful food plants. However, it is more likely that the dandelion seeds were uninvited "guests" in some unsuspecting immigrant's baggage.

2-3 WHAT'S A NATIVE?

Look at the plants pictured in Figure 2–3 below and Figure 2–4 on page 25. Do either of these grow in your neighborhood? If so, you must live in one of the two locations shown on the maps. These two plants, the saguaro cactus and the alpine sunflower, have such specific growing needs that they cannot grow anywhere else.

Two Specialized Plants

The hot, dry deserts of Arizona are home to saguaro cacti. Imposing and sturdy as this plant may appear, it is actually quite fragile. It has very specific light, temperature, soil, and water needs. If you change any of these abiotic factors even slightly, the plant may die.

Figure 2–3 A saguaro grows about 2 cm per year. Do you think these plants are very old? The map shows where saguaro cacti grow.

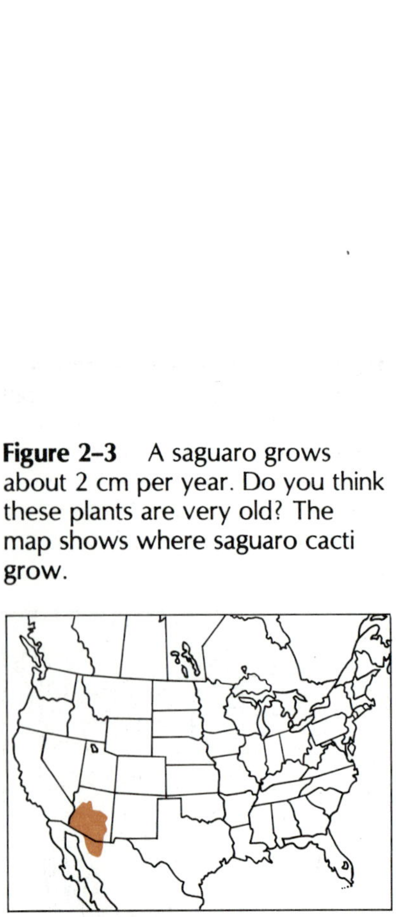

These needs are obvious when the saguaro is still a seed. The Sonoran desert in Arizona receives the exact amounts of summer rains and warmth needed for the seeds to sprout. The California deserts won't do—they get their rains in the winter when temperatures are too low to allow the seed to grow.

When the seedling is less than an inch high, it grows two long, sharp spines. If it weren't for this protection, the cactus seedling would immediately become a meal for a hungry animal. These long spines will continue to protect each saguaro cactus throughout its life.

There are many other features that make this plant well-suited to its desert life. One of these is its ability to store large quantities of water in its heavy stalk. During long dry spells, it can use its water reserves. The long ridges you see on the outside of the cactus are the clue to how the saguaro holds water. These ridges can flatten out when the cactus is storing water. Then, as the water is gradually used up, the ridges and grooves become deeper and the cactus body "closes" back down. This is similar to the way an accordian works when it opens and closes.

In contrast, the alpine sunflower could never survive a desert summer. Instead, look for these perky plants above treeline on tall mountains. They thrive in the harsh, cold weather found at high altitudes. Give them desert temperatures and they wither and die.

Figure 2–4 The alpine sunflower grows on the tundra. Its leaves look gray-green because of the small hairlike structures that cover them. What do you think could be the purpose of these hairs? The map shows where alpine sunflowers grow.

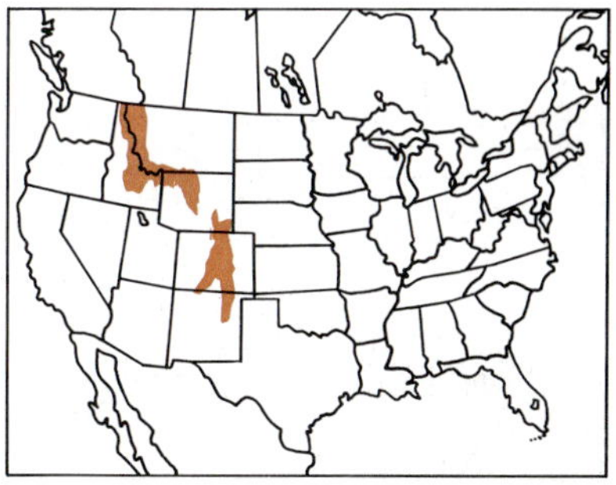

The blasts of winds at high altitudes are so strong that a tall plant would soon be blown over. Notice the short stalk of the alpine sunflower in Figure 2–4 on page 25.

The alpine tundra has a short growing season; most of the year its temperatures are below freezing. Thus, there is not enough time in the summer for this plant to complete its life cycle. Instead, for several years, it puts out only leaves. Eventually it manages to store enough food so it can emerge, put out leaves and a flower, and produce seed.

Native Plants

The saguaro and alpine sunflower are very characteristic of their environments. When you see one of them, you immediately know the abiotic conditions of that area. Not many plants are this specialized. However, all ecosystems have characteristic plants. You might want to think of them as the natives; they were here long before the European settlers arrived. Native plants have evolved right along with the environment.

However, when land is settled, things change. Trees and shrubs are cut; meadows and prairies are plowed. Buildings spring up; boardwalks, sidewalks, and parking lots appear. Dams are built; water flow changes; fields are irrigated. Pollutants build up in the air, on land, and in the water. Sometimes native plants cannot survive these changes or they may survive but only in isolated spots. Look for them along railroad tracks, in the corners of a cemetery, or in marshy spots.

Figure 2–5 The 2.5 cm (1 inch) white or pinkish flowers help identify field bindweed. How would you describe the shape of its leaves?

Non-Natives

If you go looking for the natives, you're likely to find "new natives." They're the ones that have moved in and thrived. They may have "escaped" from someone's garden, been planted along with seed for some crop, or blown in on the wind. In any case, they have been introduced into the ecosystem. These plants are not usually as picky about their growing conditions as the natives so they are more widespread. Sometimes these plants grow so well that they crowd out the natives. When this happens, they're likely to be called weeds.

The bindweed—also called wild morning glory—pictured in Figure 2–5 is an example of a pesky newcomer. Its viny stems creep along the ground, growing around everything it encounters. Farmers don't like it because it outgrows the crop plants and can even smother them. Gardeners have the same problems with bindweed. It wraps around fences and roses and tomatoes and beans.

Bindweed's long roots grow out in all directions. They can store up to three years of food. That means that even if a small piece of root is left behind when the plant is pulled, it will be able to grow again. Once the bindweed takes hold, it is just about impossible to get rid of. This plant is found throughout the United States and the world. If there are disturbed, bare patches of soil anywhere, bindweed seeds are likely to take hold.

When you are out looking at plants, you will see cultivated plants in addition to the natives and newcomers. They are plants that were planted on purpose and are being tended. They wouldn't be growing there on their own. They only survive because someone is taking care of them. (They may need more water or shade than is typical for the area or perhaps they need people to keep them bug-free.) In any case, without people they might be able to survive for a while; but probably they would eventually die.

"Reading" the Environment

If you know some of the plants in your ecosystem—even if it's only a few of them, you can start to learn "to read the environment." You will realize that an area has not been disturbed much if there are lots of natives growing on it. Looking at the natives will also help you make predictions about the local growing conditions.

ANALYSIS QUESTIONS

1. You read about two plants whose characteristics fit their particular environments.
   a. What is one of the saguaro's characteristics that makes it able to survive long, dry spells?
   b. What characteristic of the alpine sunflower makes it able to survive the strong tundra winds?
2. What does the term *native plant* mean?
3. The reading says that people change the land when they arrive.

a. List 3 of the ways that are mentioned in the reading.
b. List 2 other ways that people affect the environment.
c. Explain one thing you have done today that affects the environment.

4. Here are several definitions of a weed. Read them and then explain which one you think is best.

   A weed is

   a troublesome plant.

   an ugly plant.

   a plant that is growing where you don't want it.

   a plant that we haven't learned to appreciate.
5. List at least 3 advantages to having weeds around.
6. Think about the plants that you saw on your study site.
   a. Which kind was the most abundant?
   b. Do you think it is a native plant, an introduced plant, or a cultivated plant?
7. Try to name
   a. 3 examples of weeds.
   b. 3 examples of cultivated plants.
   c. 3 examples of natives.
8. If you wanted to find plants that are native to your area, where would you look?

TAKE A STAND

Poaching is a problem in the Sonoran desert. Poachers steal the saguaros to sell to gardeners. Since these plants are so fragile, poachers should be heavily fined or jailed for at least 5 years.

- Do you **agree** or **disagree**?
- What **comments** would you like to make?

EXTENSION ACTIVITIES

1. Learn more about the Sonoran desert—home of the saguaro—or the alpine tundra. Both of these ecosystems are fascinating places. They have highly specialized plants and animals that can survive their harsh growing conditions. Find out how these plants and animals are suited to their environment. Write a report or make a poster to show what you learn.
2. Find out more about bindweed. Look it up in gardening and plant books or talk with someone in a gardening center about it. Find out if it is considered a problem in your area. If so, ask where you could see it. Try to find a plant to bring to class after you obtain permission to pick it. Learn how to control this hardy newcomer.

2-4 PLANT IDENTIFICATION GUIDES

Having someone tell you about a plant is the easy way to learn about it; but suppose you want to learn about a plant on your own. Would you be able to do so? No matter where you live, there are books that can help you identify plants. Some of them are easier to use than others, but they all provide the same basic information.

- What can you learn about a plant from a plant identification book?

Use a plant identification book to look up some local plants. After you have read about a few plants, you will be able to answer this question.

MATERIALS

2 plant identification guides
worksheet: ***Plant Identification Guides*** (Figure 2–6)

PROCEDURE

Follow this procedure using one of the plant books and the worksheet. When you finish with one book, repeat the procedure with the second one.

Part A. Learning About the Book

1. No matter which book you use, you should always know its title and the name of the author. Look at your book to find this information and record it on your worksheet.
2. It is usually helpful to know when a book was written. Sometimes a plant's name changes so it may have two different names depending on whether you are looking in a new book or in an old one. Record the copyright date of your book on the worksheet. (The copyright date is usually on the back of the title page or on the title page itself.)

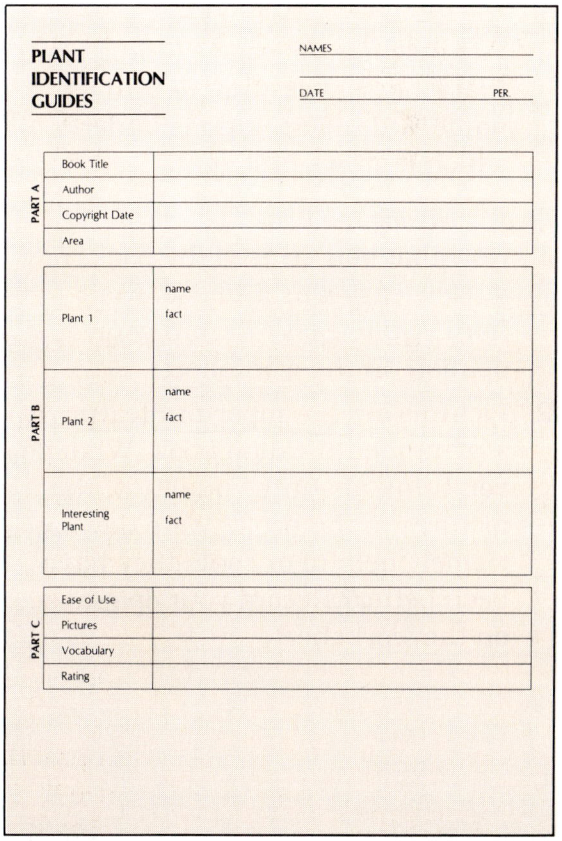

PLANT IDENTIFICATION GUIDES

NAMES

DATE PER

PART A	Book Title	
	Author	
	Copyright Date	
	Area	
PART B	Plant 1	name fact
	Plant 2	name fact
	Interesting Plant	name fact
PART C	Ease of Use	
	Pictures	
	Vocabulary	
	Rating	

Figure 2–6

3. Plant identification books are usually written for certain parts of a country. You can usually get a clue about where the book is meant to be used from its title. If not there, check the cover, title page, and introduction. Record the area for which the book is intended on your worksheet.

Part B. Looking Up Plants

1. Choose one of your local plants to look up in the plant book.
   a. Record the name of the plant on the worksheet.
   b. Look up the plant and read about it. Record one new piece of information that you learned about this plant on your worksheet.
2. Look up a second local plant. Record the same things for this plant that you did for the first one.
3. Flip through the pages. Find a plant that looks interesting. Record its name and the conditions under which it grows best.

Part C. Rating the Book

1. Now that you've used the book, rate it. How easy was it to use? Very hard? Quite easy? Record your opinion on the worksheet.
2. Pictures are an important feature in a plant identification book. What do you think about the quality of the pictures in the book? Were they extremely helpful or could they have been better? Record your opinions on the worksheet.
3. Some plant books use a lot of technical vocabulary and are hard for a beginner to read. Others are quite easy to understand. How would you describe the vocabulary in this book? Easy, medium, or difficult? Again, record your opinion.
4. Overall, what do you think of the book? Record your opinion.

Remember to repeat this procedure with a second book. Look up the same two plants in the second book that you did in the first one so you can compare the two books.

CONCLUSION

Now that you've spent some time with two plant identification guides, look back at the focusing question on page 29. Write an answer that includes at least 5 types of information you can learn about a plant from a plant identification guide.

EXTENSION ACTIVITY

Check out a plant identification guide to use at home. Look through the pictures to see if there are any plants you recognize; if so, look up their names. Then study the appearance of a plant that should be common in your area. Later take a walk around the neighborhood or a local park to look for it.

2-5 A PLANT STUDY

Each plant and animal provides clues about the ecosystem in which it lives. The more plants and animals you study, the more you will learn about the environment; but even a single plant may reveal information about weather conditions, local animals, and, of course, more about the plant itself. It doesn't take a lot of training to interpret some of nature's clues, but it does take careful observation.

- What can you learn about your environment by closely inspecting a single plant?
- What evidence can you find that a plant interacts with animals or with other plants?

Spend several minutes observing a plant in order to find the answers to these questions.

MATERIALS

pencil and paper
tape measure
colored pencils

PROCEDURE

Work alone or with a partner.

1. Choose a plant to study. Any plant will do; but the larger and older the plant is, the more you can learn from it.
2. Take pencil, paper, and a tape measure with you when you visit your plant. Plan to spend **at least 20 minutes** making your observations. (Twenty minutes is not very long. Even after 20 *hours* of observation you still would not have seen all there is to see.)
3. Refer to the Plant Observation Guide on page 32 while you study your plant. Record at least 15 observations on your paper.
4. Use the colored pencils to help you analyze your observations.
   a. Use a *brown* colored pencil to circle the observations you made that involved animals.
   b. Put a *blue* check mark by your most interesting observation.
   c. Put a *red* check by your best observation—something you noticed that might be easy to overlook.
   d. Use a *yellow* pencil to underline your observations that involved something other than sight. Underline anything you smelled, felt, or heard.
5. Answer the analysis questions.

ANALYSIS QUESTIONS

1. Think about the life history of the plant you observed.
   a. Estimate the age of the plant.

Plant Observation Guide

Location. Sketch a map showing the location of the plant.

General appearance. Draw and color your plant as it appears today.

Physical characteristics.

Record some of the plant's measurements, such as height and circumference of main stalk or trunk. You might also want to count the number of main branches or the number of leaves on a stalk.

If the plant has leaves, sketch one. Comment on the leaf size, color, and texture.

Other characteristics. (Use your senses to describe the plant.)

Describe the odor of the plant. Smell different parts, such as stems, leaves, bark, and flowers.

Record textures of the plant parts.

Describe sounds the plant may make.

Animal life.

Look for signs of animal life on and around the plant. Describe the evidence you see.

Speculate about the types of animals that use this plant.

List any animals that you find on the plant.

Other plants. (Describe how this plant "interacts" with other plants.)

Describe any plants that grow on your plant.

Measure the distance between your plant and its nearest neighbor.

Describe the plant that is most abundant near your plant.

Record how many plants of the same kind as yours are within sight of your plant.

Abiotic influences.

Look for signs of abiotic influences on the plant—from such things as strong wind, hail, heavy snow, or drought.

Feel the dirt around the plant. Dig down a few inches. Comment on the moisture content of the soil.

b. How much longer do you think this plant could live? Explain.
c. How do you think the appearance of this plant will have changed six months from now?

2. Animals use plants for food, for shelter, and for a variety of other reasons. Look at the items on your list that involved animal life.
   a. What evidence did you find that your plant was used for food? List *all* the clues you observed.
   b. What evidence did you find that your plant was used for shelter?
   c. For what other reasons did an animal use your plant?
3. What are 3 reasons someone might want to plant a plant like the one you observed?
4. Look at your list to see if you made any observations that involved weather or other abiotic factors. If so, what could you tell about the abiotics by looking at your plant?
5. The procedure includes the statement "the larger and older the plant is, the more you can learn from it." Explain why this is true.

CONCLUSION

Write a paragraph that answers the two focusing questions asked on page 31.

TAKE A STAND

Even though the large, healthy, old trees in a forest are valuable sources of lumber, they should not be cut down.

- Do you **agree** or **disagree**?
- What **comments** would you like to make?

EXTENSION ACTIVITIES

1. Make careful observations of your plant again in 3 months or 6 months. Compare your notes from the two visits. What did you notice during the second visit that you did not see the first time?
2. Design a "composite plant." That is, draw a picture that shows this plant being used by as many animals as you can put into one picture. Remember to meet the needs of a variety of animals —birds, insects, spiders, mammals, amphibians, and reptiles—in this majestic plant.
3. Dead plants are important to an ecosystem, too. Locate a standing dead tree or cactus and make a plant report about it. Try to answer these questions about the plant.
   Why might it have died?
   Is anything living in or on it now?
   Should it be cut down and hauled away or left where it is?

THE BIGGEST TREES

Tree	Place Found	Height	Trunk Circumference	Crown Spread
Common apple	Santa Fe, NM	35′	11′	44′
White ash	Palisades, NY	95′	25′4″	83′
Bigtooth aspen	Marquette, MI	132′	8′9″	67′
American beech	Ashtabula Co., OH	130′	18′6″	75′
Yellow birch	Deer Isle, ME	76′	21′	91′
Coconut	Hilo, HI	92′5″	5′	27′10″
Plains cottonwood	near Hygiene, CO	105′	36′	93′
Coastal Douglas fir	Olympic National Park, WA	221′	45′5″	61′
American elm	east of Louisville, KS	99′	23′2″	133′
Pignut hickory	near Brunswick, GA	125′	15′3″	87′
American holly	St. Mary's City, MD	50′	11′6″	39′
Western larch	near Libby, MT	175′	24′5″	37′
Bigleaf maple	Jewell, OR	101′	34′11″	90′
Red mulberry	Edmond, OK	62′	20′8″	76′
Papaya	Castello Hammock, FL	25′	1′8″	4′
Pecan	Warren County, MS	130′	23′10″	90′
Coast redwood	Humboldt Redwoods State Park, CA	362′	52′5″	74′
Rosebay rhododendron	Ellicott's Rock Wilderness Area, SC	40′	2′1″	22′

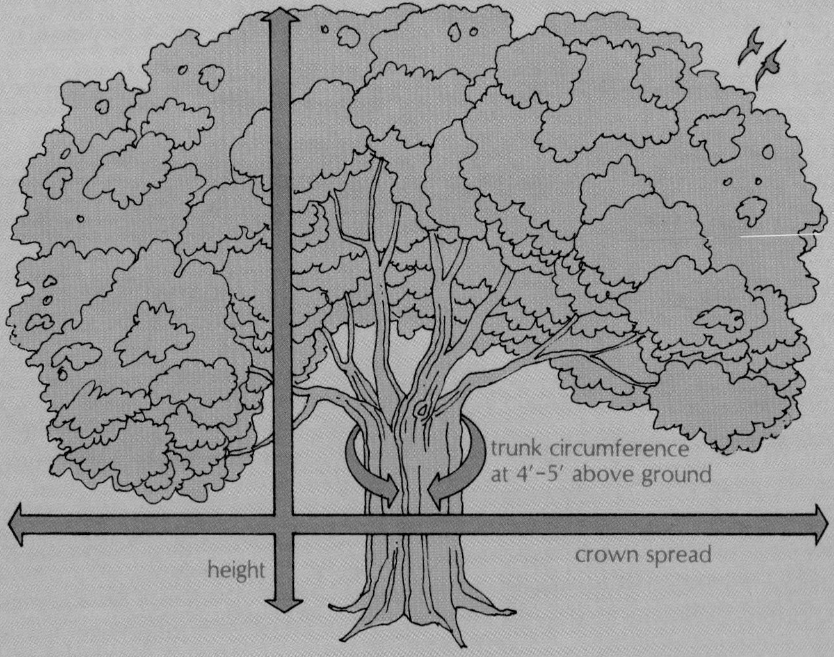

2-6 GROWERS' ALERT

SUPERIOR SEED COMPANY
580 Angiosperm Drive • Dicot, Colorado

Dear Student,

Every year we select several classes to participate in our New Products Testing Program. This year we are asking your class to help us evaluate a new strain of seed.

This particular strain is the result of years of research by our horticulture department. We now need to determine the optimal growing conditions for these seeds. We are asking you to plant and grow these seeds while carefully monitoring *one* environmental condition.

Here is a list of the questions about growing conditions that we need to have answered for these seeds.

1. Do the seeds need to be planted? If so, how deep?
2. Does the type of soil/dirt affect seed growth?
3. Will the plant grow in anything besides soil?
4. How much water is ideal for plant growth? (One, two, or three teaspoons per watering? How often?)
5. Do the seeds/plants need light? How much? (none? one hour a day? 8 hours? 14 hours?)
6. Is there an ideal temperature for growth?
7. Is growth affected by sounds--noise, music, or voices?

Please keep careful notes. We will need to know details about the growing conditions, complete growth data, and your final conclusions. This information will be used on the seed package labels.

Thank you for your time and cooperation.

The Superior Seed Company

Preparing for an Experiment

1. Decide if you want to work alone, with a partner, or in a group. Keep the things below in mind when making your decisions.
   - The experiment takes about four weeks. If you plan to work with other people, make sure you get along with them.
   - You will need to grow these seeds at home. If you work in a group, you will need to decide who will keep the seeds. Do you live close enough to each other that all of you will be able to monitor the growth of the seeds?
2. Choose one condition to study from the list in the letter from the Superior Seed Company.

Planning Your Experiment

1. You will be given a limited number of seeds to test. Plan how you will use them.

Figure 2-7 This was taken from a student's notebook. Notice that this student explained what materials to use and exactly how to proceed.

2. Write down your prediction about how your seeds will grow under the conditions you plan to use.
3. Now plan your experiment in more detail. Write down exactly what you plan to do.
4. Have your teacher check your plans and initial them.

Doing the Experiment

Time needed: about 3 weeks

Keep a daily log. Here are some things you should record.

1. Your observations.
2. Measurements. (These must be specific.) Examples:
   excellent: Leaf #2 is 3 cm long.
   Leaf #5 is 2.5 cm long.
   poor: Leaf #2 is longer than leaf #5.
   unacceptable: The leaves are different sizes.
3. An explanation of any problems you encounter.
4. An explanation of what you do to the seeds every day.

Option: You can take photographs showing your set-up and/or how the plants are growing.

Putting Together the Final Report

1. Prepare your final report. It should include the following:
   a. a title
   b. a description of the condition you are testing
   c. your prediction or hypothesis
   d. your procedure
   e. your data
   f. charts or graphs that organize the data
   g. your conclusion

h. an explanation of how your prediction compared with your conclusion

2. Your final report should also include your plan with your teacher's initials and the answers to the analysis questions.
3. Be prepared to explain your experiment to the rest of the class.

ANALYSIS QUESTIONS

1. What would you suggest that the Superior Seed Company print on their package label about the condition you tested?
2. Did you have any unexpected problems? If so, what were the problems and how did you solve them?
3. If your seeds or plants died, explain why.
4. What were some possible sources of error in your experiment?
5. If you were going to do this experiment again, what would you do differently to improve your results?

Answer the questions at the top of the next column after the class discussion.

6. What factor(s) seemed to be the most critical for growth?
7. Which factor made the least difference in growth?
8. List at least 4 other factors that could have been tested.

EXTENSION ACTIVITIES

1. Continue this experiment for another 3 or 4 weeks. Watch and record what happens to the plants. (You may need to transfer the plants to a larger pot.)
2. Repeat the experiment, making changes to improve your experiment and to give you clearer results. Again, keep detailed notes and be prepared to explain the results.
3. Read the information printed on 5 seed packages. List the types of information that appear on all the labels. Which abiotic factors are mentioned? Which factor seems to be most critical for plant growth? Decide which of these plants you would be able to grow in your area.

2-7 SOCKS TO CELLOPHANE

Plants are an important part of our lives; they're not just for decoration. Besides being producers in an ecosystem, they "produce" our lifestyles. Without the use of products made from plants, life would be drastically different. Have you ever thought about the plants in your life? How do you use plants? How many plants or plant products do you estimate that you use in a typical day?

Keep these questions in mind as you read the following story. It's about a day in the life of a girl your age. While you read, list the items that are mentioned that you think are made either totally or partially from plants.

Even though it was early on a Saturday morning when her clock radio went off, Jodi didn't mind. She never had trouble getting up on a skiing day. Leaning over, she turned off the radio on top of the nightstand, stretched, and practically jumped out of bed. Once up, she straightened the sheets, fluffed her pillow, and pulled up the bedspread. When she opened the curtains, she paused to admire the snow glistening on the branches of the tree in front of her window.

After showering and shampooing and conditioning her hair, Jodi dried off with her favorite towel. Then she put on her deodorant before returning to her room to get dressed. She'd set out her clothing the night before so she wouldn't forget anything. She loved winter sports but she didn't like to be cold. First, she put on long underwear and cotton socks. Next came a turtleneck and wool socks, followed by her ski pants. She laced her hiking boots, dabbed on her favorite perfume, and was ready for breakfast.

When she entered the kitchen, she found her brother Timmy already helping himself to bacon, scrambled eggs, and hash brown potatoes. Sliding into her chair, she reached for the bacon and eggs. Orange juice and a glass of milk completed their meal. Her parents were reading the newspaper while they sipped their coffee. They'd already finished their pancakes with margarine and warm maple syrup. (This was not their usual breakfast. On most days, they ate something that could be prepared quickly, like dry cereal and milk or oatmeal with cinnamon and sugar.)

After breakfast the family hurriedly rinsed the dishes and loaded them into the dishwasher. Since Timmy was getting over a throat infection, his mother reminded him to take his penicillin. Jodi fed the cats, then went to brush her teeth with a new brand of toothpaste. After rinsing her mouth with mouthwash, she pulled on a heavy wool sweater, grabbed a book to read in the car, and hurried out the door, making sure to lock it behind her.

Jodi slid into her place in the back seat. The car felt so clean. It had just been repainted. Then it had been waxed and the upholstery cleaned. The effort had been worth it; even the tires glistened!

This ski outing was a special occasion. For this trip, Jodi and Timmy were using the money they had earned from delivering the daily paper and mowing lawns the past summer. Jodi was especially excited because it would be the first time she used the skis she had gotten for her birthday. They had been wrapped with several rolls of wrapping paper, yards of string, and at least 3 meters of transparent tape.

The highway had recently been resurfaced with a fresh coat of asphalt so it contrasted sharply with the white snow on the hillside. When they reached the top

of the first hill, Jodi took a picture of the snowcapped peaks against the brilliant blue sky. The rest of the drive to the ski area was just as scenic.

They parked in a spot marked in yellow paint for "compact cars only." Jodi took her skis off the rack and got her down parka, gloves, wool hat, and sunglasses out of the car. Jodi's parents stopped at the lodge to buy some lip balm but Jodi and Timmy headed straight for the lift.

Jodi handed the attendant a twenty-dollar bill to pay for her lift ticket. She then fastened it to the zipper pull on her jacket. She and Timmy raced past the guide ropes and were soon riding the lift to the top of the mountain.

Jodi and Timmy got in four runs before lunch. Not bad! At the snack bar Jodi selected a ham sandwich sealed in cellophane, an apple, cookies, and a carton of milk. Timmy smothered his hot dog with mustard and pickle relish and filled his cup with hot apple cider. They each grabbed a straw and a napkin, then joined their parents at a table outside so they could all enjoy the view while they ate.

In the afternoon, they completed three more runs. Jodi couldn't believe it when she saw that her watch said 4:00. The family piled into the car for the return trip.

Once home, her father lit a fire in the fireplace while her mom warmed the chili she had prepared the day before. Jodi made hot chocolate for everyone, topping each cup with a blob of whipped marshmallow. Ending the day like this, in front of the fireplace, was part of what made skiing so enjoyable for Jodi. As she snuggled under a blue-gray wool afghan on the couch, the warmth of the fire and the crackling of the logs lulled her to sleep.

STORY ANALYSIS

1. Count and record the number of items you listed.
2. Look at the items on your list; many of them come from trees. Put a star in front of each of the tree products.
3. List at least 10 additional items that Jodi probably used sometime during this Saturday but that were not specifically mentioned in the story. Circle the items that you think may be plant products.

HOMEWORK

1. What plant products do you use? Choose one of these two assignments to help you find out.
   - Write about a Saturday in your own life. Mention the things you do, see, and eat. Then, go back and circle, or list, all of the items you mentioned that came from plants.

     or

   - Take an inventory of a room. You could use your classroom, a room at home, an office, your garage, or any other room. Be observant. List at least 50 items that are in that room.
2. Analyze your list. See if you know the kinds of plants that were needed to produce the things you listed.
   a. If you know, or think you know, the kind of plant used to make an item, record the name of the plant by the item.
   b. How many different kinds of plants did you identify?
   c. What kind of plant provided the most items?

2-8 PLANTS AS PRODUCERS

What can be made from peanuts besides peanut butter? If you concentrate, can you think of over 300 uses for peanuts? George Washington Carver did.

In the early 1900s Carver started to experiment with them using less equipment than you have in your kitchen today. He wanted to find more uses for peanuts in order to save farmers in the south from financial disaster. Since their cotton crops were doing poorly, Dr. Carver suggested that they plant peanuts instead. He realized, however, that he had to be sure that there would be a large market for the peanuts so that all the farmers would be able to sell their peanuts at good prices.

Carver spent months developing uses for peanuts. He smashed, soaked, and analyzed them until eventually he produced 30 dyes from the peanut skins alone. He also developed leather stains, wood stains, and ink. He made instant coffee, cream, skimmed milk, and butter, too. As a result of his work, we use peanuts in a variety of ways. Some of these are listed in the chart that follows.

PRODUCTS MADE FROM PEANUTS

Food Products
milk
flour
peanut butter
margarine
vegetable oil
cheese
mayonnaise
meal
candies

Made from Shells
insulation filler
floor sweeping compound
buffing for steel mills
wallboard

Non-food Products
shaving cream
glues
paper
ink
plastics
salve
cosmetics
shampoo
shoe polish
dyes
lubricating oils
metal polish
fertilizer
fodder for livestock

Plant Uses

Peanuts are just one example of a plant that we use for food as well as in industry. There are dozens of other such multipurpose plants. When we think of plants like these, it is obvious that plants are the producers of the world. They provide us with food, medicines, and drugs; with energy for heat; and with the raw materials for many of the items that we use on a daily basis.

Plants provide all of the animals on earth with the food necessary for survival. Humans, however, eat relatively few kinds of plants. Of the 500,000 plant species known, six provide most of our food. These six crops are cassavas, corn, potatoes, rice, wheat, and yams. (See Figure 2–8.) Think about the foods that line the

Figure 2–8 Top: This is a cassava field in Ecuador. Bottom: In the United States, wheat is much more abundant than cassava.

grocery store shelves; they are either plants themselves, are made from plants, or are meat and dairy products that come from animals that survive on plants. Can you think of any foods you eat that do *not* have a plant source?

In winter, people worldwide use many tons of firewood to keep warm. Additional energy comes from coal, natural gas, and other petroleum products. Where did these come from? From plants, of course. These were the trees, flowers, and vines that grew eons ago. After they died, these plants began the long, slow process of turning into the energy-producing products that we use today.

Aspirin, one of the most widely used painkillers, is naturally found in willow trees. Long ago, people would chew the bark and leaves of certain willows in order to get relief from pain. Today, pure forms of the active chemicals in aspirin can be made in laboratories. Like aspirin, many of our other medicines and drugs originally came from plants but are now artificially made. One can't help but wonder how many more drugs await discovery, still "hidden" within the fruit or leaves of a plant.

The list of plants used in industry is long. Natural rubber and waxes come from tropical species. Fibers for clothing, string, twine, and canvas come from plants, the most widely used being cotton. Dyes, perfumes, plastics, soaps . . . the list goes on and on. Without plant products, our lives would be significantly different.

Natural Values

Aside from the dollar worth, plants have scenic value, too. Can you imagine a lake without grass or a picnic site with no green plants? Natural areas are chosen as vacation spots by many. Backpacking, camping, hiking, and animal watching can all be enjoyed in state and national parks. There the native plants and animals are preserved for all to enjoy. Even those people who don't look to the wilderness for recreation tend to chose areas rich in plants for their relaxation. Think about your area. Where are the popular weekend spots? City parks, green belts, and golf courses attract crowds of people. Would these places be as popular without plants?

Figure 2-9 The scenic value of plants is obvious in our national parks. What are the dominant plant types seen here in Yellowstone?

It is becoming very important to think about the balance in nature. How many plants are needed to keep our ecosystems healthy? Within cities and towns, how much concrete should there be compared with "green space"? Where do we set aside land for plants? What kinds of plants should be there—"useable" plants, decorative plants, or wild plants? There are no right answers to any of these questions but, with more and more people living on the same amount of land, the questions are becoming more critical. Have you ever thought about how many acres of land must be cleared in order for you to live each year? How much of the pavement in your community "belongs" to you?

ANALYSIS QUESTIONS

1. What are 5 examples of products made from peanuts?
2. Think about the 6 plants that provide most of our food.
   a. List them.
   b. Underline the ones you eat at least once a week.
   c. Circle the ones you have never tasted.

> **TAKE A STAND**
> Peanut plants are more valuable than saguaros.
> - Do you **agree** or **disagree**?
> - What **comments** would you like to make?
>
> Rare plants should be protected, even if it costs money to do so.
> - Do you **agree** or **disagree**?
> - What **comments** would you like to make?

3. Trace the following foods back to their plant source.
   (example: eggs ⟶ chickens ⟶ grains)
   a. cheese
   b. spaghetti noodles
   c. hamburger
4. Here is a list of 12 products that you probably use.

comb	door knob	towel	softball bat
pencil	milk carton	milk	pillow
book	toothpaste	water	quarter

   a. Which of these items come from animals?
   b. Which come from plants?
   c. Which items are made from something other than plants or animals?
5. List 5 things you enjoy doing in your spare time.
   a. Circle each one that requires plant products.
   b. Put a star by each one that is made more enjoyable if living plants are around.
6. Parking lots are necessary but they are seldom attractive.
   a. What are 3 characteristics of an attractive parking lot?
   b. What are 3 characteristics of an ugly parking lot?
   c. What are at least 3 costs of planting a tree in a parking lot?

EXTENSION ACTIVITIES

1. Learn more about George Washington Carver; the story of his life is fascinating. He was born a slave but was orphaned almost immediately. The white couple who raised him tried to help this black child face a world of discrimination. Ask a librarian to help you find a book about his childhood. Compare his life with your own. Would you want to trade places?
2. Find a recipe that uses cassava. Decide if you would like to prepare it for your family or for the class. If you would like to bring the prepared food to class, talk to your teacher about cost, amount to make, and a convenient time to bring it in.

Figure 2–10 George Washington Carver in his laboratory at Tuskegee Institute, Alabama

2-9 A SEAWEED EXTRACT

Ink from peanut skins? Plastics from trees? The plant origins of some products such as a wooden pencil or a loaf of bread are obvious; but there are hundreds of other things that we seldom associate with plants. We are surprised when we learn of their plant origins. For example, would you think of toothpaste as coming from the sea? While it doesn't come directly from the sea, one of its important ingredients does come from seaweed.

- What kinds of substances come from seaweed?
- How can you extract a substance from seaweed?

In this activity, you will extract an important substance from seaweed. After following this procedure and reading about uses of seaweed, you will be able to answer these questions.

MATERIALS

large test tube
test tube holder
graduated cylinder
test tube rack
masking tape
safety goggles
hot water bath
small piece of Irish moss (a type of seaweed)
reading: ***Substances from Seaweed*** (page 49)

PROCEDURE

Work with a partner.

Part A. Starting the Extraction

1. Copy the data table below onto your own paper.

APPEARANCE OF IRISH MOSS

timing	**sketch** (actual size)
At start	color ______
After 2 minutes	color ______

2. Bring the large test tube, test tube holder, and test tube rack to your desk.
3. Use masking tape to label the upper part of the test tube with your and your partner's names.
4. Get one small piece of Irish moss. It should be 1 to 1.5 cm long.
   a. Blow off any sand that is clinging to it.
   b. Fill in the first part of your data table.
5. Drop your piece of Irish moss into the test tube.

a. Add about 10 ml of very hot tap water to the test tube.
b. Wait 2 minutes and then fill in the second part of your data table.

6. **Put on your safety goggles.** Then put your test tube in the hot water bath and leave it there for about 20 minutes.
7. While you are waiting, read ***Substances from Seaweed*** starting on page 49. Then answer analysis questions 1–4.

Part B. Finishing the Extraction

1. **Put on your safety goggles.** Then, using your test tube holder, take your test tube out of the water bath and place it in the test tube rack to carry it back to your desk.
2. Observe the extract—the liquid in the test tube—and the piece of Irish moss.
   a. Fill in the second part of your data table.
   b. Answer the rest of the analysis questions.
3. Clean up.
   a. Pour your extract into the large beaker that your teacher has labeled "Irish Moss Extract." Try to keep the seaweed in the test tube. You may want to use your pencil to hold the seaweed back.
   b. Throw away your piece of Irish moss.
   c. Wash your equipment and put it away.

ANALYSIS QUESTIONS

1. Describe the changes you observed in the Irish moss after it had been in hot water for 2 minutes.
2. Define the word *seaweed.*
3. What 3 important substances are extracted from seaweed?
4. You are using Irish moss in this experiment.
   a. What substance are you extracting?
   b. List 5 products in which it is used.
5. Describe the way the Irish moss looks after 20 minutes in the hot water bath compared with the way it looked after it had soaked for 2 minutes.
6. Use three of your senses to observe the *extract.*
   a. Describe the way it looks.
   b. How does it smell?
   c. Once it cools a little, stick your finger into it. How does it feel?
7. It would be fun to taste chocolate pudding made from your extract. List at least 2 reasons why this is *not* part of the procedure.
8. Now that you have closely observed your extract and read about several uses of seaweed extracts, think of at least 2 additional ways the extract could be used. Be creative!

CONCLUSION

Look at the focusing questions on page 47. Imagine that your parents ask you these same two questions tonight. Write down what you will say to them.

EXTENSION ACTIVITY

Look for seaweed additives in foods you have at home. Agar, algin or alginate, xanthan gum, and carrageenan are used as thickeners in a variety of products. Make a collage of the labels from the foods that have these additives.

Substances from Seaweed

Seaweed* is a source of several substances that are used in food and industry. When these substances absorb water, they thicken and cause mixtures to gel, or stabilize. For example, oil and vinegar mixed together for salad dressing normally separate on standing. But when a seaweed substance is added, the liquids remain evenly suspended for a longer period of time. Seaweed substances do not provide nutrients, nor do they help preserve food.

The three most frequently used substances are algin, agar, and carrageenan. Each one has a variety of uses.

*Seaweed is not a scientific term. It is a word that refers to any plant that lives in the ocean. Green, red, and brown algae are types of seaweed.

Algin

Algin prevents ice crystals from forming in frozen desserts and makes them taste smoother and creamier. Algin is also used to thicken shampoo, lotions, and cosmetics. Without algin, water-based paints would separate and be less "spreadable." Film is coated with a thin layer of algin. If your dentist has made an impression of your teeth, algin was used to make the mold.

Algin is found in the cell walls of seaweeds such as kelp. Kelp is a type of large, brown algae that grows from the bottom of the ocean. It grows rapidly, perhaps as much as 70 cm (about 2 feet) a day. If the water temperature and depth permit, kelp can grow as tall as 67 meters (200 feet)! Kelp is harvested in southern California in specially equipped boats that cut, or "mow," the top 130 cm (4 feet) of the plants. This allows more sun to reach the kelp and encourages the rest of the plant to grow. (See Figure 2–11.)

Figure 2–11 Divers must be careful not to get tangled in kelp beds.

Agar

Agar is used in making marshmallows, jelly candies, ice cream, and sherbet. The gel around canned meat may contain agar; it is also the main ingredient in some laxatives. Laboratories grow microorganisms on agar. If you've had a throat culture tested for strep, the cells from your throat were placed on agar to grow.

Agar is extracted from several kinds of red algae. (See Figure 2–12.) Most agar is produced in Japan.

Figure 2–12 Red algae

Carrageenan

Carrageenan has many of the same characteristics as agar and algin. In addition, it works particularly well in the production of milk products. This makes it useful as an additive in pie fillings, puddings, chocolate milk, and cream cheese. Gravies, cough syrup, shaving cream, deodorants, and water-based paints may also contain carrageenan.

Carrageenan comes from seaweeds such as Irish moss. Irish moss is really not a moss; it is a red algae that grows along the coast of New England and eastern Canada. Irish moss is gathered from boats by using long-handled rakes. It can also be collected where it washes up on the beach. (See Figures 2–13 and 2–14.)

Figure 2–13 This Irish moss plant is about 3 cm tall.

A Sea Crop

Seaweed crops must be tended and cared for similarly to the way we care for crops that grow on land. As a result of human actions, however, some sea crops have become endangered. Pollution, the destruction of predators, and careless harvesting can greatly reduce the supply of these valuable resources.

Figure 2–14 These mossers are unloading their Irish moss harvest at a wharf in Nova Scotia.

2-10 PLANT CELLS

Like all living organisms, plants are made up of cells. And, like most of these, plant cells are not visible to the naked eye. They can, however, be seen with a microscope. The **cell wall** that surrounds the cell is usually the easiest to find. The **cytoplasm** is the jellylike substance within the cell that contains all of the other cell structures.

- In which plants can *you* observe cells?
- Do green plants have green cells?

Look at plant cells through a microscope in order to answer these questions.

MATERIALS

microscope
microscope slide
cover slip
water in a dropper bottle
colored pencils
worksheet: ***Plant Cells*** (Figure 2–15)
pieces of plants
elodea
onion, tomato, or cucumber
green leaf (blade of grass, weed, tree leaf, pine needles. . . .)

PROCEDURE

Work with a partner.

1. Since you will be making your own microscope slides, you will need to know how to prepare a **wet mount.** Read through the procedure in the chart to the right to make sure you understand what to do.
2. Take turns preparing the wet mounts as you follow the procedure on the worksheet.

WET MOUNTS

1. Place a *very thin* piece of tissue on the microscope slide. **Light must be able to pass through the tissue in order for you to see anything.**
2. Drop 2 or 3 drops of water onto the tissue.
3. Set the cover slip at an angle on the slide.

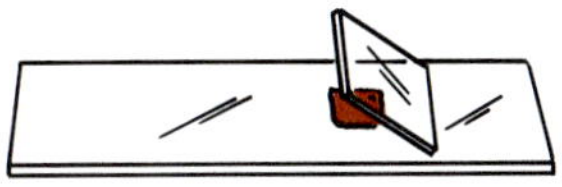

4. Carefully lower the cover slip to cover the tissue.
5. If the cover slip does not rest flatly on the slide, tap it gently with your eraser to smash the tissue. (Be careful not to crack the cover slip. Your tissue sample may be too thick if it does not flatten easily. If so, you will need to start over.)

ANALYSIS QUESTIONS

1. What was the most difficult part of doing this lab?

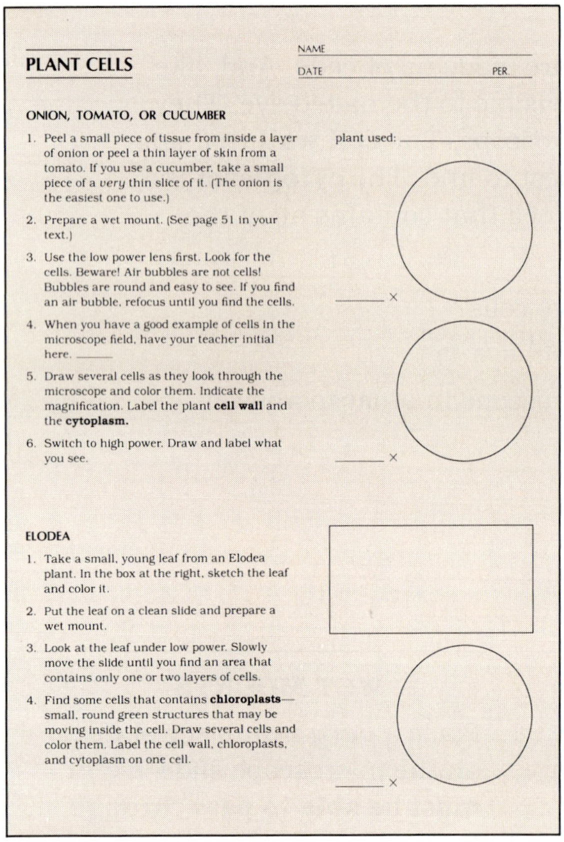
PLANT CELLS

NAME ____________________
DATE ______________ PER. ______

ONION, TOMATO, OR CUCUMBER

1. Peel a small piece of tissue from inside a layer of onion or peel a thin layer of skin from a tomato. If you use a cucumber, take a small piece of a *very* thin slice of it. (The onion is the easiest one to use.)
2. Prepare a wet mount. (See page 51 in your text.)
3. Use the low power lens first. Look for the cells. Beware! Air bubbles are not cells! Bubbles are round and easy to see. If you find an air bubble, refocus until you find the cells.
4. When you have a good example of cells in the microscope field, have your teacher initial here. ______
5. Draw several cells as they look through the microscope and color them. Indicate the magnification. Label the plant **cell wall** and the **cytoplasm.**
6. Switch to high power. Draw and label what you see.

plant used: ____________

______ ×

______ ×

ELODEA

1. Take a small, young leaf from an Elodea plant. In the box at the right, sketch the leaf and color it.
2. Put the leaf on a clean slide and prepare a wet mount.
3. Look at the leaf under low power. Slowly move the slide until you find an area that contains only one or two layers of cells.
4. Find some cells that contains **chloroplasts**—small, round green structures that may be moving inside the cell. Draw several cells and color them. Label the cell wall, chloroplasts, and cytoplasm on one cell.

______ ×

Figure 2–15 (Worksheet, page 1)

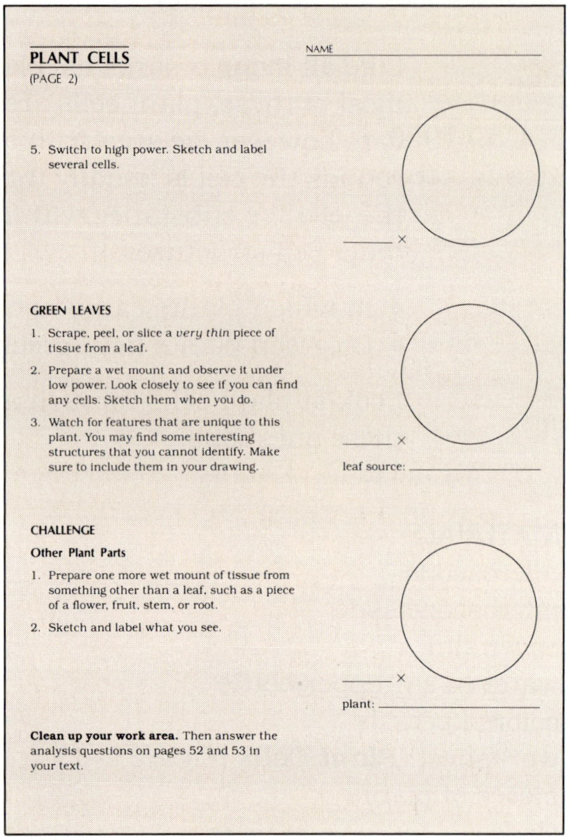
PLANT CELLS
(PAGE 2)

NAME ____________________

5. Switch to high power. Sketch and label several cells.

______ ×

GREEN LEAVES

1. Scrape, peel, or slice a *very thin* piece of tissue from a leaf.
2. Prepare a wet mount and observe it under low power. Look closely to see if you can find any cells. Sketch them when you do.
3. Watch for features that are unique to this plant. You may find some interesting structures that you cannot identify. Make sure to include them in your drawing.

______ ×

leaf source: ____________

CHALLENGE

Other Plant Parts

1. Prepare one more wet mount of tissue from something other than a leaf, such as a piece of a flower, fruit, stem, or root.
2. Sketch and label what you see.

______ ×

plant: ____________

Clean up your work area. Then answer the analysis questions on pages 52 and 53 in your text.

Figure 2–15 (Worksheet, page 2)

2. Think about each type of plant cell you looked at.
   a. Which were easiest to see?
   b. Which were hardest?
3. Animal cells do not have cell walls, only plant cells do. What do you think is the function of a cell wall?

CONCLUSION

Write a sentence or two to answer the focusing questions on page 51.

EXTENSION ACTIVITY

Look at cells of the same plant from 5 different places. For instance, if you choose an apple tree, you could look at cells from the apple peel, apple "meat," apple stem, upper surface of leaf, lower surface of leaf, bark from trunk, bark from young twig, layer of tissue inside of bark, or flower petal. Look closely. Are all of the cells the same or do they differ from one another? Illustrate your observations.

2-11 CELL WALLS AND CHLOROPLASTS

In 1950, two brothers were digging peat near Silkeborg, Denmark. When they reached a depth of 3 meters (9 feet), their spades touched something that didn't feel like peat. Peering down, they found themselves gazing at a face. The eyes were closed, the lips were lightly pursed, and stubble was on its chin. Continuing to clear the peat, they soon found that the body was intact. They contacted the police, but the officials did not recognize the man as anyone on a missing persons list.

Realizing that this corpse might be old, the police decided to turn it over to the archaeologists. They identified it as a bog body; in this case, the corpse of a man who died more than 1500 years ago. The body organs, including the digestive tract, were well preserved. By analyzing the stomach contents, the scientists found that the dead man's last meal was a gruel of grains and seeds—the remains included microscopic bits of barley, linseed, flaxseed, and knotgrass.

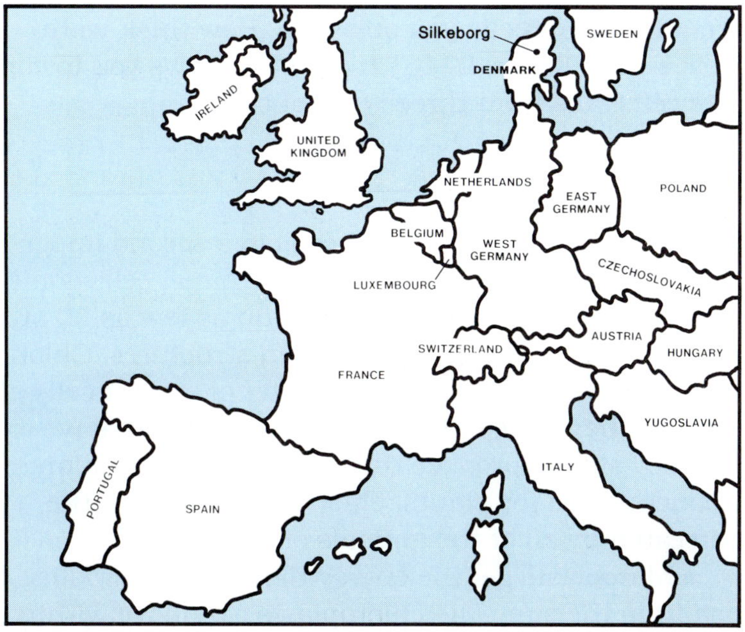

Cell Walls

As you can see, only plant cells remained in the bog man's stomach; no animal cells were found. That could be because he didn't eat meat. However, it is also possible that any cells from animal tissue had been completely digested. This is because plant cells have cell walls; animal cells do not. Cell walls make plant foods harder for humans to digest. When a person eats a meal that contains both plant and animal cells, the animal cells break down first. The plant cells remain identifiable even after they've been in the stomach for several hours.

A **cell wall** is like a sturdy box that supports and protects the cell. It is made of long cellulose molecules. Humans cannot digest cell walls because we don't produce the enzymes needed to break down cellulose. Instead, cellulose moves right through our digestive tracts. Nutritionists refer to this undigestible plant material as "fiber." Plant-eating animals such as cattle, deer, and horses cannot break down cellulose either. However, they have bacteria in their digestive tracts that can digest it.

The strength and thickness of cell walls vary. Plants that need lots of support have heavy cell walls. For instance, the cells in a tree trunk have thick walls; they provide the support the tree needs to stand upright. Many water plants are made up of cells with thin walls. The water provides the support they need. Think about the plant cells you observed. How thick were their walls? (Figure 2–16 on page 55 allows you to compare cell walls from three types of cells.)

Chloroplasts

Plant cells differ from animal cells in a second important way—they have chloroplasts. Animal cells don't. A single green plant cell may contain as few as 20 or more than 100 of these small green structures. Chloroplasts are the food production center of a plant cell. Because they have chloroplasts, plants do not have to eat; they simply produce their own food. In addition to producing food for plants, chloroplasts produce the food eaten by all of the animals of the world.

Chloroplasts get the energy they need to produce food from the sun. The chloroplasts can move within

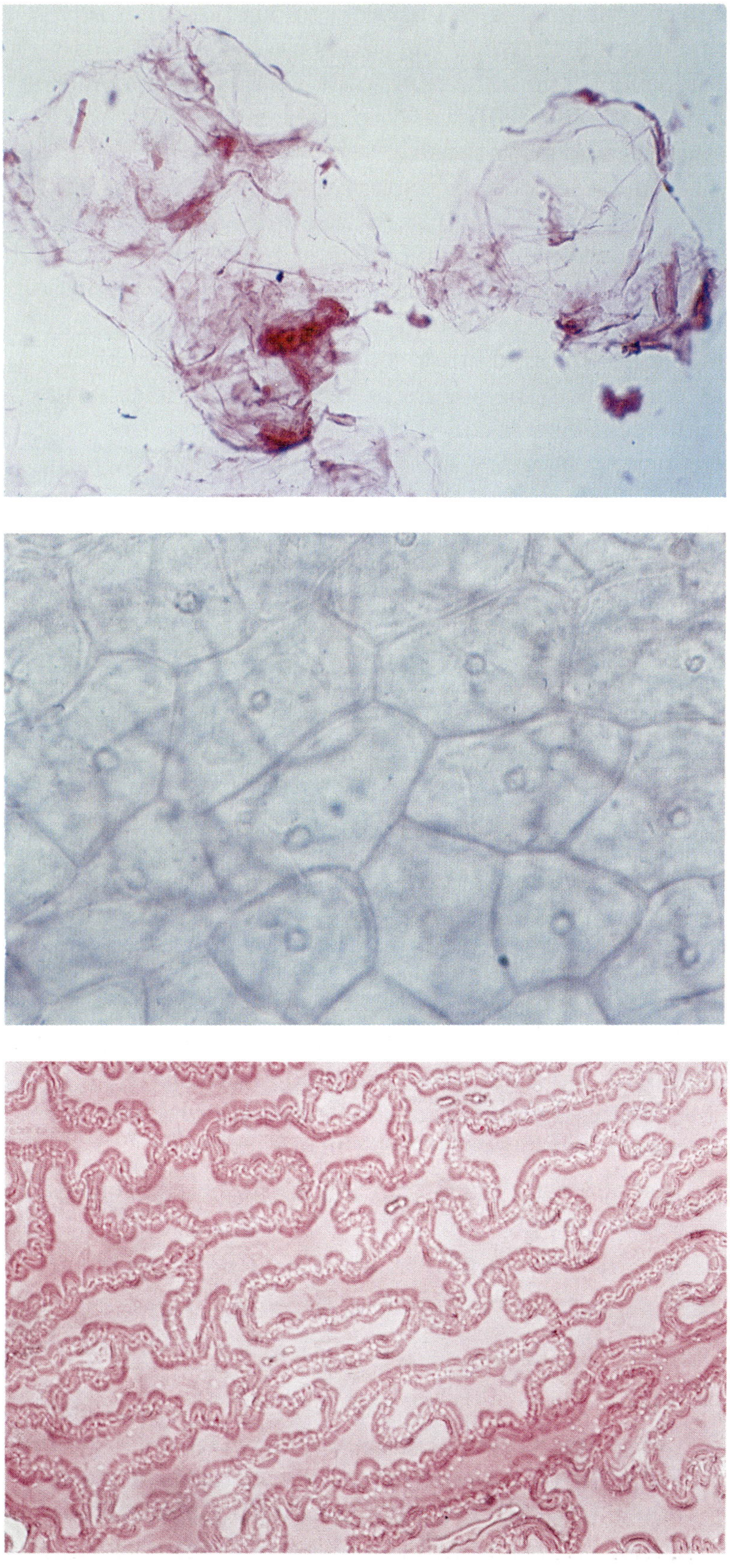

Figure 2–16 The insides of many fruits are pulpy. Top: These are apple pulp cells. They are rounded and very thin-walled. Middle: Green pepper pulp cells aren't quite as rounded and have slightly thicker walls. Bottom: In contrast, the skins of chili pepper are made up of jigsaw-puzzle shaped pieces with thick cell walls.

the cytoplasm of a cell in order to position themselves to catch more rays. If the sun is too bright, they can turn away. The solar energy allows the cell to combine carbon dioxide with water to make sugar. This process is called **photosynthesis.**

Once the cell has a supply of sugar, it can use it in three general ways. First, it may store the sugar for use later. It does so by producing starch. (When we eat food such as potatoes, we are eating food that the potato plant has stored.) Second, the cell may use the sugar molecules to build other cell structures. For instance, by connecting the sugars in long chains, the cell can make cellulose. It can then use the cellulose to strengthen its cell wall. Third, it may break down the sugar. This releases the energy which the cell needs to stay alive.

Oxygen is also produced during photosynthesis. In fact, all of the oxygen in the air is produced by chloroplasts. Even though we use plants to produce so many items we use in our homes and businesses, plants are most important in their role as producers of sugar and oxygen. Because of plants, all other organisms can survive.

ANALYSIS QUESTIONS

1. What 2 structures are found in plant cells, but not in animal cells?
2. Define the term **photosynthesis.**
3. Explain why plants are called the "producers" of an ecosystem.
4. When you burn wood, what do you think is burning—plant cells, cell walls, chloroplasts, and/or cytoplasm? Explain your answer.
5. List at least 3 ways your life would be different if your skin cells contained chloroplasts.
6. Why do you think that a tree does not need blood?

TAKE A STAND

Every Life Science class should plant one tree to help clean up the atmosphere and beautify their neighborhood.

- Do you **agree** or **disagree**?
- What **comments** would you like to make?

EXTENSION ACTIVITIES

1. Get some pencil shavings from the pencil sharpener. Put a thin piece of shaving on a microscope slide, add a drop of water, and cover with a cover slip to make a wet mount. Look at your slide under the

microscope. Sketch and label what you see. What part(s) of a plant cell remain in wood?

2. Find out more about the bog men. Read the article in the June 1977 issue of *Reader's Digest* entitled, "Who Killed the Bog Men of Denmark? And Why?" or *The Bogman* (1987) by Don Brothwell. Look for an explanation as to why these people were so well preserved in peat.

Plant a Tree

The Environmental Defense Fund would like to see everyone planting trees. No, they do not own tree nurseries. They are concerned about our air. They know that more and more carbon dioxide is building up in the atmosphere. They also know that carbon dioxide can increase the earth's temperature. Trees can absorb this extra carbon dioxide. They hope to clean up the air by planting trees.

CONSUMERS

Is this animal part of your environment? What do you think it likes to eat?

3–1 MORE THAN MAMMALS

Producers produce food not only for themselves but also for all of the consumers of the world. Who are the consumers? Animals, of course. But who are the animals? You have lived with them all of your life. There are animals in your home, your backyard, and your community. You've read about them, seen them on television, and studied them in school. How much do you really know about animals?

- What does the word **animal** mean to you?
- How many kinds of animals live on earth?

You will find out answers to these questions by naming and categorizing some animals with which you are familiar.

MATERIALS

notebook paper, 2 pieces

PROCEDURE

Work with a partner.

Part A. Working in Pairs

1. Fold and tear 2 pieces of notebook paper into 16 squares each so that you will end up with 32 pieces of paper that are all about the same size.
2. Write the name of one kind of animal on each piece of paper. Be as specific as you can. For example, instead of writing *bird*, write *sparrow*.
3. Now, look over the 32 kinds of animals that you have recorded on the pieces of paper and sort them into groups. Put similar kinds of animals together.

Part B. Gathering Class Data

1. Do step *a* and have your partner do step *b*.
   a. Add the names of "your" animals to the groups your teacher describes.
   b. Study the data table your teacher gives you.
2. Together with your partner, fill in the "Predictions" column on your data table.
3. Answer analysis questions 1 and 2.
4. Fill in the rest of the data table during the class discussion. Then answer analysis questions 3 and 4.

ANALYSIS QUESTIONS

1. Think about the 32 animals you listed.
   a. On a scale of 1 to 5, how difficult was it for you to list 32 kinds of animals?
   very hard 5 4 3 2 1 very easy
   b. If you had more time, do you think you could have listed 100 kinds of animals?
   c. For which group did you list the most kinds of animals?
2. Look at your predictions.
   a. Which group do you think has the smallest number of kinds of animals?
   b. Which group do you predict is largest?

c. How many kinds of animals do you predict live on earth?

3. Think of the animals that the other students listed.
   a. Were there some animals listed for every animal group?
   b. For which group did your class identify the most animals?
4. Look at the actual numbers of kinds of animals for each group.
   a. What is the smallest animal group?
   b. What is the largest?
   c. About how many kinds of animals live on earth?

CONCLUSION

Look at the focusing questions on page 59. You will find that you answered them when you answered the analysis questions so, in conclusion, answer this question instead: "What do you think was the most interesting thing you learned by doing this activity?"

HOMEWORK

Make a graph to illustrate the numbers of kinds of the various animal groups that live on earth. You could make a bar graph, a pie graph, or a pictorial graph —letting one animal drawing represent a thousand kinds of animals.

TAKE A STAND

Of the money available to protect endangered animals, more should be spent on mammals than on other kinds of animals.

- Do you **agree** or **disagree**?
- What **comments** would you like to make?

3–2 ANIMAL INVENTORY

Even though we humans share this earth with millions of other kinds of animals, we seldom see more than a few different species. Why? Usually it's because we don't look closely enough. For example, in your backyard there may be dozens of different animal species.

- Which type of animal is most abundant in your local environment?
- Where in your environment are you likely to find the largest variety of animals?

Do animal inventories at two sites to help you answer these questions.

MATERIALS

sweep net
small, clear plastic bag
data page: ***Animal Inventory*** (Figure 3–1), two copies

PROCEDURE

Work with a partner.

Part A. Getting Ready

1. Read the entire procedure so both you and your partner will know what to do when you are outside.
2. You will need to take animal inventories in two different sites—one that is managed and one that is not. Your teacher will assign the sites. Make sure you understand exactly where your sites are located.
3. Look over the data page. Is there anything on it that you do not understand? If so, ask your teacher about it *now*.
4. Discuss and answer the first analysis question with your partner.
5. When you are outside, be quiet. The less noise and movement you make, the more animals you are likely to find.

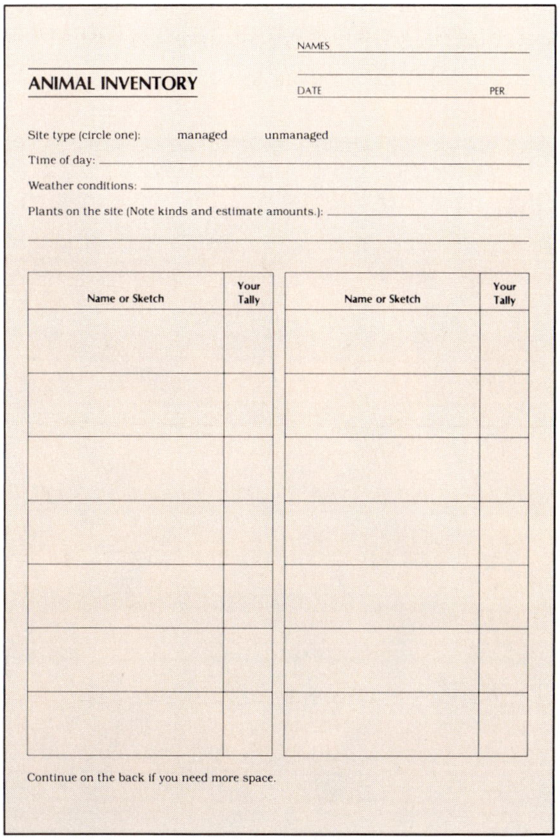

ANIMAL INVENTORY

NAMES

DATE PER

Site type (circle one): managed unmanaged

Time of day:

Weather conditions:

Plants on the site (Note kinds and estimate amounts.):

Name or Sketch	Your Tally	Name or Sketch	Your Tally

Continue on the back if you need more space.

Figure 3–1

Part B. Doing the Inventory

1. Go directly to one of your study sites. Sit down. Look around. Fill in the top part of one of the data pages.
2. While you are looking around, keep track of any kinds of animals you happen to see. Record each kind on your data page. If possible, list them by name. If you don't know the name of an animal, just describe or sketch it. Include characteristics such as size, color, number of legs, whether or not it has wings, and its speed of movement.
3. Have one person use the net to sweep your study area for insects while the other person observes and keeps track of the materials.
   a. Swing the net using back and forth movements. Take a step with each swing. Take a total of 16 steps forward.

   b. Quickly grasp the top of the net so nothing can escape. Empty the net into the plastic bag.
   c. Quickly sketch each little animal on your data page.
   d. Count the number of each little animal and record the tally.
   e. Empty the plastic bag so that you can release the animals back into their environments.
4. Take another look around. Record any additional animals that you observe.
5. Switch roles so that each of you has a chance to fill in a data page and inventory a study site. Use the second study site this time.
6. Go back to the classroom. Put away your equipment and answer the rest of the analysis questions.

ANALYSIS QUESTIONS

1. Make the following predictions.
   a. Where will you find the greatest variety of animals, in the managed area or in the unmanaged area? Why?

b. How many animals do you predict that you will find at each site?
c. What do you predict will be the most abundant animal you find?

2. What was the most difficult part of taking an animal inventory? Why?
3. Look at your data for the managed area.
   a. How many kinds of animals did you find?
   b. What was the most abundant animal you found?
   c. What was the most unusual animal you found? Describe it if you don't know its name.
4. Look at the data for the unmanaged area.
   a. How many kinds of animals did you find?
   b. What was the most abundant animal you found?
   c. What was the most unusual animal you found? Describe it if you don't know its name.
5. Which area had the largest variety of animals, the managed one or the unmanaged one? Why do you think this is so?
6. Copy the chart below onto your own paper. Use it to summarize the number of *kinds* of animals that you found in each area.

	Managed	**Unmanaged**
birds mammals amphibians reptiles insects spiders other		

7. Which kind of animal was most abundant
   a. in your inventory sites?
   b. worldwide?
8. As a human, which of the two areas do you prefer? Why?
9. How do humans affect the variety of animals found in an area?

CONCLUSION

Write a paragraph that answers the two questions on page 61 and summarizes what you learned by doing this activity.

EXTENSION ACTIVITIES

1. Repeat this activity under different weather conditions, at different sites, or in another week or two. Predict what you will find; then take the inventory. Compare your data with the data you got during class.
2. Learn more about your local insects. Use an insect key to identify the 5 most abundant kinds of insects that were found in each study area.
3. Study the local birds. Set up a bird feeding station and keep track of the kinds of birds you see. Call your local nature group or Audubon Society for suggestions on the types of bird feed to use and for hints on identifying birds.

Insect or Spider?

Many people think that every tiny, crawling critter is an insect. That's not true. Consider, for example, spiders. They are not insects. However, like insects, they have jointed legs and several main body parts. Here are some clues you can use to tell them apart.

Number of legs: Spiders have eight legs. Insects have six.

Number of body parts: Spiders have two main body parts. Insects have three.

Wings and antennae: Spiders don't have wings or antennae. Most insects have both.

golden garden spider

eastern lubber grasshopper

3–3 NATIVE ANIMALS

Obviously, all species of animals do not live in or near your community. Just as your area has a characteristic assortment of plants, there are also specific animals that are most likely to live in your environment. These animals require specific foods and abiotic conditions in order to survive. As with plants, the more you know about your local animals and their needs, the better you will understand your environment.

- What are five examples of animals that are native to your area?
- What can these animals tell you about your environment?

Listen to a slide talk about your native animals to learn the answers to these questions.

PROCEDURE

1. Make a chart like the one below.
2. Fill in your chart while you watch the slides. Take neat, complete notes so they will be helpful to you later.
3. When the slide show is over, look back over your notes and, for each animal, ask yourself these questions:
   Would I be able to identify this animal if I saw it? What characteristics would help me recognize it?
   What does this animal eat? What else does it need in order to survive?
   What is the most interesting thing I learned about this animal?
4. Answer the analysis questions.

ANALYSIS QUESTIONS

1. Look over your chart.
   a. How many animals did you learn about?
   b. How many of these are mammals?
   c. How many are insects?
2. Which animal in the slide show do you think is the most interesting? Why?
3. List 5 locations in your area that you think might provide suitable living conditions for these animals.
4. Which of these animals do you think you could identify?

CONCLUSION

When you write your conclusion to this activity, make sure to answer the two

Animal	Identifying Characteristics	Basic Needs	Other Information

focusing questions on page 65. Also include a few sentences that explain how you felt about this activity.

EXTENSION ACTIVITIES

1. Learn more about one of your native animals so that you can teach others about it. Using what you learn, you could prepare a poster for your classroom. Another activity could be to write a story as if you were the animal. You could explain what you eat, how you spend your day, what frightens you, who "your friends" are, and anything else that makes your life interesting.
2. Put together slides about another animal for your teacher to use when he or she presents the slide show again. Write the narrative to go along with the slides. It should include important information about the animal.

Not a Native

The perky house sparrow is a familiar sight in most major American cities. These sparrows seem to be happiest when living by people; they prefer making their nests by occupied buildings. House sparrows who live in the cities seldom move more than a mile or two from their birthplace. However, sparrows who live in more rural areas may flock to nearby fields during the summer months.

House sparrows are not native to North America. They were brought over from Europe in the 1850s by farmers who wanted the birds to eat insects. Many immigrants welcomed this friendly reminder of home.

3–4 ANIMALS IN AN ECOSYSTEM

Animals are found throughout the world—inhabiting every ecosystem from those that are hot and dry to ones that are wet and cold. They live in, on, and above the ground. They thrive on the tallest mountains and in the depths of the oceans. They may creep, run, fly, swim, or barely ever move. They may produce millions of young every year or only one or two in a lifetime.

However, all these animals have one thing in common; they must eat in order to get the energy they need to survive. It does not matter if an animal is a toad, a clam, a slug, a human, or anything else; each kind of animal must eat. To say it another way, **every animal is a consumer.**

Basically, there are three kinds of consumers—those who eat plants, those who eat animals, and those who eat both plants and animals.

Three Kinds of Consumers

Animals that eat plants are called **herbivores.** There are thousands of species of herbivores. They include

Figure 3–2 Beavers and caterpillars are herbivores. What do you think they eat?

everything from tiny, leaf-eating insects to large mammals. Herbivores are also known as **primary consumers.** You may be an herbivore if you avoid eating meat and other foods that come from animals. (Two examples of herbivores are shown in Figure 3–2 on page 67.)

Some other animals seldom eat plants but live by eating other animals. While you may have heard them called **carnivores**, they are also known as **secondary consumers.** A trout is an example of a secondary consumer since it lives off smaller fish and other little

Figure 3–3 Carnivores, such as praying mantises and octopi, eat meat. What kinds of meat do you think these animals eat?

Figure 3–4 Bears and sponges are omnivores. Their diets include both plants and animals.

water animals; so are all spiders and snakes. More typical examples are usually mammals such as mountain lions, wolves, and bobcats. (See Figure 3–3 on page 68.) Do you know any people who are carnivorous? Probably not. A person who lived on a diet of nothing but meat would not be healthy.

Those animals who eat both plants and animals are called **omnivores.** Most people are omnivores, and so are other mammals such as pigs, bears, raccoons, and rats. Some sea animals—such as sponges, clams, and oysters—filter the water, eating anything they happen to catch. Since they filter out both microscopic plants and animals, they, too, are omnivores. (See Figure 3–4.)

Carnivore, herbivore, omnivore . . . all of these words come from Latin. If you understand the Latin bases, it is easier to remember their definitions.

vorus = to devour or eat greedily
carn = flesh
herb = grass
omni = all

By referring to this list, you can see, for instance, that *carnivorous* means "to eat flesh (meat) greedily."

Which Kind of Consumer?

Since all animals are consumers, every animal can be labeled as a carnivore, an herbivore, or an omnivore. It sounds as if it would be easy to classify an animal based on what it eats; but it isn't. We seldom know enough about an animal's diet to know what kind of consumer it is. For example, what would you call a robin? A carnivore since it eats worms? Actually, a robin eats berries, fruits, and other plants in addition to worms and other soil animals so it is an omnivore.

The more you learn about animals and their environments, the more you will be amazed by the complex interactions that go on not only between animals but also between the animals and other components of their ecosystem.

ANALYSIS QUESTIONS

1. Considering your diet, which one of these terms best describes you? Explain.

primary consumer	herbivore	omnivore
secondary consumer	carnivore	

2. Copy this list of animals and indicate if each is an herbivore, a carnivore, or an omnivore.

horse	wolf	deer
canary	frog	garter snake

3. Give 3 examples of each of the following types of animals. Think of other examples than those that have been mentioned in this text.
   a. primary consumer   b. secondary consumer

4. Using the Latin bases on page 70 as your guide, what does *omnivorous* mean?
5. Think of at least 2 other words that come from the Latin base *carn* (flesh). Write down these words and their definitions.
6. If you consider every kind of animal on earth, which do you think is more numerous—the primary consumers or the secondary consumers? Defend your answer.
7. Mountain goats live atop high mountain ridges in Canada, Alaska, and a few other states. They are herbivores who graze upon such plants as low shrubs, daisies, strawberry blossoms, grasses, ferns, and mosses. Imagine that it has been a particularly long and harsh winter. Months of blizzards and bitter winds have made it impossible for the mountain goats to find anything to eat. One starved animal comes across a mouse that had recently died. Will he eat it? Explain your answer.

HOMEWORK

Consider the consumers that live in your own environment. Learn more about 3 of them. Look them up in several references or call someone who may know about these particular animals. Find at least 6 examples of plants and/or animals that each one eats and at least 2 examples of animals that eat each of them. Then decide if each of these consumers is an herbivore, a carnivore, or an omnivore.

Insect Diets

About one-half of all insects are primary consumers; their diets consist only of plants. About one-fourth are scavengers; they feed on decaying plant and animal material. The other fourth are secondary consumers; they get their energy by eating other animals. While some of these insects rely on birds, mammals, reptiles, and other fairly large animals for food, most of them feed mainly on other insects.

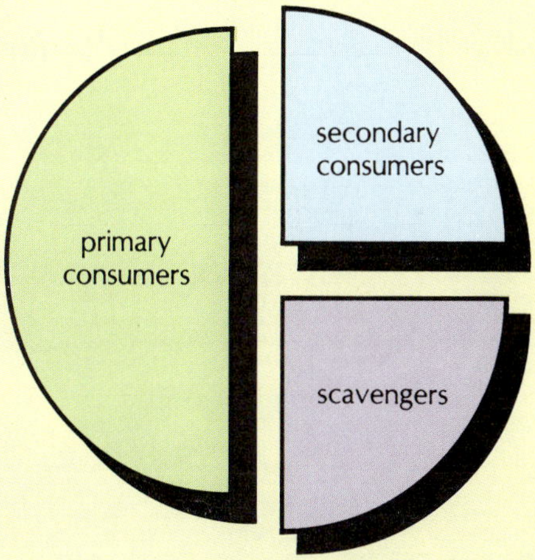

3-5 WHO EATS WHAT?

Look closely at any ecosystem and you will see animals eating. The herbivores will be eating plants and the carnivores will be eating other animals. If you watch for a long period of time, you may end up making a series of observations such as these: "An owl eats skunks; skunks eat mice; mice eat the seeds on grass." If you put all of this information together, you can make a **food chain.** Each organism can be thought of as a link in that chain.

However, from your observations, you would also realize that owls don't always eat skunks. In fact, they're more likely to eat mice or even small birds. Likewise, skunks don't live on mice alone. They also eat frogs, small rabbits, and eggs. When all of these eating relationships are put together, a **food web** is formed. A food web shows how energy is transferred from one organism to another in an ecosystem.

- What is the hardest part of making a food web?
- What can you learn about an environment by making a food web?

Answer these questions after you have worked on a food web that includes at least 12 organisms from your local environment.

MATERIALS
colored pencils or markers

PROCEDURE
Work in a group with two or three other students.

Part A. Getting Ready
1. Ask one person in your group to be the recorder and entitle one piece of paper "Local Organisms."
2. As a group, brainstorm a list of at least 15 *producers* that grow in your local ecosystem. Be sure to include native plants. The recorder should list the organisms on the paper.
3. Now brainstorm a list of at least 15 *consumers* that live in your local ecosystem. Again, remember to include native animals. (These should be wild animals, not pets.) The

recorder should also record the organisms.

4. Follow these steps to choose 12 organisms to use in your food web.
   a. Look over your list and, as a group, decide on at least 3 producers to include.
   b. As a group, choose at least 6 consumers from your list.
   c. If you have not selected at least 12 organisms, choose as many more as you need.

Part B. Drawing a Food Web

1. Ask someone in your group to draw the food web. Choose someone who can write clearly and neatly to label the web.
2. Have this person start by writing the names of the 12 organisms all over a new sheet of paper.
   a. As a group, decide which of the organisms are primary consumers. Circle them with yellow.
   b. Circle the secondary consumers with red.
   c. Circle the producers with green.
3. Draw arrows between the organisms to show where each organism gets its energy. Since these arrows show the movement of food energy, they should point from the food to the consumer. (See Figure 3–5.)

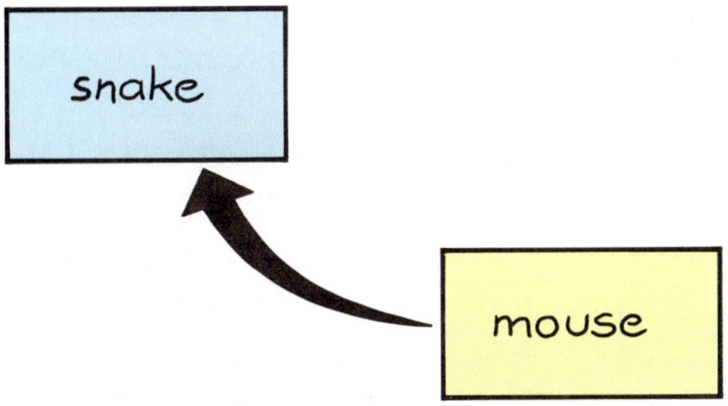

Figure 3–5

4. Most organisms eat more than one kind of food. Draw as many arrows as you need to show all of the eating relationships.
5. Finish your food web.
   a. Put a title on your food web.
   b. List at least 2 abiotic characteristics of this ecosystem on your page. (You may need to look back on page 15 in the reading entitled ***Ecosystems*** to help you remember what abiotic characteristics might include.)

CHALLENGE

Use a brown colored pencil to add the word **decomposers** to your food web. Make brown arrows to show how they fit into the web.

ANALYSIS QUESTIONS

1. What is the difference between a food chain and a food web?
2. Which organisms in your food web don't eat anything? Why?
3. Which organisms in your food web aren't eaten by anything? Why?
4. Do you think there is an organism in your food web that is more important than any of the others? Explain your answer.
5. Consider the organisms in your food web. In this ecosystem, which do you think is
   a. most abundant? Why?
   b. least abundant? Why?
6. Use the small, imaginary food web in Figure 3–6 on page 74 to answer these questions.
   a. What kind of an animal could the jorb be?

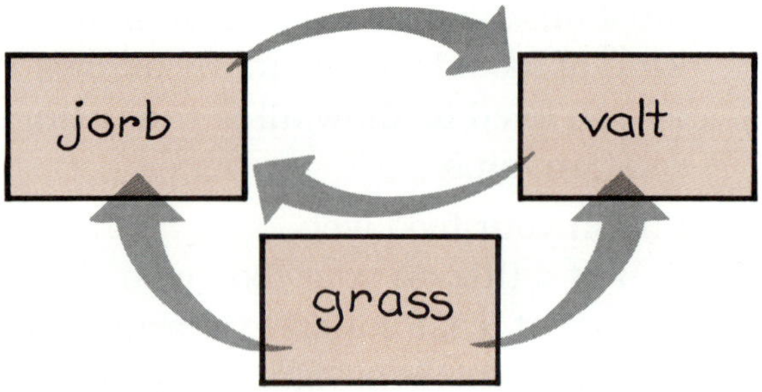

Figure 3–6

b. What could the valt be?
c. Do you think this situation exists in any real ecosystem?

7. Look at the food web in Figure 3–7. Imagine that all of the mice just died of a viral infection.
   a. How will the coyotes be affected?
   b. How will the sunflowers be affected?
   c. What might happen to the dandelion population?
   d. How would the prairie dog population be affected?
8. List at least 4 ways that humans could affect the food web in Figure 3–7.

CONCLUSION

Write short paragraphs to answer each of the focusing questions on page 72.

EXTENSION ACTIVITIES

1. Turn your food web into a work of art! Add illustrations. Make it into a mobile or a diorama. Design a bulletin board entitled "The Local Food Web" for your classroom.
2. *"Benjamin Mac Donald was following a mouse. The fact that he was doing so was nothing out of the ordinary for Ben; he often followed mice. For that matter, he followed birds, too, . . . and ground squirrels and snowshoe rabbits and anything else if he got half a chance."*

   So begins *Incident at Hawk's Hill* by Allan W. Eckert. It is the story of a shy 6-year-old who is adopted by a badger. The story is based on a true incident that took place in the summer of 1870. Read the book and then ask yourself if this could really have happened. You will learn quite a bit about badgers from the book; use the information to make a food web for badgers.

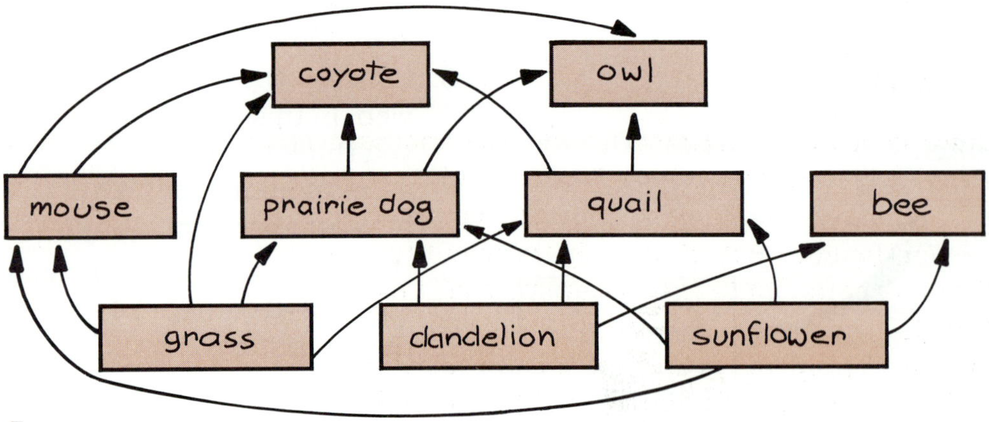

Figure 3–7

3-6 PREDATORS

"Little pig, little pig, let me come in!" cried the big bad wolf.

The little pig was very frightened but he acted brave and replied, "Not by the hair of my chinny chin chin."

So the wolf said, "I'll huff and I'll puff and I'll blow your house in."

So he huffed and he puffed and he blew the house in and. . . .

So goes the familiar childhood story of "The Three Little Pigs." You've probably heard this story when you were young and, like millions of other children, you probably cheered for the pigs and booed the wolf. And yet, the wolf was only doing what he needed to do in order to survive. This is a classic example of a predator-prey relationship.

Predators and Prey

A **predator** is any animal that kills and eats another animal in order to survive. Lions, hawks, and crocodiles are predators; so are robins, garter snakes, and trout—even though we don't usually think of them as such. They all catch and eat other animals as a part of their diets. The animal that is hunted is called the **prey.** A robin's prey may be an earthworm, while a frog's is likely to be a fly; a trout's might be mosquito larvae.

Some animals can be both a predator and prey—it depends upon the situation. A frog, for instance, is a predator when it is feeding on an insect but it becomes prey when it is eaten by a heron.

Survival

The predator hunts to survive. If it does not catch its prey, it will die. We often forget this when we watch one animal stalk another. We may cringe when we see a lion pounce on an unlucky zebra during a television program or we get upset if we see a blue jay steal a baby warbler from its nest. But what would happen if the lion never did catch a zebra or the blue jay never succeeded in raiding a nest? (See Figure 3–8.)

Figure 3–8 What do you think each of these predators is eating? Left: emerald tree boa; Top right: purple heron; Bottom right: chameleon.

Figure 3-9 Robber flies survive by sucking the body juices out of bees, beetles, and other insects.

Ecosystem Balance

Predation—the preying of one animal on another—is not only essential for the survival of the predator, it also plays an important role in the balance of an ecosystem. Predators help control the size of their prey populations. Without predation, prey populations would grow so quickly that the ecosystem would suffer. For example, in the early 1900s, the ranchers in the Kaibab plains near the Grand Canyon killed many of the local wolves and coyotes. Without these predators, more deer survived. After several years, there were so many deer that food became scarce and thousands of deer starved to death.

Predators and prey need each other to keep their populations in balance. Predators usually capture the old, crippled, sick, or very young animals. Very rarely are healthy adults caught and killed. In this way, only the strongest and smartest animals are left to reproduce. Over long periods of time, predation actually improves the health of the prey population.

Changing Attitudes

People have not always recognized the importance of predators. For years, we hunted and feared the large predators and did not even realize that the small ones existed. Recently, attitudes have changed. Laws have been written to protect predators such as wolves, cougars, bears, and bobcats. Gardeners are using predatory insects to help control the populations of other insects. Land managers and farmers realize that snakes and hawks can help them control the rodent populations. These signs are encouraging. They mean that predators will be protected so they will remain an important part of our environment.

Think about your own attitude towards predators. Do you tend to feel sorry for the prey or see predators as villains? Where did you get your attitude? Consider for whom you would cheer—the predator or the prey—if the story of the three pigs read like this:

> The worried wolf thought, "I must find food for my children. They haven't had anything to eat for two days. The little one was too weak to stand up this morning."
>
> The wolf soon came across a very silly pig who was sunning himself on a large flat rock. . . .

ANALYSIS QUESTIONS

1. Write a definition for each term.
   a. predator
   b. prey
2. Which of the following types of animals could also be called predators? Explain.
   omnivores herbivores carnivores
3. Copy the examples of animal pairs listed below.

rat, rattlesnake	caterpillar, leaf
ladybug, aphid	pelican, carp
clover, bee	fox, plum
ant, anteater	bobcat, ground squirrel

   a. Cross off the pairs that are *not* examples of predator-prey relationships.
   b. Underline the prey in each of the remaining pairs and circle the predator.

4. Do you eat meat? If so, do you think that *you* are a predator? Explain your answer.
5. Think of at least 2 examples of animals that can be called both a predator and prey.
6. Many people have studied the effects of poison when used to kill prairie dogs. In one study, it was found that when the prairie dog population dropped, the ferret population dropped, too. Give *at least 3* reasons why the ferret population may have dropped.
7. Two neighbors are having an argument. Mr. Burnett wants Ms. Quinn to keep her cat in the house. He feeds birds and does species counts for the Audubon Society. He has watched Ms. Quinn's cat kill two birds in three days. Ms. Quinn says that it's a cat's instinct to hunt. "Cats have always hunted birds; my cat is just doing what is natural." She refuses to keep her cat locked in the house.
   a. With whom do you agree?
   b. If you were asked to settle this argument, what would you suggest?

TAKE A STAND

People should be allowed to hunt predators such as skunks or rattlesnakes without needing to purchase a hunting license.

- Do you **agree** or **disagree**?
- What **comments** would you like to make?

EXTENSION ACTIVITIES

1. Rewrite the story of "The Three Little Pigs" so children will cheer for the wolf instead of the pigs.
2. Think of at least 4 examples of predators that are native to your environment. Find out if these animals still roam wild there today and if there are any regulations about hunting them.
3. Read *The Summer of the Falcon* by Jean Craighead George and meet 13-year-old June Pritchard and her young sparrow hawk, Zander. Her twin brothers have raised and trained hawks and owls and help June train her own bird. When you read, keep track of what these birds of prey eat and who their enemies are.

Figure 3–10 A kestrel. These birds used to be called sparrow hawks; but that is misleading since they eat mainly grasshoppers, not sparrows.

3–7 COYOTE CONTROL

When homesteaders moved west, they plowed fields, fenced land, and brought in farm animals. The changes they made in the environment upset the native animals, especially the predators. The coyote, for example, was here long before the settlers arrived. For thousands of years, coyotes had roamed the open lands; but ranchers soon claimed that the coyotes must go. Coyotes, they said, were killing their livestock.

These complaints continue today. The government tried to help ranchers and farmers control coyotes by setting out traps and poisons; they have even hunted coyotes from airplanes. The coyote "problem," however, hasn't been solved. For this reason many states have passed laws about coyote control.

- Why is the coyote problem difficult to solve?
- What do you need to consider when you evaluate a possible solution to an environmental problem?

Keep these two questions in mind while you decide how you would vote on a proposed coyote control law. You will be "listening" to four people to help you make your decision.

MATERIALS

one of four viewpoints (pages 82–85)

PROCEDURE

1. There are three main parts to the proposed coyote control law. Read the contents of the proposed law below and then answer the first analysis question.

   The proposed law would
   - outlaw the use of poisons for predator control.
   - provide money for research to find more humane methods for controlling predators. (This might include a coyote-repellant for sheep and better herding techniques.)
   - pay ranchers for livestock lost to coyotes or other predators.
2. Your teacher will divide the class into four groups. Each group will be assigned one viewpoint to read.

Figure 3–11 A coyote

a. Get into your groups. (You may need to rearrange your chairs in order to talk with each other.)
b. Read these role descriptions and decide who will take each one.
   group leader—keeps group on task and discussion going
   recorder—takes notes during group's discussion
   timekeeper—makes sure group finishes on time

3. Each person in your group should read quickly and silently through the viewpoint that has been assigned. (This should take about 3 minutes.)
4. Have each person reread the viewpoint, but this time read more slowly and concentrate. Then answer analysis questions 2–5.
5. In your group, use the following questions to discuss your viewpoint. The recorder should take notes.
   What reasons does this person give for supporting the law?
   What reasons are given for you to vote *no* on the law?
   What questions would you like to ask this person?
6. During the class discussion, listen to the viewpoints of the three other people in the text. Ask questions to make sure that you understand how they feel about the proposed law.
7. After the class discussion, answer the rest of the analysis questions.

ANALYSIS QUESTIONS

1. Which part of the proposed law do you think is most important? Why?
2. Why do you think the person whose viewpoint you read is interested in the law?
3. What part(s) of the law do you think this person is most interested in?
4. What does this person think of coyotes?
5. According to this viewpoint, what do coyotes eat?
6. Do *you* think the proposed law should be passed? Why?
7. How do you think the law could be improved?
8. Describe what you think either a cattle or a sheep rancher will have to say about the coyote situation ten years from now if this law passes.
9. Think of at least 2 questions you have about coyotes—about their behaviors, their diets, their young, or anything else you might want to know.

CONCLUSION

Write a paragraph to answer each focusing question on page 80.

EXTENSION ACTIVITIES

1. Call your state division of wildlife or department of natural resources to find the answers to these questions:
   Do coyotes live in your state? If so, where?
   Are they a problem in your state? If so, where and why?
   What are the laws about killing coyotes?
   Can the poison 1080 be used legally in your state?
2. Do some research on the poison 1080. Find out how it works and how it is used. Decide whether or not you think it is harmful to the environment.

From a Sheep Rancher's Viewpoint

"You want me to tell you what I think about the proposed law? I think it's awful. We've got a real coyote problem here. That law would put me out of business. Let me tell you just how serious the coyote problem is in my area.

"I'm a sheep rancher. There are several of us in this area. Each year we all lose sheep to the coyotes. Last week I lost four sheep in one night; they all had coyote teeth marks on them. To make it worse, the coyotes only ate part of the sheep. That's what they always do; they seem to kill for the sake of killing. That's one reason why we hate coyotes. We work hard to raise our sheep, and then coyotes kill them. It's pretty upsetting to see baby lambs with teeth marks on their necks.

"This last time I called government officials for help. They set out poison for the coyotes but they didn't use enough. I've lost three more sheep since then, so I'm going to put out my own poison. Some people say the poison is killing other wildlife, such as bobcats, mountain lions, bears, ferrets, and badgers. Some of these animals might eat the poison, but mostly the coyotes do. Fences don't keep coyotes out and traps won't work. Coyotes can sense a trap and won't come near it. What am I supposed to do?

"If this law passes, it will be the end of sheep ranching. I'm already losing about 8% of my sheep to coyotes. If we continue to lose sheep, we'll go out of business. There'll be less lamb for people to eat and less wool to make clothing. With such a shortage, lamb and wool will get so expensive that no one will be able to afford them. I won't be able to stay in business.

"I know there are a lot of people who don't like killing coyotes. I wish there were another way, but I don't like coyotes killing sheep. Even if the bill passes, many ranchers are going to get poison and do their own coyote-controlling."

From a Cattle Rancher's Viewpoint

"I want you to know that as a cattle rancher, I'm totally in favor of this law. At one time I thought that killing coyotes was a sensible thing to do, but I've changed my mind. Let me explain why.

"It's true that coyotes affect my cattle business, but they help me more than they hurt because they eat a lot of rodents and rabbits. When there are fewer coyotes, there are more gophers and rats. The rodents eat the grain I store for my cattle and the rabbits eat the grass my cattle need for grazing. So if coyotes kill a few calves, I figure that's a small price to pay for keeping the rodents and rabbits from eating the cattle feed. Besides, gophers make holes in the pastures and the cattle can break their legs in the holes. Fewer gophers mean fewer holes.

"The poisons and traps that some people have been putting all over the range are a bigger problem than the coyotes. I've lost two dogs in the last three years. I tracked both of them in the snow and found them dead not far from some poisoned meat that was meant to kill coyotes. Luckily I found the dogs and buried them before some cougars got killed from eating the poisoned dogs.

"I've heard some sheep ranchers say, 'The only good coyote is a dead coyote.' Coyotes do seem to kill a lot of their lambs, but killing all those coyotes hasn't helped much. The government trappers have convinced the sheep ranchers that killing coyotes is the only way to handle the problem. They know that if the law passes, they won't have a job; they want to continue spreading poison and trapping and shooting coyotes. You'd think that sooner or later they'd admit that they aren't solving the problem. Coyotes seem to be smarter than people in some ways; they have to be in order to survive.

"Coyote-killing makes me mad because it's doing more harm than good. There has to be a better way to control coyotes and other predators. I'm glad that the proposed law would set aside money for research on this problem."

From a Conservationist's Viewpoint

"I belong to the National Conservation Club. I'd like to explain our position on the proposed law that would restrict the use of poisons for controlling coyotes and other predators.

"We're especially concerned about the poison 1080 which is now used to kill coyotes. No one knows how many animals have been killed by eating the poisoned bait which is set out for coyotes. Most ranchers have lost at least one dog to 1080. Hundreds of mountain lions, badgers, eagles, hawks, and other predatory animals probably have been killed by it, too. In addition, when it rains, some of the poison from the bait may dissolve and be carried into water supplies. Once there, no one knows just how long 1080 remains active.

"We've heard that sheep ranchers are against this law; they want to continue killing coyotes. They seem to forget that, in 1900, before predator control programs, there were 50 million sheep in this country. Today, after over 90 years of poisoning, shooting, and trapping coyotes, there are only 16 million sheep. Efforts to control coyotes haven't helped the ranchers because coyotes aren't the main problem; the main problem is economic.

"Prices for sheep have dropped over 30 percent in the last 20 years. In addition, ranchers' expenses have increased nearly five times. Even though coyotes have not caused these problems, they are easy to blame. It's easier to kill coyotes than to kill increasing expenses and competing industries.

"The effort to control coyotes has not only failed to help sheep ranchers; it has hurt them. Sheep ranchers don't seem to understand that coyotes eat a variety of foods including rabbits and rodents. When the coyote populations decrease, the number of rabbits and rodents can increase. Rabbits eat mainly grass. When there are too many rabbits, livestock have a hard time finding enough grass to eat. Rodents often eat grain that has been stored. An increase in rodents and rabbits costs the ranchers money.

"I urge you to vote *for* the proposed law. Killing coyotes has not helped them so let's try a new approach."

From a Trapper's Viewpoint

"I'm concerned about the proposed changes in the coyote-control law. I've been a trapper for the U.S. Fish and Wildlife Service for the past 15 years. One of my main duties is to kill coyotes when a sheep rancher requests help. I don't enjoy killing coyotes, but I see the need for it.

"In my territory, there used to be 35 sheep ranchers; now there are only 17. The rest of them quit or switched to raising cattle because it isn't very profitable to raise sheep anymore. The ranchers blame their losses on the coyotes who kill their lambs.

"The ranchers depend on us to help them protect their sheep. There are several methods that we can use; let me mention a few of them. Sometimes we shoot coyotes from airplanes and helicopters but this is pretty expensive. Besides, coyotes often come out at night, and it's hard to spot them from the air.

"We also do quite a bit of trapping, but I don't like this method because other kinds of animals can get caught. Then, too, a coyote may just be injured in a trap. Injured coyotes can't run fast enough to catch rabbits and rodents so they become 'killer' coyotes, stalking lambs that don't run very quickly. Trapping just makes more problems for the rancher.

"Poisoning is the best method we have to get rid of coyotes. We set out rabbit or sheep meat that contains 1080. This poison kills the coyote in a hurry. True, 1080 can kill some other animals if they eat the poisoned bait, but the little bit of poison in the body of a dead coyote won't hurt a cougar or other animal that eats the remains. I don't see how the poison we use really hurts the environment.

"I think outlawing poisons would be a mistake. Right now we're running a sensible control program. We haven't caused any animals, including coyotes, to become extinct. If you vote to outlaw poisons, sheep ranchers may break the law. They may do their own poisoning, and try to wipe out all the coyotes. I don't want to see this happen."

3–8 A SURVIVOR

The pups are restless and hungry. It's feeding time but the mother coyote hasn't made her usual mid-morning visit. The pups whine, nip at each other, and scramble to the front of the den, only to run clumsily to the back again. They stop when they hear a soft whine. It's the voice of the father coyote. He's been standing at the mouth of the den, cocking his head from side to side, and listening to the scratching and yipping of the pups inside. Again, he whines. Even though the pups have never heard this command before, they know that it means "Come on out!" The five pups tumble out into the sunshine, eager and curious about the world.

If you were to see these coyotes, you might think that you were looking at a family of dogs. A coyote looks a lot like a shepherd-collie mix. Like medium-sized dogs, coyotes are from 15- to 24-inches high at the shoulder and they weigh from 25 to 30 pounds. Their fur color varies depending on where they live. Coyotes native to woodlands tend to have dark fur, allowing them to blend with shadows. Those that live in the deserts have lighter colored coats which makes them hard to see against the sand.

Figure 3–12 What do you think this young coyote pup is pondering?

A Coyote Diet

Coyote pups learn to hunt by stalking insects. At first they think this is a new game and that grasshoppers, crickets, and butterflies are amusing toys. It isn't long, however, until they realize that the "game" is serious and can mean survival. They will soon venture out on their own, constantly on the watch for food and danger.

Many studies have been done to find out what a coyote eats. Most of these have involved analyzing the

Food Items in a Coyote's Diet

Rodents and other small mammals **63–70%**

This group includes mice, cottontails, gophers, ground squirrels, meadow voles, skunks, porcupines, beaver, hares, marmots, and woodrats. Many of these prey animals are also considered to be pests by ranchers and farmers.

Deer, livestock, and other large mammals **5–16%**

In addition to deer, this group includes antelope, sheep, elk, and cattle. Coyotes do not usually kill these animals; they eat the carcasses of animals that have died from other causes. When they do take live prey, they usually look for unfit individuals (the sickly, old, or wounded) or the very young.

Insects and other invertebrates **7–21%**

This could include moths, butterflies, caterpillars, crickets, beetles, spiders, dragonflies, and anything else that may catch a coyote's attention.

Vegetable matter **4–8%**

This varies considerably depending on the location and time of year. Coyotes eat apples, corn, watermelon, berries, grasses, and other vegetation.

Birds, reptiles, and amphibians **3–4%**

Ground-feeding birds—such as juncos, robins, road runners, and sparrows—and frogs, toads, snakes, lizards, tadpoles, and salamanders may all become coyote prey.

Miscellaneous items **2–6%**

This includes anything that's easy to find—string, rags, orange peels, shoe leather, gunny sack pieces, paper wrappers. When starved, a coyote will eat just about anything.

contents of dead coyotes' stomachs; others are based on examining coyote droppings. All of the studies have shown that coyotes eat whatever is available. Exactly what each eats depends on the season of the year and where the coyote lives. As with all wild animals, the coyote must struggle for its survival—the preferred food may not be available when hunger strikes.

The results of a typical study might look like those shown in the chart on page 87. As you can see, small mammals make up most of a coyote's diet. However, this varies considerably from one coyote to another. Some coyotes are mainly vegetarians, relying almost totally on plants for food; others are strict carnivores, eating only meat. Occasionally, a coyote will discover that livestock are easy prey. They may then become problems for ranchers, since they watch for vulnerable sheep and cattle. Because of these problem animals, some people believe in strong coyote-control programs.

People Versus Coyotes

A variety of methods have been used "to control" the coyote population and protect the livestock. Many of these methods are directed towards *all* coyotes and not just the problem ones. Some of these methods include the following:

Direct shooting is one of the oldest methods; rewards have been offered for every animal killed.

Aerial hunting may be used by flying over pastures where coyotes have been killing livestock.

M-44 is a small device that is scented to attract coyotes and then buried in the ground. A poison (sodium cyanide) is released into the mouth of the animal that digs it up.

1080 is a poison that is sometimes used in bait. Because of the controversy that surrounds this poison, it is legally controlled. (See the table on page 89 for more information about 1080.)

Coyote-proof fencing uses high voltage wires to prevent coyotes from reaching livestock.

Corralling livestock at night reduces the opportunities for predator attacks.

Guard dogs living with sheep can protect them from predators.

Figure 3–13 In other countries for 3,000 to 4,000 years, dogs have been used to guard sheep and goats.

The Poison 1080

The poison 1080 has been used in a variety of ways to control predators. It is especially effective against coyotes, wolves, and dogs. There have been many studies on the amount of poison required to kill various animals. This amount is referred to as the **lethal dose.** Not all studies give the same results. The data from one study is shown in the table. As you can see, to some extent, the lethal dose depends on the size of the animal.

This has been a "popular" poison since it is odorless and tasteless when used in bait. There is no antidote or treatment for a person or animal that has eaten this poison.

Lethal dose (milligrams)	Animal
4.6	badger
48.0	bear
1.4	black vulture
2.3	bobcat
.1	cat (domestic)
4.5	chicken
.5	coyote
.5	dog (domestic)
.6	fox
8.0	golden eagle
5.6	great horned owl
5.6	hawk
48.0	human (150 lbs.)
.1	magpie
.5	marten
.5	mink
76.9	opossum
.4	prairie dog
7.7	raccoon

Lambing in sheds rather than on the open range protects the newborns from bad weather as well as from predators.

Herding by trained herders can cut down on livestock loss from all causes.

Lithium chloride laced bait can be used to train predators to stay away from livestock. It is a nonpoisonous chemical that makes animals sick.

Rodent Control

Coyotes are an important part of the ecosystem. Since they eat so many rabbits and rodents, they help control the number of these small animals. Over the years, a natural balance develops between the rodent and coyote populations. When the coyote population goes up, the rodent population goes down—causing the coyote population to drop. Then the rodent population goes back up again.

For example, in 1946, the ranchers who owned property around the National Bison Refuge in western

Figure 3–14 Coyotes are known for their mournful evening howls. This is a true sound of the wild.

Montana asked the United States government to destroy the coyotes living inside the refuge. They claimed that coyotes who were bred in the refuge moved out onto their land and killed livestock. As a result of the ranchers' request, the government trapped and poisoned almost all the coyotes inside the refuge. Consequently, the rodent populations grew rapidly, eating the grass the bison normally ate. The bison were starving; half of them had to be killed so the remaining bison would have enough food. When the killing of coyotes was stopped, the coyote population increased. Then the rodent population dropped and the bison again had enough grass to eat.

Now that we recognize the importance of coyotes to the ecosystem, we are doing a better job of protecting them. Once people accept the coyote, they may be more willing to accept other predators and preserve other ecosystems.

ANALYSIS QUESTIONS

1. Consider the food coyotes eat and decide if they are carnivores, herbivores, or omnivores.
2. Make a food web that includes the coyote and at least 12 of the foods it eats. Add arrows to show as many eating relationships as you can.
3. Look at the descriptions of methods used to control coyotes.

   a. Which of them would definitely affect only problem coyotes and not the other coyotes in an area?
   b. Which of them do not involve killing coyotes?
4. Read the table on page 89.
   a. How many milligrams of 1080 does it take to kill a coyote?
   b. What other animals would be killed by the same amount (or less) of 1080 that kills a coyote?
5. If you find a dead lamb in your pasture, list 4 reasons for its death.
6. What happened when the coyotes in the National Bison Refuge were killed?
7. Have you ever seen a coyote? If so, where?
8. If you were in a meeting that concerned methods of coyote control, what would you support doing?

EXTENSION ACTIVITIES

1. If you are interested in learning more about coyotes, read *Wonders of Coyotes* by Sigmund A. Lavine. From this book, you can find out more about the raising of coyote pups, Native Americans' attitudes toward coyotes, or the history of coyote control.
2. Read about wolves and take notes. Look for ways that wolves and coyotes are alike and then consider their differences. Why are coyotes surviving despite human pressure and yet wolves are endangered?

Figure 3-15 What similarities do you see between this wolf and the coyotes pictured earlier? What differences?

3-9 HOW MANY COYOTES?

Imagine that you are talking to some ranchers about their coyote "problem." They tell you that they don't mind a few coyotes but that recently there have been too many of them. "Too many." How many are too many? Is there a limit to the number of coyotes that can live in an area?

- Would it make any difference if there were 2, 10, or even 40 coyotes living on a ranch? Why?
- What limits the number of organisms that can live in a given area?

Do an activity in which you simulate a coyote population to help you answer these questions.

MATERIALS

data page: ***How Many Coyotes?*** (Figure 3–17)

PROCEDURE

Part A. Getting Ready

1. Read the information below and study Figure 3–16.

   Until recently, coyotes lived only in the open country and grasslands of western North America. Now they can be found throughout the continental United States as well as in much of Canada and Mexico.

Figure 3–16 The coyote range in North America

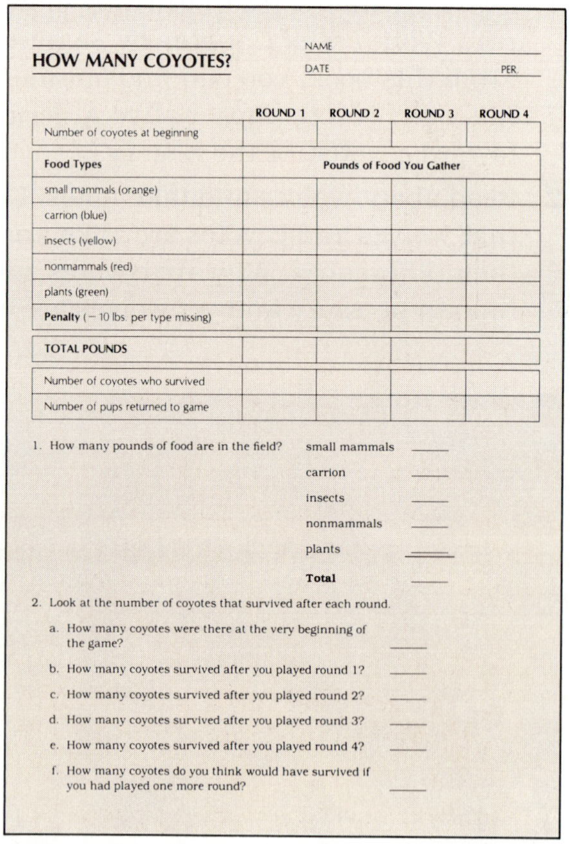

HOW MANY COYOTES?

NAME ______

DATE ______ PER. ______

	ROUND 1	ROUND 2	ROUND 3	ROUND 4
Number of coyotes at beginning				
Food Types	**Pounds of Food You Gather**			
small mammals (orange)				
carrion (blue)				
insects (yellow)				
nonmammals (red)				
plants (green)				
Penalty (− 10 lbs. per type missing)				
TOTAL POUNDS				
Number of coyotes who survived				
Number of pups returned to game				

1. How many pounds of food are in the field? small mammals ____ carrion ____ insects ____ nonmammals ____ plants ____ **Total** ____

2. Look at the number of coyotes that survived after each round.
   a. How many coyotes were there at the very beginning of the game? ____
   b. How many coyotes survived after you played round 1? ____
   c. How many coyotes survived after you played round 2? ____
   d. How many coyotes survived after you played round 3? ____
   e. How many coyotes survived after you played round 4? ____
   f. How many coyotes do you think would have survived if you had played one more round? ____

Figure 3–17

2. Read the rules in the next section before you go outside to do the coyote population simulation.

Part B. Doing the Simulation

1. "Coyote food" is scattered around the field. You, as a coyote, can choose a "den" anywhere around the field.
   a. Write your name on your data page and use it to mark your den. (If a pencil is not heavy enough to hold the paper in place, use a rock or book to weigh it down.)
   b. Once you have identified your den, stand by it.
2. You will all be coyotes, but you will not all be exactly equal.
   - One young male coyote was trotting across a ranch last week when he put his foot in a trap. As a result, he has a broken leg. This coyote must hunt by hopping on one leg.
   - The other special coyote is a mother with two young pups. She must gather twice as much food as other coyotes.
3. When the teacher says to start, *walk* through the field to gather food.
   a. When you find one thing that looks appetizing, pick it up and return to your den with it. (Coyotes actually would not return to their dens; they'd eat on the spot.)
   b. After "eating" an item, return to the field to find something else to eat.
4. A round in the simulation ends when all the food has been gathered.

NOTE

There are five different kinds of food represented by five different colors. In order to have a balanced diet, you must eat from each of these food "groups." Coyotes, like people, need a variety of foods in order to stay healthy.

5. Before you start, you need to know a little more about coyote behavior.
   - Snatching food from the crippled coyote is natural; wild animals are seldom generous.
   - Coyotes do not steal from each other's dens.
   - Coyotes seldom fight. If they did so, they could be injured and unable to gather sufficient food; then they would starve.
6. After one round, use your data page to calculate the number of pounds of food you collected. (The numbers on the colored papers/sticks represent pounds of food.)
   a. Record the number of pounds of small mammals you collected by adding up the numbers on the orange papers. Total the number of pounds in the same way for each of the other food types.
   b. For every type (color) of food you're missing, penalize yourself 10 pounds.
7. A coyote needs 40 pounds of food to survive for ten days.
   a. If you collected less than 40 pounds, you died. "Dead coyotes" must go to the "graveyard."
   b. Any coyote with at least 30 extra pounds of "food" can have a pup. When this happens, an "adult" coyote from the graveyard should

return to the game. The most pups that any one coyote can have is three.

8. When you have played four rounds, finish filling in the data page and answer the analysis questions.

The object is not to see who "wins." There is no winner. The object is to see how many coyotes are able to survive in this area.

ANALYSIS QUESTIONS

1. If a coyote needs 40 pounds of food to survive for 10 days, how many coyotes should have been able to live in this area? (Use the total amount of food you calculated on the data page to answer this question.)
2. Since the amount of food never changed, why did the number of surviving coyotes change each time you played the game instead of staying the same?
3. If there is enough food for one mother coyote *or* two pups, who do you think should get the food—the mother or the pups? Explain your answer.
4. List at least 3 ways that this simulation is like a real coyote population in a real ecosystem.
5. List at least 4 ways that the real situation would be different from this game.
6. If you wanted to figure out how many coyotes could live in an area, you would have to analyze the food supply; but you'd also have to consider other things. List at least 5 other factors that could affect the size of a coyote population.

CONCLUSION

Write a paragraph that answers both of the focusing questions on page 92.

3–10 THERE'S A LIMIT

You would never see 50 squirrels living in someone's backyard—unless, of course, it was a very large yard. Nor would you ever hear about 5 adult bobcats living on an acre of land. Why? Because there is a limit to the number of organisms that a given amount of land can support. This limit is called **carrying capacity.**

Exceeding the Limit

Imagine what would happen if 50 squirrels were forced to live in an average-sized yard. Since they aren't used to living so close to each other, there would be lots of fights; some animals would be hurt or even killed. Of course, the squirrels wouldn't fight all the time; they'd pause to eat in order to keep up their energy. Soon, every seed, bud, berry, and edible leaf would be gone. Some animals would weaken and die. The starving squirrels would quickly devour the newly dead. Disease might strike.

Obviously, this yard cannot support 50 squirrels; it does not have a carrying capacity for that many squirrels. And now, since so much of the vegetation has been eaten and the growth "damaged," the yard has a carrying capacity for even fewer squirrels than it had in the beginning. Based on this example, we can say that carrying capacity is the number of animals *of a specific kind* that can live in a given area *without damaging it.*

Figure 3–18 Red squirrel

Limiting Factors

There is always something that will limit the number of animals that can live in a given area. **Food** is one of the most obvious limiting factors. There has to be enough food to feed the animals year round—not just during the summer months. If there isn't, the animals may need to move from season to season. For instance, during the summer months, deer and elk may live high in the mountains. However, when winter arrives and deep snows cover the grasses and other edible plants, they may move down to the lower valleys and plains to graze.

Figure 3–19 How would you describe this alligator's habitat?

In some areas, **water** is critical. This is especially true in the west. Animals may guard their water supply, not allowing others to settle nearby.

Space is also important. This is related to each animal's size and need for food—a grizzly bear needs much more room than does a shrew. However, there's more to the need for space than the physical requirements. Some animals have personalities that need more space than others. They may feel crowded if other animals are around.

Shelter can be another limiting factor. Animals need it for protection from enemies while they are sleeping and for raising their young. Shelter means different things to different animals. A spider may be satisfied with a crack in a rock; some owls may not be able to survive unless they can find a hole at the right height in a specific kind of a dead tree. A pile of brush can provide cover for a chipmunk, but an alligator wouldn't consider living anyplace but in water.

Changing Capacity

Carrying capacity is not an absolute number. It changes from season to season and from year to year. When people move into an area, carrying capacity often changes for most of the native plants and animals. Their habitats are destroyed by the bulldozers and backhoes. As a result, some animals no longer have enough room to hunt; others may be disturbed by noise or activity. Consequently, they may need to leave; but where are they supposed to go? The neighboring areas probably already have reached their carrying capacity.

For years people did not give much thought to the wildlife when they moved into an area. Now we hear more concern for the native plants and animals. Natural areas are often set aside; highways and developments are planned so they will have minimal effects on the land.

As our population grows, it is more and more important that we plan for wildlife. If we don't, what will their future be?

ANALYSIS QUESTIONS

1. What does the term **carrying capacity** mean?
2. List at least 4 things that can limit the size of a population.
3. What do you think could happen in an area where the number of grasshoppers exceeds the carrying capacity?
4. Over 12 million people live in Shanghai, China. The government provides 2- or 3-square meters of living space for each person.
   a. Do you thing Shanghai has reached its carrying capacity? Explain.
   b. Do you think the Chinese population could continue to grow until the entire country is as populated as Shanghai?
5. Do you think there is a carrying capacity on earth for humans? Explain.

TAKE A STAND

There should be laws to limit the number of children people can have.

- Do you **agree** or **disagree**?
- What **comments** would you like to make?

EXTENSION ACTIVITY

Find out more about the size of the human population. Here are some questions for you to research.

How many people live on Earth?
How many live in the United States?
What do people think will be the world's population in the year 2000?
What do the terms *birthrate* and *death rate* mean?
Which countries are growing the fastest?

3-11 A WILDLIFE PROBLEM

Can you think of any place on Earth where people have *not* affected the environment? Is there anywhere that animals can live completely free from human influence? Probably not. Directly or indirectly we have made our presence felt everywhere. And, as the human population grows, we will continue to affect the wildlife. As we do so, we run into problems. Most of these problems do not have easy answers.

- Why are wildlife problems often hard to solve?
- What types of things need to be considered when solving a problem that involves wildlife?

After you have worked at solving a wildlife problem, reconsider the answers to these two questions.

MATERIALS

Parts I, II, and III of A Problem

PROCEDURE

Work on a committee with three other people.

1. Look over the schedule below and make sure you understand what you will be doing.

SCHEDULE

Part I – 10 minutes
- read the problem
- follow the directions
- record your responses

Turn in Part I and get Part II.

Part II – 10 minutes
- read the problem
- follow the directions
- record your responses

Turn in Part II and get Part III.

Part III – 10 minutes
- read the problem
- follow the directions
- record your responses

Turn in Part III.

2. Talk with the people on your committee and decide who will assume each of these roles:
   reader—reads the information to the group
   monitor—reads the section called "Your Task" and keeps the group on task
   recorder—writes the group's responses
   runner—returns and gets the parts of the problem when the teacher calls time
3. All members of the committee should sign the report to indicate that everyone agrees with what is written.

CONCLUSION

Look at the focusing questions above. Talk over these questions within your committee and then write a one-paragraph response to each of the questions.

TAKE A STAND

Middle school students are too young to be involved in making decisions that affect wildlife.

- Do you **agree** or **disagree**?
- What **comments** would you like to make?

EXTENSION ACTIVITY

Learn more about a wildlife problem that is occurring in your own state. There may be a population control problem, an animal becoming extinct, or an animal whose habitat is being destroyed. Talk to people. Ask about the advantages and disadvantages of various solutions. Let them know what you think should be done about the problem.

Figure 3–20 Oil spills cause multiple problems for wildlife.

CHAPTER 4

DECOMPOSERS

Which of these things will decompose the fastest?

4–1 SPECIAL TERRARIUMS

(Designed for a Clean-Up Crew)

Consumers and producers are the obvious parts of an ecosystem. We can all list dozens—maybe even hundreds—of different plants and animals, but very few people know much about the "clean-up crew"—the scavengers and decomposers who quietly go about their business, efficiently changing dead tissues back into raw materials. They don't do this overnight; it takes time. Some things can be broken down in a week or so, others take months or years, while yet others will be around forever.

- Which kinds of things decompose the fastest?
- What conditions are best for decomposer action?

Construct a terrarium for decomposers and scavengers and observe this natural clean-up crew at work in order to answer these questions.

MATERIALS

clear plastic or glass container
ruler
tape for labeling
paper towel
data page: ***Special Terrariums*** (Figure 4–2)
litter
- 2 leaves
- fruit slice
- 3 pine needles
- piece of paper, about 3-cm square
- piece of plastic, about 3-cm square

soil—sterilized or sifted garden
organisms—sow bugs, earthworms, or other kinds

PROCEDURE

Work in a group with two or three other students.

Part A. Assembling the Terrarium

1. Get your litter and observe it closely. On your data page, describe what it looks like. You will be inspecting this litter every 8 to 10 days to see how it changes over time.
2. Your teacher will tell your group which terrarium number to set up. Use the chart on the next page to decide what kind of soil to use.

TERRARIUM PREPARATION CHART

No.	Type of Soil	Organisms to Add
1	sterilized	none
2	sifted garden	none
3	sterilized	8 sow bugs
4	sterilized	6 earthworms
5	sterilized	6–8 other "bugs"
6	sifted garden	8 sow bugs and 6 earthworms
7	sifted garden	none + NO WATER

3. Get the container for your group. Put the litter in the container against the glass or plastic surface so you can see it. Gradually add the soil to hold the litter in place. (See Figure 4–1.)
4. Refer to the chart to see what you need. Carefully bring the organisms to your table. Measure them and record their sizes on the data table. Put the organisms in your terrarium.

Figure 4–1 Terrarium set-up

NAMES ______

DATE ______ PER ______

SPECIAL TERRARIUMS

Terrarium Number: ______ Contents: ______

	Describe Contents	Predict Appearance After 4 Wks.	1st Observation (Days)	2nd Observation (Days)	3rd Observation (Days)
organism(s) (kind, size, number)					
2 leaves (size, kind, appearance)					
3 pine needles (length, appearance)					
1 fruit slice (size, kind, appearance)					
paper (size, kind, appearance)					
plastic (size, appearance)					
other:					

Figure 4–2

5. Water your terrarium UNLESS you have terrarium #7. It should be moist but not soggy. Cover it securely but make sure there is some way for the organisms to get air.
6. Use tape to mark the locations of your litter.

Terrarium #4

Darwin Baca
Stacy Lau
Mark Everly

Period 5

Figure 4–3

a. On the tape, write the name of the litter and, if you feel ambitious, a sketch of what it looks like.
b. Place the piece of tape *near* the piece of litter—not over it.

7. Label your terrarium with its number, the names of everyone in your group, and your class period. Place it where your teacher shows you. (See Figure 4–3 on page 102.)
8. Clean up your work area.
9. Make some predictions. What do you think each piece of litter will look like after it's been in your terrarium for four weeks? Fill in your predictions on the data table.

Part B. During the Waiting Time

1. Every few days, check to see if your terrarium needs water—unless you have terrarium #7. It's important to keep your terrarium moist but not too wet.
2. Observe your terrarium three times—after about 10 days, after 20 days, and after 30 days. Record your observations each time. (See Figure 4–4.)

Part C. Making Final Observations

1. Record your final observations on the class data table.
2. Observe the other terrariums. See if you can find any of the inhabitants. Notice the appearance of the litter in each terrarium.
3. Clean up your terrarium—or add more food and keep your decomposers alive. Your teacher will tell you where to put the sow bugs and earthworms so they can be returned to the environment.
4. Answer the analysis questions.

ANALYSIS QUESTIONS

1. Why did the procedure say to sift the garden soil?
2. Which kind of litter changed the most? Describe how it changed.
3. Which kind of litter changed the least?
4. Which terrarium(s) showed
   a. the least change in litter? Why?
   b. the most change in litter? Why?
5. Look at the drawings in Figure 4–5 on page 104. Drawing A is of the terrarium when it was first put together. Drawing B shows what it

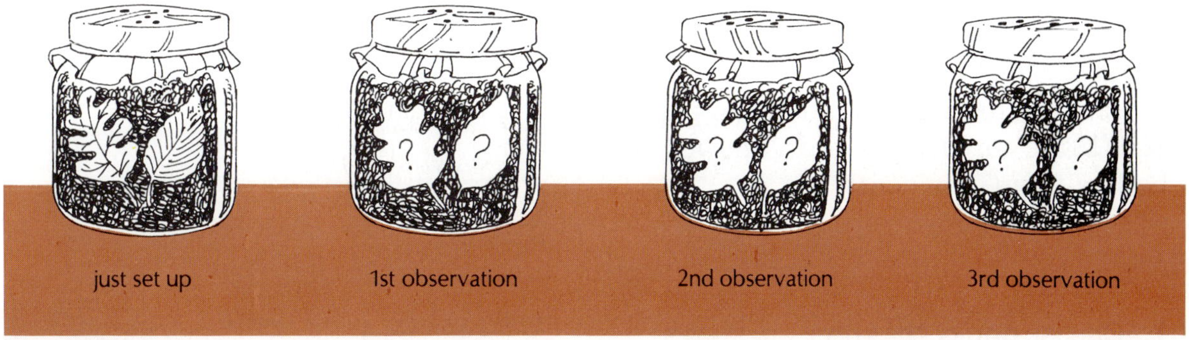

Figure 4–4

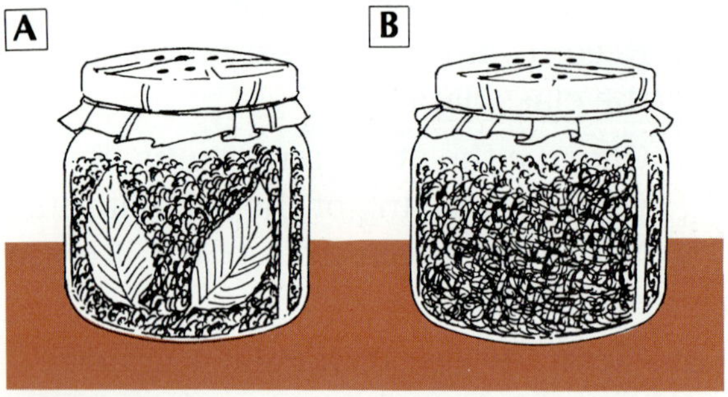

Figure 4–5

looked like 4 weeks later. No organisms were added but it was watered once a week. Since no organisms were added, what caused the litter to decompose?

6. Think about the litter that did not change. These are the "foods" that the decomposers and scavengers did not like.
   a. List at least 3 other types of litter that you don't think decomposers could use for food.
   b. How are these items alike?
7. Which of these places provide good habitats for decomposers? Explain.

garbage dump	edge of swimming pool
gravel pit	under lilac bush
edge of pond	
under a cactus	

8. If you wanted to start another decomposition terrarium to test other organisms, where would you look for decomposers? Why?

CONCLUSION

Look at the focusing questions on page 101. Make sure that you can answer both of them confidently.

EXTENSION ACTIVITIES

1. Choose one of the three questions near the top of the next column. Then set up your terrarium so you will be able to answer that question.
   What kind of litter is eaten the fastest?
   What kinds of litter are not broken down within two months?
   How long does it take a grasshopper carcass to decompose?
2. Sometimes decomposers go to work when we don't want them to. Food spoils because decomposers—especially fungi and bacteria—start to break it down. Choose a food, such as bread, and read the labels to find one that has preservatives and one that does not. Place each slice in a small plastic bag with some moist dirt and observe it for several days. Does the preservative affect the decomposing time?
3. Compare the work of the clean-up crew in your terrarium with the clean-up crew that lives in a garden or under a bush. Put together 3 more sets of litter and place each in a different container. Use a nylon bag, a mesh bag, and a plastic bag. Bury the 3 bags in one hole. Separate the bags from each other by several centimeters of dirt. Cover the bags with about 5 cm of soil. Wait 2 to 6 weeks and then dig them up. Compare the contents of the bags. Explain your observations. How do these results compare with what happened in your terrarium?

4–2 LITTER LIFE

There's a clean-up crew at work in every ecosystem. You can find these organisms in a desert, in a swamp, or under a bush. In fact, the decomposing twigs and leaves under a bush, called **litter**, are very good habitats for scavengers and decomposers.

- "Who" makes up the clean-up crew in your local litter?

Search litter for scavengers and decomposers to answer this question.

MATERIALS

6 clear plastic vials with lids or closable plastic bag
data page: ***Litter Life*** (Figure 4–7)

PROCEDURE

Part A. Hunting for Litter Critters

Work with a partner. You will be searching through litter to find scavengers and decomposers. Read these directions before you go on your search. This is an outside activity; your teacher will tell you where to work.

1. Work quickly. Litter life may be hard to catch.

Remember! These are living animals. Handle them gently!

2. You and your partner will need 6 plastic containers for what you find:
   - In 1 container, place something that is in the process of decomposing.
   - Use 4 containers for litter critters—only put one critter in each container.

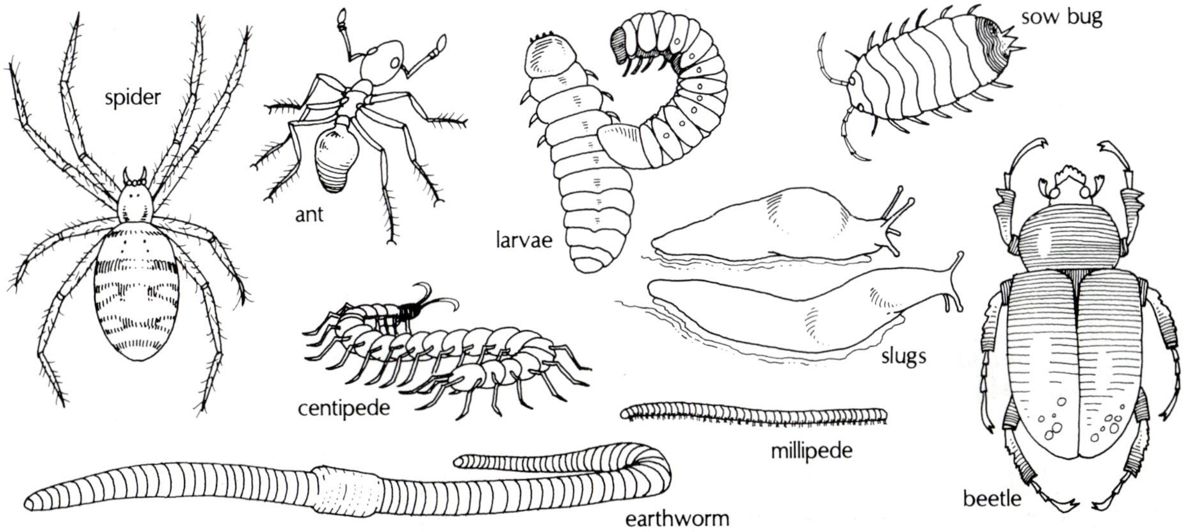

Figure 4–6 Examples of some animals you are likely to find in litter

- Try to find fungi at work. A fungus, such as a mold, may look like white branching threads in the soil. All fungi are easily disturbed. If you are lucky enough to find one, carefully place it in the last container.

3. Try not to disrupt the area. Litter is on the surface of the ground so there is no need to dig. Replace any clumps of litter, rocks, or branches that you disturb.
4. Stop after you have filled your containers or when your teacher calls time, whichever comes first. You will have about 10 minutes for your search.

Part B. Inspecting the Critters

This may be done outside or in the classroom.

1. Look at the 4 critters you collected. Sort them according to the categories at the top of page 107:

LITTER LIFE

NAMES ____________

DATE ____________ PER ____

	Total No.	No. of Kinds	Sketch and color or describe appearance (color, shape, size, number of legs, and so on)
earthworms			
sowbugs			
ants			
slugs			
larvae			
beetles			
spiders			
millipedes/ centipedes			
bugs with wings			
other critters			
fungi			
decomposing plants and animals			

Figure 4–7

earthworms	beetles
sow bugs	spiders
ants	bugs with wings
slugs	millipedes/ centipedes
larvae	
other organisms	

2. Your teacher has set up stations for each kind of organism. Place each container at the appropriate station.
3. Inspect the organisms at each station. Record your observations on the data table.
4. Return the organisms to their habitat.
5. Answer the analysis questions.

ANALYSIS QUESTIONS

1. Look at your data table.
   a. Which critter was most abundant in your study area?
   b. Which one was hardest to find?
   c. Describe the most interesting one.
   d. How many kinds of organisms did your *class* find?
2. If you were a litter critter, describe what your idea of a perfect home would be.
3. Write a sentence that describes decaying leaves. Include at least 4 adjectives in your description.
4. Imagine that you are the teacher.
   a. Would you have students do this activity? Explain.
   b. Where would you take your class to look for litter life?
5. Each of the following events could harm the critters who live in litter. Explain why.
   a. removing all leaves from under bushes
   b. keeping the soil around the bushes very wet
   c. spraying the bushes with insecticides
   d. letting trash accumulate under bushes
6. Many litter organisms are scavengers. Fungi and bacteria are decomposers. What do you think is the difference between a scavenger and a decomposer?

CONCLUSION

Write several sentences describing the clean-up crew you found in this ecosystem.

TAKE A STAND

Pesticides that could kill decomposers should be banned for use by home-owners and farmers.

- Do you **agree** or **disagree**?
- What **comments** would you like to make?

EXTENSION ACTIVITIES

1. Check one spot in different weather and different times of the year and look for other kinds of litter critters. What kind of weather is best for litter critters? In what time of year are they the most abundant? Check different habitats. Where do you find the greatest variety of critters?
2. Read the report on ***Earthworms*** on page 108. Write a similar one about some other form of litter life.

Earthworms

An earthworm's body is divided into segments. Each segment has several stout bristles that help the animal move through the soil. It may be hard to tell which end of the worm is the front since its head is so small, and it doesn't help to look for eyes since worms don't have any. Instead, look for the swelling that is behind the head. This swelling looks like an enlarged segment of the worm but actually it's a structure called the *clitellum.* The clitellum secretes a cocoon for the fertilized eggs. This structure is found only in adult worms.

When an earthworm is cut in two, it never makes two worms. If the tail is cut off, the worm can grow a new one; but the tail fragment dies. If the cut is between its head and clitellum, the worm will die. The only way to make more earthworms is by reproduction. Each worm has both male and female body parts. Thus, a worm can produce both eggs and sperm.

An earthworm's body is essentially two tubes, one within the other. The inner one is the digestive system. The worm's dark coloring is due to the food in this tube. Earthworm "manure" looks like pellets. These pellets contain lime which helps enrich soil.

Earthworms are an important part of an ecosystem. By loosening the soil, they make spaces for air and water to get to plant roots. They also turn under decaying plant and animal matter; this helps fertilize the soil. Earthworms may accidentally plant seeds by pulling them underground along with the leaves they drag into their holes for insulation.

On Your Own

Dig up a small area of soil and find some worms.

How long is the biggest worm you can find? the smallest?

Can you find adults and young?

Look for pellets. What size are they?

Gently slide your fingers forward and then backward along the worm. Can you feel the bristles?

Set the worm on a smooth patch of soil. Watch how it digs a hole. Describe its movements.

4–3 THE CLEAN-UP CREW

Nature's clean-up task is a large operation. It is handled by thousands of different organisms ranging from microscopic bacteria to muscular lions. Each organism has its own role in clearing up dead vegetation, animal wastes, carcasses, and the things we throw away every day. This clean-up crew includes both scavengers and decomposers.

Scavengers

Scavengers begin the cleaning-up process by eating dead matter. Large scavengers, such as hawks, eagles, coyotes, hyenas, lions, and sharks feed on dead animals, but they also hunt. In a food web, these animals are both secondary consumers and scavengers. Some of them have gotten bad reputations from their scavenging habits. For instance, a hawk found eating a lamb is likely to be accused of killing the lamb—when actually it is only taking advantage of a "free meal."

Other meat-eaters never kill to eat; they are true scavengers. Vultures are one example of such an animal. They feed exclusively on dead and decaying carcasses. Even if they wanted to eat fresh meat, they probably couldn't. Their bills and feet are too weak to capture and kill live prey. One type of vulture is shown in Figure 4–8.

Figure 4–8 Reportedly, turkey vultures locate carrion by smell as well as by sight. The range map shows the areas where turkey vultures live.

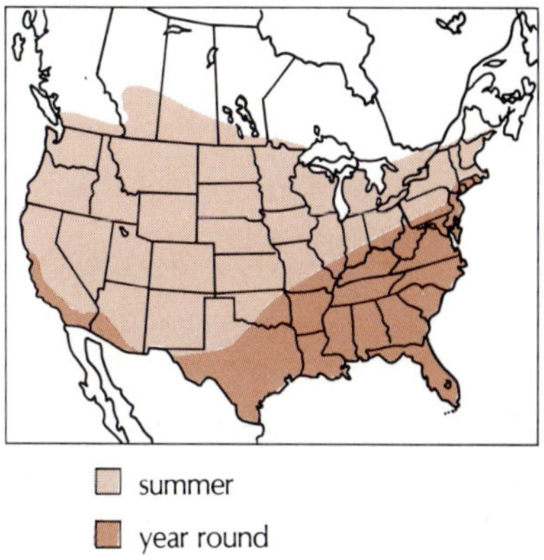

Figure 4–9 A single rotting apple may contain as many as 100,000 roundworms. Without these worms, it would take longer for apples to decompose.

Figure 4–10 Dung beetles lay their eggs in dung. When the eggs hatch, the developing larvae feed on it. Beetles are important in recycling waste.

Most scavengers are much smaller than mammals and birds. Slugs, beetles, sow bugs, tadpoles, flatworms, and larvae are examples of small scavengers. Their small size allows them to crawl through the tiny spaces in rotting plant and animal matter. Litter is a perfect habitat for them since it is moist, dark, and full of hiding places and it provides a ready source of "appetizing" food. (See Figures 4–9 and 4–10.)

Scavengers produce waste of their own in the form of droppings and their own dead bodies. These, too, must be further broken down by other scavengers and decomposers.

Decomposers

The **decomposers**—bacteria and fungi—take care of the final step in cleaning up. They convert litter and waste back into raw materials so these can be recycled. You seldom see the decomposers at work because they are so small. Most of them are microscopic. In one teaspoon of rich soil, there may be 3 billion bacteria, a million microscopic animals, and 200,000 algae and fungi. All these organisms are living off the materials and nutrients in the soil.

Sometimes, when conditions are right, you may see the fruiting body of a fungus. This fruiting body is called a mushroom. Next time you see a mushroom, look at where it's growing. It may be on a decaying log or on a pile of cow manure. (See Figure 4–11 on page 111.) If it's growing on a grassy area, think about what

Figure 4–11 Top: This tiny fungus is about 1 cm wide. It only grows on dead wood, helping old twigs and branches to decay. Try pronouncing its name — it's fun to say: *Scutellina scutellata* (Scū′ tell ī na scū′ tell ă tă). Bottom: Most *Coprinus* grow on dung. These are common mushrooms so look for them when you are in a pasture.

might be *under* the ground. There may be a network of dead roots, layers of decomposing leaves, or a long-buried log. In any case, the mushroom only represents a very small part of the fungus. Most of it is a large network of fibers that you may never see.

Decomposers don't actually "eat." They release enzymes into the rotting material to break down large molecules into small ones. (See Figure 4–12.) They can then absorb the small molecules as their source of food. Once the decomposers have finished their jobs, logs, carcasses, and vegetation are returned to the soil in the forms of water, minerals, carbon dioxide, and other nutrients.

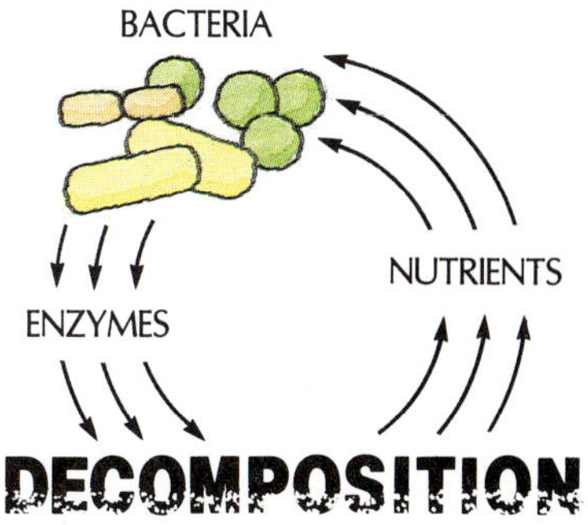

Figure 4–12 Bacteria release enzymes that release nutrients from rotting material.

Figure 4–13 Forest ecosystems, like all ecosystems, have decomposers.

On Land and in the Water

Scavengers and decomposers are at work in every ecosystem. You can find them in a desert, a rain forest, on top of a mountain, and at the bottom of the ocean. Without its clean-up crew, an ecosystem couldn't survive.

ANALYSIS QUESTIONS

1. Define the term **scavenger.**
2. Think of examples of scavengers.
    a. List 3 large scavengers that are mentioned in the reading.
    b. List 3 of the small scavengers that were mentioned.
    c. List at least 2 other animals that you think are scavengers.
3. What is meant by the term **decomposer?**
4. Since we have scavengers, why do we need decomposers?
5. You've learned about predators and scavengers.
    a. How are these two groups alike?
    b. How are they different?

6. Look at Figure 4–13 on page 112. Describe this ecosystem by listing
   a. at least 3 of its producers.
   b. at least 3 consumers you would expect to find.
   c. at least 2 of its scavengers.
   d. 2 decomposers that you know live here.
   e. one of its abiotic characteristics.
7. Here are 10 examples of litter.
   a. Put the examples into 3 groups: easy to decompose, decomposes slowly, very difficult to decompose.

glass jar	petunia plant	walnut shell
dead worm	aluminum can	car engine oil
tree branch	squirrel feces	chicken bone

   b. In general, what kinds of things do you think decompose easily? Why?
   c. In general, what kinds of things do you think are hard to decompose? Why?
8. What are some examples of food for decomposers and scavengers that *you* supply?

EXTENSION ACTIVITIES

1. Make a scavenger trap. Bury a container so it's even with the surface of the ground. Put some raw meat in the container. Cover it with a board but make sure there's space for the scavengers to get in. Check your trap twice a day and record what you find. (See Figure 4–14.)

Figure 4–14

2. Make a spore print. (Mushrooms don't produce seeds; they make spores.) When you find a young mushroom—not one that's dry and shriveled, place its cap on a piece of paper. Leave it undisturbed for a day. Pick up the cap and you will find that you have made a spore print. Try different kinds of mushrooms; you'll get spores of different colors.
3. Discover the clean-up crew at work in water. Use 3 containers: a nylon stocking, a mesh bag, and a plastic bag. Put a set of litter in each container. Close the containers securely and attach them to a stake. Stick the stake into a pond or lake so the containers are completely covered by water. Check the containers every week for four or five weeks. Record the changes you observe in each container.

4–4 DECOMPOSERS AND HUMAN WASTE

Humans, like all other animals, produce body waste. In addition, we manufacture items that become waste. Waste clutters our daily lives. How do we get rid of all this waste? Some of it is packed in sacks and hauled off to local dumps or landfills. Other wastes are washed into our sewers.

Figure 4–15 shows three main entry points for water into a sewer system: from homes, from water runoff into gutters, and from industry, schools, and hospitals.

On your own paper, list at least 30 waste products that could get into the sewer systems. Include wastes from each of the sources pictured. You should be able to think of more waste products than just those shown in the drawing. (Save your list, you'll need it again later.)

Figure 4–15 Along with water, there are many other things that can get into our sewer systems.

As you can see, there are lots of things in the sewers besides water so this water should not be emptied into our lakes and rivers. What happens to wastewater?

Water Treatment—Naturally

For years, many sewer lines drained into nearby streams and rivers. The plants and animals living in the streams were expected to "clean up" the water. The scavengers and decomposers thrived when given so much extra food. Under such ideal conditions, they reproduced quickly.

As their populations increased, they used up the oxygen in the water. Without oxygen, the fish, frogs, and other organisms died. Meanwhile, more sewage continued to flow into the streams. Even the scavengers and decomposers couldn't handle all of the waste so it began to accumulate and spread downstream.

In the 1960s, people became extremely concerned about the "death" of our streams and lakes. Laws were passed requiring that all towns clean up their wastewater before allowing it to reenter the natural waterways. (See Figure 4–16.)

Figure 4–16 Top: A healthy lake has plenty of oxygen. Bottom: When nutrients are added, plants and microorganisms grow rapidly. They use up oxygen, causing fish and other lake organisms to die.

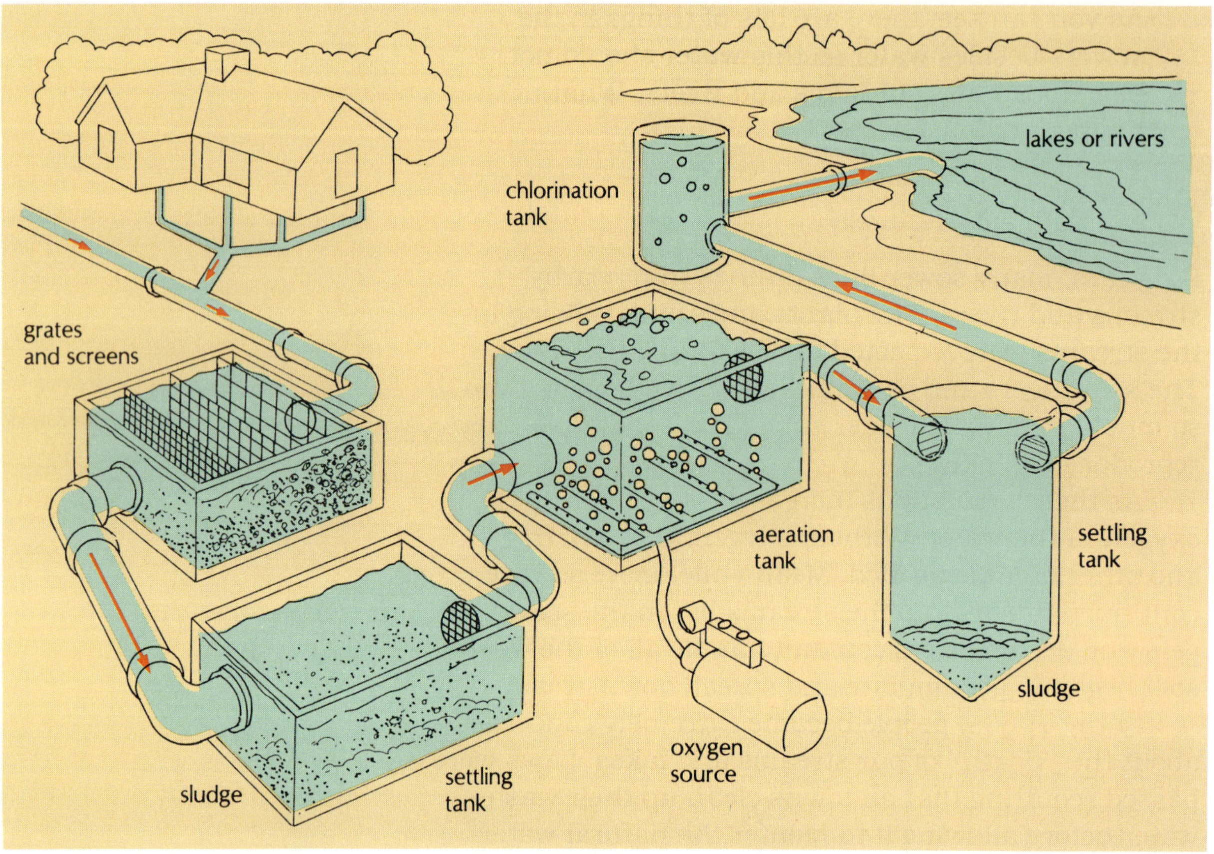

Figure 4–17 Wastewater treatment occurs in a series of steps.

Wastewater Treatment Plants

Now, most wastewater flows from the sewer lines into treatment plants. The clean-up process in the treatment plant is similar to what occurs naturally—but at a faster rate. Wastewater treatment takes place in a series of steps.

First, grates and screens filter out the largest objects. Water then enters a large tank where sand, dirt, and other solids settle out. These solids are called **sludge.** At this point, the water may look a little better; but it still isn't clean.

In the next step, decomposers are put to work. The nutrient-rich wastewater becomes a feast for millions of bacteria, fungi, and other microorganisms. They can live on the small particles of animal and plant material as well as on some of the chemicals that remain in the water. They are provided with plenty of oxygen to use while they break down our waste products. After several hours, the water moves on and is again allowed to settle. More sludge accumulates at the bottom of another settling tank.

The water that leaves this second settling tank is usually chlorinated to kill the extra bacteria and then returned to the streams. Many phosphates, nitrates, salts, radioactive materials, and pesticides still remain in the water. Some treatment plants are set up to remove these pollutants, but most are not. While not drinkable, the water that leaves the treatment plant is much cleaner than when it entered.

Meanwhile, the sludge that was collected in the settling tanks must be treated to destroy harmful microorganisms and then hauled away. It may be taken to a landfill or dried and used as fuel. Some treatment plants are set up to recycle the sludge so it can be returned to the land as fertilizer.

All of the steps in this process are illustrated in Figure 4–17 on page 116.

A Final Word

You have learned that we, as consumers, could not survive without producers. Even though it is not as obvious, we also depend on decomposers. In fact, they are so important in breaking down our wastes that we raise decomposers in these special treatment plants just for that purpose. Imagine what our world would be like if there were no decomposers!

ANALYSIS QUESTIONS

1. In general, what kinds of waste are removed from the water at each of these points in a wastewater treatment plant?
   a. by screens and grates
   b. in settling tanks
   c. by the decomposers
2. What pollutants are likely to remain in the water after it has been treated?
3. What is **sludge?**
4. Divide a sheet of notebook paper into 5 sections. Label the sections as follows:

   Is screened or filtered out
   Settles out
   Is decomposed
   Remains in the water and flows into our rivers
   ? ? ?

CHALLENGE

For homework, find out where your community's wastewater is treated and where the water goes when it leaves the treatment plant.

TAKE A STAND

People should be fined if they dump things into sinks or gutters that decomposers cannot break down.

- Do you **agree** or **disagree**?
- What **comments** would you like to make?

In order to cut down on the dangerous pollutants in our water, Life Science students should memorize a list of chemicals that should not be poured down the drain.

- Do you **agree** or **disagree?**
- What **comments** would you like to make?

Now look at your list of waste products. Decide what would happen to at least 20 of the products. Then list each in the appropriate section on your paper. If you don't know what happens to an item, put it under the question marks.

5. List at least 3 things *you* could do to decrease the amount of wastewater you produce.

EXTENSION ACTIVITIES

1. Draw a cross section of your own house. Include all of the appliances that drain into the sewer system. What do you think are the 10 most hazardous substances that you dump into your sewer system?
2. Call the nearest wastewater treatment plant. Explain that you have some questions about how wastewater is treated. When someone offers to help you, take careful notes and report your findings to your class. Here are some questions to ask:
   - How much water is treated at this plant each day?
   - How complete is the treatment process?
     - Are wastes removed by filtering and then by a settling process?
     - Are decomposers used for further clean-up?
     - Are there any third-level steps taken to remove phosphates, salts, or similar pollutants?
   - What is hardest to remove from wastewater?
   - What happens to it?
   - How much sludge is produced?
   - Where does the water go when it leaves the treatment plant?

Figure 4–18 A sewer treatment plant

CHAPTER 5

ABIOTIC FACTORS

How are humans affecting this environment?

5-1 ABIOTICS ONLY

The producers, consumers, decomposers, and scavengers make up the living or **biotic** part of every ecosystem. Ecosystems also have a non-living or **abiotic** part that must be considered. (The words **biotic** and **abiotic** both come from the Greek language. *Bio* means "life" and *a-* means "not" or "without." Putting this together, you can see that **abiotic** means "not living.")

Prairie Grassland

How do you identify an ecosystem when you see it? If you are like most people, you probably look at the biotic parts of the ecosystem to make your decision. For example, look at the prairie grassland shown in Figure 5–1. Even if it weren't labeled, how would you know that this is a prairie? Both the large open expanse of grasses and flowers and the clump of cottonwoods in the distance are helpful clues. The consumers shown are also typical of a prairie.

Could you identify the prairie ecosystem if all of the producers and consumers were removed? Your only

Figure 5–1 A prairie ecosystem

Figure 5–2 The abiotics of a prairie ecosystem

clues would be the abiotic ones. Figure 5–2 shows what this prairie looks like without plants or animals. Would you have been able to identify it?

Other Ecosystems, Abiotics Only

Figure 5–3 below and Figure 5–4 on page 122 show the abiotic conditions for two additional ecosystems. None of the plants and animals are shown in either one. All

Figure 5–3 Abiotics of a second ecosystem

Figure 5–4 Abiotics of a third ecosystem

that is pictured are the abiotics. What ecosystem do you think you are seeing in each drawing?

We often overlook the abiotic components of an ecosystem, but they are very important. Abiotic factors such as water, temperature, light, and soil type determine what can live in a given area. In this chapter you will study a few abiotic factors to see how and why they affect living organisms.

ANALYSIS QUESTIONS

1. Explain the meaning of the word **abiotic.**
2. Write a sentence that uses the words *abiotic* and *plants.*
3. Decide if the following characteristics are abiotics or not. Answer by using *yes* or *no.*
   a. amount of rainfall
   b. number of days that the temperature goes below freezing
   c. height of the tallest plants
   d. number of decomposers in the soil
   e. type of soil
   f. type of predators
   g. average wind speed
4. Look back at Figure 5–2 on page 121.
   a. What are 2 clues that might tell you this is a prairie ecosystem?

b. What other ecosystems could this be?
c. Where could this *not* be?

5. Use either Figure 5–3 on page 121 or Figure 5–4 on page 122 to answer the following questions:
   a. What are at least 3 abiotic characteristics of this location?
   b. What kind of a place—what ecosystem—do you think this is?
   c. What are 2 examples of producers that would be found here?
   d. What are 2 examples of consumers that would be found here?
   e. What other type of ecosystem could this be?
6. Several abiotic conditions are described below. For each one, choose the value that you think best matches your ecosystem.
   a. coldest winter temperature
      below −10° −10° to 10° above 0°
      11° to 35° above 36°
   b. altitude
      0–3000′ 3001′–6000′
      6001′–9000′ above 9000′
   c. annual rainfall
      0–10″ 11″–20″ 21″–30″ more than 30″
7. Imagine that the abiotic conditions in your neighborhood have changed suddenly. You are now getting many more extremely windy days than you had before. How do you think this wind would affect
   a. the plants? (Be specific.)
   b. the animals? (Be specific.)
   c. the decomposers?
   d. you?

EXTENSION ACTIVITIES

1. Do you live in the desert? by a seashore? on the tundra? Look up your ecosystem in an encyclopedia. List what it tells about the abiotic factors of your ecosystem. Look it up in another encyclopedia. Does this second reference tell you anything new?
2. Start with a copy of either Figure 5–3 or Figure 5–4 and turn it into a picture of a complete ecosystem. Sketch in the producers and consumers. Try to include an example of a decomposer or scavenger, too.

5-2 SOIL PERMEABILITY

Plants and animals need water in order to survive; but just because it rains or snows does not mean they will get the water. Water seeps quickly into some soils but may immediately run off the surface of others. **Soil permeability** refers to how easily water can soak into and flow through the soil.

- Which part of your school grounds has the soil with the lowest permeability?
- What affects soil permeability?

After you measure the soil permeability at four different sites on your school grounds and think about your results, you will be able to answer these questions.

MATERIALS

soup or vegetable can—10½ oz size—with both ends removed
tomato juice can, 6 oz size
water (in bucket or large plastic jar)
stop watch or watch that shows time in seconds

PROCEDURE

Work with a partner.

Part A. Getting Ready

1. On a full sheet of paper, prepare a table like the one below so you can record your data and observations.

Site #	Site Location	Predictions	Site Observations	Water Entry Time
1				
2				
3				
4				

2. Decide on 4 sites on the school ground where you will test the soil's permeability. Try to choose spots that you think will have different soil characteristics. Describe the general location of these sites on your data table.
3. Make some predictions.
   a. Write "most" on the data table by the site that you think is most permeable. (In other words, where do you think water will soak in the fastest?)
   b. Write "least" by the site that you think will be least permeable.
4. Read through Part B of this activity. Make sure you understand what you will be doing when you are outside.
5. Get your materials so you will be ready to go outside.

Part B. Measuring Soil Permeability

1. Go to your first site. On your data table, describe the soil at the site. Mention such things as the size of

the particles, the plant life present, the moisture content, and whether the soil is packed or loose.

2. Push the large can 1 to 2 inches into the ground. (See Figure 5–5.)

Figure 5–5

3. Fill the juice can with water and then pour the water into the large can.
4. Record the number of seconds it takes for all of the water to be absorbed by the soil—the entry time.
5. Repeat this procedure at 3 other sites.
6. Return to your classroom, put away your materials, and answer the analysis questions.

ANALYSIS QUESTIONS

1. In your own words, explain what the term **soil permeability** means.
2. Describe the characteristics of the soil at the site that was most water permeable.
3. Describe the soil characteristics at the site that was least water permeable.
4. What type of surface do you think would be completely *im*permeable?
5. List at least 2 reasons why some areas are more permeable than others.
6. Think about the procedure. List at least 2 things that could cause someone's data to be wrong.
7. Think about a soil that is highly permeable.
   a. Why might this be an advantage?
   b. How could it be a disadvantage?
8. Name at least 2 things you think could happen to the water after it soaks into the ground.

CONCLUSION

Check to make sure you learned something about soil permeability by answering the two questions on page 124.

5–3 SOIL DRAINAGE

Both rainfall and soil permeability affect the available water supply; but that's not all there is to it. The soil's ability to **hold** water must also be considered. This is called **water retention.** For example, a soil may be highly permeable and have low water retention. This means water will soak quickly into the soil but then it will drain right out.

- How can you test a soil's water retention?
- What type of soil can retain the most water?

Test permeability and water retention of several types of soil in order to answer these questions.

MATERIALS

filter paper (3 pieces)
funnel
beaker
beaker with 100–120 ml water
graduated cylinder
ring with ring stand
paper towels
3 dirt samples—clay, sand, and pea gravel
clock or watch with second hand
data page: ***Soil Drainage*** (Figure 5–8)

Figure 5–6

PROCEDURE

Work with a partner.

Part A. Getting Started

1. Get your materials and set up the apparatus as shown in Figure 5–6.
2. Fold a piece of filter paper and place it in the funnel. (Your teacher will show you how.)
3. Label 3 paper towels as shown in Figure 5–7.

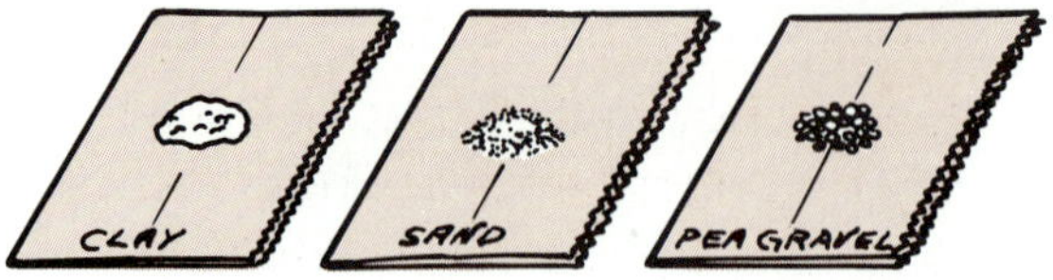

Figure 5–7

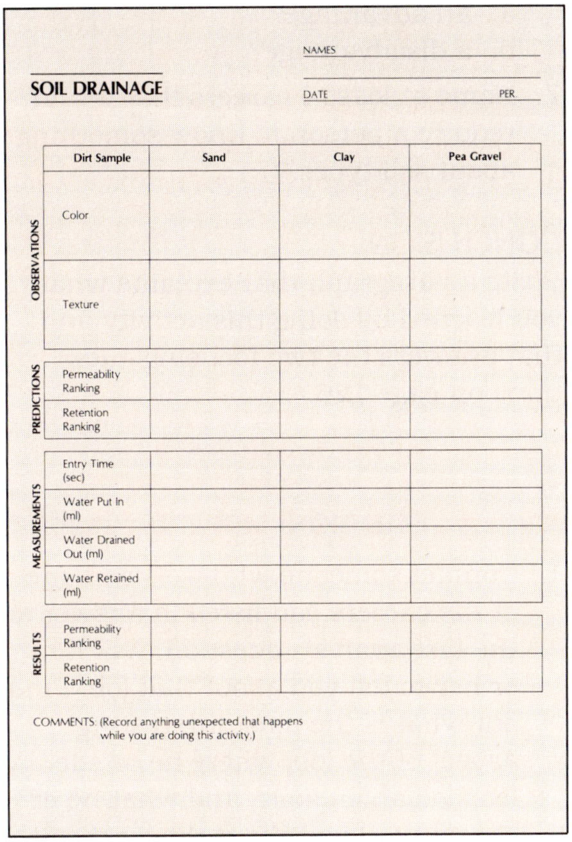
SOIL DRAINAGE

NAMES

DATE PER.

	Dirt Sample	Sand	Clay	Pea Gravel
OBSERVATIONS	Color			
	Texture			
PREDICTIONS	Permeability Ranking			
	Retention Ranking			
MEASUREMENTS	Entry Time (sec)			
	Water Put In (ml)			
	Water Drained Out (ml)			
	Water Retained (ml)			
RESULTS	Permeability Ranking			
	Retention Ranking			

COMMENTS: (Record anything unexpected that happens while you are doing this activity.)

Figure 5–8

4. Empty one vial of each dirt type onto the appropriate paper towel. Carefully observe each sample. Note its color and texture. (Is it fine? grainy? smooth? full of rotting sticks?) Record your observations on the data page.
5. Predict how permeable each dirt sample will be. Rank the samples from 1 to 3. Use "1" to indicate the one that you think will be the most permeable and "3" for the one that will be the least. Record your predictions on the data table.
6. Predict how well the samples will hold—retain—water. Rank them from 1 to 3. Use "1" to indicate the one that will hold the most water and "3" for the one that will hold the least. Again, record your predictions.

Part B. Doing the Soil Tests

1. Carefully empty the dirt from one paper towel into the funnel.
2. Use the graduated cylinder to measure 15 ml water. You are now ready to measure permeability and retention.
3. Pour the water into the soil sample. Note the number of seconds between the time when you start pouring and when there's no more water resting on top of the dirt. (This is the **entry time.**) Pour quickly but do not let the water overflow out of the funnel. Record the entry time on your data table.
4. Measure the amount of water that drained into the beaker. Record this amount on the data table.
5. Empty your dirt sample into the "used" buckets. Do not mix up the samples! (See Figure 5–9.)
6. Throw away the used filter paper and replace it with a new one.
7. Repeat this procedure for each dirt sample.

Figure 5–9

8. Look at your results and use them to record the actual rankings on your data table.

ANALYSIS QUESTIONS

1. In your own words, explain the difference between the terms **water retention** and **soil permeability.**
2. Consider the permeability of the clay, gravel, and sand samples. Into which sample did water soak
   a. the fastest?
   b. the slowest?
3. Look at the data for the sample that retained the most water.
   a. Which one was it?
   b. Why do you think it could hold the most water?
4. Look at the data for the sample that retained the least water.
   a. Which one was it?
   b. Why do you think it held the least amount of water?
5. Scott and Amanda were comparing their data on the clay samples. Scott found that his sample retained 7 ml of water; Amanda's sample retained 12 ml. Give 3 possible reasons for the differences in their data.
6. If a soil retains a lot of water, how can that be
   a. an advantage?
   b. a disadvantage?
7. Name at least 4 careers that would require a person to know something about soil types.

CONCLUSION

Write a paragraph that explains what you learned by doing this activity and that answers the two focusing questions on page 126.

EXTENSION ACTIVITIES

1. Contact someone who works in one of the careers you listed in answer to the last analysis question. Ask him or her to tell you about that job. Find out what is the most interesting aspect of the job, where he or she learned about soils, and what he or she can tell you about the local soil.
2. Look up "soil" in a book on houseplants. What kind of soil is recommended? What does the book say about watering and drainage? If you were going to use one of the soils you tested in class to grow a houseplant, which one(s) would you use? Why?

5-4 SOIL: AN ABIOTIC FACTOR?

Soil. What is it? You might say "soil is dirt," but what does this mean? To find out, you could inspect some soil. You could go out with your shovel and start to dig, pausing to inspect each shovelful that you lift out of the growing hole.

First, you might have to move aside some branches and twigs. (Are they part of the soil?) Then you might remove some grasses and leafy plants. With the tip of your shovel, you could scrape away some damp, rotting leaves and other decaying matter that you cannot identify. A centipede and a few insects may scurry away as you disturb them. (Are these part of the soil?)

Then, you carefully lift out a small amount of the next layer. It's dark and damp; it smells "earthy." As you poke around, you find roots, fragments of leaves, and twigs that crumble under your touch. An earthworm quickly pulls back when your finger comes too near.

Down you dig another few centimeters. It still looks dark and feels damp but you don't see much sign of rotting plants. You do see plenty of evidence of plant life though. There is a network of fine roots lacing this layer along with a few larger roots.

Your shovel drops a little the next time you dig. It has broken through a mole's tunnel but you can be sure the mole is nowhere in sight. That mole had scurried away when it felt the vibrations of your approaching footsteps. Bend down and inspect this layer. Still pretty dark, but not quite as black. Roots. Practically no rotting leaves. A slightly grittier feeling between your fingers. Is this soil?

As you dig, you continue to notice change. Fewer roots. Fewer insects. No rotting leaves. A lighter color. Pebbles. Small rocks. Depending on the type of minerals, the dirt might look beige, reddish, or yellowish. You may hit a hard rocky layer or you may dig through lifeless dirt for 5, 10, or more meters. Is this soil?

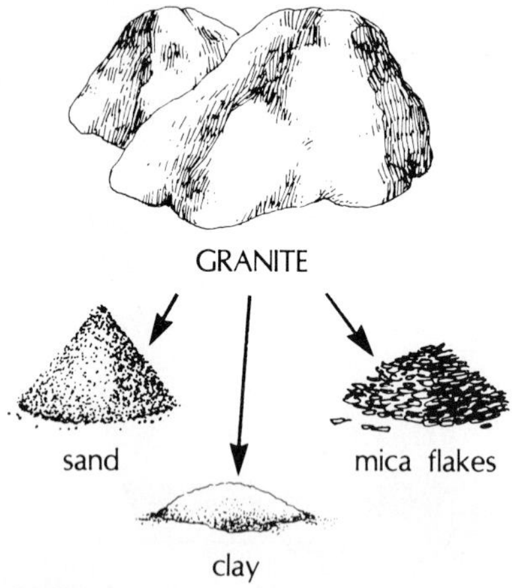

Figure 5–10 When granite weathers, it produces sand from the quartz, clay from the feldspar, and mica flakes.

Soil Content

Soil is a combination of plants, animals, gases, water, dirt particles, and minerals. Without the plant and animal life, it wouldn't be soil; it would be dirt. Depending on where you live, you may encounter any one of a variety of different soils. The soil that you just "dug" through was obviously not from a desert. It contained too much moisture and had too much evidence of plant life. This area had not been disturbed either; you could recognize distinct layers as you were digging.

The type of soil found in an area depends on several factors. One factor is the type of rocks and minerals that are found locally. The gradual breakdown of the rocks into smaller and smaller particles gradually forms dirt. Different types of rocks form different dirts. For example, quartz breaks down to form sand; feldspar helps form clay. (See Figure 5–10.)

Climate is a second factor that affects soil type. An area that is cool and moist generally has more plant life than does a hot, dry place. Therefore, the soil that forms in the moist area will contain more decaying material than will the soil in the dry area. The local plants and animals also affect the soil. For instance, earthworms churn the soil, allowing more air to get deeper into the soil than usual. This can change the chemical makeup of soil.

Particle Size

Soil particles vary in size. The smallest particles are called **clay.** In order to see a clay particle, you would need to use an electron microscope. **Silt** particles are a little larger; they feel like powder between your fingers. You can see **sand** grains with your naked eye and study them with a low power microscope. What we often call sand is actually gravel. Any particle larger than 2 mm in size is **gravel.** As you just learned, soil contains a mixture of these particle sizes.

Name	Particle Size
clay	less than .002 mm
silt	.002–.05 mm
sand	.05–2.0 mm
gravel	more than 2 mm

Figure 5–11 Prickly pear cactus grows better in a sandy soil than in clay. Why do you think this is so?

The size of the particles is important. Different size particles have different amounts of surface area to which water can cling. Fine particles have lots of surface area and can hold lots of water. The amount of space between the particles is also important. If the spaces are very large, water will run right through the soil. Clay, for example, has lots of surface area and very small spaces between its particles. What happens to water when it is poured over a sample of clay?

Interactions

Not only do plants and animals affect the soil, the soil itself affects which plants and animals can live in a given area. Plants, such as alders and willows, grow best in soil that holds a lot of water. A cactus does best in sandy soil which drains quickly. (See Figure 5–11.) Other plants may require specific minerals for growth. In turn, the forms of animal life will depend on both the local plant life and soil conditions. A prairie dog cannot build tunnels through mud and a slug will not survive on a granite outcropping. How is your life affected by your local soil conditions?

ANALYSIS QUESTIONS

1. Define the word **soil.**
2. What is the difference between soil and dirt?

CHALLENGE

Sketch the appearance of each type of soil particle as it looks when viewed through a microscope. (Remember to record the magnification.)

3. What are 3 factors that affect the type of soil found in an area?
4. The title of this reading is "Soil: An Abiotic Factor?" Why is a question mark included in this title?
5. Explain why clay soils and gravelly soils retain different amounts of water.

EXTENSION ACTIVITIES

1. Try to locate an area that has "native soil." Look for a spot that has not been plowed or bulldozed and that has not had anything dumped on it. (If a site has lots of native plants, that also might be a clue that it has native soil.) Do some soil permeability tests in this area. How do your results compare with the results from your school ground?
2. Analyze some native soil to see what size particles it contains. Scrape away the rotting vegetation and take about 2 tablespoons of soil. Place the soil in a large test tube that is about two-thirds full of water. Put a cork on the test tube and shake it thoroughly. Set the test tube someplace where the soil particles will be undisturbed. After the soil has settled, describe what you see.

CHAPTER 6

ECOSYSTEMS

How many reasons can you think of for recycling aluminum cans?

6-1 ECOSYSTEM SURVEY

Every ecosystem has its own unique combination of producers, consumers, decomposers, and abiotic factors. This variety of climates and living organisms makes the world an interesting place in which to live.

- What are two helpful sources of information about an ecosystem?
- Of the four components that make up an ecosystem, which is easiest to find information about? Which is the hardest?

Answer these questions after you have made a poster about an ecosystem and have learned about other ecosystems from your classmates.

MATERIALS

reference books
large piece of paper or poster board
variety of other materials: magazines, paints, markers, yarn, scissors, colored paper, glue, and whatever else you need to make your poster

PROCEDURE

Part A. Making a Poster

Work alone, with a partner, or in a small group.

1. Each ecosystem is a special place. Choose one that interests you.

tundra	coniferous forest
desert	deciduous forest
ocean	tropical rain forest
marsh	freshwater pond
grasslands	your own town or city
coral reef	some other ecosystem

2. Make a poster that captures the spirit of the ecosystem you chose. Make it colorful, interesting, and informative. Include these items on your poster:
   a. *5 producers*—Try to choose examples that are unique to this ecosystem.
   b. *5 consumers*—Again, choose examples that are unique to the ecosystem or very common in it. Give examples of what each one eats and indicate if it is a primary or secondary consumer.
   c. *3 decomposers or scavengers*—If you learn something special about one of these organisms, include that on your poster, too.
   d. *5 abiotic factors*—Be specific about the conditions. In other words, don't include "snow." Instead, include how much snow and when it falls. Make sure to include average annual rainfall and average temperature.
   e. *locations*—Give some examples of where in the world this ecosystem is found.

ECOSYSTEM: ____________________

Producers	Consumers	Decomposers	Abiotic Factors
1.	1.	1.	1.
2.	2.	2.	2.
3.	3.	3.	3.

Other interesting facts:

Figure 6–1

3. Become an expert on the ecosystem you have chosen. You, or your group, will have three or four minutes to tell the rest of the class about your fascinating ecosystem.

Part B. During Class Presentations

1. Now is your chance to become an ecosystem authority. Listen closely while your classmates explain their ecosystems. For each presentation, record the name of the ecosystem and at least 3 examples of each of its four components. One method for organizing your notes is shown in Figure 6–1.
2. Ask questions if you do not understand something or if the presenter forgets to provide you with some necessary information. Take good notes! You will be tested on these ecosystems. This is your chance to find out what you will need to know.

CONCLUSION

Conclude this activity by answering the two questions on page 134.

EXTENSION ACTIVITIES

1. Islands are "contained" ecosystems. San Nicolas Island, off the coast of southern California, is the setting for *Island of the Blue Dolphins* by Scott O'Dell. It is the story of a young girl struggling to stay alive while she waits to be rescued. Read this book as though you, too, had to survive on this island. Which producers grow there? Which consumers are mentioned? How would you describe the abiotic conditions? Look for clues about the local scavengers and decomposers. What is one important way that an island ecosystem differs from other ecosystems?
2. A space travel capsule needs to be a mini-ecosystem. Design a mini-ecosystem for a trip to Pluto. Consider these questions when making your plans.
   - Who will be the producers?
   - Who will be the consumers?
   - How will you control your abiotic factors?
   - How are you going to handle all of the different waste products?

6–2 ECOSYSTEM MAP

Perhaps you already know where the ocean ecosystems are located, but you may want to know where to find a tropical rain forest. You also may want to explore a grassland in Africa. Would you know where to find one? This information is readily available if you know where to look. There are maps that show the major ecosystems of the world.

- Which major ecosystem covers the most area in the United States?
- In which major ecosystem do you live?

Color a map of the major ecosystems found in North and Central America to answer these questions.

MATERIALS

map: ***Ecosystems of North and Central America*** (Figure 6–2)
crayons or colored pencils (gray, green, yellow, orange, blue)

PROCEDURE

Read the following steps and color the appropriate part of the map when directed.

1. *Neatly* label these places on your map:
   Canada
   the Caribbean Islands
   Central America
   Mexico
   the United States—label Alaska and Hawaii separately
2. On the map, find about where you live and mark that spot with a star.
3. Look at the very northern part of Canada and Alaska. This area is tundra.
   Color the tundra and the appropriate box in the key a very light gray.
4. Forests are found throughout most of Canada, the eastern United

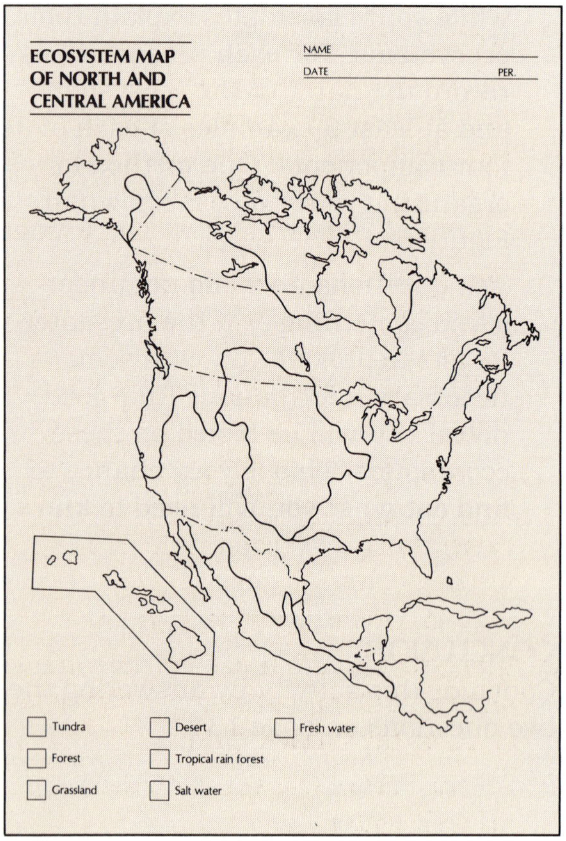

Figure 6–2

Figure 6–3 These pictures show an Alaskan tundra and a tropical rain forest.

States, and the west coast of the United States. The western parts of both Mexico and Central America are also forest.

Color these areas and the appropriate box in the key a light green.

5. The western United States and northern Mexico are desert.

   Color the desert and its box yellow.

6. The central part of the United States and a small area of southern Canada are grassland.

   Color the grassland and its box orange.

7. Hawaii, the Caribbean Islands, and the eastern edges of Mexico and Central America are tropical rain forests.

   Color these areas dark green.

8. The Great Lakes and the three lakes shown in Canada are fresh water.

   Color them dark blue.

9. All the rest of the areas on the map are salt water.

   Shade them a light blue.

ANALYSIS QUESTIONS

1. Which ecosystem is
   a. the largest?
   b. the smallest?
2. In which ecosystem would you expect to find the greatest variety of organisms? Explain.
3. Name at least 2 smaller ecosystems that you might find within a large grasslands ecosystem.
4. In which large ecosystem do you live?
5. There are two places in the United States where three of the major land ecosystems come together. Where are these two locations? Try to use the names of the states if you can. Otherwise, just describe where these two areas are.
6. If you could live in another ecosystem for six months, which one would you choose? Why?

CHALLENGE

Where else in the *world* would you find

a. a tundra?
b. grasslands?
c. a desert?
d. a tropical rain forest?

Check your answers in an encyclopedia or in some other type of reference.

CONCLUSION

Now that you have completed this activity, look back at the two questions on page 136. Write a short paragraph that answers both questions.

TAKE A STAND

An insect-infested marsh near the edge of a city should be filled for building sites.

- Do you **agree** or **disagree**?
- What **comments** would you like to make?

It's not important for Life Science students to learn about all types of ecosystems.

- Do you **agree** or **disagree**?
- What **comments** would you like to make?

EXTENSION ACTIVITIES

1. Compare a map of the rainfall of North America with your ecosystem map. How do the two compare?
2. Ask your social studies teacher for a blank map of the world. Use reference books to look up **biomes**—large ecosystems are often called biomes—and then locate the grasslands and the deserts of the world. Color them on your map. Which ecosystem covers more land—the grasslands or the deserts? If you feel ambitious, color the other major ecosystems on your map.
3. Some people are more aware of their environment than others; they realize that they are part of the ecosystem. Read either *Julie of the Wolves* or *Water Sky* to learn something about the Eskimo's attitude towards nature. In both of these books by Jean Craighead George, she describes life in the frozen north while telling exciting adventures in the lives of two young people.

6-3 ECO-WISE DECISIONS

You are at a remote lake on a three-day fishing trip. The fishing has been great. By the afternoon of the first day, you have already caught your limit. When you mention this to the other two people who are fishing, they laugh and open an ice chest revealing many large fish. They tell you to go ahead and keep fishing because no one ever comes here to check the catch. What are you going to do?

Well, if you were in this situation, what would you do—keep fishing or quit? You know what you'd *like* to do. It's been good fishing and it's the beginning of the trip so, of course, you want to keep fishing. However, you probably will ask yourself, "What's the *right* thing to do?"

Since you already have your limit, you really should quit fishing. You might also want to consider ways you could legally continue to fish. In addition to the law, you might think about the environment and what would be the "eco-wise" thing to do. For instance, by continuing to remove the biggest fish from the lake, what would you be doing to the local ecosystem?

- What is involved in making an "eco-wise" decision?

Keep this question in mind while you work in a small group to make a decision that affects the environment.

MATERIALS

reading: ***Eco-Decisions—Part A***
(page 141)
Eco-Decisions—Part B

PROCEDURE

Work in groups of four.

1. Get in your group and decide who will be the recorder. (Suggestion: assign this role to the person with the shortest last name.)
2. Read the situation entitled "Styrofoam Trash" on page 141.
   a. Discuss what you would do if you were in this situation.
   b. Try to agree on one option. Have the recorder write down the group's reasons for this choice. If you cannot agree on what to do, the recorder should record both choices and the reason(s) for the disagreement.
   c. Ask your teacher for part B of "Styrofoam Trash."
   d. Again, discuss what you would do and have the recorder write down your reasons.
3. Read the other two situations and choose one of them to discuss. Follow the same procedure that you did for "Styrofoam Trash."
4. As a group, answer the analysis questions and then write your conclusion to this activity.

ANALYSIS QUESTIONS

1. Look at the choices that are given for "Styrofoam Trash" on page 141.
   a. Which one is easiest to do?
   b. Which choice is the best for the environment?
2. Look at the choices that are given for "A Dead Bird" on page 141.
   a. Which one is easiest to do?
   b. Which choice is the best for the environment?
3. Look at the choices that are given for "Pollution Control" on page 142.
   a. Which one is easiest to do?
   b. Which choice is the best for the environment?
4. Why do you think these situations were given to you in two parts instead of giving you the whole situation all at once?

CONCLUSION

Look back at the focusing question on page 139. Think about what you need to do first when making a decision that can affect the environment. What would you do next? Then what? Write a paragraph that explains the steps that are involved in making an "eco-wise" decision.

EXTENSION ACTIVITIES

1. Discuss one of these situations tonight with your family. Listen to what they have to say. What kinds of things do they consider? Do they think of any additional options?
2. Do some research on styrofoam. Find out why some styrofoam products have been banned in some cities. Has your city set any limits on the use of styrofoam?

Eco-Decisions—Part A

Styrofoam Trash

You have just heard that styrofoam contributes to environmental problems. It litters the streets and highways; it fills dumps. Gases that pollute the air are produced when it is made. You think about this when you go to a fast food restaurant. Do you

- forget about it and enjoy your meal?
- order your meal but tell the people you are with about the problems caused by styrofoam?
- ask them not to give you the styrofoam and explain that you'd prefer a paper wrapper?
- write to the president of the company suggesting that they change their packaging?
- go someplace else to eat?
- ???

A Dead Bird

You and a friend are target shooting when a bird perches on top of the target. Before you realize what is happening, your friend shoots the bird. An hour later, when you are leaving, someone tells you that a bird has been shot and asks if you know anything about it. Do you

- deny knowing anything about it?
- admit your friend did it?
- take the blame yourself?
- make up a story accusing someone else?
- say nothing, but call later with an anonymous phone tip?
- ???

Pollution Control

Your parents own a large factory. They are concerned about the pollution it creates so they asked a committee to review the situation. The committee reports that the factory is barely under the legal limits. In fact, this factory is the biggest polluter in the community. The equipment necessary to stop the pollution would cost so much that your parents would not be able to buy the new, larger house with a swimming pool and sauna they have been considering. Do you hope your parents will

- add the equipment and forget the house?
- forget about the pollution?
- buy less expensive equipment that will stop half of the pollution?
- wait a few years to see if the cost of the equipment comes down?
- sell the factory?
- ? ? ?

6-4 ENVIRONMENTAL RESPONSIBILITY

A paper plate or a glass one?
Canvas shoes or leather ones?
A plastic bag or a paper one?
A shower or a bath?
Chicken or pizza?

All these decisions affect the environment. When you make your decisions, do you try to make the choice that will be best for the plants and animals as well as for yourself? Possibly not. You may not even realize that your decisions have environmental effects. Think about some of your more routine decisions—what to wear, how much homework to do, what to eat, how to spend your free time. As you get older, you will be making even bigger decisions that can have even bigger effects on the environment.

"Why," you may be wondering, "should I think about the environment when I make a decision?" Why not? You live on this earth and share it with several *billion* other people as well as with hundreds of billions of plants and animals. As the human population grows, we are using up more and more of the earth's resources and producing more and more waste. As a result, we hear more about shortages, crime, pollution, and problems with waste disposal. It can be discouraging. You may feel helpless when it comes to these problems; but there are many things you can do that will make a difference. You may be doing some of them already. Take the survey on page 144. You will find out just how ecologically aware you are.

After you take the survey, think about it again and answer these questions:

Which thing in the survey do you think is easiest to do?
Which do you think is hardest?
Which ones can affect wildlife?
What 5 other suggestions can you think of for "living lightly" on this world?

Don't stop here. Take your stand for the environment. Be aware of the issues and problems. That doesn't mean you must read every article and watch

TAKE A STAND

It's more important for adults to assume environmental responsibility than it is for students.

- Do you **agree** or **disagree**?
- What **comments** would you like to make?

Eco-Awareness

How ecologically do you live? Take this survey to find out. Give yourself 5 points for each statement that applies to you.

1. I recycle aluminum cans.
2. I eat all the food I take.
3. I do not litter.
4. My showers are shorter than 5 minutes.
5. I turn off the lights when I leave a room for longer than two minutes.
6. The volume on my radio is turned low.
7. I quickly close the door behind me in cold weather.
8. When possible, I walk or ride a bike instead of asking someone to drive.
9. I turn off the water while brushing my teeth.
10. I wear my clothes until they are worn out or outgrown.
11. I am aware of extra packaging on items I buy, such as fast foods, tapes, and toothpaste.
12. I donate items that I no longer use.
13. I stay on paths and trails when asked to do so.
14. I do not pick wild flowers.
15. I sometimes read about environmental issues in the newspapers or in nature magazines such as *National Wildlife, World,* and *National Geographic.*

BONUS

- I feed the birds. (5 points)
- I've planted a tree or shrub. (5 points)
- I pick up litter when I see it. (10 points)
- I've called or written to my representative or senator about an environmental issue. (15 points)
- I've written a letter to the editor of a newspaper about an environmental issue. (15 points)

IF YOU EARNED	YOUR SCORE IS
over 60 points	Excellent! Keep up the good work.
40–55 points	Good, but you can do better.
20–35 points	Fair—you are just beginning to think about the environment; work a little harder at it.
below 20 points	Wake up! You live in this world too; help it out!

every television program that's on the environment. It does mean you should take some time to learn the facts and to consider what you can do to make a difference.

EXTENSION ACTIVITY

Make a list of ecological gift ideas. Every time you give a gift, you make a decision. Whether or not the gift is "eco-wise" could be one of the factors you consider. Try to list at least 20 gifts that would not add to our environmental problems and might even help.

UNIT 2

BODY STRUCTURE

CHAPTER 7

BONES

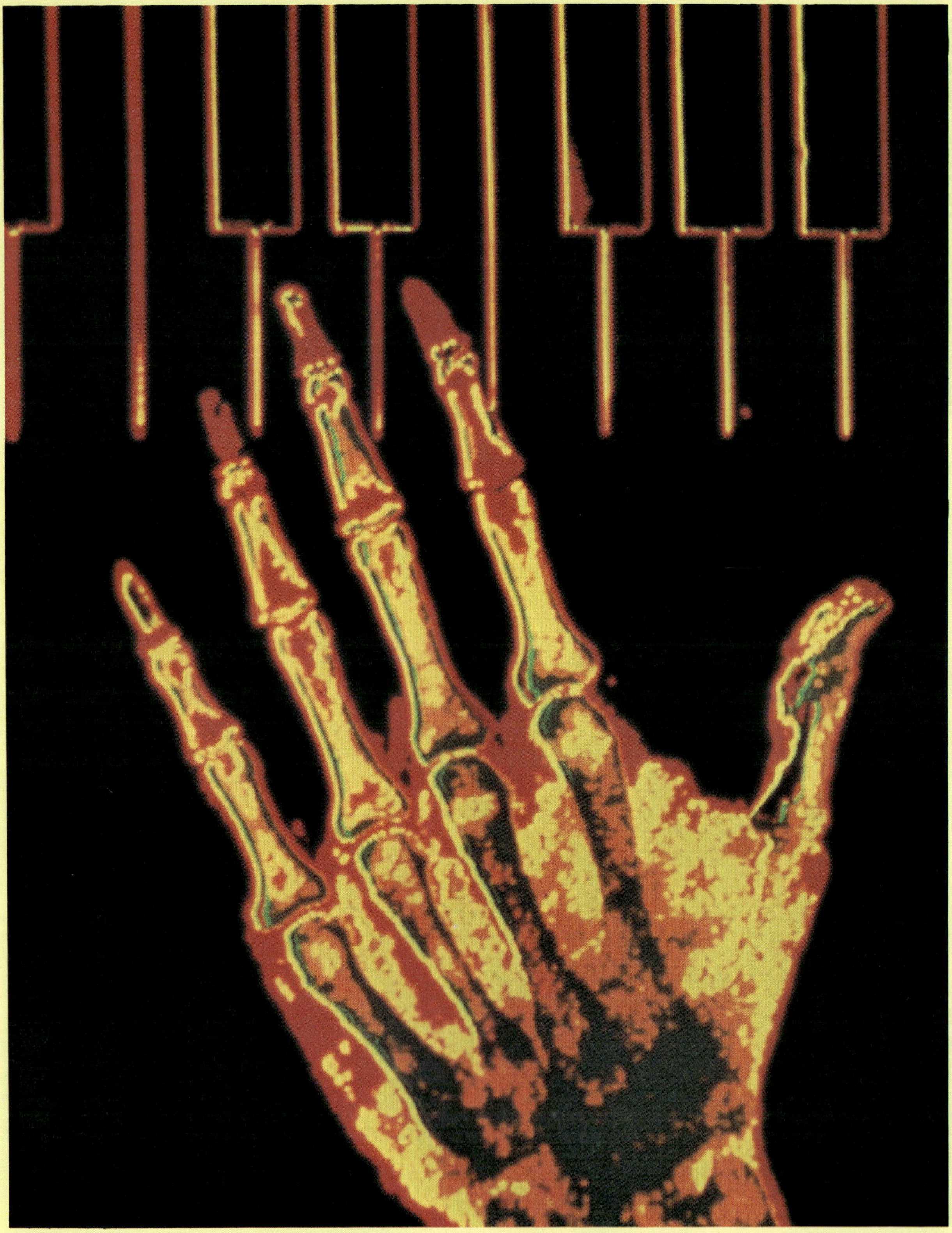

Joints provide the flexibility needed to play a piano. How many joints do you think are in a hand?

7-1 WHOSE BONES?

Imagine that you've been hiking in the woods for several hours when you finally reach your destination, a beautiful mountain lake. Immediately you feel as if the lake belongs to you. No one else is around. The grasses are not trampled and there is no litter to be seen. It's so quiet; all you hear is the wind in the trees, the chatter of a squirrel, and the occasional quack of a duck.

Walking more slowly now, you scan the lake shore, looking for a place to rest. Ahead of you some whitish objects catch your eye. Are they flowers? rocks? As you get closer, you see that they aren't flowers or rocks. Then you wonder if perhaps they could be pieces of driftwood. No! They're bones! Bones of various sizes and shapes are scattered all around the area. Curious, you pick up the nearest bone. As you begin to examine it, you notice that some of the other bones look very different from the one in your hand. One by one you examine several of the bones. You wonder, "How did these bones get here? What animals are they from?"

- What can you learn about an animal by observing one of its bones?

Test your powers of observation by studying several real bones. Imagine that you have picked up these bones from the lake shore. Keep this question in mind as you inspect the bones.

MATERIALS

bones—4 different ones
data page: ***Whose Bones?*** (Figure 7–1)

PROCEDURE

This activity requires you to think quickly so be prepared.

1. Work in a group with two or three other students. Your group will have four bones to inspect.
2. Inspect the bone that your teacher gives your group.
   Fill in the appropriate column on the data table while you inspect the bone.
   Your group will have about 6 minutes to finish with this bone.
3. When your teacher calls time, pass your bone to another group and get a new bone for your group to inspect. Complete the data table for this bone. Remember, you have only 6 minutes to work with this bone.

Handle the bones with care. Bones are strong, but they can be fragile too!

4. Repeat this procedure with the third and fourth bones.
5. Answer the analysis questions on your own paper. Use complete sentences and be prepared to discuss your answers.

ANALYSIS QUESTIONS

1. List at least 3 things all the bones had in common.
2. List at least 3 ways that the bones differed from each other.
3. What functions could these bones have served?

Figure 7–1

WHOSE BONES

NAMES ______________ ______________
______________ ______________
DATE ______________ PER. ______

	BONE W	BONE X	BONE Y	BONE Z
1. List at least 4 features about this bone that might give you clues about the animal.				
2. Interpret these clues. What do they tell you about the animal?				
3. Where in the animal's body do you think you would find this bone? Give a reason for your answer.				
4. What could have been a function of this bone?				
5. What animal do you think this bone is from? Give a reason for your answer.				
6. What other animal(s) could this bone possibly be from?				

Figure 7–2 This paleontologist is studying bone and skull artifacts in a lab.

4. What kinds of things can you tell about an animal by looking at one of its bones? Copy this list of characteristics. Then circle each one that you could determine by looking at that animal's bones.

size	the sounds it made
eye color	whether it is male or female
age	
weight	if it were sick or healthy
diet	
intelligence	if it lived in water or on land
favorite food	
	if it could fly

5. List at least 3 careers in which you might need to recognize bones.
6. If you really did find an assortment of bones on a lake shore, you probably would wonder how those bones got there. Give at least 2 possible explanations.

CONCLUSION

Write a short paragraph that answers the focusing question on page 149.

EXTENSION ACTIVITY

1. Analyze another bone. Maybe you have a bone at home that you would like to learn more about or your teacher may have an interesting one at school. Inspect the bone closely to learn everything you can about it. Consider the same kinds of things you did during this activity.
2. Learn more about one of the careers you listed in question 5. Find an article that describes this career or interview someone who works in this field. With your teacher's permission, invite this person to speak to your class.

7-2 ANIMAL SKELETONS

Bones fit together to form skeletons. Skeletons provide the support that is vital for the survival of many animals. Think of some animals that are supported by bones. Then think about some animals that do *not* have bones. How do you think their bodies are supported?

In addition to bony support, there are two other important types of support systems. As you read about the three types, consider what life would be like if you had one of the other two types of support systems. Think about what problems you would have and what you could do more easily.

Bones Inside

When you study bones, you are studying parts of an **internal skeleton.** Humans and many other animals have internal skeletons; that is, their skeletons are *inside* their bodies. (See Figure 7–3.)

Figure 7–3 Snake eels and chameleons have internal skeletons.

Every bone fits in a specific place in a specific kind of skeleton and each bone is adapted to do a different job. With the help of the muscles, some bones allow for movement. Other bones protect vital organs. Together the bones that make up an internal skeleton support the body and give it shape.

"Bones" Outside

Lobsters, crabs, and grasshoppers are examples of animals with **external skeletons.** These animals do not have bones; instead they have a hard outer covering, as can be seen in Figure 7–4.

There are advantages and disadvantages to having an external skeleton. This hard covering is tough and water-resistant. (The shells of sea animals are so strong that they can even hold up under the continuous battering of waves.) External skeletons also protect their "owners" from drying out in the sun. However, animals with external skeletons do not grow as easily as do animals with internal skeletons. In order to grow, an animal such as a crab must shed its existing skeleton and wait for a new one to harden. During this time it is left defenseless. Luckily, animals with external skeletons do not remain "skeletonless" for very long.

Figure 7–4 Barnacles and lobsters have external skeletons.

Figure 7–5 Jellyfish and green anemones have neither bones nor shells.

No Bones

Some animals, such as those in Figure 7–5, have neither bones nor hard coverings. Instead, they are supported by the pressure of fluid within their bodies.

This type of support system has its advantages and disadvantages, too. Worms, for example, definitely find it easy to fit through small spaces. And whoever heard of a worm having a broken bone? Two disadvantages are that these animals can never grow very large and are easily squashed.

ANALYSIS QUESTIONS

1. What is an advantage to having
   a. an internal skeleton?
   b. an external skeleton?
   c. no skeleton?
2. What is a disadvantage to having
   a. an internal skeleton?
   b. an external skeleton?
   c. no skeleton?
3. Decide if each of these animals has an internal skeleton (I), an external skeleton (E), or no skeleton (0).

porcupine	clam	slug
tape worm	spider	turtle
frog	snake	mosquito

4. What 3 "jobs" do bones do for animals who have internal skeletons?

Look at the drawings of the four internal skeletons in Figure 7–6 to answer questions 5 through 8.

5. For each skeleton, identify the animal that skeleton came from. Use the letters A, B, C, and D in your answers.
6. List or sketch 3 bones, or structures, that you can observe on all four skeletons.
7. List 1 characteristic that makes each skeleton different from the others. For example: the bird has a beak, none of the others do.
8. Look at each skeleton on Figure 7–6. If it has a lower or back leg, tell how many bones make up that leg. (Don't count foot bones.) Use the letters A, B, C, and D in your answers.

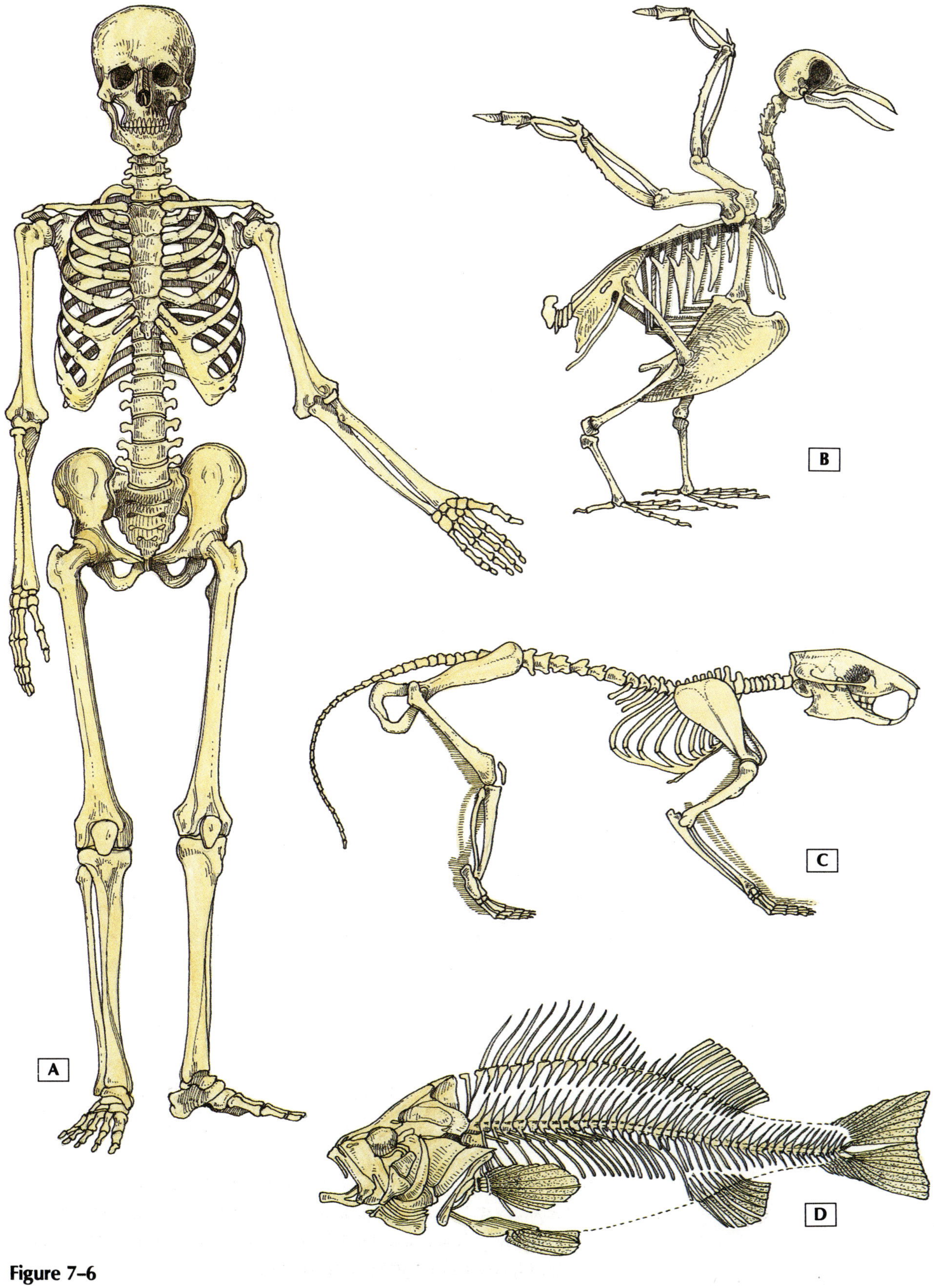

Figure 7–6

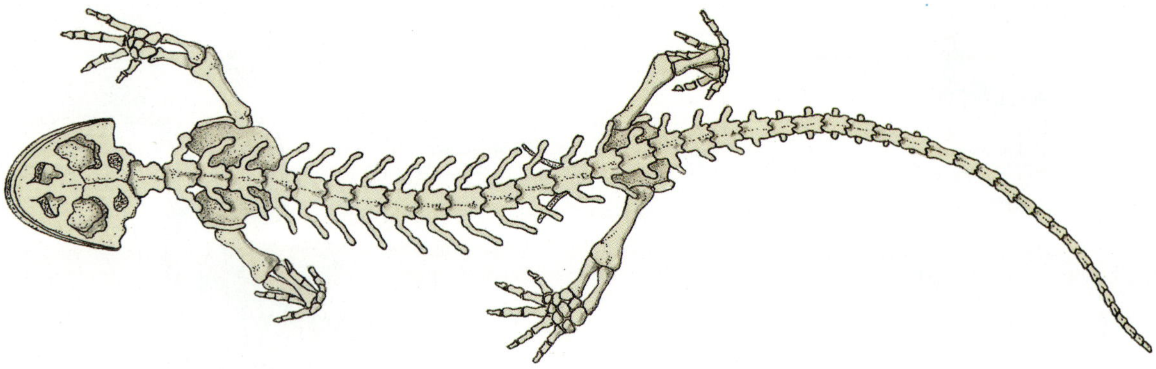

Figure 7–7

9. Look at the "mystery skeleton" in Figure 7–7. Describe at least 5 things that you can tell about this animal by looking at its skeleton. How many other things can you list?

ON YOUR OWN

Imagine that you are a worm, a clam, or some other animal that does not have an internal skeleton. Write a creative paragraph or two about your life as this animal.

EXTENSION ACTIVITIES

1. Visit a natural history museum or another place that has a bone collection in it. Inspect the bones or skeletons and try to answer some of these questions.

   Do any skulls have nose bones?
   What do tail bones look like?
   Are there any animals that *do not* have ribs?
   About how long is the longest bone on display?
   What skeleton looks the strongest? the most fragile?
   What does the most interesting bone you can find look like?
   What clue would help you recognize an upper leg bone if you found a pile of bones? (Look at the upper legs of at least four animals before you answer this question.)
   Are there bones in an elephant's trunk?
   Do deer or elk skulls have antlers?
   Do dinosaurs have knee caps?

Figure 7–8 A prehistoric mammoth skeleton

2. Put together a skeleton. Start with a chicken from a store or a pile of bones that you found somewhere. If you have to clean the bones first, choose one of these three methods to try.
   - Scrape, cut, or pull as much meat off the bones as possible. Set the bones in the sun to dry.
   - Set the bones on top of an ant hill and let the insects clean off the meat. You may have to cover the bones with a wire screen so dogs don't drag them away.
   - Boil the bones until the meat falls off.

   If you want the bones to look even cleaner, soak them for a few days in water with 2 tablespoons of liquid bleach.

   Finally, use pictures of skeletons to help you assemble the bones. The feet will be very difficult because they contain so many tiny bones.

7-3 THE HUMAN SKELETON

Your internal skeleton is made up of 206 bones. How many can you name? Actually, it is not important to memorize the names of all those bones, but it is helpful to know the names and functions of a few of them. As you read, look at Figure 7–9 on page 159.

The skull is made up of two parts, the **cranium** and the **mandible.** The cranium protects your brain. It is made of more than 20 bones that have fused together. The mandible is the largest and strongest facial bone. Your dentist may talk to you about your mandible when you have your teeth examined.

The vertebral column supports the skull and body. It is made up of **vertebrae.** These bones are separated by discs that act as miniature shock absorbers. The vertebral column is not perfectly straight; it curves naturally at the neck and lower back. If it curves too much or in the wrong way, it may need to be corrected surgically.

The **clavicles** are important for posture because they brace the shoulders. They also help hold the **scapulas** in position. Do you remember what the rat scapula looked like? The shape of yours is very similar. Its broad, flat shape provides a large surface area for the attachment of muscles that move your arm.

The rib cage is made of the **sternum** and the **ribs.** The ribs are attached to the vertebrae and curve forward to attach to the sternum. (Can you feel where your ribs and sternum meet?) The rib cage has to be tough to protect your heart and lungs. However, it also has to be flexible. Inhale deeply, then exhale. What happens to your rib cage?

The **pelvis** is a broad and strong bone that supports the body. Archaeologists can often tell the sex of a skeleton by looking at the pelvis. A female's pelvis is usually broader than a male's. Why do you think this is so?

The **humerus, radius,** and **ulna,** as well as the **femur, tibia,** and **fibula,** are long bones. Because they are shaped like hollow cylinders, these bones can take enormous pressure. Even though they are very strong,

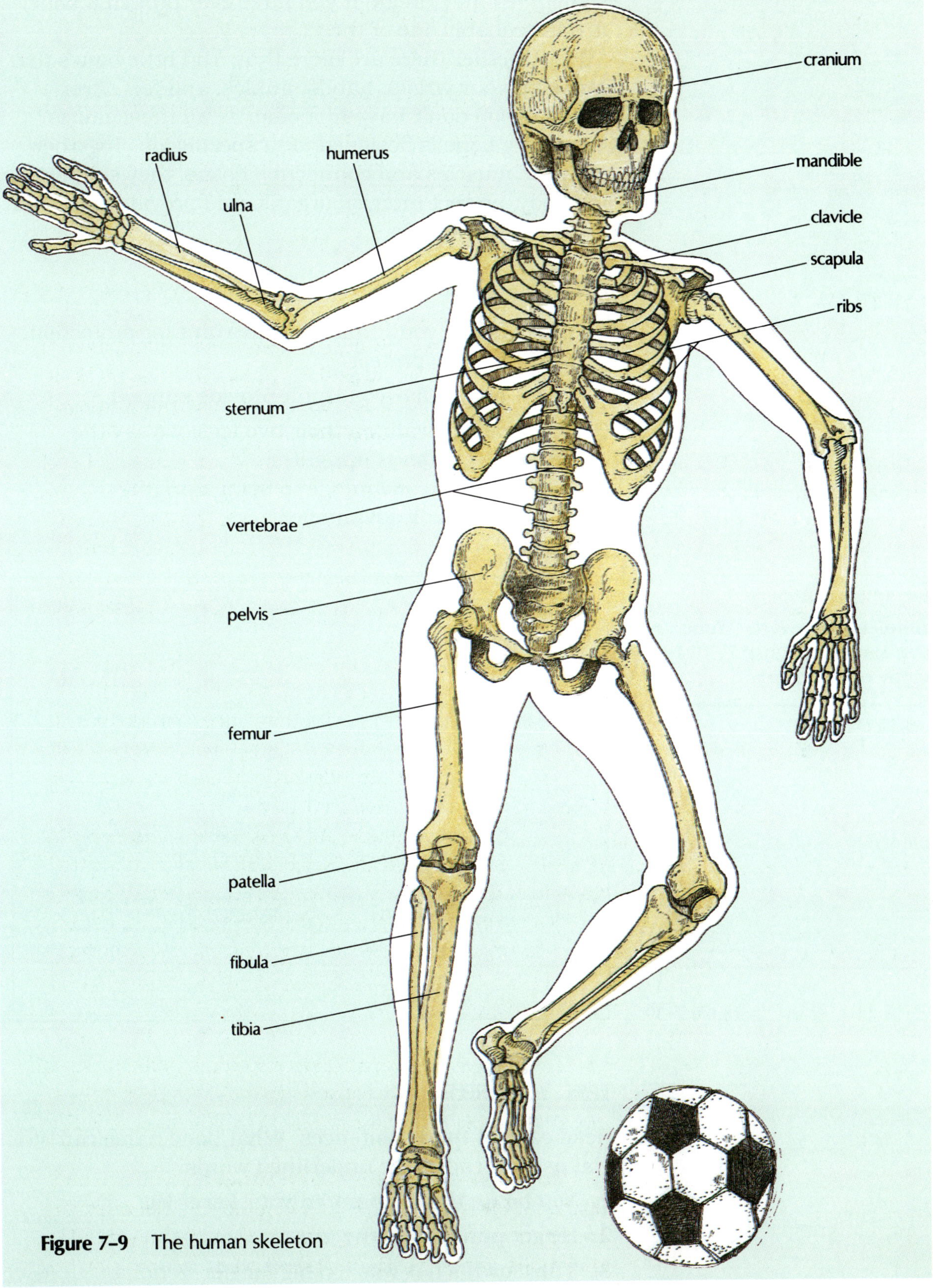

Figure 7–9 The human skeleton

sometimes they break. If you have ever broken a bone, it was probably one of these.

Altogether there are more than 100 little bones that make up your wrists, hands, ankles, and feet. Aren't you glad you don't have to memorize all their names?

Every bone is designed for a specific job. Together with the muscles and connective tissue, they support the body, protect internal organs, and provide places for muscle attachments.

ANALYSIS QUESTIONS

1. List 2 bones you think are important for protection. What do they protect?
2. List 3 bones that you think provide support.
3. Since people walk on their two legs, we say the human skeleton is upright.
   a. What is an advantage to being upright?
   b. What is a disadvantage?

ON YOUR OWN

Complete the chart, "Your Own Bones." Figure 7–10 is a copy of the chart.

YOUR OWN BONES

NAME ____________

DATE ____________ PER. ____

TIBIA How long is your tibia? ______	**HUMERUS** How long is your humerus? ______	**STERNUM** How long is your sternum? ______	**PELVIS** How wide is your pelvis? ______
MANDIBLE How wide is the widest part of your mandible? ______	**CLAVICLE** How long is one of your clavicles? ______	**VERTEBRAL COLUMN** How long is your vertebral column? ______	**VERTEBRAE** How many vertebrae can you count? ______
ULNA How long is your ulna? ______	**FEMUR** How long is your femur? ______	**RIBS** How many ribs can you count on one side? ______	**RADIUS** How many ends of your radius can you feel? 0 1 2
SCAPULA Find the best position to put your arms so you can feel your scapula.	**FIBULA** Where can you feel your fibula? ______ ______ ______	**CRANIUM** How big around is your cranium in centimeters? (What is its circumference?) ______	**PATELLA** Measure your patellas. Which statement is true? ___ The right one is bigger. ___ The left one is bigger. ___ They are exactly the same size.

Figure 7–10

TEST YOURSELF

Read each of these sentences. What bone name can you use instead of the underlined words?

1. You broke the *big bone* in your lower leg.
2. He got punched in the *jaw.*
3. The clown's hat was too big for her *head.*

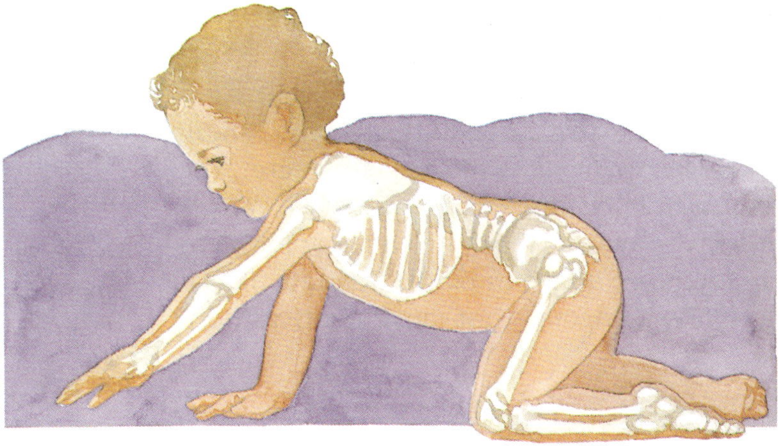

4. Tom's posture is beautiful because he keeps his *back* so straight.
5. Joanne rested the *side of her lower arm* on her desk.
6. The fighter pounded his *chest* and let out a yell that could be heard for blocks.
7. The baby is crawling on its *kneecaps*.

EXTENSION ACTIVITIES

Long bones are shaped liked cylinders. A hollow cylinder is surprisingly strong. Roll a sheet of paper loosely to make a cylinder, keeping it together with a rubber band, as shown in Figure 7–11. Stand it on end. Try supporting different objects on top of the cylinder. Will it support a pencil? a paperback book? a textbook? two textbooks? Use one sheet of paper and try to form a shape that would be stronger than a cylinder.

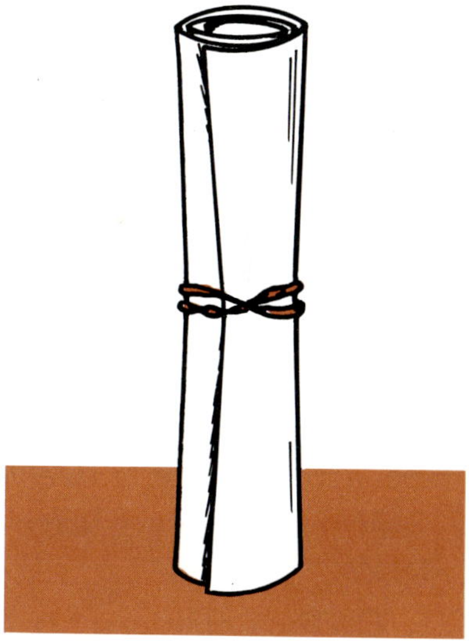

Figure 7–11

7-4 JOINTS JOIN

Bones form a skeletal framework that supports the body. When muscles are added to this framework, movement can occur. These motions are possible only because we have **joints.** Wherever two bones meet, there is a joint.* Most joints allow for some movement while still holding the bones in place. In a few places, however, bones are so firmly connected that movement cannot occur.

- Which bones that make up a joint can you identify on x-rays?
- How does the structure of a joint relate to its function?

There are three parts to this activity—a worksheet that you complete while moving your own joints, x-rays on which you can observe joints, and a reading for you to complete. After finishing all three parts of the activity, you will be able to answer the two focusing questions.

MATERIALS

x-rays: shoulder and skull
worksheet: ***Joints Join*** (Figure 7–12)
reading: ***Joint Structure*** (page 165)

PROCEDURE

Part A. Human Joints

Use the worksheet to record your data for this part of the activity. As you can see, arrows point to twelve joints on a skeleton. There are three lines by each arrow.

1. On the top line, write the name of the joint. If the joint does not have a name, put a dashed line (---) in the space.
2. On the second line, record the number of bones that meet to form that joint. If you want to test your memory, list the names of the bones that form the joint.

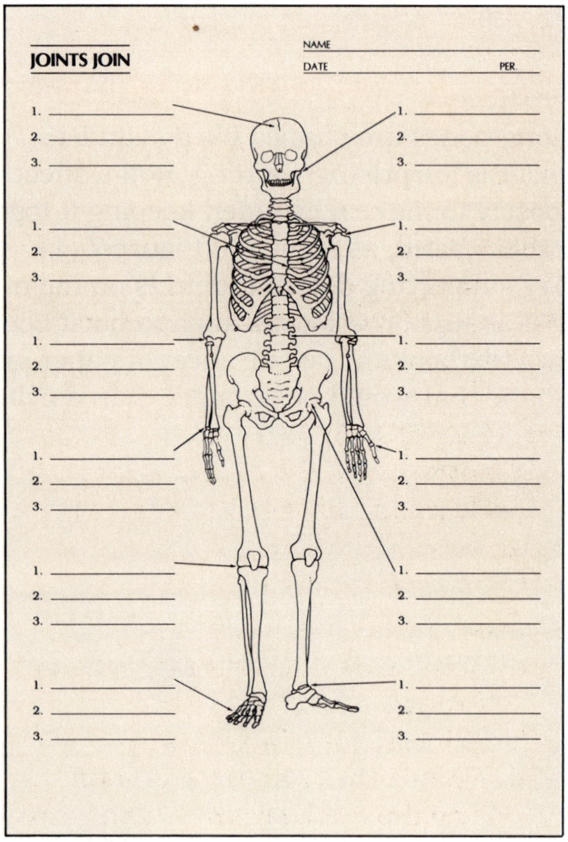

Figure 7–12

*Since your skeleton is made up of *206 bones*, how many *joints* do you think are in your body?

3. Use the third line to rate the ease with which you can move your own joints. Here is the rating scale you should use.
    ++++ moves easily and in several directions
    ++ movement is limited
    0 does not move

Part B. X-Rays
Radiologists use x-rays to see and "read" what is going on inside a person's body. It takes a great deal of training to learn how to read an x-ray. In some ways, x-rays are like negatives of photographs. Dense, heavy structures look white while spaces appear black.

Look at the 2 x-rays in the packet and answer analysis questions 1–3.

X-rays can easily be damaged. Make sure your hands are clean before you touch them. Always hold x-rays by the edges.

Part C. Joints
Read ***Joint Structure***, starting on page 165, and then come back and answer analysis questions 4–7.

ANALYSIS QUESTIONS

1. What color do you think each of the parts listed below is on an x-ray?
    a. bone   c. fat
    b. muscle   d. air
2. Look at the skull x-ray.
    a. What view of this person's head are you seeing?
    b. Which bones can you see?
    c. Sketch the shape of this skull and draw the largest "joint." (Do not draw the jaw joint.)
    d. Do you think this skull x-ray is normal? If not, explain why.
3. Look at the shoulder x-ray.
    a. Which bones can you see?
    b. How many bones can you see forming the shoulder joint?
    c. Do you think this shoulder is normal? If not, explain why.
4. Look at the hinge shown in Figure 7–13.
    a. Does a hinge move in two directions (up and down *or* side to side) or in four directions (up and down *and* side to side)?
    b. List at least 3 joints that you think move like hinges.

Figure 7–13 A hinge

Figure 7–14 A ball and socket

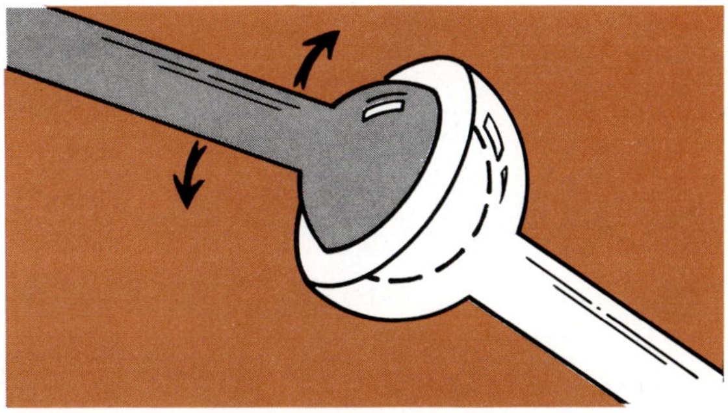

5. A ball and socket is one way to attach two structures together and still allow for movement—an example is shown in Figure 7–14.

Name 2 joints that work like a ball and socket.

6. What 3 features found in joints make movement possible?
7. Which joint do you think is most easily injured, the hip or the shoulder? Explain.

CONCLUSION

Think again about the questions at the beginning of this activity. Write a short paragraph to answer the second question. Use examples to help explain your answer.

EXTENSION ACTIVITIES

1. Contact your local hospital or medical clinic. Ask to talk to someone who works with x-rays. Explain that your class is studying bones and joints. Ask if you could have any discarded x-rays to examine. Look at each x-ray and do the following:
   a. Decide which part of the body you are seeing.
   b. Try to identify the bones.
   c. Inspect the structure of any joint you see.

   After you have had time to inspect the x-rays, you may want to ask questions about what you are seeing.
2. Inspect some joints on animals that have external skeletons. You may want to begin with a lobster shell or a grasshopper. Look closely! You may want to use a magnifying glass. Consider these questions during your inspection.

   How many joints can you find?

   In which directions do these joints move?

   How would you describe each joint?

   How are these joints *like* your joints?

   How are they different?

"Wing Joints"

Ball and socket joints, hinge joints, elbows, hips . . . these joints are important for movement in animals with *internal* skeletons.

What about animals with *external* skeletons? Lobsters have to be able to move too, so do insects. In fact, not only do insects have to be able to move, they have to do so very quickly.

A physiologist from Finland calculated how fast certain insects beat their wings in a complete cycle. Butterflies move their wings at 5 cycles per second. Mosquitoes were faster, going at 587 cycles, and the tiny midge beats its wings at the incredible rate of 1,046 cycles per second!

Joint Structure

Most joints in the body allow for some movement. However, joints must also hold bones firmly together. Isn't it amazing that one structure can serve both these functions?

Movable Joints

Consider one of your movable joints for a minute. It really does not matter which one you think about—wrist, knee, thumb joint; any moveable joint will do. How many times do you think you bend that joint in an hour? in a day? in a month? You will never oil it and, yet, no matter how many times you use it, you expect it to last your entire life. Very few machines are that efficient!

Joints have three unique features that make smooth, easy movements possible: cartilage, ligaments, and fluid. Two of these are shown in Figure 7–15.

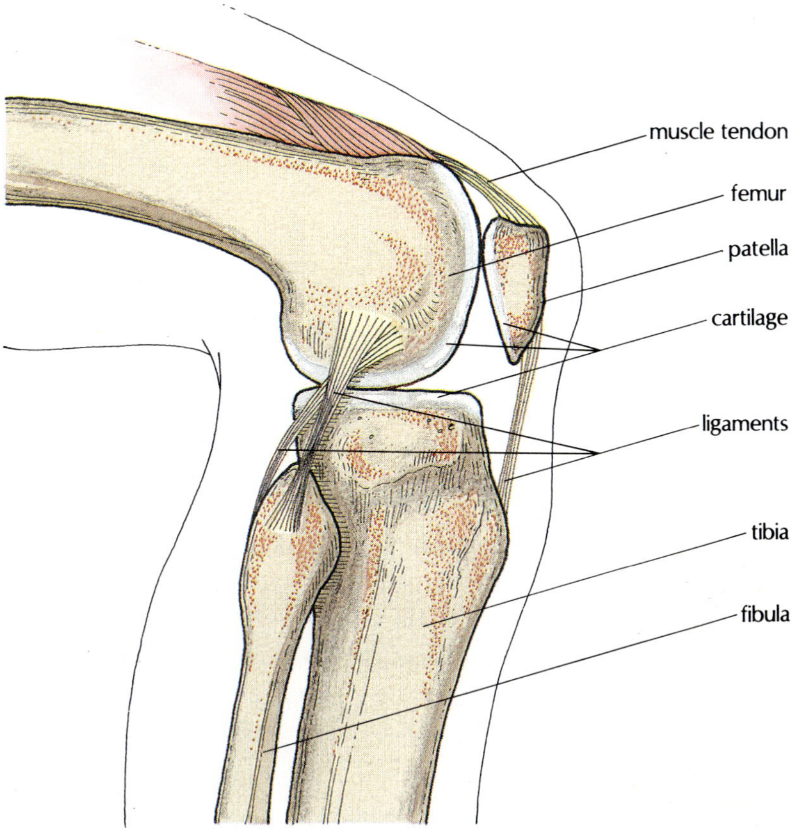

Figure 7–15 A knee joint

Cartilage is an extremely smooth and shiny tissue. It not only protects bones, but lets them glide easily past one another as they move. If the ends of the bones were not covered by cartilage, they would soon wear out.

Ligaments are strong, fibrous bands that connect the bones that make up a joint. They hold that joint firmly together and, at the same time, allow for some movement.

Fluid lubricates joints. This fluid is thick, slippery, and rich in the nutrients—food—that cartilage cells need. Without fluid, cartilage would dry out and bones would grate against each other as they move.

In order for a joint to function smoothly, all three of these components must be present. When people have a joint injury, they usually have damaged the bones, ligaments, or cartilage that make up that joint. The treatment they receive will depend on exactly what has been injured.

Immovable Joints

Unlike movable joints, immovable joints do not contain fluid, nor are the ends of the bones that make up an immovable joint covered by cartilage. They are not held together by ligaments either; instead, a very strong tissue connects the bones. The tough fibers in this connective tissue are so strong that they do not allow the bones to move.

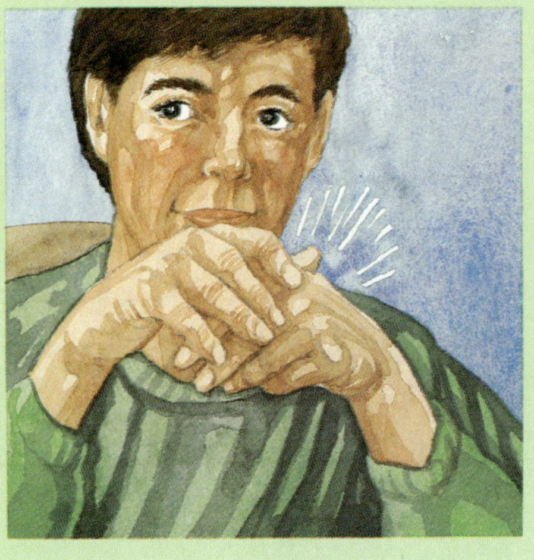

Do You Crack Your Knuckles?

If you do, you might want to think about stopping. When the bones are pulled apart, a gas bubble forms in the joint. As pressure on the joint continues, the gas bubble collapses—making a popping sound. The continued cracking of a joint can lead to arthritis and other joint problems. If you crack your joints, are you prepared for joint problems?

7-5 JOINTS, GROWTH, AND INJURIES

Joint injuries are common. At some time in your life, you are likely to injure one of your joints. You may already have experienced the sudden pain of a ''turned'' ankle. Some joint injuries can have serious results because they can permanently damage joints. In children and teenagers, certain joint injuries even can affect bone growth.

A Dislocation

One of the most common joint injuries is a dislocation, and one joint most likely to be dislocated is the shoulder. Three bones join to form the shoulder—the clavicle, the scapula, and the humerus. Feel these three

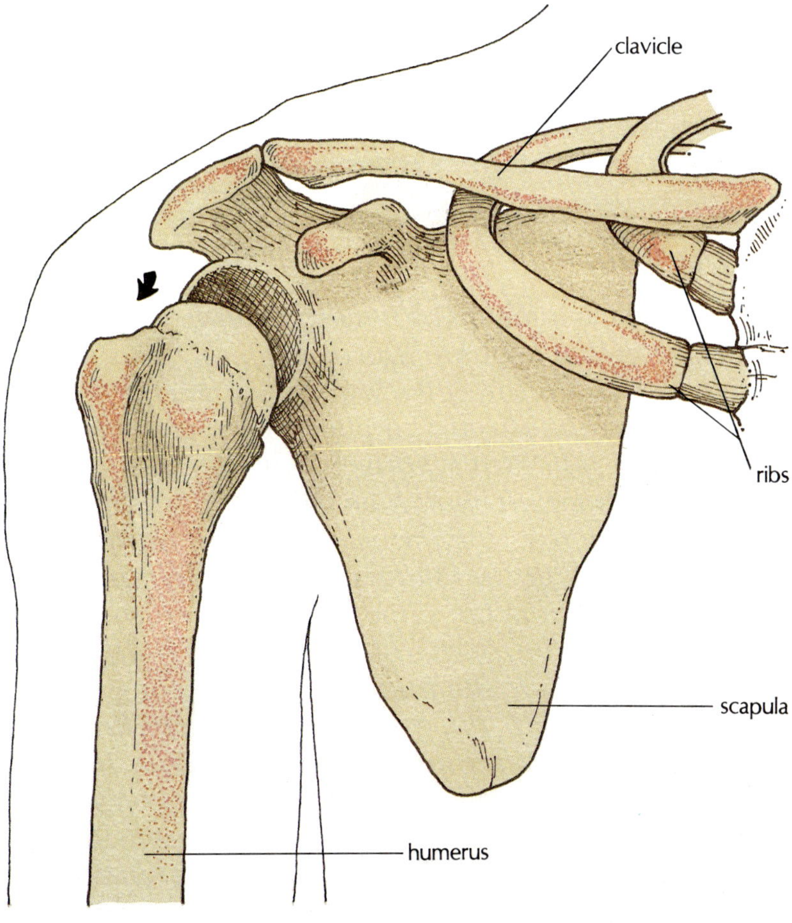

Figure 7-16 When a shoulder is dislocated, the humerus is not in its socket.

bones on your shoulder. You can feel your scapula best in two places—where it meets your clavicle and on your back by your shoulder. Lots of muscles attach to the bones at the shoulder joint so you cannot feel the ends of the bones.

Sometimes the upper end of the humerus slips out of its socket. When this happens, the shoulder is said to be dislocated. Once a person has been injured this way, it can easily happen again because now the socket has been damaged.

Other joints can also be dislocated. Dislocations occur when joints are forced to move farther than they are supposed to. However, sometimes a joint can become dislocated for no apparent reason.

Treatment for Dislocation

If you think you have dislocated a joint, do not let just anyone try to "fix it." This could damage the joint even more. Simply protect the injured area as well as you can and go to a doctor or hospital right away.

A dislocated joint should be repositioned by a trained person as soon as possible. The longer it is out of place, the more that joint will swell. When this happens, it is more painful to reposition.

Figure 7–17 A femur has several growth centers. Which ones do you think could be damaged when someone has a hip injury?

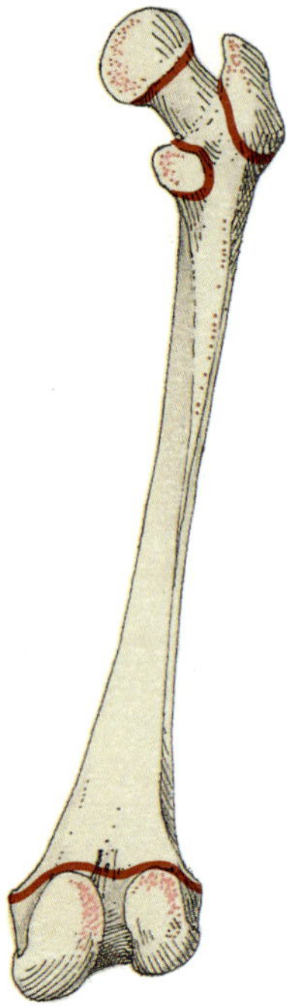

Growing Bones

When a joint is injured, the bones, cartilage, and/or ligaments of that joint are injured. If one of the bones is damaged as a result of the joint injury, growth of that bone may be affected because a bone grows from growth centers located at both of its ends.

When a bone is growing, its growth centers are softer than the rest of the bone. Bones continue to grow until the growth centers harden. The growth centers do not all close down at the same time; they harden in a predictable order.

Because you are still growing, most of your growth centers are still soft. The growth centers in the bones that form the elbow are among the first to close down. They usually do so between 13 and 16 years of age. The age range means that these centers may harden in one person at 13 while they still may be growing in another person until age 14, 15, or even 16. Each person's

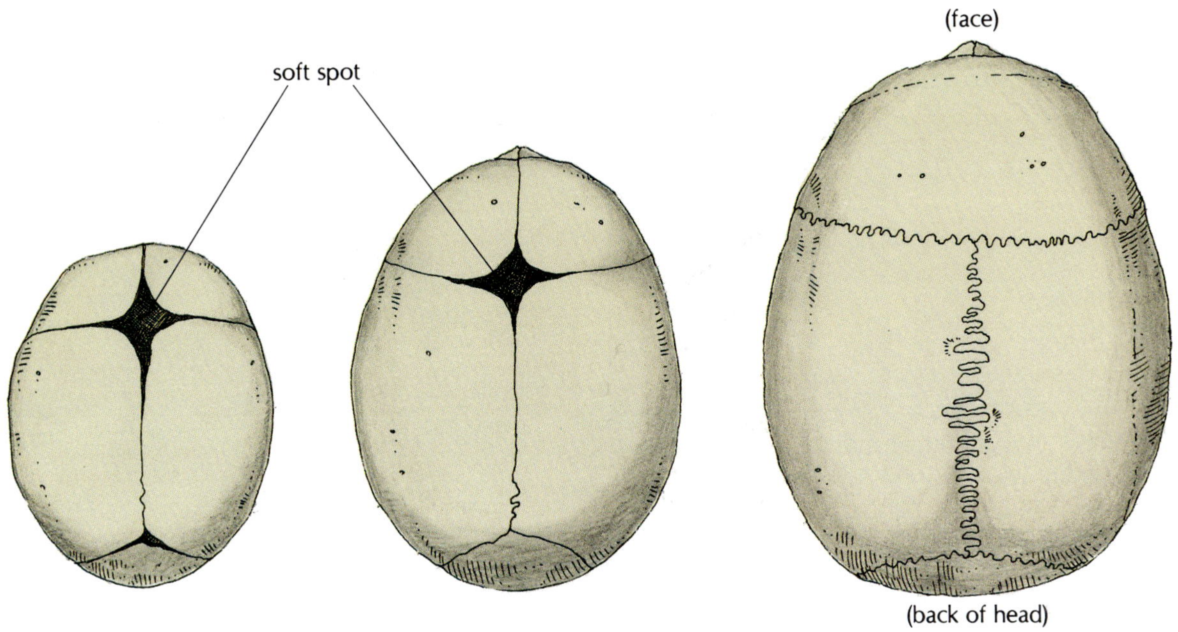

Figure 7–18 As a baby grows, the soft spots in the skull are gradually filled in with bone.

bones grow and harden according to his or her own timetable. In general, the second growth centers to close down involve the bones that form the hip joint. Among the last bones to harden are those that form the wrist. Most bones are fully hardened by the time a person is 25 years old.

Skull Growth

Cranial bones grow along their edges in the cranial joints. On a newborn baby, the cranial bones are separated by tough tissue. This tissue can be felt in the area we call **soft spots.** As a baby grows, the cranial bones grow and fill in the tissue areas. Eventually the bones join and the top of the skull becomes solid. (See Figure 7–18.)

Bone Growth and X-Rays

Bones show up well on x-rays because they contain calcium. Because the softer growth centers are not yet hardened, they appear fuzzy on x-rays. This allows a doctor to use an x-ray to check a child's bone growth. The four x-rays in Figure 7–19 on page 170 show the development of bones in a hand and wrist.

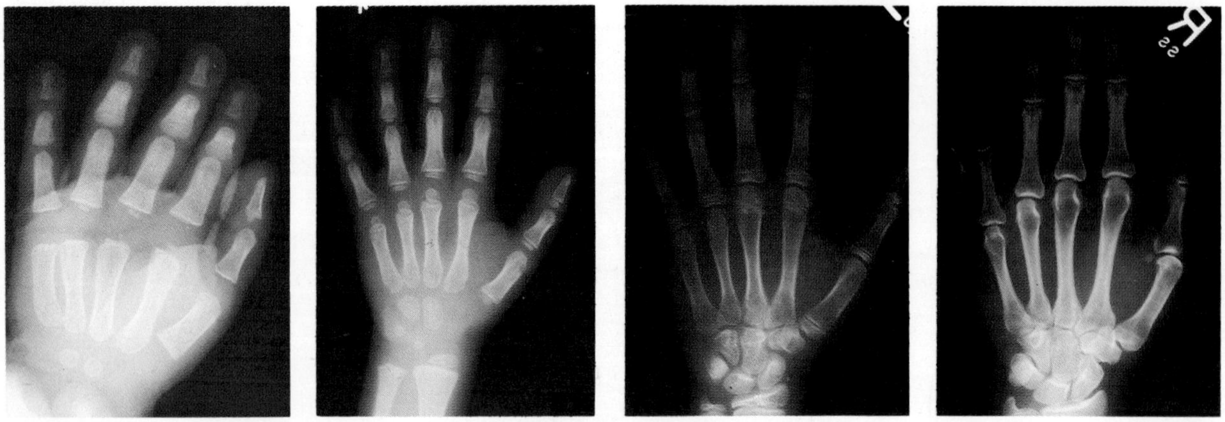

Figure 7–19 X-rays of the hands can be used to tell a person's age. From left to right, the following ages are shown: 18 months, 5 years, 12 years, 17 years.

TEST YOURSELF

1. What are two advantages for a baby not having fused skull bones at birth?
2. Estimate the age of the two people whose x-rays are shown in Figure 7–20. You can compare these pictures with the ones above to help you solve this problem.

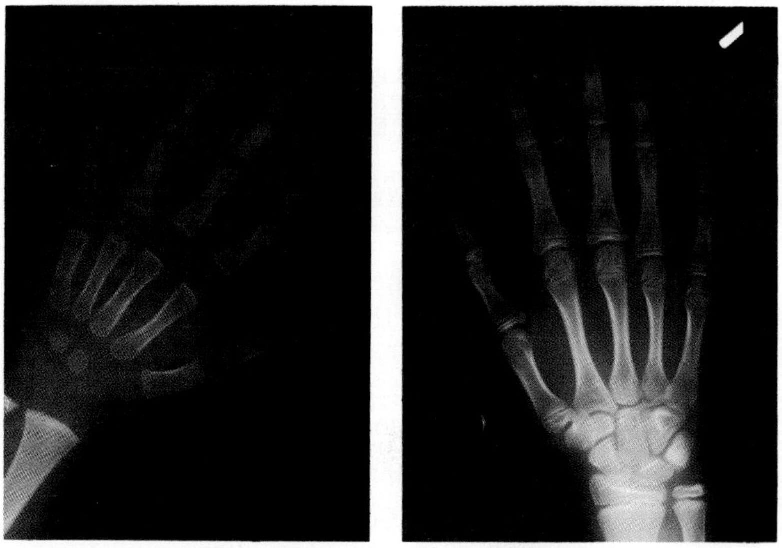

Figure 7–20

7-6 BONING UP

If you were describing a bone, you might say something like this, "A bone is hard, white, and dry and it has very tiny cracks all over its surface." This description would fit an old, dry bone from a skeleton. However, it would not fit a "fresh" bone. We are much more familiar with "dead" bones than we are with living ones such as those in our own bodies.

- How many differences can you observe between a dead bone and a living bone?
- What is inside of a long bone?
- What does a bone cell look like?

Inspect a section of a "fresh" bone and a microscope slide of bone to help you answer these questions.

MATERIALS

paper towel
probe or toothpick
beef bone, cut
microscope
prepared slide: bone
data page: ***Boning Up*** (Figure 7–21)
reading: ***Bones and Blood*** (page 174)

PROCEDURE

Work with a partner. Record your findings on your data page.

Part A. Bone Inspection

1. Obtain a fresh bone section from your teacher. Imagine what it looked like before it was cut. Think about what a whole fresh bone would look like and how it would compare with an old dry bone. Record your thoughts by completing the sentence on the data page.
2. Sketch your bone in the space provided at the side of the data page.
3. Use Figure 7–22 on page 172 to help you identify the four main parts of a bone. You may be able to find two types of bone—**compact bone** and

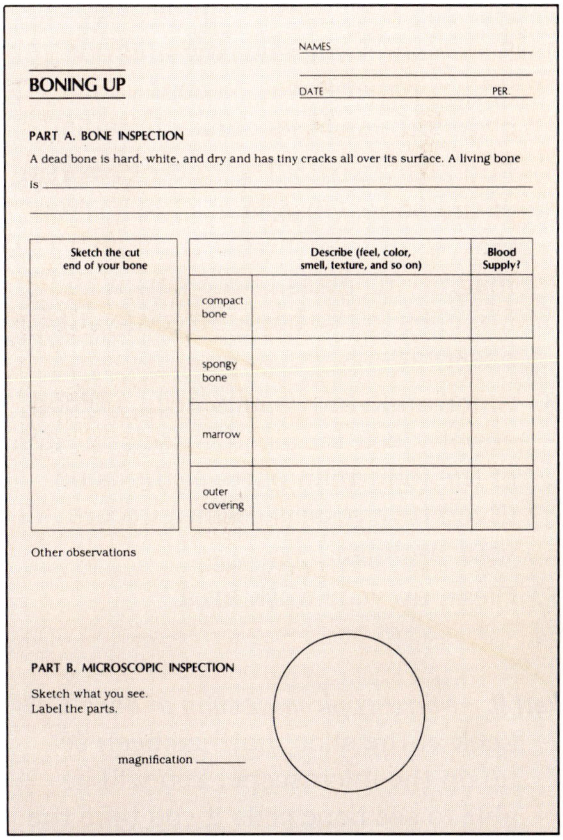

BONING UP

NAMES ______

DATE ______ PER. ______

PART A. BONE INSPECTION

A dead bone is hard, white, and dry and has tiny cracks all over its surface. A living bone is ______

Sketch the cut end of your bone		Describe (feel, color, smell, texture, and so on)	Blood Supply?
	compact bone		
	spongy bone		
	marrow		
	outer covering		

Other observations

PART B. MICROSCOPIC INSPECTION

Sketch what you see.
Label the parts.

magnification ______

Figure 7–21

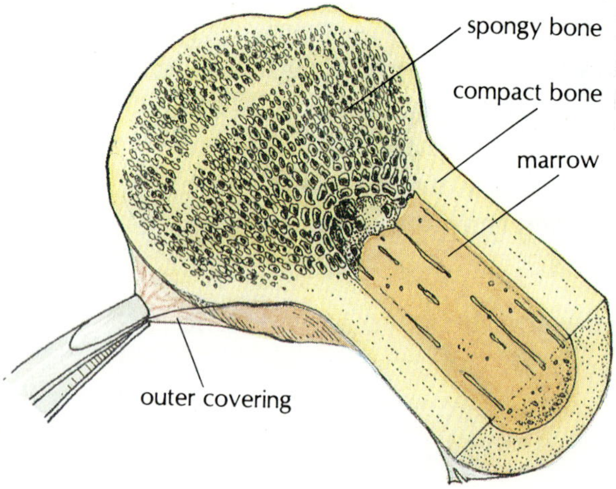

Figure 7–22 Cross section of a bone

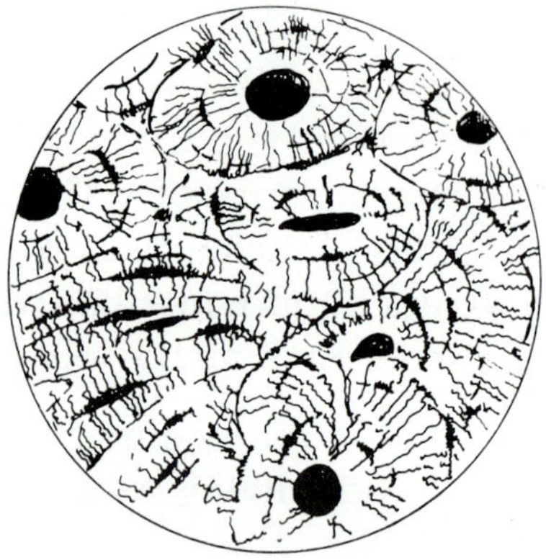

Figure 7–23 Microscopic view of bone

spongy bone—on your piece of bone. (Not all bones have spongy bone. You may need to look at it on some other person's bone.) Record your observations about these two types of bones on the data page.

4. Scoop out some **marrow** and put it on a paper towel. Record your observations of its characteristics.
5. Pull away part of the **outer covering** of the bone. Observe it closely. (You may want to inspect it through a magnifying glass.) Note how it attaches to the bone. Record your observations.
6. Since all living tissues need a blood supply, look for blood vessels in each of the four main parts of the bone. (Even tiny vessels count.) Record your findings.
7. Clean up your work area.

Part B. Microscopic Inspection of Bone Cells

1. Look at the slide that has been set up on the microscope. You will be able to see bone cells if you take the time to focus your microscope.
2. Sketch what you see on the data page. Label the parts you can identify. (Figure 7–23 may help you.)

Part C. Reading and Writing

Read ***Bones and Blood,*** starting on page 174. Write the answers to the analysis questions on the back of your data page. The information on the front of the data page will help you answer the questions.

ANALYSIS QUESTIONS

1. Did the bone you inspected have more spongy bone or compact bone?
2. List at least 4 differences between living bones and dead ones.
3. All of the things listed below can be found in a *living* bone. Copy the list and circle each one that is also found in a *dead* bone.

outer covering	marrow
bone cells	spongy bone
tiny blood vessels	compact bone
minerals	water

4. Which parts of the bone—compact bone, spongy bone, outer covering, marrow—had a blood supply?

5. Is a fresh bone mostly living cells or minerals? Explain.
6. Why do bones need a blood supply?
7. Imagine that you have broken your arm. Your little brother wants to know why it takes so long for the bone to heal. What would your explanation be?

CONCLUSION

Reread the three focusing questions on page 171. Answer them using complete sentences.

EXTENSION ACTIVITIES

1. Take a long, thin uncooked bone—like a rib or chicken leg—and remove the muscles and other tissues. Soak it for a week or so in a dilute acid. (When using a 5% solution of nitric acid, plan to let it soak for three to five days. Vinegar, a weaker acid, also works but, when using it, you will need to wait a

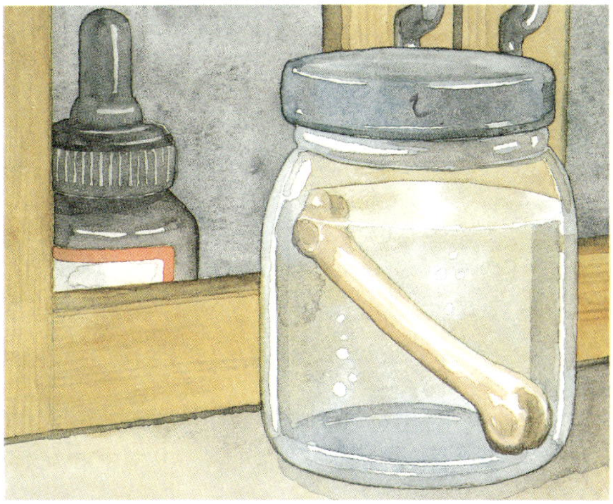

couple of weeks.) Record your observations.
2. Sometimes people need to take calcium tablets. Call a doctor, nurse, or dietician to find out answers to these questions.
   Why do pregnant women usually need to take calcium tablets?
   How can a doctor tell when someone needs more calcium?
   Is it possible to get too much calcium?

Bones and Blood

Bones are alive! In order to stay alive, they need nutrients and oxygen. Because they are alive, bones can grow, and hurt, and heal. There are four parts to bones. (See Figure 7–24.)

Parts of Bones

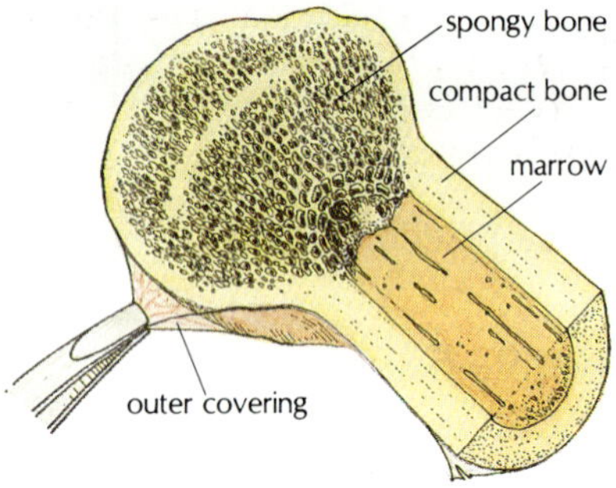

Figure 7–24

A bone's **outer covering** is thin and tough. It provides a place for muscle attachment and contains bone-forming cells. When these cells divide, the bone grows a bit wider.

Directly under the outer covering is a layer of **compact bone.** Next to this is a layer of bone that is full of spaces. Although this is called **spongy bone,** it is just as strong as the compact bone. (If a skeleton were all compact bone, it would be very heavy.)

Within the center of many bones is **marrow.** Marrow contains large amounts of fat and blood-forming cells. Yes, most blood cells are formed inside of bones; they are not made in the heart or in any of the blood vessels. Blood cells are shuttled out of the bone and into circulation by the blood vessels that are around and inside the bone.

Every Ten Years

Bones need a steady supply of blood so their cells can stay alive. As blood circulates, it brings nutrients and oxygen to the bone cells. Even though these cells are "trapped" within bone, they are very active. What do they do? They constantly tear down and rebuild bone. In fact, every ten years your bone cells tear down and rebuild your complete skeleton! Your skeleton is being remodeled even as you are reading this page.

A Broken Bone

A blood supply is also important for healing broken bones. Without blood a broken bone would not mend. The better the supply, the faster an injury will heal.

When a bone breaks, all the little blood vessels break, too, and a pool of blood collects around the

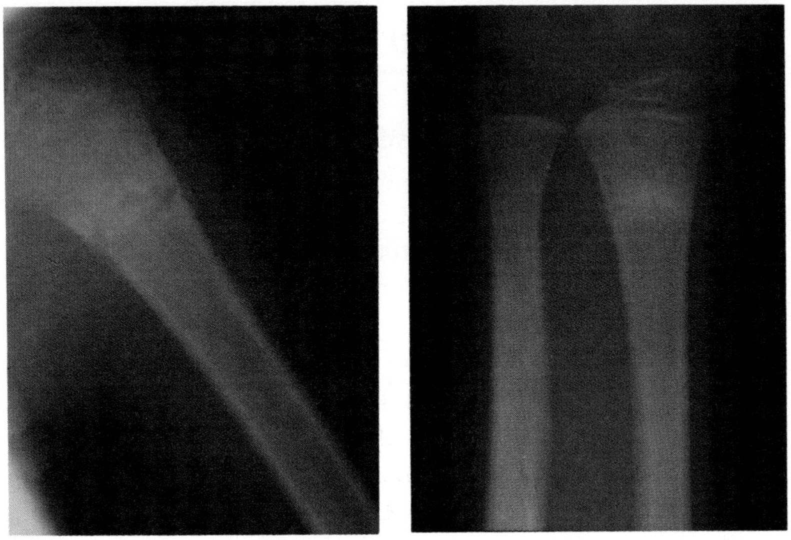

Figure 7–25 Look at these two x-rays. Both bones have been broken. Look for the location of each break.

broken ends of the bone. This is the first step in healing. New bone cells need this blood to grow. As they grow, they deposit minerals around themselves. Eventually, they are surrounded by a hard bony case. Once this has happened throughout the area of the break, the bone will be healed. Here is an approximate time line for healing.

immediately	—	blood collects around break
1 week	—	immature bone cells appear
2 weeks	—	bone cells start to deposit minerals
6 weeks	—	bone has regained strength
1 year (or longer)	—	completely healed

In a well-nourished, healthy young person, healing can occur so completely that it may be impossible to tell that the bone was ever broken.

7-7 WHAT DO BONES NEED?

Rickets is a bone disease that affects children more often than it does adults. Children who have rickets have weak bones. Some bones, such as the leg bones, may bend under the weight of the body and may even become permanently misshapen.

During the 17th and 18th centuries, many children developed rickets. This was the era of the Industrial Revolution. For the first time, large factories were built; they produced clouds of black smoke that hung in the air for days. Many children worked in these factories and lived nearby; their homes, with very small windows, were built along crowded, narrow streets. These "city children" were the ones who developed rickets. As you might guess, doctors were perplexed by the increase in the number of children who developed rickets. They wondered what could be causing the increase.

Imagine that you are trying to solve this problem. How might you proceed? First, read about two experiments and then answer the two questions on the next page.

Figure 7-26 Early factories produced lots of smoke.

- What experiment could you set up to find out what causes rickets?
- What are some possible causes of rickets?

Smoke Causes Rickets? NO!

First, you might wonder if the smoke itself could cause rickets. You could test this idea by examining rats that were raised where there was a lot of black smoke in the surrounding air. You would find that these rats do not get rickets. Monkeys raised under similar smoky conditions do not get rickets either. Based on these observations, you would probably conclude that smoke by itself does not cause rickets.

An Experiment

There is a second important clue to consider; the children in Northern Europe were not exposed to as much sunlight as children had been in the past. They were living in buildings that let in very little light and they were working inside during the day instead of playing outdoors. Then, even when they were outside, much of the sunlight was blocked by smoke from the factories.

Considering the above information, you might decide to test the following possibility:

"Sunlight prevents rickets."
or, to put it another way,
"Lack of sunlight causes rickets."

In order to test this possibility, imagine that you are observing an experiment using several groups of laboratory rats. Each group of four rats is being kept in a separate cage. The living conditions are the same for each group of rats except for the amount of sunlight they get. All of the rats are fed plain rat pellets without any vitamins or minerals. The results of this experiment are summarized in the following table.

A RAT EXPERIMENT—DATA TABLE

Rat Group	1a	1b	1c
Sunlight	12 hours	6 hours	0 hours
Diet	rat pellets	rat pellets	rat pellets
Results	no rickets	no rickets	rickets

ANALYSIS QUESTIONS

Answer these questions to see how well you understand this experiment.

1. Think about the experiment.
   a. How many rats were needed to do this experiment?
   b. What living conditions do you think were the same for each group of rats?
   c. What conditions were different for each group?
   d. What was the idea being tested in this experiment?
2. What can you conclude from this experiment?
3. A student says rats need at least six hours of sunlight to stay healthy. Do you agree? Explain.
4. Think of another set of living conditions to test.
   a. What conditions would you use to raise the rats?
   b. What would you be testing by using these conditions?
5. Some people think it is cruel to use laboratory animals for scientific experiments. Do you think this experiment was cruel? Explain.

A Second Experiment

The results of the first experiment indicate that sunlight has something to do with rickets. Considering that, here is another important fact: sunlight causes a person's skin to produce vitamin D. "Perhaps," you might wonder, "rickets result when rats do not get enough vitamin D." You might now want to do a second experiment to test this idea.

Imagine that you set up another rat experiment. The living conditions you tested and the results are shown below.

A RAT EXPERIMENT—DATA TABLE

Rat Group	2a	2b
Sunlight	0 hours	0 hours
Diet	rat pellets with no vitamin D	rat pellets high in vitamin D
Results	rickets	no rickets

ANALYSIS QUESTIONS

6. What was tested in this second experiment?
7. What conclusion can you make?
8. Consider *both* experiments. List 2 ways you could prevent rickets in rats.
9. Predict whether or not each of these conditions would give rats rickets.
   vitamin D and 0 hours of sunlight
   vitamin D and 12 hours of sunlight
   vitamin C and 12 hours of sunlight
   vitamin C and 0 hours of sunlight
10. Rats are not like people. What would you need to do before you could apply these findings to humans?
11. Now that you know something about causes of rickets, reread the information about rickets on page 176.
    a. List 3 ways people could have tried to prevent rickets during the Industrial Revolution.
    b. Explain which solution you would try first.
12. Would you volunteer for an experiment that could either harm or help you? Why or why not?

CONCLUSION

Write several sentences to answer each of the focusing questions on page 177.

EXTENSION ACTIVITIES

1. Analyze your own diet for calcium and vitamin D. Find out how much you are getting of each of these nutrients. Which foods are good sources of vitamin D and calcium?
2. Other bone problems can develop when something is missing from a person's diet. Find out what osteoporosis is and how it can be prevented. Do more research to learn what other bone problems are preventable by changing a person's diet.

7–8 "MILK BUILDS STRONG BONES"

You have probably heard someone say, "Drink your milk; it's good for you" or "Milk will make your bones grow." What do you think? Is milk really important for bone growth? Read the following information so you can answer this question yourself.

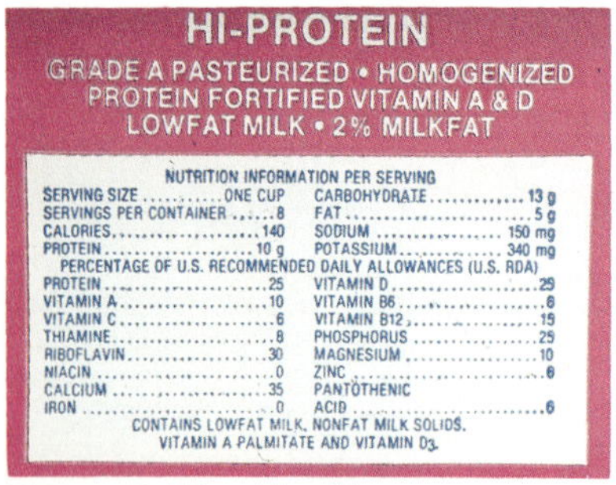

Figure 7–27 Nutrition information from 2% milk

Milk Ingredients

Read the ingredient panel on the milk carton in Figure 7–27. It gives information about some nutrients that are essential for healthy bones. Two of the most important ones are calcium and vitamin D. What percent of the recommended daily amounts of these two nutrients do you get from an 8-ounce glass of milk?

Calcium

Calcium is one of the main components of bone. Without calcium, bones would not be very strong. They would break easily and they would probably become deformed.

Bone cells take calcium out of blood and deposit it in the bone around them. You may wonder how calcium gets into the blood in the first place. Like other nutrients, it is absorbed out of the contents of the small intestine into the blood. This is not an easy process. Here is where vitamin D is important! Vitamin D increases the amount of calcium that is absorbed.

Vitamin D

As you can see on the label in Figure 7–27, milk supplies a large amount of vitamin D. However, vitamin D is not a natural milk nutrient; it has to be added to milk. This addition is important since it would be hard for you to get enough vitamin D from other foods. Vitamin D is found naturally in fish but in very few other foods. In addition to food sources, vitamin D is made by the skin. Sunlight changes a chemical found in skin to vitamin D.

THINK ABOUT IT

Why is vitamin D added to milk instead of to orange juice, pop, or some other food?

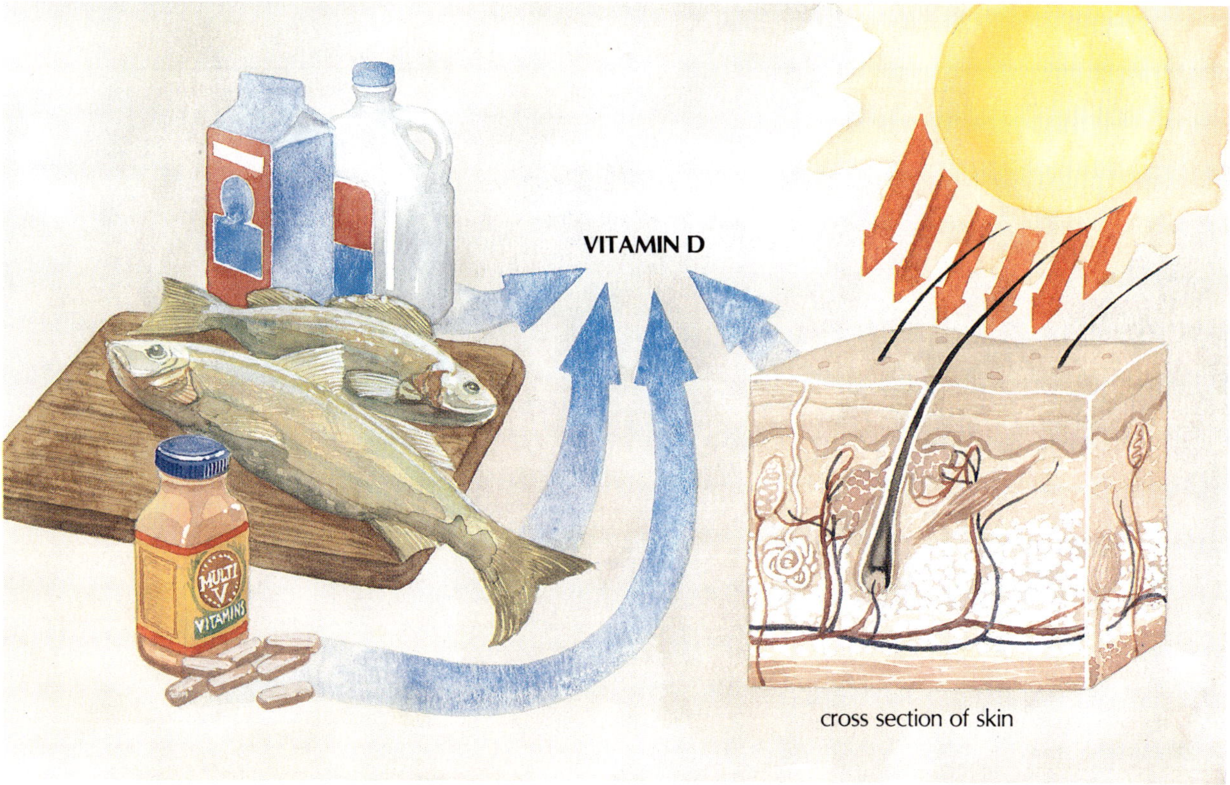

Figure 7–28 Sources of vitamin D

Now see if you can explain why either vitamin D or sunlight is important for the prevention of rickets.

Other Nutrients

In addition to vitamin D and calcium, several other nutrients are important for bone development. A well-balanced diet provides all of the nutrients most people need to have healthy bones. Bone growth is also influenced by hormones and exercise.

CHAPTER 8

MUSCLES

Which muscles do you think are used when playing tug-of-war?

8–1 CHICKEN WINGS

There is more to a body than bones. Bones provide support but other tissues* are needed to hold the bones together and allow for movement. Look at your hand, for example. What tissue besides the bones do you think are found there?

It is difficult to look at the tissues in your body, but you can observe them in other animals. For instance, you can learn about some relationships between body tissues by dissecting a chicken wing.

- What tissues make up a chicken wing?
- Which tissues are involved in movement?

Keep these two focusing questions in mind while you dissect a chicken wing.

MATERIALS

chicken wing
dissecting pan
scissors
soap
paper towels
plastic bag
dissection guide: ***Chicken Wings*** (Figure 8–1)

PROCEDURE

Work in a group with one or two other students.

1. Choose one person in your group to read the dissection guide aloud. As each step is finished, have this person record the observation in the space provided.
2. Work quickly and carefully. Make sure that each person in your group can see what is going on. As you work, check the observation tables

*A tissue is a group of similar cells that have a similar function; examples are bone tissue, blood tissue, and muscle tissue.

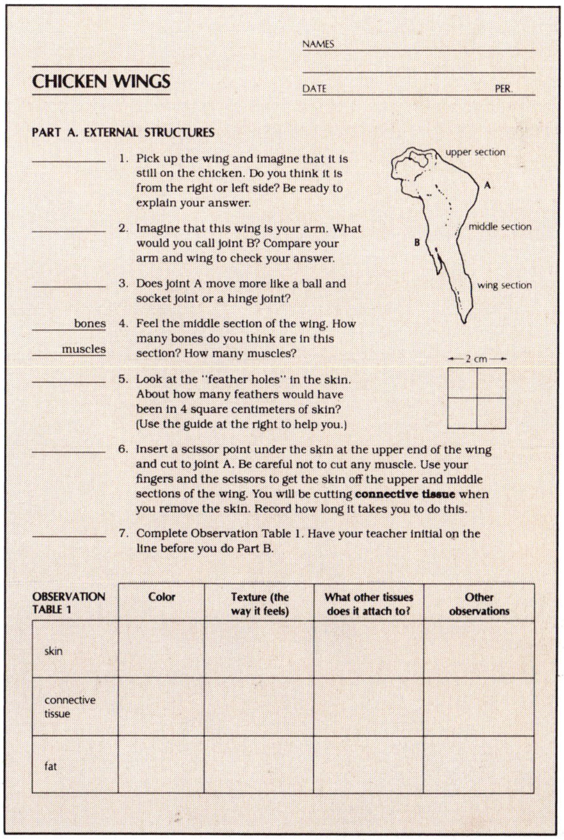

NAMES ______

CHICKEN WINGS

DATE ______ PER. ______

PART A. EXTERNAL STRUCTURES

______ 1. Pick up the wing and imagine that it is still on the chicken. Do you think it is from the right or left side? Be ready to explain your answer.

______ 2. Imagine that this wing is your arm. What would you call joint B? Compare your arm and wing to check your answer.

______ 3. Does joint A move more like a ball and socket joint or a hinge joint?

______ bones ______ muscles 4. Feel the middle section of the wing. How many bones do you think are in this section? How many muscles?

______ 5. Look at the "feather holes" in the skin. About how many feathers would have been in 4 square centimeters of skin? (Use the guide at the right to help you.)

______ 6. Insert a scissor point under the skin at the upper end of the wing and cut to joint A. Be careful not to cut any muscle. Use your fingers and the scissors to get the skin off the upper and middle sections of the wing. You will be cutting **connective tissue** when you remove the skin. Record how long it takes you to do this.

______ 7. Complete Observation Table 1. Have your teacher initial on the line before you do Part B.

OBSERVATION TABLE 1	Color	Texture (the way it feels)	What other tissues does it attach to?	Other observations
skin				
connective tissue				
fat				

Figure 8–1 (Dissection guide, page 1)

NAMES ______________________________
____

CHICKEN WINGS
(PAGE 2)

PART B. INTERNAL STRUCTURES

______ 1. Gently loosen the connective tissue around one large muscle in the middle section of the wing. Each end of the muscle narrows to form a **tendon.** Where does the other end of a tendon attach?

______ 2. Finish loosening this muscle from the other muscles but do not cut the tendons. Gently pull on the muscle. Try to demonstrate what happens to a living chicken when this muscle contracts. Can you?

______ 3. Slowly but firmly pull the muscle away from the bone. Draw a sketch to show where it tears.

______ 4. How many muscles did you predict would be in the middle section of the wing? (Part A, Step 4)
______ How many muscles *are* there? Remove them one at a time while you count.

______ 5. Check here when you have completed Observation Table 2.

OBSERVATION TABLE 2	Color	Texture (the way it feels)	What other tissues does it attach to?	Other observations
muscle				
tendon				

______ 6. How many bones did you predict would be in the middle section of the wing?
______ How many bones are there?

Figure 8–1 (Dissection guide, page 2)

NAMES ______________________________
____

CHICKEN WINGS
(PAGE 3)

______ 7. Remove the tissues from around joint A so you can observe the joint. **Ligaments** hold bones together. How many ligaments can you count at this point?

______ 8. Try to pull the bones apart at joint A. Don't twist! Just pull them straight apart. Can you do it?

______ 9. If you did not succeed in pulling the joint apart, twist it apart and inspect the **cartilage** at the ends of the bone. Think of one word to describe cartilage.

______ 10. Can you find an outer covering on these bones?

______ 11. Check here when you have completed Observation Table 3.

OBSERVATION TABLE 3	Color	Texture (the way it feels)	What other tissues does it attach to?	Other observations
bone				
ligament				
cartilage				

______ 12. Have your teacher initial here when you complete step 11.

______ 13. Write your initials here when you have thoroughly cleaned your table, dissection tools, and hands with soap and water.
______

Optional

Predict what you would find *inside* of a chicken bone. ______________________________
____

Ask a technical education teacher to cut open a chicken bone for you. What do you see?
____
____

Figure 8–1 (Dissection guide, page 3)

in the dissection guide to see what needs to be recorded.

3. If your group does not finish, save your wing in a plastic bag. Write your names and class period on the bag and put it in a refrigerator.
4. Thoroughly clean your hands, desk, and tools with soap and water.

CAUTION

Salmonella bacteria live in raw chicken. These bacteria can cause food poisoning so make sure to wash your hands and equipment with soap and water.

CONCLUSION

Reread the focusing questions on page 183. Answer each question in complete sentences.

EXTENSION ACTIVITIES

1. Analyze a complete chicken. Dissect it the same way that you did the chicken wing. Compare the ways a chicken uses its wings and its legs. How does the structure of the leg differ from that of the wing to allow for their different functions?
2. Tendons connect skeletal muscles to bone. Look for tendons on the back of your hand. Wiggle your fingers. How many tendons can you count? The largest tendon in your body is called the Achilles tendon. You can feel it by pressing on your lower leg just above your heel. What is the purpose of this tendon? Find at least 2 other places where you can easily feel some of your tendons.
3. Buy a lobster claw, clam, or a crab at a supermarket. Try to find muscles in this animal. Describe what they look like. How do muscles attach to an external skeleton?

8–2 MAJOR MUSCLES

Mike Franzinni tripped down several stairs and landed on his arm, jarring his shoulder. His dad thought Mike might have dislocated his shoulder so he rushed him to the their doctor's office. After gently feeling Mike's shoulder, the doctor decided that the location and amount of pain indicated that a muscle had been injured. The doctor explained that Mike had strained his **deltoid.** Mike was told to put an ice pack against the muscle and to not use his arm for awhile.

Injuries like this one happen frequently. The results make people very aware of their muscles. In this particular case, Mike suddenly needs to learn something about one of his muscles. He needs to know exactly where the deltoid is and what movements it controls.

People involved in athletics also are interested in specific muscles because they often want to improve their strength or endurance. Whether because of an injury or an interest in athletics, many people want to know the names and locations of their major muscles.

Muscle Names

Altogether the human body contains more than 600 muscles. Every one of them has a name. The location of the deltoid as well as that of several other major muscles is shown in Figure 8–2. The **biceps** and **triceps** are the two muscles people are most likely to call by name but the other muscles are equally important. The largest muscle of all is the **gluteus maximus.** You are probably sitting on your gluteus right now.

On your own body, locate each of the muscles shown in Figure 8–2. Try to contract and relax each muscle so you can feel it move. Memorize the names and locations of at least 4 of these muscles.

Muscle Action

The muscles pictured in Figure 8–2 are all called **skeletal muscles** because they attach to the skeleton.

Figure 8–2 Major muscles

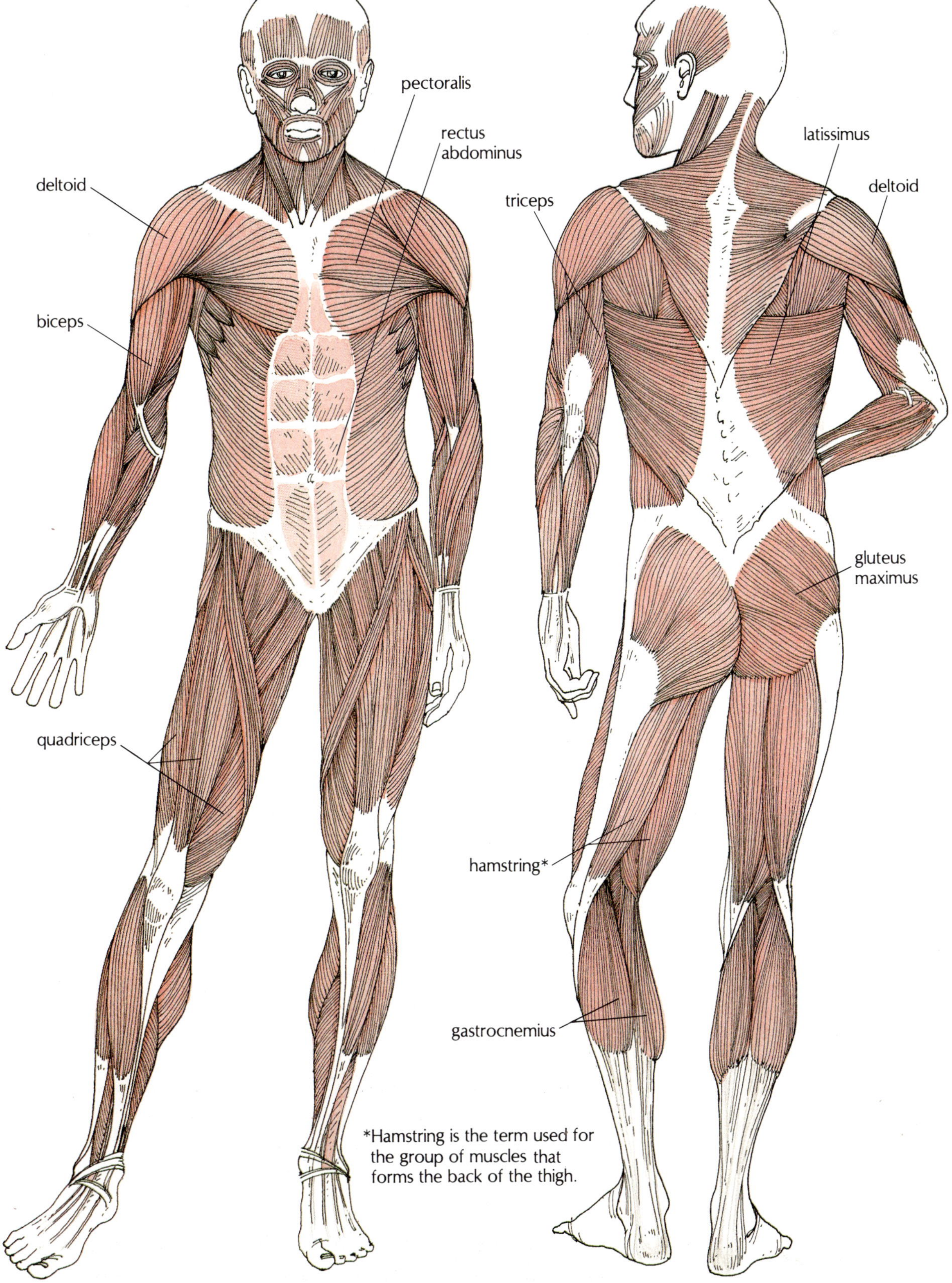

*Hamstring is the term used for the group of muscles that forms the back of the thigh.

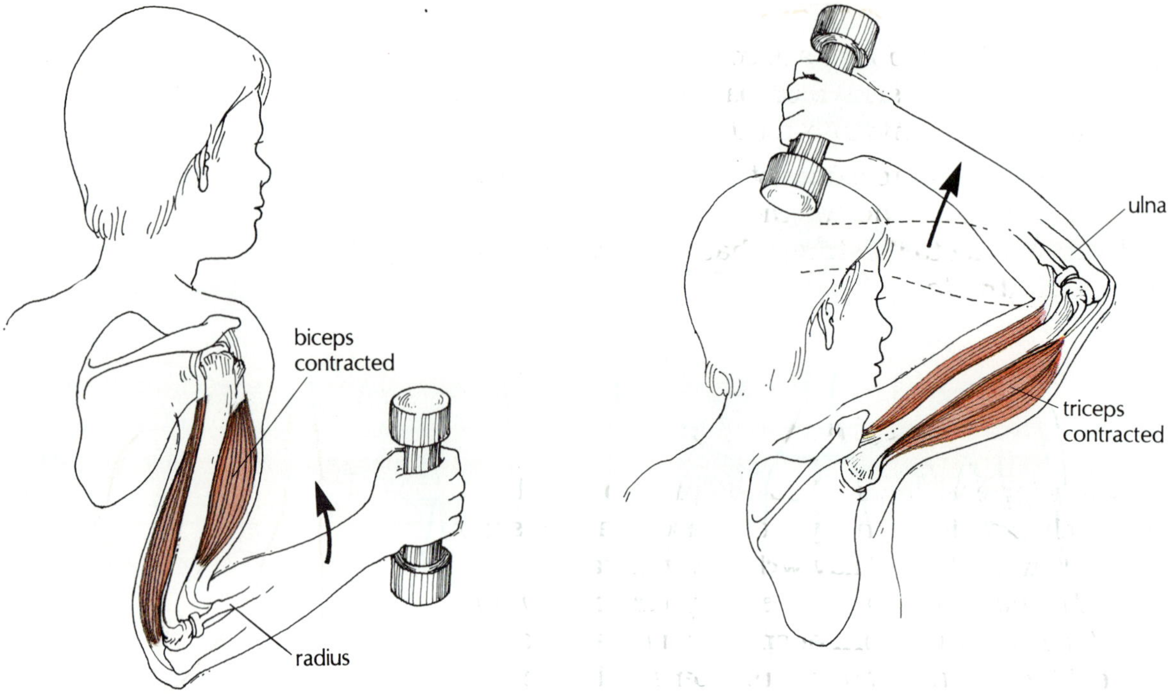

Figure 8–3 Muscles work in pairs. The biceps and triceps are an example of a muscle pair.

However, they do not hook directly to bones; instead, they are attached by **tendons.** If you dissected the chicken wing, you saw how strong a tendon attachment is.

Skeletal muscles cause movement when they shorten or **contract.** You can tell what movements a muscle controls if you know where its tendons attach. Figure 8–3 shows where the biceps attaches to the radius and the scapula. When the biceps contracts, it pulls on the radius and causes the elbow to bend.

In order to straighten the arm, other muscles are needed. The biceps cannot straighten the arm because a muscle cannot "push"; all it can do is pull. Here is where the triceps is needed. It attaches to the scapula and humerus at one end and to the ulna at the other. When it contracts, it pulls on the ulna and straightens the arm. (See Figure 8–3.)

Now, remember Mike's problem—he needs to rest his deltoid. Try to figure out what motion this muscle controls. One end of the deltoid is a broad tendon that attaches along the clavicle. Point to this bone on your own body and imagine the muscle that is attached there. The other end of the deltoid attaches to the humerus. Find this area on your arm.

Imagine that the deltoid is contracting. What motion does it cause? What motions should Mike avoid?

EXTENSION ACTIVITIES

1. Determine what motions are controlled by each of the major muscles shown on page 00. Make a chart that summarizes this information.
2. Construct a model to demonstrate the action of one particular muscle. Include the tendons and bones. Use your model to illustrate what happens when the muscle contracts.

Face Muscles

Try to wiggle your ears. You will probably find yourself exercising many face muscles at the same time. Your forehead may wrinkle, your jaws clench, your eyebrows flutter, or your nose twitch.

There is a network of small muscles in your face. They control your expression and help communicate your mood—even if you don't want them to! Facial movements are unique to humans. Most animal faces are immobile. Have you ever seen a giraffe laugh or a snake smile?

Try the facial muscle exercises shown here. They are fun to do and even more fun to watch. While you do them, try to figure out which facial muscles you are using. Think about your facial expressions. What parts of your face do you use when you look angry? happy? in pain? scared?

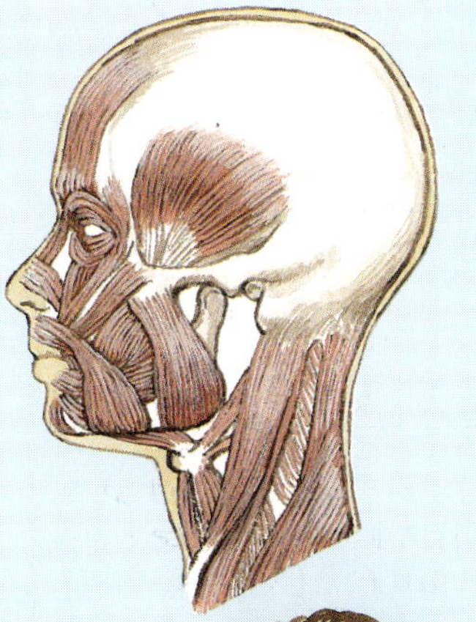

pull scalp back

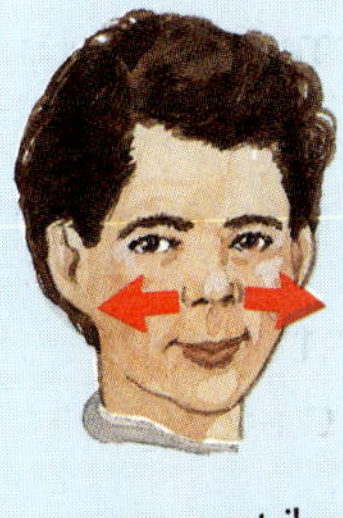

open nostrils

shut nostrils

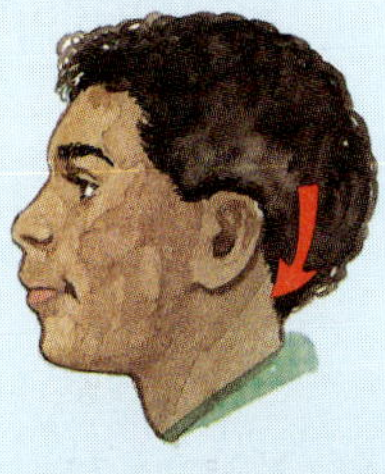

pull scalp down

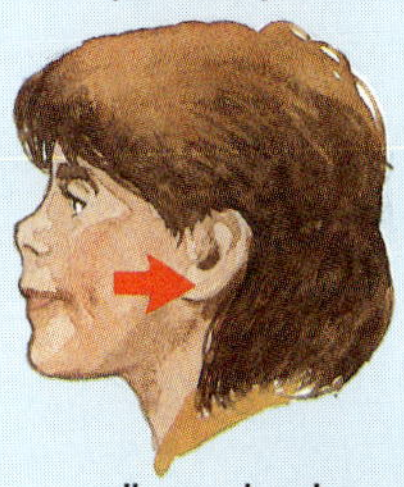

pull ears back

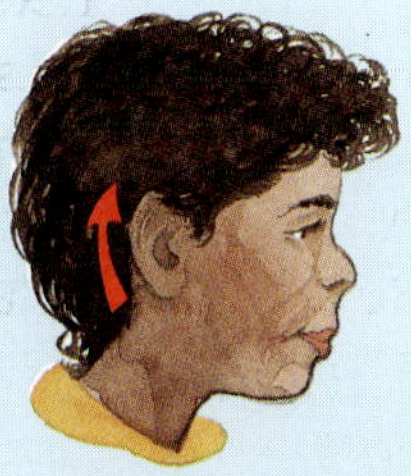

raise ears

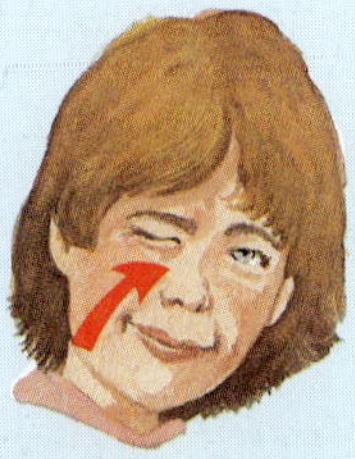

wink one eye, then the other

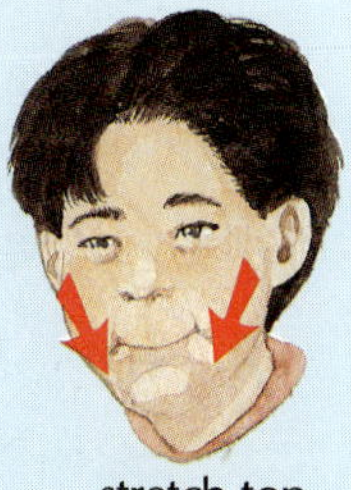

stretch top lip down

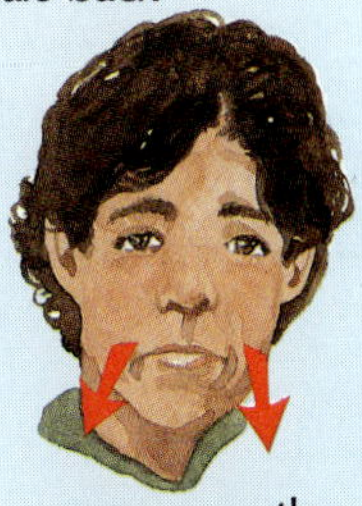

move mouth corners up & down

8–3 MUSCLE FITNESS

Even if people *do not* know the names and locations of their muscles, they should know about muscle fitness. People often assume that fit muscles mean strong ones. Actually, there are three parts to muscular fitness—strength, endurance, and flexibility.

Strength

You have probably talked about the strength or weakness of a muscle. When you talk about muscle strength, you are talking about the amount of work that a muscle can do in a single try. What activities require strength?

Endurance

Could you climb ten flights of stairs without having your leg muscles "twitch"? If so, you have good leg muscle endurance. Endurance is a muscle's ability to work for a long period of time without getting tired. What activities require endurance?

Flexibility

How far can you bend, twist, and stretch? This range of movement is called flexibility. A person may be quite strong and still have very poor flexibility. What activities would be easier when a person has worked on flexibility?

How fit are your muscles? Are your muscles strong? enduring? flexible?

8–4 MUSCLE STRENGTH AND SIZE

If you could check the strength of every muscle in your body, you would find that some of your muscles are much stronger than others. They may be naturally stronger or they may have become stronger because you used them more. Sometimes you may want to know how strong a muscle actually is. Did you know that there are ways to check the strength of specific muscle groups?

- How strong are the muscles in your lower arm and hand?
- How does strength relate to arm size?

In the following activity you will learn the answers to these questions by testing the strength of the muscles in your lower arm and hand with a grip meter and measuring the size of your lower arm.

MATERIALS

grip meter
tape measure
reading: ***Muscle Size*** (page 194)

PROCEDURE

Part A. Grips and Graphs

1. Grip strength
   a. Get a comfortable grip on the grip meter as shown in Figure 8–4.
   b. Test your grip 3 times.
   c. Record all three scores and circle your best one.
2. Arm measure
   a. Measure in centimeters the circumference of your lower arm at its biggest point as shown in Figure 8–5 on page 192. Measure the same arm that you used to test your grip strength.
   b. Record your measurement.
3. Plot your grip reading and arm measurement on the class graph. Your teacher may want you to copy this graph onto your own paper.

Figure 8–4 Sit down when using the grip meter.

Figure 8–5 Measuring arm circumference is an important part of this activity.

4. Help your teacher decide where to draw a line on the class graph to show the way strength relates to arm size. (When you do this, you are choosing the "best fit" through the points.)

Part B. Reading and Questions

In this activity you are thinking about muscle size. Did you know that muscles can shrink? Read ***Muscle Size,*** starting on page 194, and then return to this page and answer the analysis questions.

ANALYSIS QUESTIONS

1. Stacy's grip measurements were 17kg, 9 kg, and 20 kg.
   a. Which measurement should she record?
   b. Give 2 reasons why her measurements weren't all the same.
2. Look at the class data.
   a. What is the range in grip strength for your class?
   b. What is the average grip strength?
3. Consider arm strength and size. What was the arm circumference of the person with
   a. the highest grip strength?
   b. the lowest grip strength?
4. Based on your class data, explain how grip strength relates to arm size.
5. Two people have exactly the same arm circumference but their grip strengths are very different. One of them has a very strong grip and the other has an average grip. Explain how this could happen.
6. When a person's muscles get stronger, what 2 things happen that cause the muscles to look larger?
7. How can you strengthen one of your muscles?
8. Think of at least 1 example of a muscle that is better developed in one animal than it is in another. Think of an original example; don't use one from ***Muscle Size.***

EXTENSION ACTIVITIES

1. Reanalyze the class data. This time separate the data for males and females. Make two lines on the graph—one for males and one for females. How do the two lines compare?
2. Practice! Improve your grip! Every morning and evening, squeeze a tennis ball or racquetball 20 to 30 times or more. Do this for two

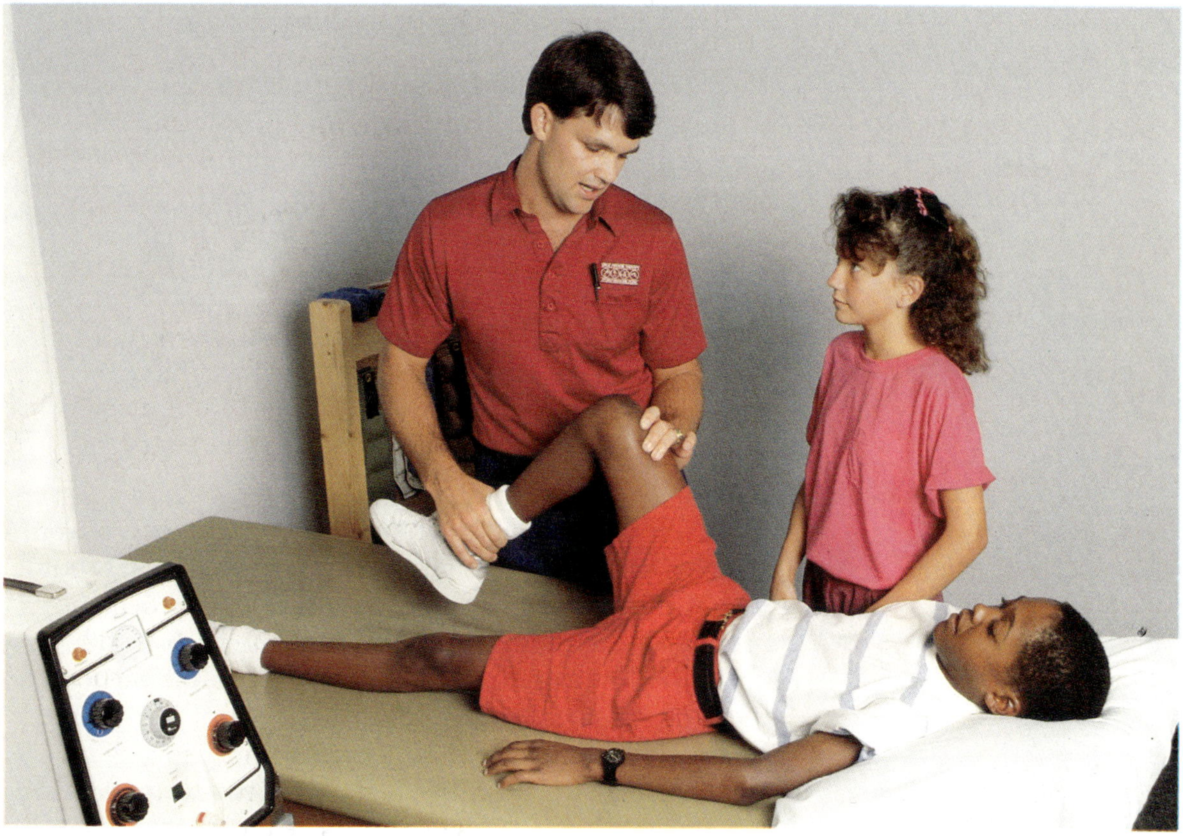

Figure 8–6 Athletic trainers work in school settings as well as with professional sport teams.

weeks. Retest your grip strength. Did it change? Why or why not?

3. People who have been injured may need the help of a physical therapist or an athletic trainer. These specialists can help people work to regain strength, motion, flexibility, or coordination. Talk to one of these specialists and visit a clinic if you can. Find out which exercises can help a person regain or improve the function of a specific muscle group. For example, you might ask what exercises would be recommended for a person whose leg or back has been in a cast for some time. Also find out how to test specific muscle groups.

Muscle Size

"Come on! Pull! You can do it!" Chin-ups usually are easy, but not many people can do chin-ups with some type of weight on their backs. It may even be hard for them to do a chin-up without extra weight. Why? Because some of us lack the necessary upper arm strength to do this type of muscle work. **Strength** is the amount of work that a muscle can do in a single try.

Big Muscles

If you have ever tried to build up the size of one of your muscles through exercise, you may have noticed that it takes time to do so. At first, the muscle may become stronger but still not look any different. In fact, the strength of a muscle can increase quite a bit without affecting its size. But eventually, with continued effort, the muscle will begin to look larger.

This growth is not because new cells form. All the muscle cells a person will ever have are present at birth. Instead, a muscle enlarges because its individual cells get larger. When a muscle works, it needs oxygen; the only way it can get oxygen is from the blood. The more a muscle works, the more oxygen it needs—so more blood vessels form. This too causes the muscle to grow larger.

Strong Muscles

Exercise is the only safe way to increase the strength and size of a muscle. No special diet or magical food can turn a scrawny body into a muscular one. The "Popeye fitness plan" will not do it. Neither spinach, protein powders, nor any other food will develop muscle strength.

Sometimes people try drugs called steroids to help them "bulk out." That's dangerous. Steroids do increase muscle size and strength but they can also ruin a person's health. Steroids may cause liver and kidney damage, shrinkage of the testes, severe mood swings, heart disease, and cancer. They can also stunt growth by closing off the growth centers at the ends of the

Figure 8–7 Many articles have appeared in magazines warning people about the dangers of steroid use.

bones. Steroids are not a good choice when it comes to developing muscles.

So what can you do if you want strong muscles? Exercise and eat a balanced diet!

Shrinking Muscles

Muscles can grow, but they can shrink, too. They do so when they aren't used. This can happen fairly quickly. For instance, if you break your arm, it is put in a cast. When the cast comes off, your arm often looks smaller. Why do the muscles in that arm shrink?

When muscles are not used, the individual cells become smaller. Eventually the whole muscle is smaller. This is called "muscle wastage."

Muscle wastage is a problem for space travelers. Astronauts do not experience the normal pull of gravity on their bones and muscles. After a long stay in space, their muscle size would decrease and their bones would become less dense if they did not exercise in space. Fortunately, once the muscle is used again, the muscle cells enlarge and the entire muscle returns to normal size.

Figure 8–8 Astronaut in space shuttle doing one of a series of exercises

Very Large and Very Small

Some muscles, such as a penguin's flight muscles, have changed size over hundreds of thousands of years. These muscles are now quite small compared with those of most birds. You may be thinking, "So what? Penguins don't fly, so they don't need flight muscles." That's right, but these flightless birds did not suddenly appear. Eons ago there was probably a group of birds who spent quite a bit of time on the ground. Over hundreds of generations, these birds may have spent more and more time walking and less time in flight. Birds who were born with smaller flight muscles had better and better chances to survive. In fact, there was probably an advantage to having small flight muscles. Eventually, the descendants of these birds became what we now know as penguins.

An elephant's trunk is its upper lip and nose. It is made up of about 40,000 muscles and tendons. These muscles have developed over hundreds of thousands of years. An elephant now can control its trunk muscles so well that it can pick up a single blade of grass!

Figure 8–9 Top right: Emperor penguin. Bottom left: African elephant.

8–5 MUSCLE CONTROL

Wink. Stand up. Sit down. Clap. Wiggle your toes. These are complex muscle movements, and yet you can do them easily because *you* control your muscles.

Nerve Connection

Your brain tells your muscles what to do. Of course, the brain doesn't "talk" directly to a muscle. Instead, a nerve carries the message from the brain to the muscle. When the message reaches the end of the nerve, certain chemicals are released. These cause the muscle to contract. After contracting, the muscle quickly relaxes until it receives another message to contract. Figure 8–10 illustrates this nerve muscle interaction.

Figure 8–10 A muscles contracts when it is triggered to do so by nerves.

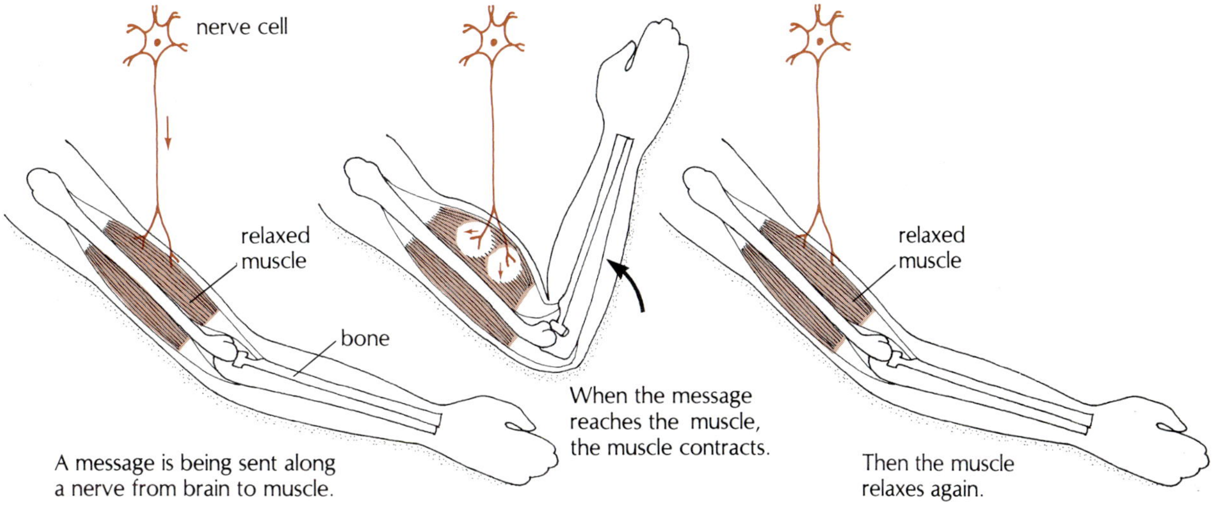

Tired Muscles

Sometimes, your muscles do not "listen" to you. You may have noticed this if you have tried to do an exercise such as sit-ups. Maybe you wanted to do one more sit-up. Your brain may have been saying, "Come on! One more time!" but your abdominal muscles could no longer respond. The muscles were tired; they had worked beyond their limit of endurance.

A muscle tires because waste products build up in its cells. Eventually, these wastes reach such high

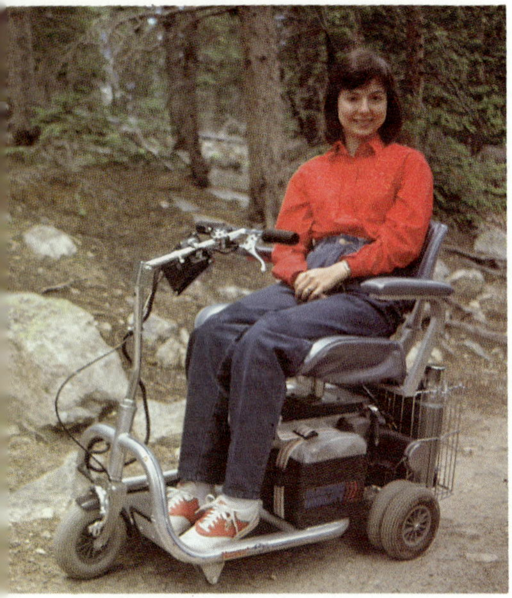

Figure 8–11 Teenager hiking on mountain trail

levels that the muscle stops working. It will not contract even if a message from the nerve tells it to do so. Once this happens, the muscle has to rest before it can work again. It has to get rid of the waste products and rebuild its supply of oxygen.

Loss of Control

Luckily, the loss of muscle control from overuse is only temporary, but sometimes people lose control of some of their muscles forever. There are several ways this can happen. For instance, injuries and disease can damage nerves. When a nerve is damaged, it cannot communicate with the muscle. Thus, the muscle will not contract. It is said to be paralyzed.

Certain diseases can cause people to lose muscle control. You may have heard of some of these—muscular dystrophy, multiple sclerosis, Parkinson's disease. Some of them affect the brain, others affect the nerves, and still others involve the muscles themselves. Scientists are working to discover the exact causes and cures for each of these disorders.

Sometimes, when an operation is taking place, anesthesiologists want to paralyze a patient's muscle temporarily; they do not want it to contract while the surgeons are working. In this case, drugs are used to block the chemicals that nerves normally release. Since these chemicals cannot signal the muscle to contract, it cannot move.

Tetanus

Figure 8–12 *Clostridium tetani* (magnified 500 times)

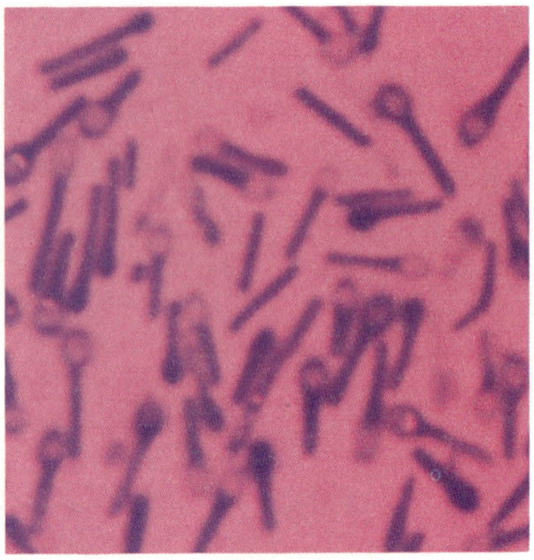

Tetanus is another example of a disease that causes a form of paralysis. It is caused by a type of bacteria that lives in soil. Figure 8–12 shows a microphotograph of this bacteria. A person may develop tetanus if these bacteria enter a cut or open wound. Puncture wounds from objects such as nails and thorns are of most concern because they are difficult to clean.

Once inside the body, the tetanus bacteria produce a chemical that makes it impossible for a muscle to relax. A person with tetanus develops rigid muscles and painful spasms. If these spasms develop in the throat and chest, breathing becomes impossible and death may follow. Tetanus results in death for about six out of every ten people who develop it.

There have been very few cases of tetanus in the United States recently because most people have been vaccinated against the tetanus bacteria. But, since the vaccination does not last a lifetime, it must be repeated at regular intervals. The usual ages for vaccination are shown in the table below.

TETANUS VACCINATION SCHEDULE

Age	Type of Vaccination*
2 months	DTP
4 months	DTP
6 months	DTP
18 months	DTP
4-6 years	DTP
13-16 years	DT
every 10 years	DT

*DTP Diphtheria/Tetanus/Pertussis (whooping cough)
DT Diphtheria/Tetanus

When did you get your last tetanus shot?

ANALYSIS QUESTIONS

1. How do your muscles "know" when to contract?
2. What does the word *paralysis* mean?
3. What are 4 things that can cause paralysis?
4. What is tetanus?
5. We often hear parents talk about taking young children to get their "shots," but what about the parents? Do they need tetanus shots? Explain your answer.
6. Describe at least 4 things that happen between your nervous system and muscles when you climb one stairstep.

EXTENSION ACTIVITY

With your parents' help, check your own medical records. When did you get your last tetanus shot? When do you need your next one?

8-6 MUSCLE WORK

Squeeze a grip meter. Do 60 sit-ups. Stretch. These three activities involve three aspects of muscle fitness—strength, endurance, and flexibility. In each case, muscles are working and this work takes energy. Where does this energy come from?

You have had experience with working muscles. For instance, think about what happens when you jog. Your leg muscles must contract and relax continuously for quite awhile. As you jog, your breathing rate increases and your heart beats faster. You start to feel warm. Why do these things happen?

Inside the Cell

When you breathe faster, more *oxygen* enters your body. Your heart starts to beat faster so that the oxygen can be delivered more quickly to the cells that make up each working muscle. The oxygen is used by the cells to break down *stored food.*

As the food is broken down, *heat, carbon dioxide,* and *energy* are produced.* The heat makes you feel warm—maybe even hot and sweaty. Since you are already breathing harder to take in more oxygen, you can get rid of the carbon dioxide at the same time. And, most importantly, the energy that is produced allows the muscles to keep working so you can keep jogging. Figure 8–13 shows what happens inside a working cell when plenty of oxygen is available.

*The actual chemical reaction is more complicated but this gives you the basic idea of what is happening.

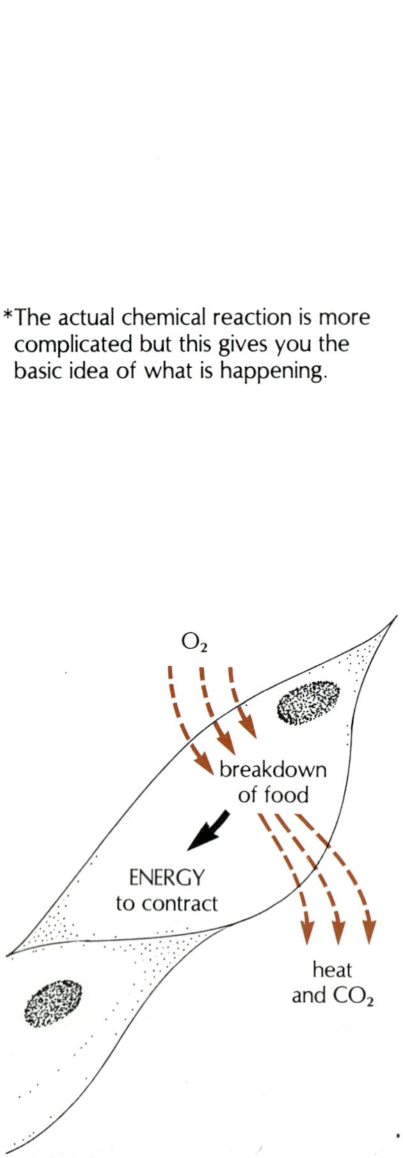

Figure 8–13 What does a muscle cell produce when plenty of oxygen is available?

Sore Muscles

Now, imagine that you have been jogging for four miles. You decide to run a fifth mile and then cool down by walking home briskly. The next morning you wake up and discover that you have sore muscles. Why?

If there is not enough oxygen available to break down the food stored in the muscle, the cells get energy in another way. They use different chemical reactions to break down the food. The waste produced by these reactions can make muscles feel sore when they build up inside the cells. The diagram in Figure 8–14 shows what happens inside a cell when not enough oxygen is present.

So when you jog and run more than usual, your muscles use up oxygen at a faster rate than you can replace it. Your muscles feel sore because they have accumulated waste products. If you continue to jog daily for several weeks, new blood vessels would form in your muscles to insure that they could get plenty of oxygen.

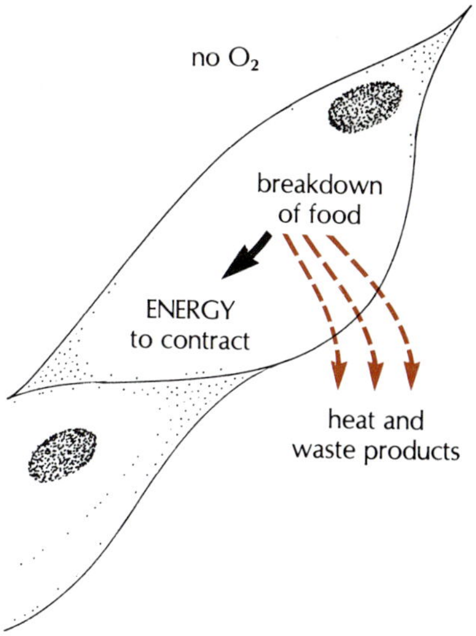

Figure 8–14 What does a muscle cell produce when it runs out of oxygen?

Aerobic Exercise

As you can see, oxygen is important for muscle contraction. When muscles get enough oxygen during exercise, the exercise is said to be **aerobic.** (Aerobic means "with oxygen.") People can exercise longer when their cells have plenty of oxygen. Aerobic exercise is recommended for losing weight, keeping the heart healthy, and improving fitness.

Aerobic exercises keep the muscles and heart working at a steady pace. An aerobic exercise does not overwork the body, but it must put some demand on the muscles. A slow walk is not really exercise at all. A faster walk might provide some exercise, but it would not be aerobic. A brisk walk or jog would get the heart rate up and could be an aerobic activity. A run or a dash is probably too fast to keep oxygen supplied to the muscles so neither one is an aerobic activity. Aerobic exercises do the most benefit if they are continued for at least twenty or thirty minutes. Ten minutes in the morning and ten in the afternoon would not provide continuous activity. The muscles must be kept working for at least twenty minutes at one time. As you might imagine, exercises that improve endurance are likely to be aerobic.

ANALYSIS QUESTIONS

1. What 2 things do muscle cells use to produce energy?
2. Why does a person breathe harder during exercise?
3. Why do your muscles feel sore after you do exercises some of the time but not all of the time?
4. Why are aerobic exercises better for the heart than other kinds of exercise?

THINK ABOUT IT

The more you know about how your muscles work, the better decisions you will make about keeping fit and exercising safely. Think about how one of your own favorite activities improves muscle fitness. Decide if it improves strength, endurance, or flexibility. Is it aerobic?

Figure 8–15 Jogging is an aerobic activity.

5. Decide which of these activities could provide aerobic exercise.

riding a bike for 1 hour	doing a backstroke for 900 yards
lifting weights	motorboating
pole vaulting	doing 100 jumping jacks
playing golf	walking briskly for 4 miles

6. What aerobic activities do you do?

EXTENSION ACTIVITY

Call a local recreation district, YMCA/YWCA, or health/fitness club and ask if they offer aerobics classes. If they do, explain that you are studying about fitness and would like to observe a class. When you visit, observe the type of exercises they do. Look for answers to some of the following questions:

Are there times for warm-up and cool down?

Does the instructor explain which muscles receive a workout from each exercise?

How can people tell if they are exercising aerobically?

Ask the instructor to explain the differences between other exercise classes and the aerobics classes. Find out what training the instructor received in order to teach these classes.

8-7 ANALYZING ACTIVITIES AND INJURIES

You can apply what you have been learning about muscles and fitness by analyzing an activity. For instance, you could ask, "Is this activity aerobic?" or "Does it require strength?" If you are trying to improve your physical fitness, this analysis will be helpful to you. At the same time, you need to be aware of the injuries that occur the most frequently during various types of exercise.

- Do you know what some of the most common sports-related injuries are?
- Would you know what to do if you or one of your friends has one of these injuries?

Keep these questions in mind while you analyze several activities.

MATERIALS

chart: ***Analyzing Activities*** (Figure 8–16)
reading: ***First Aid Guide*** (page 206)

PROCEDURE

Work with a partner. Complete the chart as you follow this procedure. Before you start, talk with your partner and decide on 3 activities to add to the first column of the chart.

NAMES ____________
____
DATE ____________ PER. ____

ANALYZING ACTIVITIES

Activity	1. Muscle fitness required*			2. Muscle(s) most likely to reach "stop" point	3. Amount of energy required**	4. Aerobic?	5. Likely injuries	6. I do this activity. (my initials)
	Strength	Endurance	Flexibility					
bicycling								
softball								
swimming								
reading								
jogging								
weight lifting								
bowling								
gymnastics								
car racing								

*Use this scale for part 1 only.
3 – very important
2 – important
1 – may help a little
0 – not important

**Use this scale for part 3 only.
3 – a great amount
2 – quite a bit
1 – some
0 – almost none

Figure 8–16

Figure 8–17 Dancers must be flexible so they can move easily. They need to have the endurance to dance the length of a performance. Depending upon their routine, they need strength in specific muscle groups.

1. With your partner, decide how important strength, flexibility, and endurance are for performing each activity on the chart. Use this rating scale to record your decisions.
   3—very important
   2—important
   1—may help a little
   0—not important
2. Different activities use different muscles. For each activity listed, indicate which muscle would tire the fastest; that is, which muscle would reach the "stop" point first. Be specific when you can. For example, write "biceps and triceps" instead of "upper arm muscles."
3. Some activities are much more tiring than others. Rate each activity according to the amount of energy it requires. Assume a person does his or her best at the activity for 30 minutes. Use this rating scale.
   3—a great amount
   2—quite a bit
   1—some
   0—almost none
4. Under normal circumstances, do you think each activity would be a good aerobic exercise? Answer **Yes** or **No** for each one. (Look on page 201 if you need to review the meaning of **aerobic.**)
5. Muscles, bones, and joints are more likely to be injured than other parts of the body.
   a. Read "Common Aches, Pains, and Injuries" on page 206. It gives information about some common complaints. You may have experienced some of these yourself.
   b. Decide which injury is most likely to occur for each activity on the chart. List these in the "Likely Injuries" column on the chart.
6. Complete the last column of the chart. Initial the activities that you do *at least one time each month.*
7. Answer analysis questions 1–5.
8. Aches and pains are the body's way of signaling that something is wrong—something may be injured. Many injuries will heal by themselves, but they will heal faster with proper care.

Read "RICE—Treatment for Common Injuries" in the First Aid Guide on page 207 and answer analysis questions 6–9.

ANALYSIS QUESTIONS

1. Fill out this information about one of your favorite activities.
   a. name of activity
   b. type(s) of muscle performance it requires
   c. muscle(s) it exercises the most
   d. amount of energy it requires
   e. 3 common injuries that might occur when doing this activity
2. List one activity you could do for each of the following:
   a. shape up your legs
   b. relax
   c. lose weight
   d. strengthen your arms
   e. end up in the hospital
3. Give an example of an injury that affects each of the following:
   a. a bone
   b. a joint
   c. a muscle
   d. a ligament
   e. the skin
4. What is the difference between a sprain and a strain?
5. Which injury on the First Aid chart seems to be the most serious? Why?
6. What does RICE stand for?
7. Here is a list of the injuries from the First Aid guide. Which ones could you treat with RICE? (Look at the treatment recommended in the guide to help you answer this one.)

bruise	fracture
cramp	sprain
strain	dislocation

8. RICE recommends using ice. In the past, people often used heat when injured. What would heat do to an injury?

RICE is good to know, but remember:
Call a physician if you are not sure what to do *or if pain lasts for more than a day.*

CONCLUSION

Go back and reread the focusing questions on page 203. Think about each one. You should now be able to answer them fairly easily.

EXTENSION ACTIVITIES

1. Professional athletes are frequently injured. Read the sports pages in newspapers for two weeks. Save articles that are about a person and his or her injury. Keep track of the injuries you read about, the sport in which each athlete was involved, and the treatments that were used.
2. What does an athletic trainer do? Talk to one and find out what is involved in the job. Ask about an injury, such as Little League elbow or shin splints, and find out its causes, its symptoms, and its treatment.
3. Are you feeling really ambitious? Improve your first aid skills! The American Red Cross or your local hospital may offer classes on what to do in emergency situations. Check with them for schedules, topics, and times.

FIRST AID GUIDE

COMMON ACHES, PAINS, AND INJURIES

Bruise = bleeding under the skin *Symptoms:* may be tender to the touch; skin discolors (looks purplish) and gradually lightens	*Cause:* a blow to anywhere on the body or a fall **Treatment:* if painful, apply ice or cold cloth with slight pressure
Cramp = when a muscle contracts but doesn't relax immediately *Symptoms:* sudden muscle pain often unexpected; the muscle may feel hard	*Cause:* not drinking enough water or for unknown reasons **Treatment:* gently massage the muscle and gradually try to move it
Dislocation = a "separated" joint *Symptoms:* pain, swelling and skin discoloration at a joint; joint may look out of position	*Cause:* stress at a joint **Treatment:* carefully apply ice, protect the joint, and get medical help
Fracture = broken bone (varies from a hairline crack to a total break) *Symptoms:* extremely variable tender to touch, swelling, bruises, slight to severe pain (may be mistaken for a sprain.)	*Cause:* stress on a bone **Treatment:* avoid moving if a break seems likely; stop bleeding; get medical help
Sprain = tearing of ligaments that hold a joint in position *Symptoms:* varies from mild pain at a joint to extreme pain	*Cause:* sideways pressure on a joint **Treatment:* use an elastic bandage and an ice pack and get medical help
Strain = a "pulled" muscle (torn muscle cells) *Symptoms:* muscle feels tender and may swell; may become "stiff" overnight	*Cause:* an overstretched muscle **Treatment:* rest the muscle; bandage to provide support; ice may ease pain

*Treatment always depends upon these two considerations:
(1) What is the diagnosis? (A muscle strain? A bone fracture?)
(2) How severe is the injury? (Is it a mild sprain? A badly torn ligament?) Call a physician if you are not sure what to do or if pain lasts for more than one day.

RICE—Treatment for Common Injuries

RICE is a useful form of first aid for injuries that everyone should know. Here is what RICE stands for.

Rest. Rest the injured body part immediately! Continued use makes an injury worse.

Ice. Cold causes blood vessels to contract or shrink so less blood flows to the injured area. This helps prevent swelling.

Compression (pressure). Compression also limits swelling. Swelling is painful and can make healing take longer.

Elevation. Raise the injured part so gravity can help drain fluid. Elevate the injured part to a level above the heart, if possible.

Start RICE as soon as possible. RICE is good first aid for almost all injuries – pulled muscles, sprained ligaments, and even some broken bones.

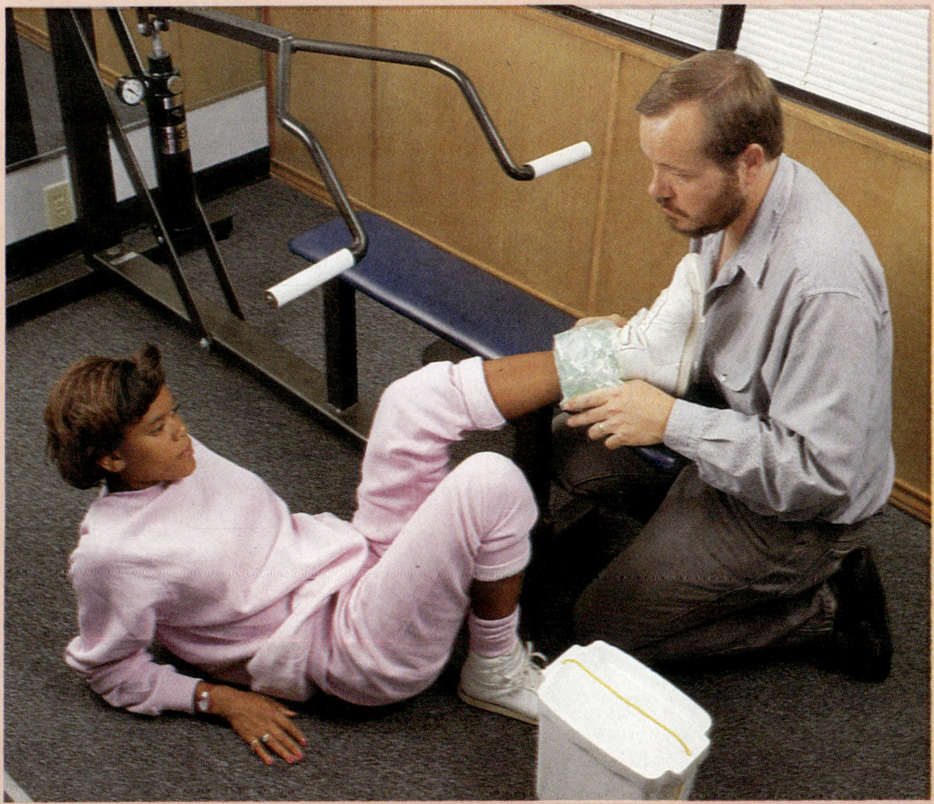

Figure 8–18 Using RICE techniques for an emergency

8-8 SMOOTH MUSCLES, TOO

Any place in the body where there is movement there is probably a muscle. Toes wiggle and foreheads wrinkle because **skeletal muscles** contract. There are also muscles in arteries, intestines, and many other body organs. When these muscles contract, they cause blood to flow in arteries and food to travel through the intestines. This second kind of muscle is **smooth muscle.**

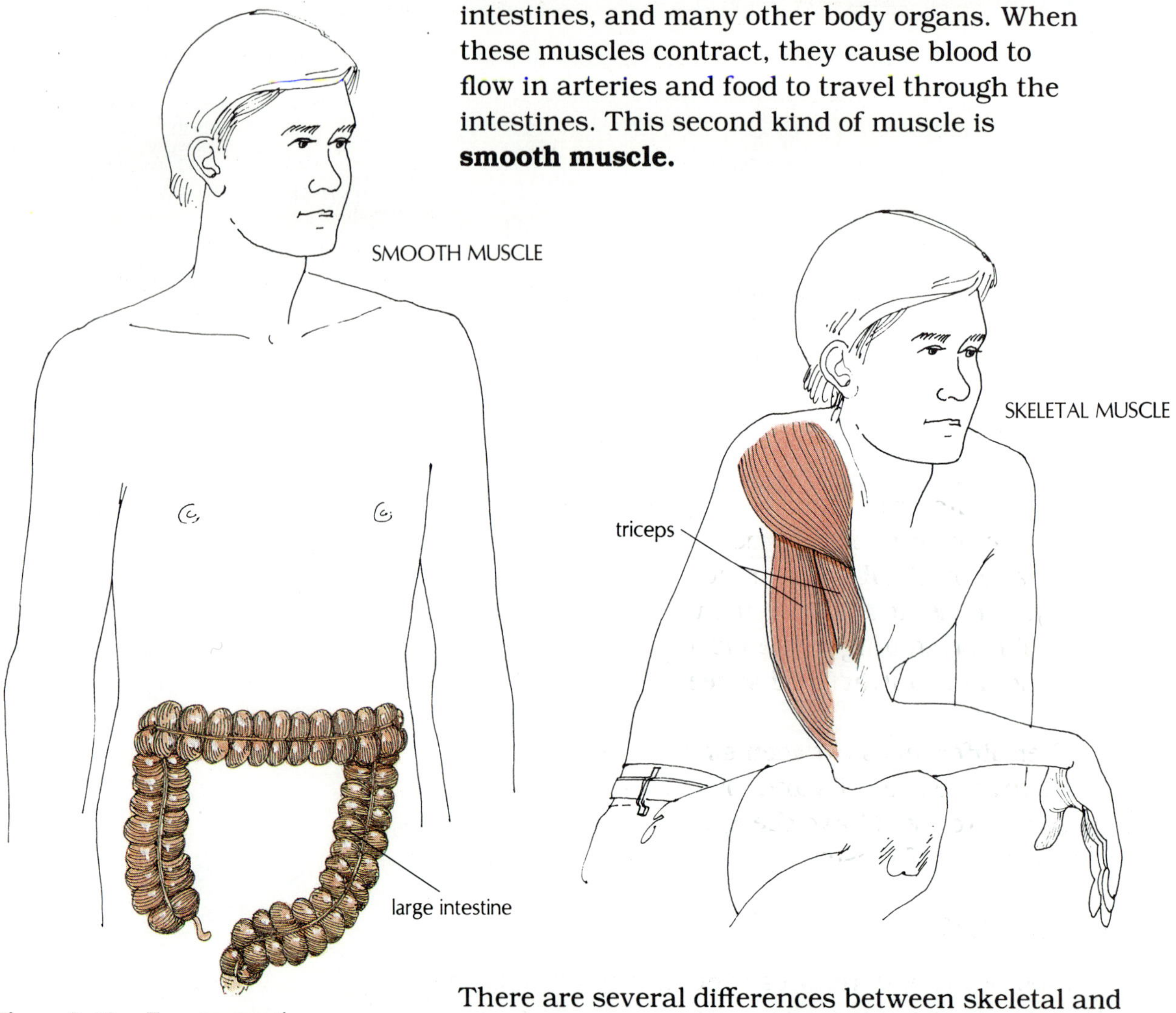

Figure 8–19 Two types of muscles

There are several differences between skeletal and smooth muscles. All these differences can be illustrated by comparing the triceps—a skeletal muscle—with the wall of an intestine—a smooth muscle. Before you read on, think about the two kinds of muscles. How many differences can you think of? (See Figure 8–19.)

One obvious difference is the location of the muscles. Skeletal muscles attach to bones; smooth ones do not. They make up the walls of many hollow body structures.

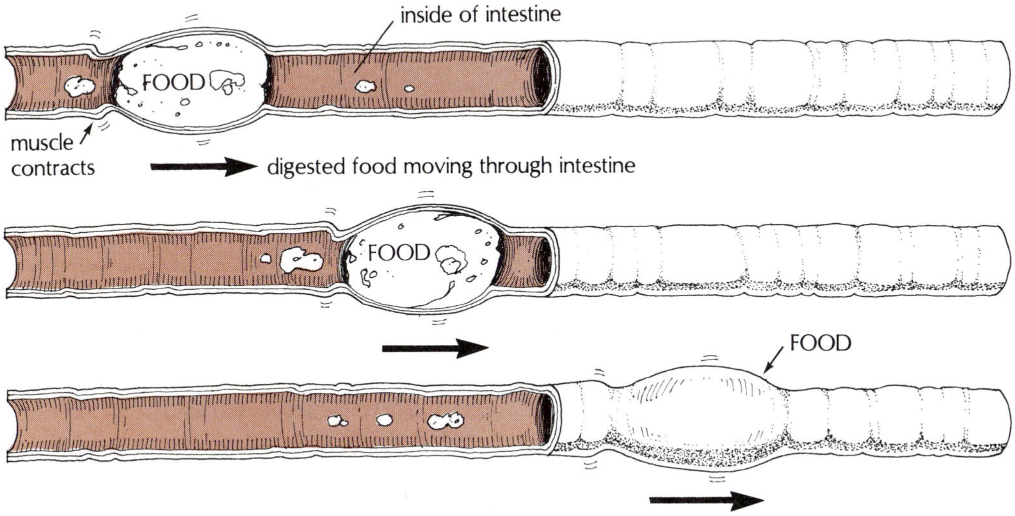

Figure 8–20 Smooth muscle contraction moves digested food through the intestine.

This leads to a second difference between the two muscle types—their actions differ. When a skeletal muscle contracts, it pulls and causes a body part to move. When a smooth muscle contracts, it "squeezes." This squeezing is what moves food through an intestine or blood through an artery.

A third difference is one of control. You control the movement of your skeletal muscles but smooth muscles contract automatically. They respond to changes inside the body. For instance, when the intestine fills, its walls stretch. The stretching of the muscle cells in the walls causes them to contract and squeeze the food along its way.

Other differences between skeletal and smooth muscles involve the cells and must be seen through a microscope. You will have the opportunity to compare these two types of muscle cells in the next activity.

TEST YOURSELF

1. Which of these structures contain smooth muscle? (Use a dictionary to look up the meaning of any word you do not know.)

foot	stomach	hip	cheeks
lips	little finger	aorta	thumb
bladder	esophagus	iris	uterus

2. In your own words, explain the 3 differences between skeletal and smooth muscles.
3. Think of an example of an activity that would exercise one of your smooth muscles.

8–9 MUSCLE CELLS

Skeletal and smooth muscles are both made up of cells. In some ways, both types of cells are alike. For instance, both can contract when the muscles work and they both need oxygen to keep working. In other ways, these cell types are not alike. There are several microscopic differences in the way they look.

- What do muscle cells look like?
- How many cell parts can you identify?
- What are some differences between skeletal and smooth muscle cells?

In this activity, you will look at microscope slides of muscle cells so you can answer these questions.

MATERIALS

microscope
prepared microscope slides—X, Y
data page: ***Muscle Cells*** (Figure 8–22)

PROCEDURE

Part A. Working from Drawings
The drawings in Figure 8–21 show what muscle cells look like through a microscope. Look at these drawings and complete part A on your data page.

Part B. Working with a Microscope
Use your microscope to look at two prepared microscope slides of muscle cells labeled X and Y. Your task is to decide what kind of muscle cell is on each slide. Be a good observer!

1. Start with either slide X or slide Y. Look at it first with low power and then with high power. Fill in the appropriate column on the data table.
2. Look at the other microscope slide, low power first. Fill in the rest of the data table while you look at the slide.

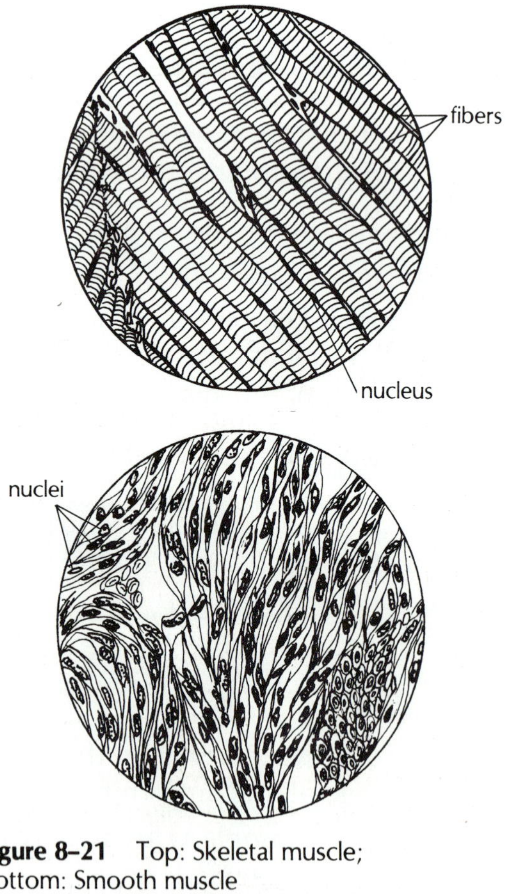

Figure 8–21 Top: Skeletal muscle; Bottom: Smooth muscle

MUSCLE CELLS

NAME ______

DATE ______ PER. ______

PART A. FROM THE DRAWINGS

	Skeletal Muscle	Smooth Muscle
Do the muscle cells look like fibers? (definitely, somewhat, or not at all)		
Do you see bands (stripes) across the cells? (definitely, somewhat, or not at all)		
How many cell nuclei do you see? (lots, some, or few)		
Other observations		

PART B. FROM THE MICROSCOPE SLIDES

	X	Y
Do the muscle cells look like fibers? (definitely, somewhat, or not at all)		
Do you see bands (stripes) across the cells? (definitely, somewhat, or not at all)		
How many cell nuclei do you see? (lots, some, or few)		
Other observations		

low power magnification ______

high power magnification ______

Figure 8–22

Part C. Analysis

Answer the analysis questions. Refer to the data you recorded.

ANALYSIS QUESTIONS

1. Look at the drawings in Part A and at your data page to answer these questions.
   a. What clues would help you recognize **skeletal** muscle cells?
   b. What clues would help you recognize **smooth** muscle cells?
2. When you looked at the microscopic slides, which cell parts could you identify?
3. Look at your notes about microscope slide X. Compare these with your notes about the drawings. What kind of muscle cell is on slide X? Explain your answer.
4. Decide which kind of muscle cell is shown on slide Y. Use the data table to help you make your decision.

5. Summarize the microscopic differences between skeletal muscle cells and smooth muscle cells.
6. Which feature of muscle cells gives you the best clue about their function? Explain your answer.
7. There is a third kind of muscle—**cardiac** muscle. It makes up the heart. Like smooth muscle, it contracts and relaxes on its own. Cardiac muscle cells do not tire easily. Describe one thing that could happen if your heart were made of skeletal muscle instead of cardiac muscle.

CONCLUSION

Answer each question on page 210. You will need to decide how long each answer should be so ask yourself, "Is one sentence enough or would two or three sentences be better?"

EXTENSION ACTIVITY

Put a very small piece of hamburger on a slide. Then observe it through the microscope. You may need to add one drop of diluted food coloring so the cells show up better. Cover the slide with a cover slip. Look for muscle cells. What kind do you see?

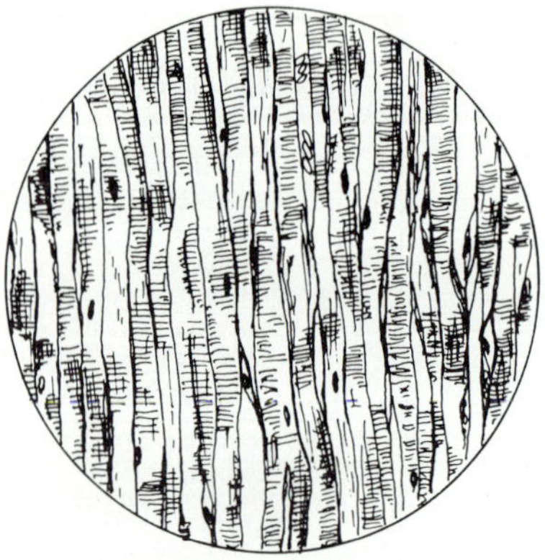

Figure 8–23 Cardiac muscle

CHALLENGE

Test your microscope skills. Look at a microscope slide of cardiac muscle cells. The drawing in Figure 8–23 shows what you might see. Observe the slide closely. Compare it with the slides of smooth and skeletal muscles. Keep these questions in mind as you work.

1. What clues would help you recognize cardiac muscle?
2. Does cardiac muscle look more like skeletal muscle or smooth muscle? Explain.

CHAPTER 9

SKIN

What differences can you see in the skins of these two people?

9–1 WHO'S LIVING ON MY SKIN?

Millions of tiny organisms share the classroom with you. They live in the pencil sharpener, float in the air, and are on every surface you touch. They can live just about anywhere because they can use just about anything for food. You don't know they are around because they are too small to see. They are **micro-organisms**. These microorganisms often settle down on your skin and try to "move in." Right now, your skin could be alive with microbes and you would not even know it.

- What do microorganisms look like?
- How can you test to find out if microorganisms are around?
- What parts of your skin are most likely to have microorganisms on them?

Find the answers to these questions by testing your own skin for microorganisms.

MATERIALS

1 plastic petri dish with nutrient agar
tape
wax pencil or marker
3 cotton swabs
data page: ***Who's Living on My Skin?*** (Figure 9–2)

PROCEDURE

Work with a partner. You will be given one petri dish with agar. Agar contains nutrients that allow microorganisms to grow.

Be careful not to open the dish until you need to. Watch out for contamination!

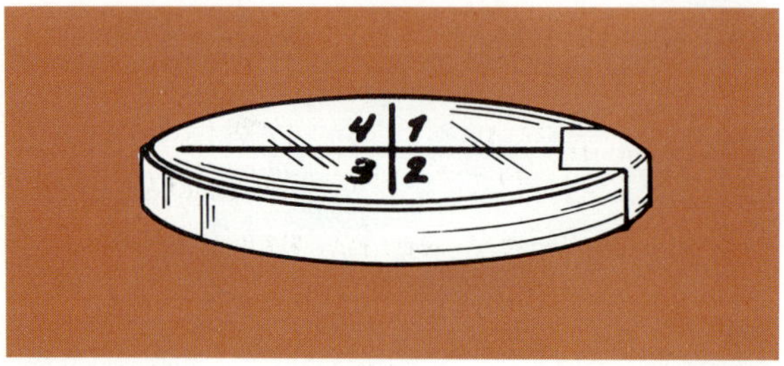

Figure 9–1

Part A. Setting Up

1. Tape the two pieces of your dish together on one side so the tape acts like a hinge. Write your names, date, and class period on the tape.
2. Use the wax pencil or marker to divide the *bottom* of the petri dish into four sections. Number the sections 1, 2, 3, and 4, as shown in Figure 9–1.
3. Decide on 3 skin areas you would like to test. Choose places such as your foot, hand, head, leg, or under one of your fingernails.
4. On the data page, record which 3 areas of your skin you will be sampling for sections 1, 2, and 3. Since section 4 will not be used to test a skin area, leave it blank.

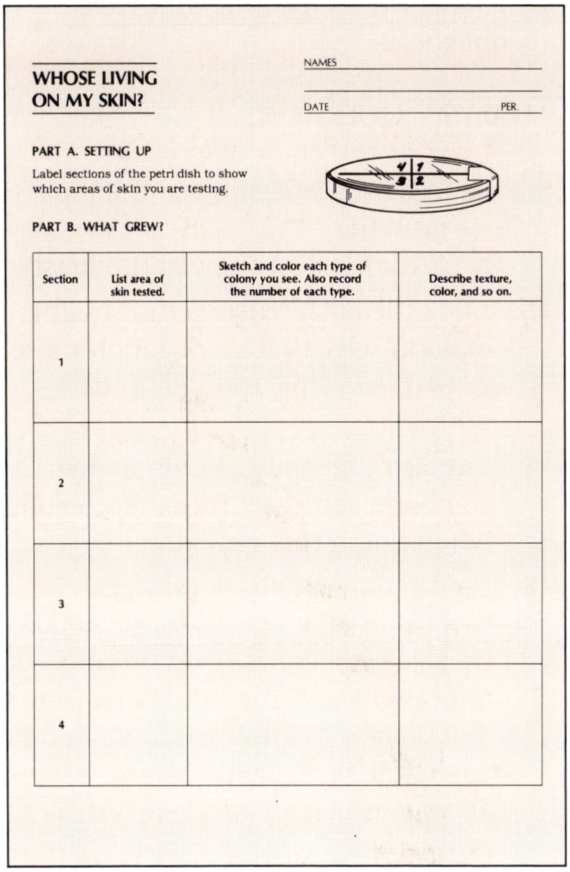

WHOSE LIVING ON MY SKIN?

NAMES ____________________

____

DATE ____________ PER. ____

PART A. SETTING UP

Label sections of the petri dish to show which areas of skin you are testing.

PART B. WHAT GREW?

Section	List area of skin tested.	Sketch and color each type of colony you see. Also record the number of each type.	Describe texture, color, and so on.
1			
2			
3			
4			

Figure 9–2

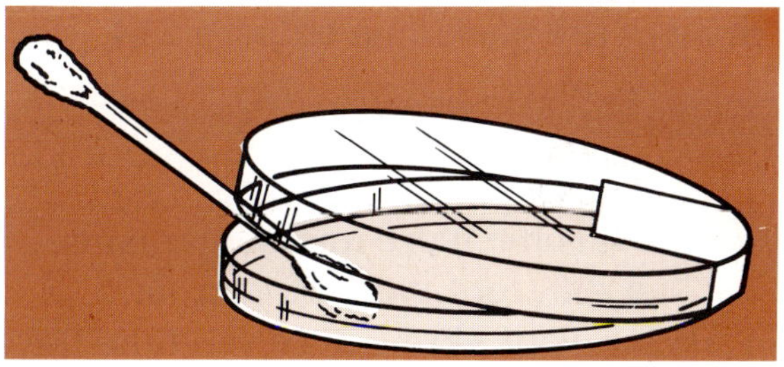

Figure 9–3

5. Follow these steps to test the first skin area you chose.
   a. Rub a clean cotton swab on that skin area.
   b. Lift the lid off the petri dish just high enough to insert the cotton swab, as shown in Figure 9–3. **Do not lift the lid completely off.**
   c. *Gently* touch the cotton swab to the agar surface in section 1. Move it around a little. Try not to break the surface of the agar.
6. Use a new cotton swab for each skin area you are testing. Follow the same procedure for sections 2 and 3. Remember, do not put anything on section 4.
7. Tape the other side of the dish so it will stay closed. Store it upside down where your teacher indicates. Look at the dish again in about 3 days.
8. Answer analysis questions 1–3. Go on to Part B when your teacher gives you that direction.

ANALYSIS QUESTIONS

1. Why is it important not to expose the agar to air?
2. Why did the instructions say *not* to put anything in section 4?
3. Which section do you predict will show the most growth? Why?

Part B. What Grew?

1. Locate your petri dish. Do not open the lid. Leave it taped closed while you record your observations.

CAUTION

Most of these microorganisms are not harmful. However, since some could cause diseases such as pneumonia or staph infections, do not take the lid completely off the dish.

Always handle bleach cautiously. It can burn skin, damage eyes, and take the color out of cloth.

Remember! Safety Counts!

2. Complete the data table under Part B on your data page. Your teacher may have you observe the microorganisms with a binocular microscope or magnifying lens while you make your sketches. Remember to record the magnification.
3. Clean up. Take your dish to the sink. Open the dish slightly and carefully squirt bleach solution into it. This will kill the microorganisms and prevent contamination. Tape the lid closed again and put it where your teacher tells you.
4. Answer the following analysis questions and then go on to Part C.

ANALYSIS QUESTIONS

4. How many different kinds of organisms grew on your petri dish?
5. How can you tell the difference between bacteria and fungi on the agar?
6. Look at the section that showed the most growth in order to answer these questions.
   a. What skin area was this from?
   b. How many different kinds of microorganisms grew in this section?
7. Which skin area showed the least growth?
8. Did your prediction come out as you thought? Why do you think it did or did not?

Part C. Class Data

1. Add your tallies to the class data table to show which skin areas you tested and how many colonies of bacteria and fungi grew from each area.
2. Answer the following analysis questions.

ANALYSIS QUESTIONS

9. Look at the class data table.
   a. Which areas showed the most growth?
   b. Which areas showed the least?
10. List at least 3 reasons that might explain why there was much more growth in some areas than in others.
11. Explain at least 3 reasons why someone could get incorrect results when doing this experiment.
12. Health care professionals are careful to wash their hands before they see a patient. This is supposed to prevent passing infections from one person to another. Explain how you could do an experiment to show if washing hands is important or not.

CONCLUSION

Write several short paragraphs that answer the questions on page 214.

EXTENSION ACTIVITIES

1. Grow some more microorganisms. Test different parts of your skin. Wash your skin before you take the sample. Is it possible to remove all microorganisms? Test skin from some animals. What kinds of growth do you get from a dog? cat? bird? hamster? fly's feet?
2. Microbiologists, virologists, and bacteriologists are all people who work with microorganisms. Contact any one of these people. They work in such places as industrial plants

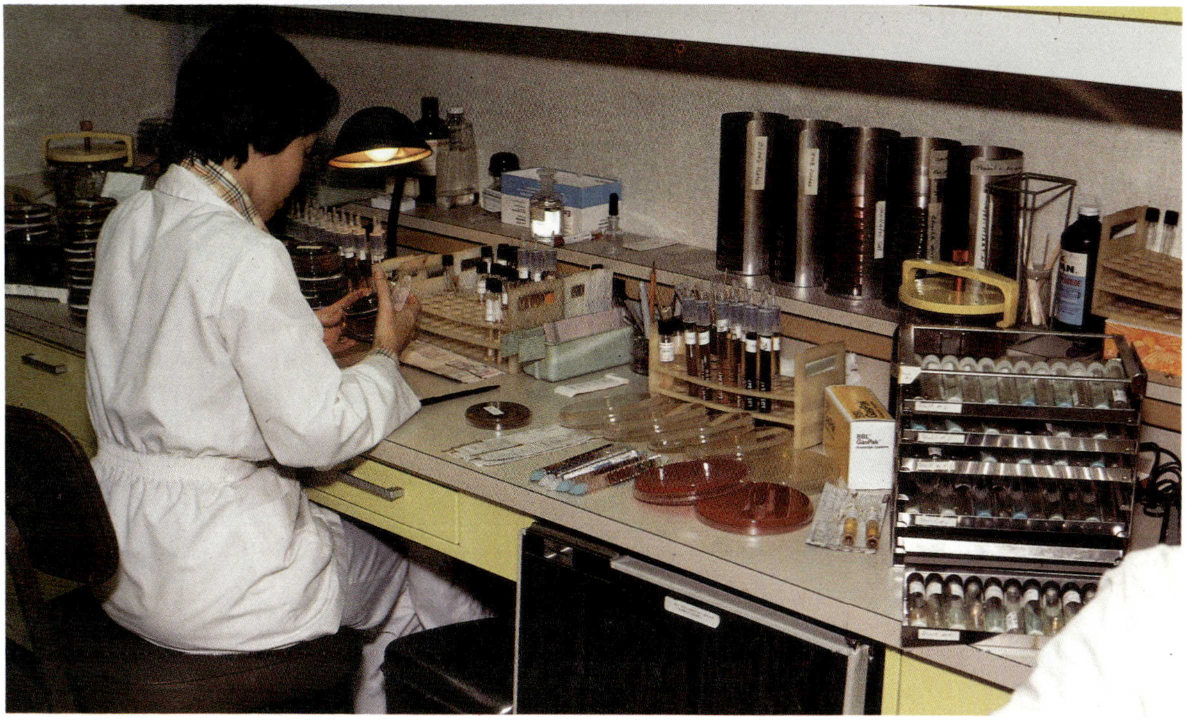

Figure 9–4 A bacteriologist in a hospital lab is working with cultures.

and laboratories, hospitals, health departments, and food packaging houses. If possible, visit one of the labs. Ask questions such as these:

What microorganisms do you check for?

Are these microorganisms bacteria, fungi, or viruses?

Could any of these microorganisms make you sick?

Why do you want to test for them?

How do you test for them?

What do you do if you find them?

3. Ignez Semmelweiss and Joseph Lister were physicians who lived in the 1800s. They were among the first people to recognize the importance of having medical professionals wash their hands before doing surgery and of keeping wounds clean. Find out how Semmelweiss and Lister discovered the importance of hand washing and how the other doctors reacted to their findings.

9-2 MICROORGANISMS

Microorganisms are much smaller than most plants and animals. Their small size makes them hard to see and study. There are several types of microorganisms. Three types are bacteria, viruses, and some fungi.

- How do you tell these three types of microorganisms apart?

Use the information and illustrations that follow to help you answer this question.

MATERIALS

prepared microscope slides:
  4 microorganisms
microscope
data page: ***Microorganisms*** (Figure 9–6)

PROCEDURE

Part A. Three Types of Microorganisms

Read about these microorganisms and look at Figure 9–5 on this page and the next one. You will need this information to complete Part B.

Fungi come in many sizes. Some of them are large enough to see without a microscope; others are single cells. Mushrooms are examples of large fungi. Yeast and molds are examples of microscopic fungi. Many of the molds will

Figure 9–5a A mushroom is a fungus.

Figure 9–5b Mold is also a fungus. Top right: Microscopic view of mold on bread. (magnification 35 times)

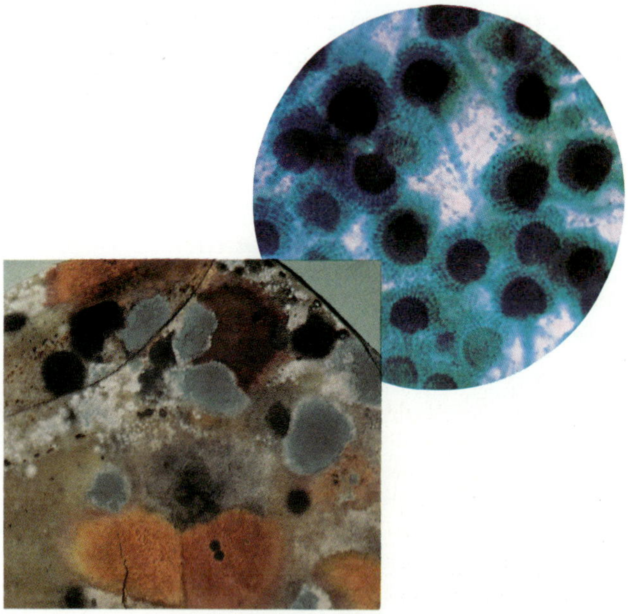

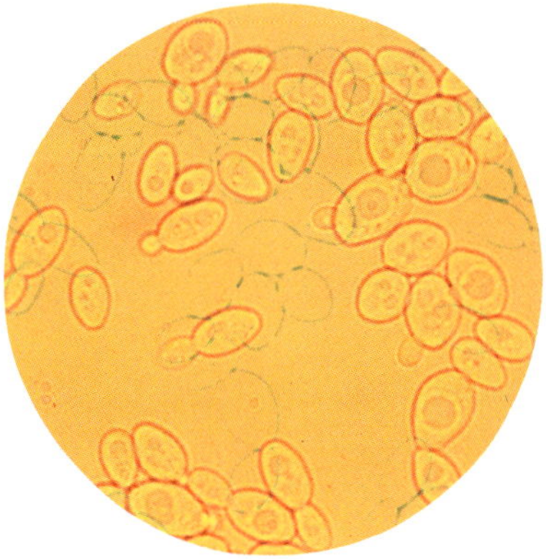

Figure 9–5c Yeast are fungi. (magnification 200 times)

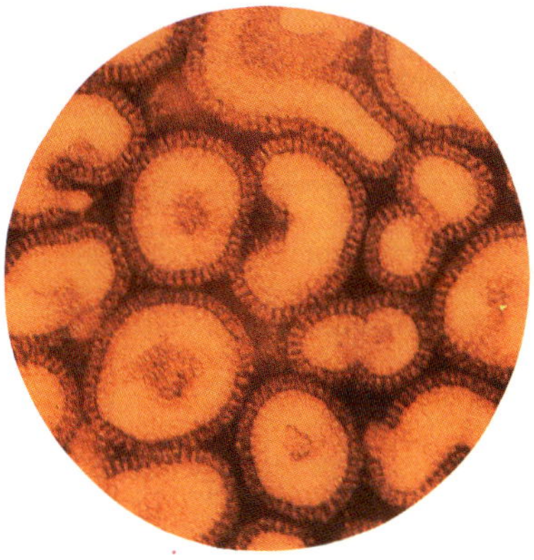

Figure 9–5e Virus causing influenza. (magnification 285,000 times)

grow on agar in a laboratory. When they do so, they look "furry." (See Figures 9–5a, 9–5b, and 9–5c.)

Bacteria are some of the simplest forms of living things. They are single cells. Bacteria may be round, rod-shaped, or shaped like a spiral. (See Figure 9–5d.) To give you a better idea of just how small these microbes are, guess how many average-sized bacteria could fit in a single drop of water. (The answer is at the end of the activity.) The tetanus bacteria and *Salmonella* are two examples of bacteria that you have read about already. When bacteria grow in groups called colonies, you can see them without a microscope. Bacterial colonies can grow on agar.

Viruses are extremely small. They are so small that they can only be seen with an electron microscope. (See Figure 9–5e.) Unlike fungi and bacteria, viruses are not cells. In fact, they cannot function or reproduce unless they are inside a living cell. For this reason, they cannot grow on agar. The common cold, flu, measles, and mumps are all examples of diseases that are caused by viruses.

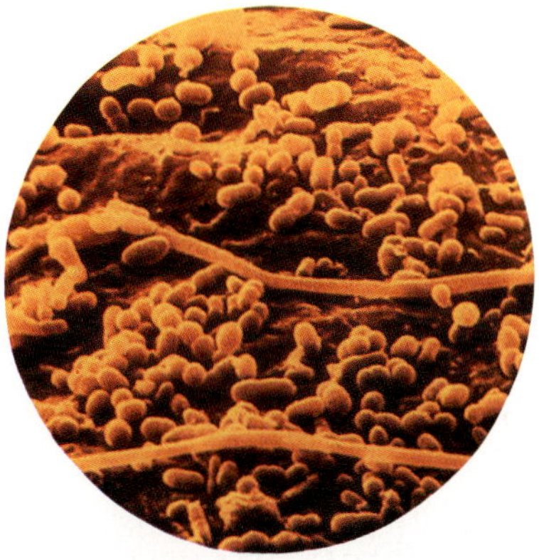

Figure 9–5d Bacteria commonly found on skin. (magnification 17,000 times)

Part B. Mystery Microorganisms

1. Answer the analysis questions. Make sure you can answer all of them before you look at the microscope slides.
2. Look at the slides of the four microorganisms. Sketch each one in the spaces provided on the data page.
3. **Use the information on page 218 and above to help you decide if each microorganism is a virus, a fungus, or bacteria. Make sure to explain your decisions on your data page.**

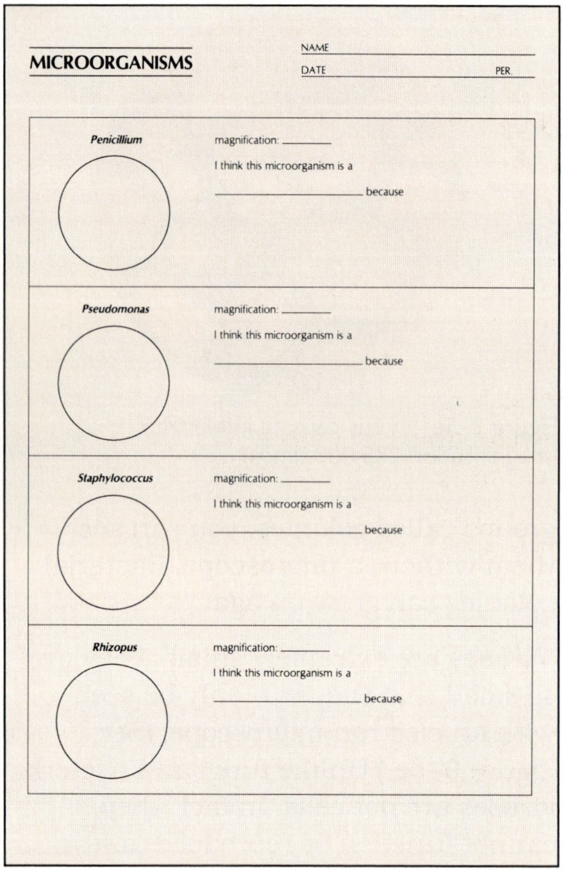

MICROORGANISMS

NAME ____________________

DATE ____________________ PER. ______

Penicillium

magnification: ______

I think this microorganism is a

____________________ because

Pseudomonas

magnification: ______

I think this microorganism is a

____________________ because

Staphylococcus

magnification: ______

I think this microorganism is a

____________________ because

Rhizopus

magnification: ______

I think this microorganism is a

____________________ because

Figure 9–6

ANALYSIS QUESTIONS

1. List the 3 types of microorganisms in order from smallest to largest.
2. What clue(s) would help you decide if you are looking at a virus, bacteria, or a fungus?
3. How many cells—0, 1, several, or thousands—would you expect to find in each of the following structures?
   a. yeast
   b. sternum
   c. virus
   d. bacterium
   e. gluteus
   f. fungus fiber
4. Which types of microorganisms can grow on agar?
5. Some people think viruses are living organisms, other people disagree.
   a. Give one reason why a virus could be considered a life form.
   b. Give one reason why a virus could be considered nonliving.

CONCLUSION

Write a paragraph that answers the focusing question on page 218.

EXTENSION ACTIVITY

Each of the mystery microorganisms plays an important role in our lives. Write a report or make a poster about one of these four microorganisms.

Answer: 2 billion (2,000,000,000)

9-3 SKIN MICROORGANISMS

The surface of the skin is a good environment for many microorganisms. It is warm, slightly moist, and provides places for them to "hide." All three types of microorganisms can live on skin. Some of them cause familiar skin problems.

Fungi

Have you ever had athlete's foot or ringworm? Both of these skin problems are caused by fungi. Skin cells provide the nourishment the fungi need to survive. Fungi must absorb food from their environment because they have no chlorophyll and thus cannot make their own food like most green plants do.

Red, itchy, flaky skin between the toes is a sign of **athlete's foot.** It is slightly contagious. People can catch athlete's foot by walking barefoot where the fungus is located. Locker room floors at a gym, fitness center, or swimming pool are a common source of infection because this fungus likes moisture. Keep your feet dry if you want to avoid athlete's foot!

Figure 9–7 Deodorants and antiperspirants can help reduce body odor.

Ringworm is not a worm; it is another skin problem caused by a fungus. Scaly, round, itchy patches form on the skin when this fungus is around. It, too, is contagious. You can catch ringworm by touch, by exchanging combs or brushes, or from a pet dog or cat.

Ringworm and athlete's foot can both be treated with **fungicides**—fungi killers. They are not serious conditions but they are unsightly. The earlier they are treated, the easier they are to get rid of.

Skin Bacteria

Millions of bacteria also live on your skin. Most of the time you are not aware of their presence but they can become embarrassing by causing body odor.

It is normal to perspire. Perspiration helps regulate body temperature. Perspiration contains water, salts, and complex chemicals. These complex chemicals are broken down to simpler ones by skin bacteria. The resulting chemical combinations can cause body odor. Some deodorants work by killing bacteria. Antiperspirants work by cutting down the amount of sweat that is produced.

Skin Viruses

Viruses can infect skin cells, too. Viral infections are difficult to treat because they live *inside* cells. Two examples of skin problems caused by viruses are cold sores and warts.

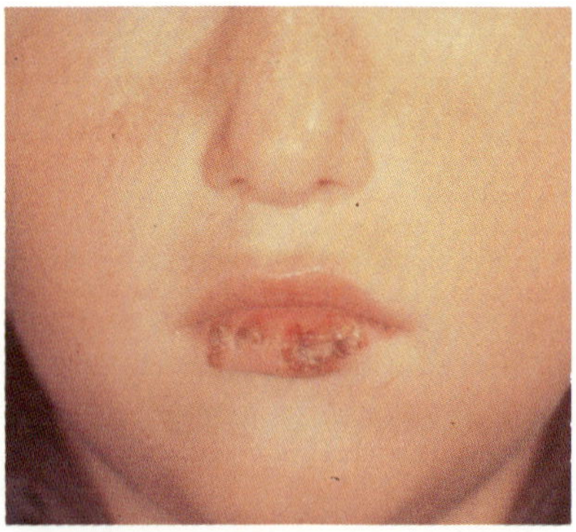

Cold sores look like blisters that form around the mouth. They start out as spots on the gums, lips, or inside of the mouth. These spots will soon swell, redden, and hurt. When a cold sore appears, the virus is active. When the infection clears up, the virus is inactive. It can be activated by stress, a cold, sunlight, or many other things. Luckily, most cold sores clear up on their own. A physician may suggest a painkiller or mouthwash to make the person feel better.

Warts are common in teenagers. There are several different kinds of warts; each kind is caused by a different virus. Some people have a low resistance to these viruses so they may feel as if they are always getting warts. These viruses form warts by causing skin cells to divide rapidly. Warts usually disappear within a few months but some take years to clear up. Several medications may be purchased in drugstores to help treat warts. They act by destroying abnormal skin cells. These medications should never be applied to the face—facial skin is too sensitive for such strong medications.

These are just a few examples of the many skin problems that are caused by microorganisms. The next time you have a skin problem, you might want to find out, "Is it caused by a fungus, a virus, or bacteria?"

ANALYSIS QUESTIONS

1. Would it be better to buy tennis shoes with or without air vents if you want to prevent athlete's foot? Explain your answer.
2. In your own words, explain why deodorants can prevent body odor.
3. What could happen if wart remover were used on the face?
4. Some infections are difficult to treat.
   a. Which is hardest to treat, an infection caused by a fungus, a virus, or bacteria?
   b. Why do you think this microorganism is the hardest to get rid of?
5. Imagine that you have a contagious skin condition. List 3 precautions you would take to make sure no one else catches it.
6. A new skin disease has just been reported. What is one way to determine if it is caused by a microorganism?

EXTENSION ACTIVITY

Go to the drugstore and find 2 medications for either athlete's foot or warts. Read the labels. List the active ingredients. Decide which medication sounds best to you. Talk to a dermatologist or your family doctor and ask what over-the-counter medications are considered effective and why.

9–4 EIGHT POUNDS OF SKIN

Skin is the largest and most visible organ of the body. An average adult's skin weighs 8 to 10 pounds. Look at Figure 9–8. How much do you think *your* skin weighs? Most of us take our skins for granted. We seldom think about all of the things the skin does, but skin is so important that if one-third of it is lost— from burns, for instance—a person will die.

Skin Functions

Figure 9–8

The main function of skin is protection. It protects our insides from harmful microorganisms, extreme heat or cold, radiation, chemicals, and other things in the environment. Besides protection, skin helps keep body temperature within a normal range. It is also a sense organ that keeps us aware of our surroundings. The sense of touch is so important that a person cannot survive without it.

Skin is a tough, waterproof, flexible covering. This amazing covering has two main layers—the epidermis and the dermis.

Epidermis

The **epidermis** is the surface of the skin; it is what you see. The top layer of the epidermis is dead cells. You wash off some of these cells every time you take a bath or shower. They even brush off on your clothes. This constant shedding keeps many microorganisms from entering the skin. A bacterium that attaches itself to a skin cell will be brushed off as soon as that cell is shed.

The bottom layer of cells in the epidermis is alive and active. These cells divide to keep producing new epidermal cells. Notice in Figure 9–9 on page 225 that the epidermis does *not* contain any blood vessels.

Several interesting structures arise from the epidermis of various animals. Hair, nails, feathers, quills, horns, and claws are all special epidermal structures.

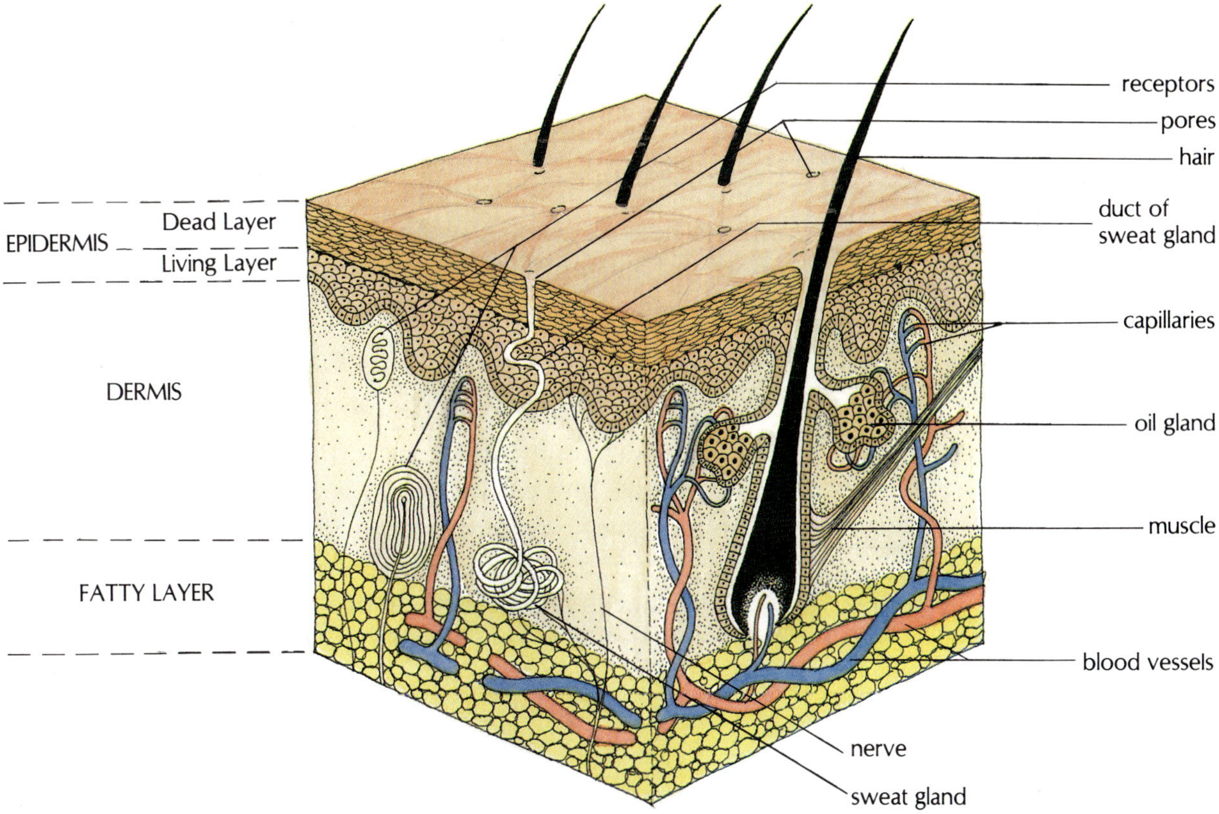

Figure 9–9 Structure of the skin

Dermis

Underneath the epidermis is the **dermis.** It is much thicker than the epidermis. It contains blood vessels, hair roots, glands, nerves, and some smooth muscle fibers.

You never *see* your dermis, but you know it is there. Oil moves out of the oil glands onto your epidermis. You can feel the oil, especially by your nose. Sweat comes up out of the sweat glands to keep your skin cool. When you cut yourself, blood may pour out of your dermis. Every time you *feel* something, you are using the nerves in your dermis. The ends of these specialized nerves are called **receptors.** See how many receptors you can find in Figure 9–9.

Fat

Beneath the dermis is a layer of fat. Fat provides insulation to keep the body warm. It varies in thickness from one part of the body to another. For instance, it is usually thick over the abdomen but it is missing from the eyelids. Premature babies have not developed this insulating fat layer. Incubators provide the environment they need until they gain enough weight and fat to keep themselves warm.

Connective Tissue

Skin is held in place by loose connective tissue. Elastic fibers in this connective tissue allow your skin to stretch.

EXTENSION ACTIVITY

If you were to buy your skin by the square foot, how many square feet would you have to buy? Find out! Wrap pieces of newspaper around your body. Have a friend trim off the extra paper. Unwrap yourself and tape the pieces together. Measure the paper. How many square feet of skin do you need to cover your body?

Sea Mammals

Polar bears have a very thick layer of fat under their skin. It's a good thing they have that extra fat! They have to survive in some very cold temperatures!

Whales, too, have a thick layer of fat under their skin called blubber. It keeps them warm and helps them stay afloat. In a large whale, the layer of blubber can be up to two feet thick!

9-5 CUTS AND HEALING

Your skin is an excellent protective barrier. Nothing can get past it into the other body tissues unless something breaks that barrier. Scratches and cuts are two of the most common skin injuries that allow dirt and bacteria to enter your body. Do you know how a cut heals? Do you know how to help the healing process?

How Deep?

You may be able to tell how deep a cut is by observing the amount of blood that escapes. If it's a scratch that does not bleed, only the epidermis has been cut. Remember, the epidermis is the upper layer of skin. It does not contain capillaries—small blood vessels. If the cut bleeds, both the epidermis and the dermis have been cut. The blood comes from the capillaries in the dermis.

Slight bleeding is normal and can help "clean" the cut. Bleeding usually stops after a few minutes because blood contains substances to form clots. When the clot dries, it becomes a **scab.** The scab protects the damaged tissues and prevents dirt and bacteria from entering the wound.

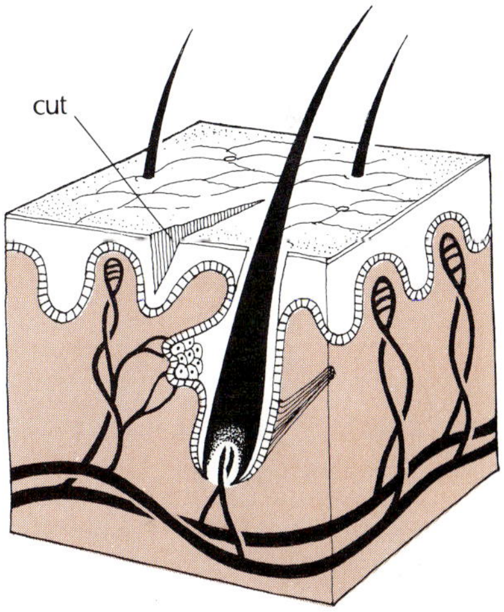

Figure 9-10 If only the epidermis is cut, there will be no bleeding.

Deep Cuts

Deep cuts may require special attention. For instance, a larger cut may leave a **scar.** The scar forms from new cells that grow to fill in the cut. Also, the risk for infection is higher when a cut is deep. In order to prevent infections and to keep the size of a scar to a minimum, a person should get medical help for deep cuts.

Here are some guidelines for when to see a doctor for a cut.

- See a doctor if the cut is *jagged* and *open*. If this type of cut is left to heal by itself, bacteria can enter and cause infection. A scar may form.
- See a doctor if the cut is *dirty*. Pieces of dirt, small rocks, or glass in the cut need to be removed.

- See a doctor for *cuts* on your *face*. These cuts may require stitches to prevent scar tissue from forming.
- See a doctor for a *puncture wound* or *animal bite*. These wounds bleed very little but allow dirt and microorganisms to get under the skin. Antibiotics and a tetanus shot may be needed.

These guidelines apply to wounds that are bleeding only slightly. **If there is lots of blood, apply pressure and call for medical help immediately!**

Minor Cuts

Most cuts are minor and bleed only a little but they, too, can become infected. To treat simple scratches and scrapes, follow these four easy steps.

1. Wash with hand soap or mild detergent.
2. Rinse with clean, running water.
3. Blot dry with a clean cloth.
4. Leave small cuts uncovered. Cover large ones with a sterile—germ-free—bandage.

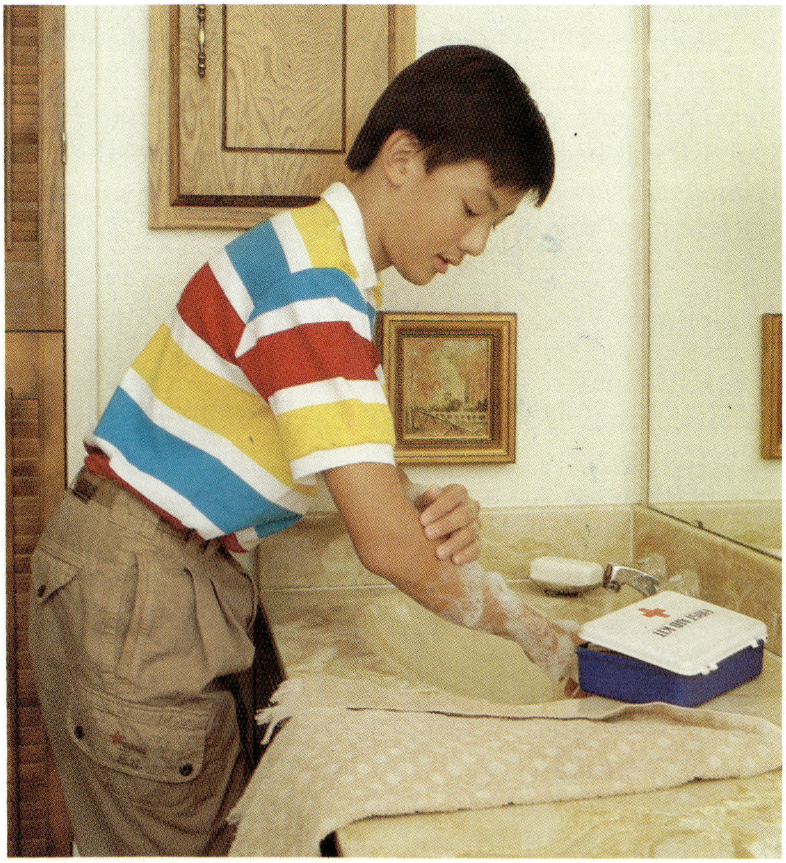

Skin Medications

Many people use skin medications on minor cuts, burns, and rashes. There are a variety of these products on the market. You may have used some of them yourself.

- What do skin medications do?
- Are there any dangers involved with using a skin medication?

Read two skin medication labels while you complete the two pages of the worksheet. Use the information on that worksheet to help you answer the two focusing questions.

CUTS AND HEALING

NAME ____________

DATE ____________ PER. ______

Use the information on two skin medication labels to fill in these two sections.

Product name ____________

Reason(s) for use ____________

Product name ____________

Reason(s) for use ____________

1. List at least 5 things that are printed on both labels. For example, an expiration date is printed on each label.

   a. ______ d. ______

   b. ______ e. ______

   c. ______ f. ______

2. Put a star (*) by the best reason for using a medication on a cut.
   a. to stop the bleeding
   b. to keep air from getting to the cut
   c. to make the cut heal faster
   d. to prevent infection
   e. to make the cut hurt less

   Cross out the things that a medication would *not* do for the cut.

Figure 9–11 (Worksheet, page 1)

CUTS AND HEALING
(PAGE 2)

NAME ____________

3. Imagine that you are redesigning the label for one of the medications so that 9-year-old children will be able to understand what it says.
   a. Copy one sentence or phrase from the label that is too hard for a 9-year-old to understand.
   b. Underline any words in the sentence that are too hard for most 9-year-olds.
   c. Rewrite the sentence so that the 9-year-old could understand it. (You may need to look up some words in a dictionary.)

4. List at least 4 warnings, or precautions, that are printed on the labels. Put a star (*) by the one you think is most important for people to know.

   a. ______

   b. ______

   c. ______

   d. ______

5. Many labels say "for minor cuts." Describe the way a minor cut would look as compared with a major one.

6. You were watching your 3-year-old nephew when he cut himself. You looked at it and decided to call a doctor. List at least 4 things you could have seen that caused you to make this decision.

   a. ______

   b. ______

   c. ______

   d. ______

Figure 9–11 (Worksheet, page 2)

9–6 SKIN COLORS

What color is your skin? The color of your skin tells a lot about you.

- It can reflect the state of your *health.*
  Are you flushed from a fever?
  Do you have a rash?
  Are you pale?
- It can reveal your *feelings.*
  Are you blushing?
  Are you "white" from anxiety?
- It can show how you *spend your time.*
  Do you have a tan over most of your body?
  Are your arms darker than the rest of your body?
- And, of course, it tells about your *ancestry.*
  Where were your ancestors from? Norway? Laos? Ethiopia? Greece? Panama?

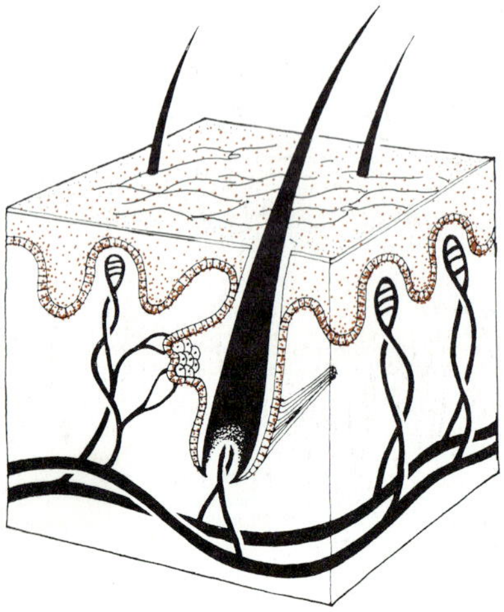

Figure 9–12 Cells that produce pigment are located in the epidermis.

What Determines Skin Color?

The anatomy of your skin determines what color it will be. Look at the diagram of skin structure shown in Figure 9–12. Remember—there are two main layers to skin, the upper epidermis and the deeper dermis.

Skin color is determined mainly by the amount of **pigment** that is present—the more there is, the darker the skin appears. The cells that produce the pigment are located in the epidermis.

Blood flow also affects skin color. The more blood there is in the capillaries, the redder, or pinker, the skin will look. (Remember, capillaries are the little blood vessels in the dermis.)

The thickness of the skin is a third factor that influences skin color. The top layer of the epidermis is dead cells. These cells may give the skin a faintly yellow hue.

Changing Colors

Some people may want to change their skin color. They may want it lighter, darker, browner, pinker, or whiter. How can they do this? They can't. They may be able to change it temporarily, but people inherit their basic skin color. A person's skin color is an unchangeable, genetic trait.

Sun Tans, Sunburns

The sun can temporarily change skin color by darkening it. The color change occurs in stages. These stages are shown in Figure 9–13.

Drawing A shows a section of skin before it has been exposed to the sun.

When skin is exposed to sun, ultraviolet light rays damage the surface skin cells. The body reacts by sending more blood to the damaged cells. This excess blood causes the skin to look red, as shown in drawing B.

The cells in the lower part of the epidermis divide to form new skin cells to replace the damaged ones. As this "new skin" appears, the red color fades. The skin looks tan because more pigment was produced and the top layer of dead cells is thicker. Look for these changes in drawing C.

After several weeks, the skin returns to its original color, as shown in drawing D. The extra dead skin cells are washed off and the new skin cells contain the "normal" amount of pigment.

Tanning is a protective reaction. When the sun's rays burn skin cells, extra pigment forms to block the rays and protect the cells. This is true for everyone, no matter whether they have inherited a very dark skin color or a very light one.

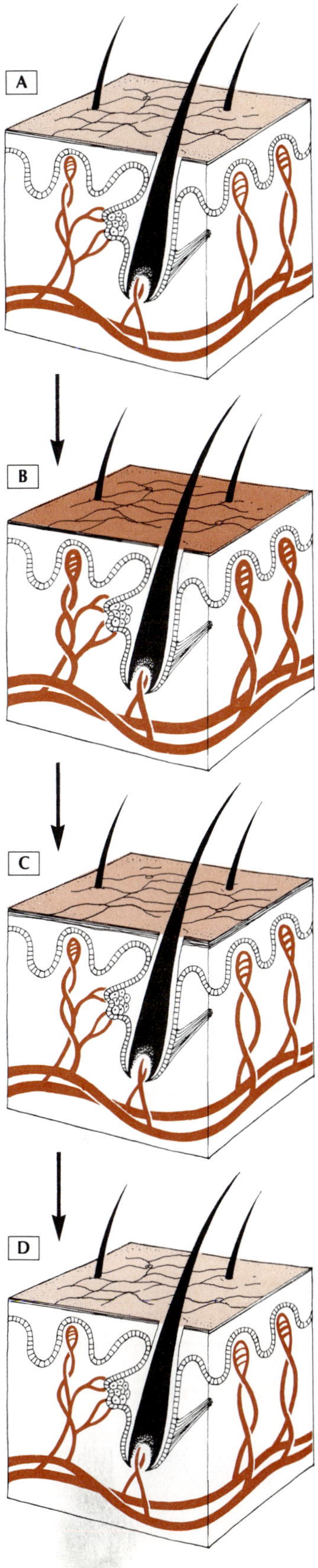

Figure 9–13

9–7 SKIN CANCER

Fire, electricity, chemicals, radiation, and the sun are all sources of heat, and all of them can cause burns. All burns are serious. They cause pain, damage tissues, and may become infected. One of the most common kinds of burns is sunburn. A sunburn results from overexposure to ultraviolet (UV) light.

Figure 9–14 All of these can be signs of skin cancer.

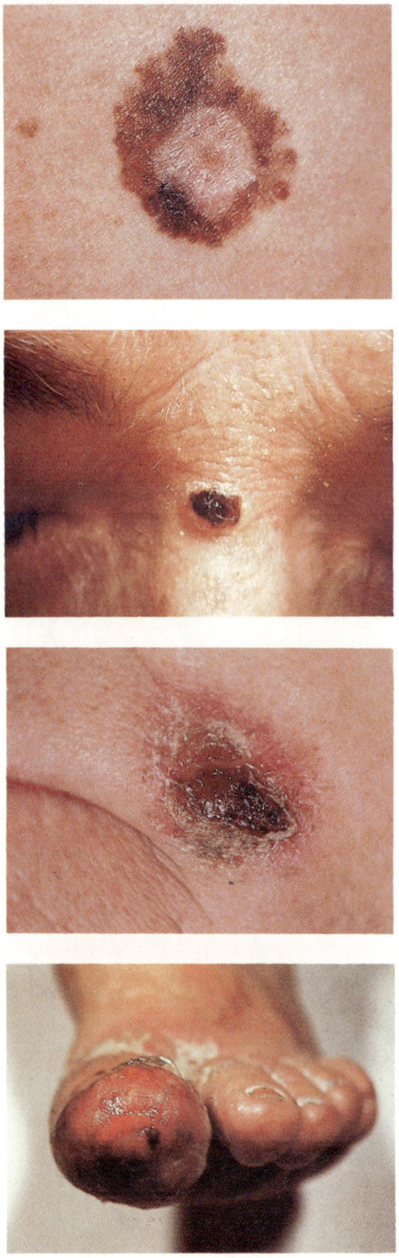

Cancer

The most serious danger from UV light is skin cancer. Anyone can get skin cancer; it doesn't matter if your skin is very light or very dark. However, the lighter a person's skin is, the higher the risks for cancer.

People do not think about this risk when they are young, but they should. Some experts believe that most of the damaging exposure to the sun occurs before a person is twenty years old. Cancer develops slowly but irreversibly; once it starts, it continues to develop.

Everyone should know the signs of skin cancer. It usually appears on areas that are not covered by clothing. The most common sites for skin cancer are the face, neck, ears, forearms, and hands. If caught early, skin cancer is curable. The signs of skin cancer are listed in the chart below and shown in the photographs in Figure 9–14.

Know These Signs of Skin Cancer

A mole, birthmark, or beauty mark that
- changes color
- increases in size or thickness
- changes in texture
- is irregular in shape

A skin growth that gets larger

A spot or growth that continues to itch, hurt, crust, scab, or bleed

A sore or wound that does not heal within four weeks, or heals and then reopens

If you ever notice any of these symptoms, see your doctor!

Protection

It is wise to protect your skin from sun damage. The very first step in sun safety is to use a sunscreen. You should choose a sunscreen according to its SPF number. SPF stands for "sun protection factor." If you usually burn after 10 minutes in the sun without a sunscreen, a SPF of 6 means you can stay in the sun 6 times longer, or 60 minutes, before burning. An SPF of 12 means you could stay in the sun 12 times longer. Many doctors think everyone should use a sunscreen of at least SPF 15. Sunscreens do not block *all* sunlight. You will still get a tan, but at a slower, safer rate, without burning.

Here are five steps for sun safety.

1. *Use a sunscreen* with at least SPF 6. Put more on every two hours, especially if you have been perspiring or swimming.
2. *Beware of sand, water, snow, and concrete.* They increase your chances of burning by reflecting sunlight.
3. *Cover up* when you will be in the hot sun for very long. Wear a hat, long sleeves, and long pants.
4. *Time your activities* for early morning and late afternoon. Avoid the most intense sun rays around noon.
5. *Avoid tanning parlors.* The UV light in tanning booths is skin damaging.

THINK ABOUT IT

Do you want wrinkly skin? Do you want to increase your chances of skin cancer?

Figure 9–15 According to the American Cancer Society, one out of every seven Americans will develop skin cancer. What could these people do to lessen their chances of skin cancer?

UNIT 3

FOODS AND DIGESTION

FOODS

If you were choosing one item from this display, what would you pick?
Why would you pick it?

10–1 HI-LO LABEL HUNT

It is easier to learn about foods than you may think. Every box or can label provides some information about the food inside that container. Some labels tell you much more than others. However, it is easy to overlook food facts since labels often are designed to catch your eye.

- What can *you* learn about a food product by reading a food label?

Answer this focusing question after you have gone on a high-low hunt through food labels to find food information.

MATERIALS

4 food labels from cans or boxes
worksheet: ***Hi-Lo Hunt List*** (Figure 10–1)

PROCEDURE

Work in teams of two or four.

1. Divide the labels among the members of your team.
2. Choose a team recorder who can write quickly and neatly. He or she will be responsible for filling in the Hi-Lo Hunt List.
3. You will have exactly 15 minutes to find the items on the list. Help each other locate the items on the labels so your group can work quickly.
4. Everyone on your team should be prepared to share the information on the list with the rest of the class when your teacher calls time.

CONCLUSION

Look back at the focusing question. As a team, write your response to the question.

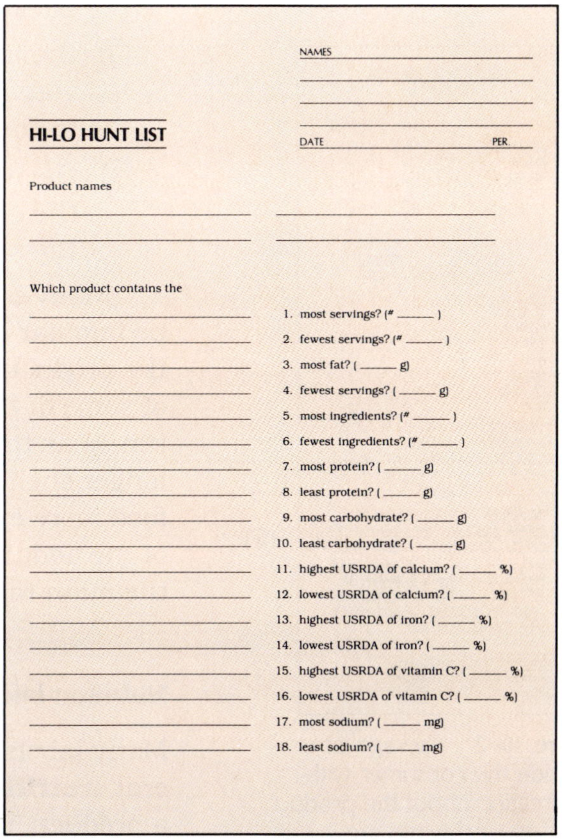

NAMES ______

DATE ______ PER. ______

HI-LO HUNT LIST

Product names

______ ______

______ ______

Which product contains the

______ 1. most servings? (# ______)
______ 2. fewest servings? (# ______)
______ 3. most fat? (______ g)
______ 4. fewest servings? (______ g)
______ 5. most ingredients? (# ______)
______ 6. fewest ingredients? (# ______)
______ 7. most protein? (______ g)
______ 8. least protein? (______ g)
______ 9. most carbohydrate? (______ g)
______ 10. least carbohydrate? (______ g)
______ 11. highest USRDA of calcium? (______ %)
______ 12. lowest USRDA of calcium? (______ %)
______ 13. highest USRDA of iron? (______ %)
______ 14. lowest USRDA of iron? (______ %)
______ 15. highest USRDA of vitamin C? (______ %)
______ 16. lowest USRDA of vitamin C? (______ %)
______ 17. most sodium? (______ mg)
______ 18. least sodium? (______ mg)

Figure 10–1

10-2 READING A LABEL

Federal law requires that all packaged foods be clearly labeled because it is important that you, as the consumer, know exactly what you are buying. In addition to the required items, many manufacturers choose to print additional information on their food labels.

①

②

NUTRITION INFORMATION PER SERVING

	IN PKG.	AS PREPARED
SERVING SIZE	1/4 BOX	3/4 CUP
SERVINGS PER PKG	4	4
CALORIES	190	290
PROTEIN	9g	9g
CARBOHYDRATE	36g	34g
FAT	2g	13g
SODIUM	420 mg	530 mg**
PERCENTAGE OF U.S. RECOMMENDED DAILY ALLOWANCES (U.S. RDA)		
PROTEIN	20	20
VITAMIN A	*	10
VITAMIN C	*	*
THIAMINE	35	15
RIBOFLAVIN	20	15
NIACIN	20	8
CALCIUM	8	8
IRON	10	10

*CONTAINS LESS THAN 2% OF THE U.S. RDA FOR THESE NUTRIENTS.

**SODIUM LEVEL WITH 1 TEASPOON SALT ADDED TO 6 CUPS COOKING WATER IS 630 mg PER ¾ CUP SERVING.

INGREDIENTS: ENRICHED MACARONI; CHEESE SAUCE MIX (WHEY, DEHYDRATED CHEESE [GRANULAR, CHEDDAR (MILK, CHEESE CULTURE, SALT, ENZYMES)], WHEY PROTEIN CONCENTRATE, SKIM MILK, SALT, BUTTERMILK, SODIUM TRIPOLYPHOSPHATE, SODIUM PHOSPHATE, CITRIC ACID, YELLOW 5 & 6 [COLOR], LACTIC ACID) ④

(M001A)

c KRAFT, INC., DIST.
GLENVIEW, IL 60025 USA ③

Figure 10–2 Package labels provide the consumer with information about the product. How many servings are in this box of macaroni and cheese?

Required by Law

People take food labeling for granted. If a label says "large pitted olives," that is exactly what we expect to find in the can. We can do so because the law requires that certain basic information appears on all labels. Each label must show the following:

① the name of the product

② the net weight (This figure gives the weight of the food, but does not include the weight of the packaging.)

③ the manufacturer's name and address (In case you have a concern about the product and want to contact the manufacturer.)

④ the ingredients (The first one listed is present in the greatest amount by weight, the last one in the smallest amount.)

Many of the ingredients listed on the label should be familiar to you since they are the foods used to make the product. However, you may not recognize the names of some of the other ingredients. These include such things as the preservatives added to give the product a longer shelf life or the food colorings used to give the food more eye appeal. Unless you have an allergy to one of the additives, you will probably be most interested in the main ingredients.

Nutrition Information

Many labels also provide nutrition information. In general, **nutrition information** tells about the nutrients in a product. The law requires that it appear on a label only if a nutrient has been added to the product. For instance, if a label claims that a loaf of bread is "vitamin enriched," you can look at the nutrition information to see which vitamins have been added.

There are six major nutrient groups. They are fat, protein, carbohydrate, vitamins, minerals, and water. These substances are needed to keep the body healthy. Some of them are used for energy. All of them are needed in order for a body to grow, repair damaged tissue, and regulate body functions. It is important to eat adequate amounts of these nutrients every day.

Fat, protein, and carbohydrates are the major nutrients; their amounts are given in grams (g) on the food label. Because we need much smaller quantities of vitamins and minerals, their amounts are given in milligrams (mg) or as a percentage.

Nutrition information is based on a single serving. Serving size is the first thing listed under nutrition information. Looking at the label in Figure 10–2 on page 238, you can see that a ¾-cup serving provides 9 grams of protein. If you ate all of the macaroni and cheese, you would get 36 grams of protein.

Keep fit and eat a Wheaties Breakfast

A breakfast of Wheaties, milk, and fruit provides an important nutritional start to your day plus energy that you need to get going.

Each 1-ounce serving contains 3 grams of dietary fiber.

NUTRITION INFORMATION PER SERVING

SERVING SIZE .... 1 OUNCE
SERVINGS PER PACKAGE 12

	1 ounce Wheaties	Wheaties plus 1/2 cup vitamin A & D fortified skim milk†
CALORIES	100	140
PROTEIN, g	3	7
CARBOHYDRATE, g	23	29
FAT, g	1	1
CHOLESTEROL, mg	0	0
SODIUM, mg	200	260
POTASSIUM, mg	110	310

THE SODIUM LEVEL IN WHEATIES HAS RECENTLY BEEN REDUCED BY 25% FROM 270 MG TO 200 MG.

PERCENTAGE OF U.S. RECOMMENDED DAILY ALLOWANCES (U.S. RDA)

PROTEIN	4	10
VITAMIN A	25	30
VITAMIN C	25	25
THIAMIN	25	30
RIBOFLAVIN	25	35
NIACIN	25	25
CALCIUM	6	20
IRON	25	25
VITAMIN D	10	25
VITAMIN B_6	25	30
FOLIC ACID	25	25
PHOSPHORUS	10	20
MAGNESIUM	8	10
ZINC	4	6
COPPER	6	6

INGREDIENTS: WHOLE WHEAT, SUGAR, SALT, CEREAL MALT SYRUP, CALCIUM CARBONATE, CALCIUM CHLORIDE, TRISODIUM PHOSPHATE.

VITAMINS AND MINERALS: VITAMIN C (SODIUM ASCORBATE), A B VITAMIN (NIACINAMIDE), IRON (A MINERAL NUTRIENT), VITAMIN A (PALMITATE), VITAMIN B_6 (PYRIDOXINE HYDROCHLORIDE), VITAMIN B_2 (RIBOFLAVIN), VITAMIN B_1 (THIAMIN MONONITRATE), A B VITAMIN (FOLIC ACID) AND VITAMIN D. FRESHNESS PRESERVED BY BHT.

Figure 10–3

Other Information

In addition to the items required by law and the nutrition information, you can find a variety of other things on the label. It may tell you more about the contents or show you a picture of the product. The manufacturer may also use the label to promote the sale of the product by including a cents-off coupon or a money-back refund. What other items have you noticed on a label?

ANALYSIS QUESTIONS

1. Which items must appear on a food label?
2. What are the 3 major nutrients?
3. What is a nutrient?
4. Use the nutrition information on the label in Figure 10–3 to answer these questions.
   a. What is the main ingredient in this cereal?
   b. How much of the cereal is considered to be one serving?
   c. Does this cereal contain more carbohydrate or more protein per serving?
   d. How much fat is in one serving?

5. In your own words, explain the difference between a nutrient and an ingredient.
6. Here is a list of 12 items. Some of them are nutrients, the others are considered to be ingredients.

flour	salt	rice	carbohydrate
vitamin C	protein	water	corn meal
iron	egg	onions	baking soda

   a. List the ones you think are nutrients.
   b. List the ones that are used as ingredients.
7. Look at the ingredients from a food label shown below. Decide which product it comes from—smoked beef jerky (a dried beef product), all beef hot dogs, or beef flavored dog food. Explain your decision.

   Beef, water, dextrose, salt, corn syrup, spices and flavorings, sodium phosphates, paprika, sodium erythorbate, sodium citrate, sodium nitrate

CHALLENGE

Here is part of a recipe for cookies. List these ingredients in the order that they would appear on a food label.

3½ cups flour
½ teaspoon baking soda
¼ teaspoon salt
1 cup shortening
1 cup sugar
1 teaspoon vanilla

EXTENSION ACTIVITIES

1. Some food labels do not provide nutrition information. You probably have a label from at least one of these products. Write to the company that manufactured the product. Ask them why they do not provide this helpful information. Have them send you a nutritional analysis of the product.
2. Look at the information printed on cat or dog food labels. In what ways is this information *like* that on "people" food labels? How is it *different?* Read the list of ingredients. Is there anything in pet food that is not in the food *you* eat?

10-3 FAT TEST

Fat is one of the three major nutrients. Our bodies need some fat in order to function normally. And yet, we often hear that there is too much fat in our diets. In order to determine the amount of fat you eat, you need to know which foods contain fat.

- How can you test for fat in foods?
- What foods contain fat?

After you have tested several foods for fat, you will be able to answer these questions.

MATERIALS

brown paper towel or a sheet of paper
fat (shortening or oil)
samples of 10 different foods
data page: ***Fat Test*** (Figure 10–4)

NAMES ______

______

FAT TEST

DATE ______ PER. ______

FOOD	PREDICTIONS Does it contain fat?	TEST RESULTS Describe appearance of paper (++, +, or −)
fat		++
water		—
apple		
celery		
cheese		
chocolate		
cookie		
hot dog		
peanuts		
potato		
potato chip		
tuna		

Figure 10–4

PROCEDURE

Work with a partner.

Part A. Getting Ready

1. Look at the list of foods on your data table. Which of these foods do you think contain fat? Discuss the foods with your partner and make some predictions. Fill in your predictions in the first column of the data table. Use a plus sign (+) if you think the food contains fat and a minus sign (−) if you think it does not.
2. Mark your paper towel into 12 sections.
   a. Label one section *fat* and another one *water*, as shown in Figure 10–5 on page 242.
   b. Write the names of the 10 foods you will be testing in the other sections.

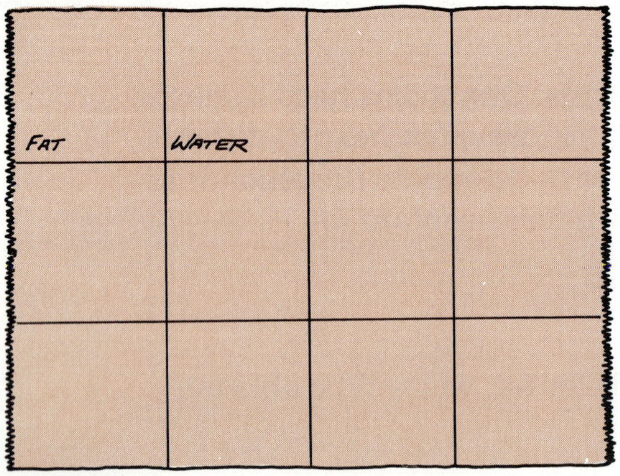

Figure 10–5

3. Twelve stations have been set up around the room. You and your partner will need to go to each station to test one of the foods. Make sure you know which station you should go to first.

Part B. Testing for Fat

1. At each station, take a *small* amount of the food to be tested. The amount you take should be smaller than a raisin.

   a. Firmly smear the food into the correct section on your paper.
   b. Brush off the remains of the food into a wastebasket.
   c. To test the water, place one drop of water in the correct section.
2. You and your partner should take turns placing the different foods on the paper towel. Clean your hands between foods or use different fingers so you don't accidently mix the foods.
3. When you have completed the 12 food tests, return to your seat, set your paper towel aside, and answer analysis questions 1–3.

Part C. Analyzing the Results

1. Watch the water spot on your paper. When it is dry, you can analyze the results of the food tests.
2. First, look at the section that you used to test fat. Hold the paper towel up to the light, as shown in Figure 10–6. Look at it from both the front and back. This shows what a very positive (++) fat test looks like.

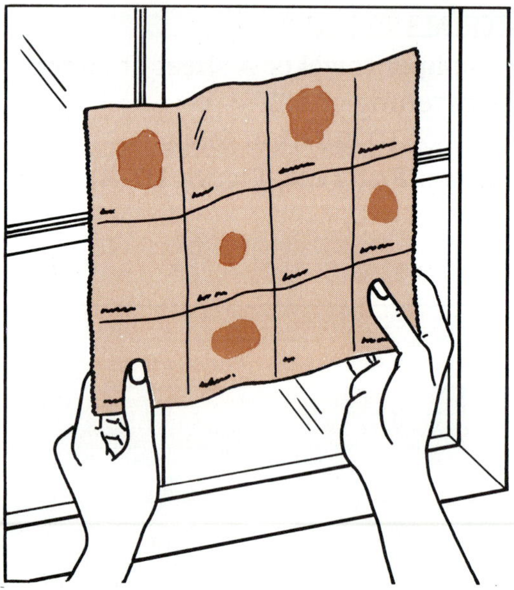

Figure 10–6

3. Look at the water section from both front and back. Since water contains no fat, this shows what a negative (−) fat test looks like.
4. Look at the food tests on the paper and record the results on your data table. Use a double plus (++) if a food contains a lot of fat, a single plus (+) for a food with some fat, and a minus (−) for a food with no fat.
5. Answer the rest of the analysis questions.

ANALYSIS QUESTIONS

1. Read the introduction again. Write down the two questions you will be able to answer as a result of this activity.
2. List at least 4 "facts" you know about fat in foods.
3. Think about the predictions you made on the data table. How did you decide whether or not something contained fat?
4. Using this fat test, how can you tell if a food contains fat?
5. Based on your results, which of the foods you tested contained fat?
6. Look at the data table. How many of your predictions were correct?
7. Look at all the foods that contained fat. Think about what they have in common. In one or two sentences, describe the kinds of food that contain fat.
8. Copy this list of foods. Then circle the ones you think contain fat.

margarine	corn
egg white	sausage
bread	lettuce
doughnut	whole milk
banana	sunflower seeds
carrot	crackers

CONCLUSION

Imagine that you are talking to another student your age who attends a different school. In writing, answer the two focusing questions as if you were talking to this other student.

10–4 FATS AND OILS

Of the six major nutrients, you will probably need to know the most about fats. Why? For a variety of reasons. Some people want to know about the fats in their food because they are concerned about a weight problem. Other people worry about the link between certain kinds of fat and heart disease. Many people want to learn more about fat so they can make wise food choices. Some fat in the diet is important, but most Americans eat too many fatty foods. Most of us need to cut down the amount of fat we eat.

Body Fat

Fat is one of the three major types of molecules found in our bodies. (The other two are proteins and carbohydrates.) Body fat serves several important functions; they are described below.

Figure 10–7

Fat molecules make up the *structure of cells*. Thus, every single cell and every part of the body contains some fat.

Fat *supplies energy*. When the body stores fat, it is storing energy for later use.

Fat *provides insulation*. The layer of fat beneath the skin serves this important function.

Fat acts as *padding*. It surrounds the heart, kidneys, and other body organs. It helps hold them in place and protects them from blows.

Fat *transports* some vitamins through the body.

Fats are used in the *production of several hormones*.

As you can see, some body fat *is* important for good health. We need some fat for insulation and padding. This fat can also be used for energy if our dietary fats are low. However, if we routinely eat too much, we will find ourselves with too much insulation and padding. That's because the body stores the fat it doesn't need right away. In addition, any extra protein or carbohydrate gets changed to fat, too.

Figure 10–8 All of these high fat foods come from plants. People are often surprised to learn that nuts are among the high fat foods.

Sources of Fat

If you are like most Americans, you are probably eating too much fat. If you want to cut back on the amount of fat you eat, you need to know which foods are high in fat. These high fat foods can come from either plant or animal sources.

Plant, or vegetable, fats are usually liquids at room temperature; we call them oils. The vegetable oils used in cooking come from corn, palms, olives, and other plants. Margarines are vegetable oils that have been "hardened" to make them solid.

You may not realize that all nuts and seeds contain lots of fat. That means foods such as peanut butter and sunflower seeds are high in fat. Chocolate, avocados, and olives are other examples of fatty foods from plants. (See Figure 10–8.)

Foods that come from animals also contain fats. Fats are found in most dairy products such as cheese, butter, milk, and cream. Eggs are another source of animal fat. One of the most obvious fats is the fat in meat, which we often trim off. All meats contain fat even if you cannot see it. As you may remember, meat is muscle and muscle cells use fat as an energy source.

Kinds of Fat

There are several different kinds of fat. Some of them seem to cause more health problems than others. When you plan a "fat-wise" diet, it helps to know which fats are the "bad guys."

Cholesterol is one fat that may cause health problems. A high level of this kind of fat is often found in people with heart disease. Health professionals worry when they find teenagers or children with high levels of cholesterol. They wonder if these young people are more likely to develop heart disease as adults.

Figure 10–9 Cholesterol is only found in animal products.

Our bodies produce some cholesterol, but we also get it from the foods we eat. Cholesterol is an animal fat; it is never found in plant foods. Egg yolks, lard, and red meats are all high in this fat. Milk products, such as cream and cheese, are also high in cholesterol. Examples of foods that contain cholesterol are shown in Figure 10–9.

People who want to avoid heart trouble cut back on foods that contain cholesterol and **saturated** fats. (Saturated fats are found in animal fats and in the hardened vegetable oils.) Healthwise, it is better to eat margarines and oils that contain **polyunsaturated** fats. (Remember: Polyunsaturated fats are okay; avoid saturated fats.)

Since plants do not produce cholesterol and most of the natural plant oils are unsaturated, most vegetable fats are "safe" to eat. However, you should avoid two of these fats—coconut oil and palm oil. Even though they come from plants, they are high in saturated fat. When you read food labels, you will find that these two oils are often used in crackers and cookies.

"Fat-Wise" Decisions

You should be asking yourself, "What can I do to make my diet more fat-wise?" You could check the fat content of your diet by adding up the number of grams of fat you eat each day. **Most 11- to 13-year-olds should**

eat no more than 65 to 110 grams of fat per day. (Or, to say it another way, fat should make up no more than 30% of the calories in your diet.) In order to follow this rule, you would need to calculate the number of grams of fat in all the foods you eat but that is not very practical to do. You usually do not list all of the foods you eat nor do you carry around a nutrient chart. So, instead of counting grams of fat, you can eat a "fat-wise" diet by following this guideline.

AVOID TOO MUCH FAT, SATURATED FAT, AND CHOLESTEROL.

Here are six recommendations made by the American Heart Association to help you follow this guideline.

1. Choose low fat versions of milk, cheese, and other milk products.
2. Trim off the fat when you eat beef, ham, or other meats.
3. Choose fish and poultry over beef or pork when you are given the choice.
4. Eat no more than 3 egg yolks per week. This includes eggs used to prepare foods such as cookies, cakes, and casseroles.

Figure 10–10 Whenever we make food choices, we should try to select low fat items.

5. Use polyunsaturated fats instead of saturated fats whenever possible.
6. Eat boiled, baked, or broiled foods instead of ones that have been fried. Especially avoid foods that are deep fat fried.

Did you read the six recommendations? Close your eyes. How many of them can you remember?

ANALYSIS QUESTIONS

1. List 6 functions of fat in the body. Now circle the 3 functions that you think will be the hardest for you to remember.
2. List 3 foods that contain animal fat.
3. List 3 plant foods that are high in fat.
4. Which of these fats should you avoid?

cholesterol	animal fats
saturated fat	polyunsaturated fat

5. How many grams of fat per day is about right for you?
6. Refer to the six recommendations for a "fat-wise" diet.
   a. Which one is easiest for you to follow?
   b. Which one is hardest for you to follow?
7. Three examples of similar foods are listed below. In each case, decide which one is the "fat-wise" choice. Explain your decisions.
   a. a pat of butter or a pat of margarine
   b. corn oil or coconut oil
   c. baked ham or fried ham

EXTENSION ACTIVITY

Ask your teacher about inviting a nutritionist or dietician to talk to your class about the diets given to heart patients. You may want to ask questions similar to the ones below.

What "eating rules" are the most important?

Why are cholesterol and saturated fats associated with heart disease?

What dietary advice would you give to students our age?

10–5 FATS IN YOUR DIET

Well? You have tested foods for their fat content. You have read about various types of fat. You know which fats may be bad for your health and you have learned some rules for making "fat-wise" food choices. Now, can you put all of this knowledge to work?

- How does your own diet rate when it comes to fats?

Analyze the foods in your food diary for their fat content to answer this question and decide if you have been making fat-wise food decisions.

MATERIALS

your food diary
orange colored pencil
chart: ***Nutrients in Foods*** (page 630)
reference materials showing nutritional values of foods

PROCEDURE

1. Use your lead pencil to do the first analysis of your food diary. Put a check mark by each of the foods that you think contains fat. Figure 10–11 shows a sample food diary.
2. Now, use nutrition charts to check the actual amount of fat in the foods you marked. Use your orange colored pencil to record the number of grams of fat found in each food.
3. Put an orange asterisk (*) by any food that contains more than seven grams of fat. These foods are your high fat choices.
4. Look over your list of foods one more time. This time put an orange circle around each food that contains cholesterol or saturated fat. Remember that this includes all foods that contain eggs.

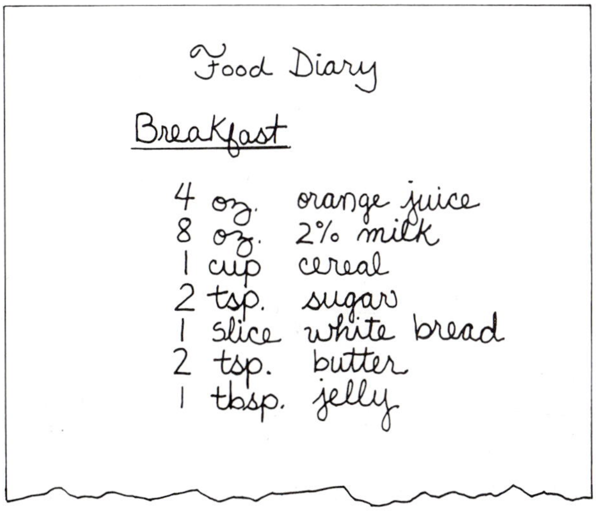

Figure 10–11

ANALYSIS QUESTIONS

1. Analyze the amount of fat you ate on the day you kept a food diary.
   a. How many grams of fat did you eat?
   b. In general, what is the suggested range for grams of fat for someone your age?

c. How did the number of grams of fat you ate compare with the suggested range?

2. List the foods you ate that contained no fat.

3. Find the 5 foods on your list that have the highest fat content. For each of these high fat foods, list a low fat food you could eat instead. One example is shown below.

High Fat	**Low Fat Alternative**
½ cup ice cream	½ cup ice milk

4. Glance over the nutrient chart in the appendix. Which food contains a very high level of fat? How many grams does it contain per serving?

5. Do you think your diet follows the dietary guideline *Avoid too much fat, saturated fats, and cholesterol?* Explain.

6. Here are some comments and questions from several students who analyzed their own food diaries for fat. Write your answer to each of these students.

   a. For dinner, Matthew usually drinks two glasses of 2% milk. He looked up milk on a food chart and found that each glassful contains five grams of fat. So now he wants to know, "Should I skip milk at dinner since I will be drinking 10 grams of fat?"

   b. Pam's mother is supposed to eat less fat so she is fixing low fat foods for the family. When Pam analyzed her meals, she found that the piece of chicken she ate for lunch contained 9 grams of fat and the serving of meat loaf she had for dinner contained 12 grams. Now Pam is asking, "Is my mother choosing low fat foods?"

   c. Jennifer was relieved to read that a teaspoon of butter only contains 4 grams of fat. She loves butter and uses about a teaspoon at a time on her breakfast muffin, lunch sandwich, and dinner vegetables and potatoes. She wants another opinion, "Since a teaspoon of butter only contains 4 grams of fat, it's okay to keep putting it on all my foods. Right?"

EXTENSION ACTIVITIES

1. Were you surprised by the fat levels of any of the foods you looked up in the nutrient charts? Were some foods higher in fat than you expected? If so, do the fat test on these foods and share the results with your class.

2. Ask a restaurant for a copy of their menu. Go through each item on the menu. Use a red colored pencil to circle the choices that would be high in fat. Use a green colored pencil to circle ones that would be low in fat. Do not circle the choices that contain average amounts of fat.

10-6 PROTEIN TEST

Of the three major nutrients, protein is the most critical because your body cannot create the protein you need. You must get your protein from foods. If you are aware of which foods provide protein, you will be able to put together a better diet. Hundreds of foods are good sources of protein. How many can you name?

- How can you test for protein in foods?
- Which foods are good sources of protein?

After doing this activity, these questions will be easy for you to answer.

MATERIALS

test tube rack
egg white
7 test tubes
safety goggles
paper towels
masking tape
dropper for water
samples of 5 different foods
nitric acid in dropper bottle
data page: ***Protein Tests*** (Figure 10–12)
chart: ***Nutrients in Foods*** (page 630)
summary table: ***The Major Nutrients***

PROCEDURE

Work with a partner.

Part A. Getting Ready

1. On your data table, record the 5 foods that your teacher tells you to test.
2. With your partner, make predictions about whether or not these foods contain protein. When you reach an agreement, record your predictions on the data table. Use a plus (+) if you think the food contains protein. Use a minus (−) if you think it does not.

PROTEIN TEST

NAMES ____________________

____

DATE __________ PER ____

Describe the reaction when nitric acid is dropped on egg white.

Food	PART A Predictions about Protein	PART B Results of Protein Test	PART C Grams of Protein per Serving
water	—		
egg white	++		

Figure 10–12

Figure 10–13

Part B. Testing for Protein

1. Put 9 or 10 drops of water in one test tube. Put a small amount of egg white—about the size of half a raisin—in another tube. (See Figure 10–13.)
2. **Put on your safety goggles**. Carefully add 10 drops of nitric acid to each of the two test tubes. Gently swirl the test tubes to mix the contents.

CAUTION

Nitric acid can be dangerous. Handle it with care. If you splash or spill any acid, *tell your teacher*. If you get acid on your skin, rinse it off immediately with cold water.

3. **Notice what happens when nitric acid is added to egg white. Most foods that contain protein will give a similar reaction.** Record the results on the data table.
4. Prepare to do the protein test on your five foods.
   a. Using masking tape and a pen, label each test tube with the name of one of the five foods.
   b. Place small amounts of each food—about the size of half a raisin—in the proper test tube. You may need to crush or crumble the foods first.
5. Take turns testing the foods.
   a. **Make sure you are wearing your safety goggles**. Then add 10 drops of nitric acid to each test tube. (See Figure 10–14.) Carefully swirl to mix.
   b. Wait 5 minutes.

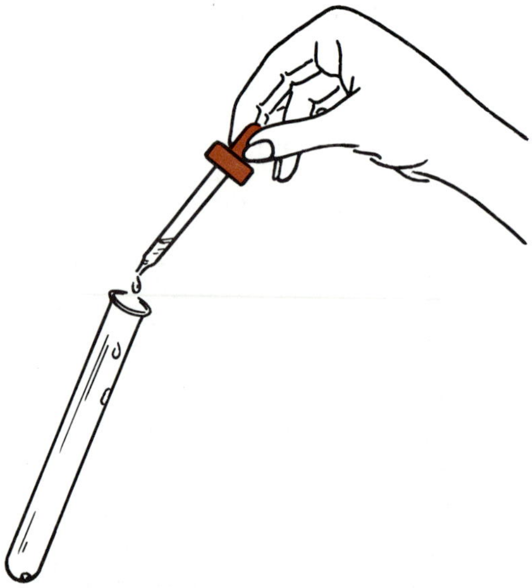

Figure 10–14

6. Record the results on the data table. Use a double plus (++) for foods with lots of protein. (They will react as the egg white did when nitric acid was added.) Use a single plus (+) for foods with some protein and a minus (−) for foods with no protein.

Part C. Finishing Up

1. Carefully dump the contents of each test tube where your teacher tell you. Rinse the test tubes with cold water, then wash them with soap. Clean up your work space.
2. Look up each of your five foods in the nutrient chart on page 630 or on a package label. Record the number of grams of protein per serving in each food on your data table.
3. Write the answers to the analysis questions on the back of your data paper.
4. During the class discussion of this activity, complete the Proteins column on the summary table.

ANALYSIS QUESTIONS

1. What happens when nitric acid is placed on protein?
2. Imagine that your lab partner spilled some nitric acid on his hand. What would you do?
3. Look at your protein test results.
   a. Which foods were high protein?
   b. Which ones did not have any protein?
4. Look at the grams of protein you listed on your data page.
   a. Which foods contained the most grams of protein per serving?
   b. How did you rate these foods on the nitric acid test?
5. Amy found that olives contain protein. Ted's results showed that olives did *not* contain protein. What are 2 possible reasons for their different results?
6. Protein is found in foods that come from both plants and animals.
   a. List at least 3 foods from plants that contain protein.
   b. List at least 3 foods from animals that contain protein.
7. If you get nitric acid on your skin or fingernails, they turn yellow and your skin will burn. Explain why the color change occurs.

CONCLUSION

Refer to the focusing questions on page 251. Write a paragraph that answers these questions. Be sure to include specific information. For example, describe the test for protein. Tell how you know if a food contains protein and give examples of foods that showed these results.

EXTENSION ACTIVITIES

1. Find 10 package labels for products that contain some protein. Write down the name of one of the products. List the 5 main ingredients in that product. Circle the ones you think provide protein. Do the same thing for the other labels.
2. Look at a world map. Choose one country that you would like to learn more about. Read about the country in an encyclopedia. Make a list of the foods that are eaten there. Circle the foods that are high in protein.

10-7 PROTEINS: BEANS OR STEAK?

Have you ever seen a photograph such as Figure 10–15? This child is listless and irritable. He does not have enough energy to walk, let alone to play. His skin may appear blotchy and wounds take a long time to heal. If he is around someone who is sick, he too is likely to become sick. The fever, vomiting, and diarrhea that go along with many childhood diseases will weaken him even more. What is wrong with this child?

Even though he may look well-fed, this child is undernourished. If you look more closely, you may be able to see that what at first appears to be fat is actually water. His system is holding fluid so he is bloated.

This child has not been getting enough protein in his diet; he has a condition called **kwashiorkor**. This condition is all too common among young children in countries where there is a food shortage. Without enough protein in their diets, people cannot be healthy.

Figure 10–15 Kwashiorkor affects millions of young children in Latin American, Asia, and Africa.

Survival

Protein is important because it does so many different things in the body. A few protein functions are listed below. You will learn more about protein functions as you study other body systems.

Protein is an important part of *cell structure.* Every cell in your body contains protein. You need a continuous supply of proteins so you can continue to produce new cells.

Besides cells, protein *makes up other structures* such as bones, hair, and fingernails.

Enzymes are also proteins. They control all of the chemical reactions that take place in the body.

One extremely important function of proteins is to fight disease. The *antibodies* that fight infections are proteins.

If you are low on fats and carbohydrates, your body can use protein *for energy.*

Figure 10–16

Enzymes, antibodies, and other protein structures do not magically appear. Our cells produce these structures by "rearranging" the proteins in the food we eat.

Protein Sources

Foods from both animals and plants contain protein. Meats of all types are good sources of protein; so are eggs, milk, cheese, and other dairy products.

Peas, lentils, kidney beans, pinto beans, soy beans, and lima beans are examples of legumes. Legumes are important sources of protein. Grains and cereals—rice, wheat, oats, and rye—as well as nuts and seeds are also protein sources. (See Figure 10–17 on page 256.) Most other vegetables and fruits are poor protein sources.

Figure 10–17 Many foods from plants are good sources of proteins. Which of these do you eat most often?

Figure 10–18 All of these are high protein foods that come from animal sources. How many of these do you recognize?

Americans eat more animal protein—meats and dairy products—than do people from other countries. These people rely more on plants for their protein. For example, a Brazilian eats an average of 50 pounds of beans a year; an average American eats about 6 pounds of beans a year.

Amino Acids

Once we eat protein, it is broken down in the digestive system. Imagine that you can look at one protein molecule with a super microscope before it is digested. You would see that it looks like a chain of smaller molecules, as shown in Figure 10–19. These smaller molecules are called **amino acids**. There are 22 different amino acids. They can be linked in thousands of different combinations. Each kind of protein is one specific arrangement of amino acids.

Our cells can rearrange these amino acids to make the proteins we need. For instance, they can change the protein in a chicken's muscle cells to protein molecules for human connective tissue.

If the proteins you eat do not contain all 22 amino acids, your body can make most of the missing ones. However, there are some amino acids your body cannot make; you can get them only from food. These are the **essential amino acids**.

Proteins from animal sources contain *all* of the essential amino acids. For this reason, they are called **complete proteins**. However, plant proteins are not complete; they are missing some of the essential amino acids. Plant foods are good sources of complete protein when they are put together in specific combinations. Since many vegetarians get all of their amino acids from plants, they must know how to balance what they eat.

By putting the right combinations of foods together, vegetarians can get all of the essential amino acids. One combination is shown in Figure 10–20. Here are other examples of good combinations:

corn tortilla with refried beans
cereal with milk
macaroni and cheese
pea soup and rye bread

In general, grains and legumes together make a complete protein; so do dairy products and eggs when they are eaten with any of the high protein plant foods.

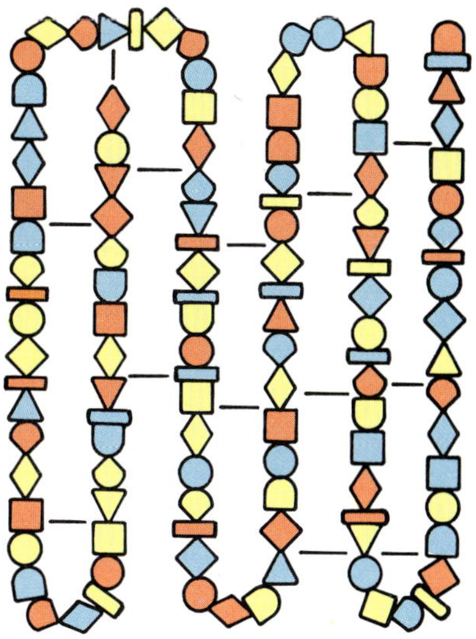

Figure 10–19 In this drawing, each different shape represents a different amino acid. Hundreds, maybe even thousands, of amino acids are linked together to form a protein.

Figure 10–20 The proteins in peanut butter with whole wheat bread provide all the essential amino acids.

How Much?

You should eat some protein each day but you probably don't have to worry that you are not getting enough protein—most Americans eat too much protein. Some people think high protein diets will make them better athletes. Actually, if you eat too much protein, cells in your body just break it down and change it to fat. Your body does not store the protein for later use.

The amount of protein a person needs changes as the person gets older. Babies need the most protein because they are growing the fastest. The recommendation for 11–14 year olds is 1.0 gram of protein for every kg (2.2 pounds) the person weighs. **Most people your age should eat 35 to 60 grams of protein a day.** (As you know, there is a big difference between 35 and 60; but there is also a big difference in people's body sizes. A small person does not need to eat as much protein as does a large person.)

Since you do not carry a nutrient chart with you, you are not going to know how many grams of protein you are eating. As with fats, you need a guideline to help you make "protein-wise" food choices. Here is a guideline for you to follow.

EAT A VARIETY OF FOODS.

If you are already following the first guideline and avoiding too much fat, just make sure that the foods you do eat are varied. If your daily diet includes some fruits, vegetables, and whole grain breads as well as dairy products and meats, you will probably be getting enough protein.

ANALYSIS QUESTIONS

1. List 5 functions of proteins.
2. Give 5 examples of foods that are
   a. good sources of protein
   b. poor sources of protein
3. What is meant by the term **complete protein?**
4. Why do vegetarians have to watch the proteins they eat?
5. Think about the protein in your diet.
   a. How many grams of protein should most people your age eat?
   b. About how many grams of protein should *you* eat each day?
   c. What guideline can you follow to make sure you eat enough protein?

6. A woman needs to eat more protein when she is pregnant than when she is not. Why?
7. Here is a lunch menu.

2 slices whole wheat bread	1 apple
1 tablespoon peanut butter	3 cookies
2 tablespoons currant jelly	1 cup whole milk

   a. How many grams of protein are in this lunch? (Use the information in the Nutrients in Foods chart starting on page 630.) Show your work.
   b. If you ate this lunch, how many grams of protein would you need to eat in your other two meals?
   c. What is one way you could make this lunch lower in fat?

EXTENSION ACTIVITY

Learn more about vegetarian diets. Find a vegetarian cookbook at home or in the library. Copy the ingredients for two main dishes. Put an asterisk (*) by each food that provides protein. Would the combinations make complete proteins? If you feel ambitious, prepare one recipe and serve it to your family.

Figure 10–21 Meatless dishes can be nutritious and add variety to your diet.

10-8 CARBOHYDRATE TESTS

Carbohydrates are delicious. They taste good raw, boiled, baked, or fried. Eat them for breakfast, lunch, dinner, or as a snack. What are carbohydrates? They are the starch and sugar foods.

- What foods contain carbohydrates?
- How can you test for starches?
- How can you test for sugars?

You will learn the answers to these questions by doing starch and sugar tests on several foods.

MATERIALS

cornstarch
corn syrup
7 test tubes
paper towels
aluminum foil
test tube rack
iodine in dropper bottle
masking tape
samples of 5 different foods
data page: ***Carbohydrate Tests*** (Figure 10–22)
summary table: ***The Major Nutrients***
Sugar Test Station
- hot plate
- beaker
- safety goggles
- test tube holder
- Benedict's solution in dropper bottle
- water in dropper bottle

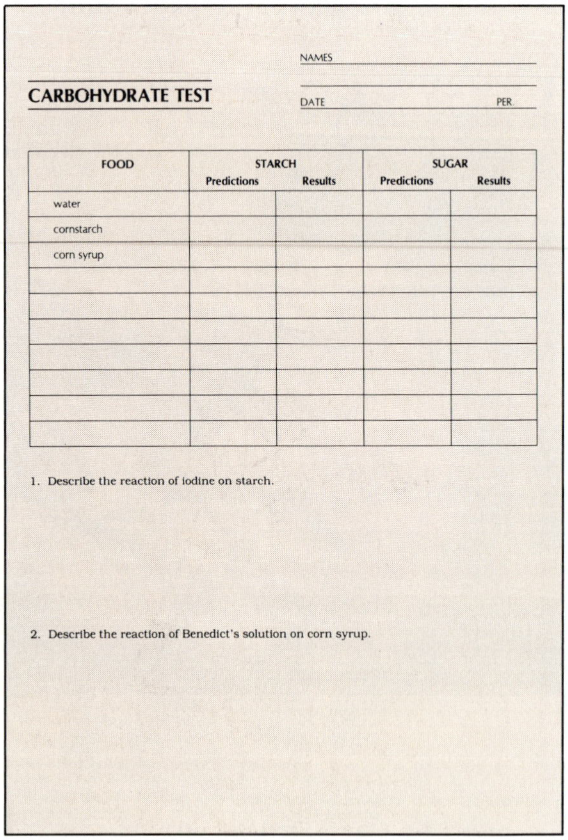

CARBOHYDRATE TEST

NAMES

DATE PER.

FOOD	STARCH		SUGAR	
	Predictions	Results	Predictions	Results
water				
cornstarch				
corn syrup				

1. Describe the reaction of iodine on starch.

2. Describe the reaction of Benedict's solution on corn syrup.

Figure 10–22

PROCEDURE

Work with a partner.

Part A. Getting Ready

1. You and your partner will test 5 foods to see if they contain sugar or starch. Record the foods your teacher assigns to you on the data table.
2. Before doing the tests, discuss the foods with your partner. Decide if

you think each food contains starch, sugar, both of these, or neither of these. When you reach an agreement, record that prediction on the data table. Use a plus (+) if you think the food contains the substance given at the top of the column. Use a minus (−) if you think it does not.

3. Your teacher will tell you which test to do first.

Part B. Testing for Starch

1. Put 3 drops of water on the foil. Put the same amount of cornstarch on a different place on the foil.
2. Add 3 or 4 drops of iodine to the water and 3 or 4 drops to the cornstarch. Answer question 1 on the data page. (See Figure 10–23.)

CAUTION
Iodine can stain your clothes. Use it carefully.

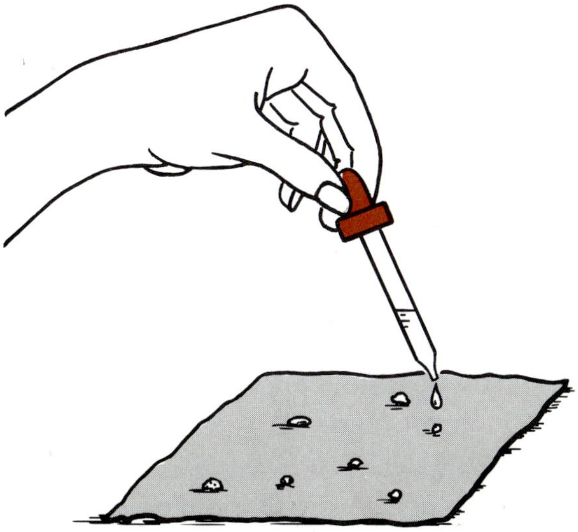

Figure 10–23

Notice what happens when cornstarch is mixed with iodine. Any food that contains starch will react in the same way.

3. Place a small amount of each of the foods you will be testing on the foil. Test each food for starch. Record your results on the data table. Use a plus (+) if a food contains starch. Use a minus (−) if it does not.
4. If the Sugar Test stations are all in use, answer analysis questions 1–3.

Part C. Testing for Sugar

(Do this at the Sugar Test station.)

1. Put 9 or 10 drops of water in one test tube. Put the same amount of corn syrup in another test tube.
2. Add 10 drops of Benedict's solution to each of the two test tubes. Gently swirl each test tube to mix the solutions.

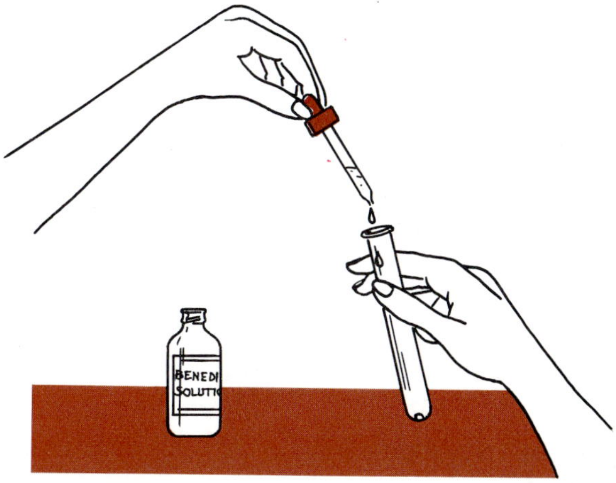

Figure 10–24

3. **Put on your safety goggles.** Heat the test tubes for about 3 minutes in a beaker of hot water as shown in Figure 10–25 on page 262. Answer question 2 on the data page.

Figure 10–25

Notice what happens when corn syrup is heated with Benedict's solution. Most foods that contain sugar will give a similar reaction.

4. Prepare to test your five foods for sugar.
   a. Label each test tube with the name of the food you will place in it.
   b. Put a small amount—about half the size of a raisin—of each food into the proper test tube.
   c. Add 9 or 10 drops of water to each one.
5. Add 10 drops of Benedict's solution to each test tube and gently swirl to mix. Heat each one as you did the other tubes.
6. Record your results on the data page. Use a double plus (++) for lots of sugar when the results are similar to the corn syrup's reaction. Use a single plus (+) when there is a slight color change and a minus (−) for a negative test (no color change).

Part D. Finishing Up

1. Wash your test tubes and clean up your work space.
2. Write the answers to the analysis questions on the back of your data page.
3. During the class discussion of this activity, complete the sugar and starch columns on your summary table.

ANALYSIS QUESTIONS

1. Look at the starch data on the data table.
   a. Which foods did you predict would contain starch?
   b. For which foods were your predictions correct?
2. List at least 6 other foods that you think contain starch.
3. Look at your data for testing water.
   a. Did water contain sugar or starch?
   b. Why did the directions say to test water for sugar and starches?
4. Look at the sugar data on the data table.
   a. Which foods did you predict would contain sugar?
   b. For which foods were your predictions correct?
5. List at least 6 other foods that you think contain sugar.
6. Tony and Pat both tested jelly for sugar. Tony got a yellow color change. Pat got a red color change. List 3 possible explanations for their different results.
7. Look at your data table.
   a. Which foods did not contain starch or sugar?

b. If they are not sugar or starch, what could they be made of?

8. Where do foods that contain carbohydrates come from—plants, animals, both, neither? Explain your answer.

CONCLUSION

Look at the three focusing questions on page 260. Answer these questions in a paragraph that summarizes what you learned doing these tests.

EXTENSION ACTIVITIES

1. There are lots of different sugars and starches. Read package labels from 10 different starchy foods or desserts. Make 2 columns on a piece of paper. List all starches found on the labels in one column and all sugars in the other. You may have to look up some of the ingredients in a dictionary to decide what they are. Keep track of any ingredients that you cannot identify. How many different starches and sugars can you find?
2. There are several kinds of sugars. Some of these can be detected with Benedict's solution, while others cannot. Test several different kinds of sugars. Read the ingredients on the package labels to help you find different ones. Which ones react with Benedict's? Which do not? Look up these sugars to find out why some react and some do not.

10-9 SUGAR, STARCH, AND FIBER

Mary was starved when she got home from the swim meet. She had eaten a candy bar right after lunch but that was hours ago. A bowl of pretzels "disappeared" while she helped get dinner on the table.

After eating three heaping spoonfuls of mashed potatoes, corn, thick slices of Italian bread, sweet potatoes, and four pieces of fried chicken, Mary felt better. Since she was getting full, she decided to skip the salad and finish her meal with a handful of cookies.

Even though she probably did not realize it, Mary ate a meal that was practically all carbohydrates. Pretzels, mashed potatoes, corn, bread, sweet potatoes, and cookies all contain starch. Sugars were in her candy bar and cookies. Mary's meal was so high in carbohydrates that she may not have gotten enough of the other nutrients she needs.

In order to make wise food choices, Mary needs to know more about nutrients. We all need to know some basic facts about fat, proteins, and carbohydrates. We should know, for instance, which foods contain which carbohydrates and how much of our diet should be made up of carbohydrates.

Carbohydrates in Foods

Many foods contain carbohydrates. We eat carbohydrates as a part of almost every meal. You have tested some foods and probably found that many of them contain a carbohydrate in the form of either sugar or starch.

There are several kinds of **sugars**. The sugar that probably comes to your mind is table sugar, but there are many other natural sugars, such as fruit sugar and milk sugar. When you read food labels, you will come across these other sugars. You can recognize the ingredients that are sugars because their names usually end with "ose." For example, fruit sugar is fruct*ose*, table sugar is sucr*ose*, and the sugar used by muscles for energy is gluc*ose*.

Figure 10–26a Foods that are high in sugar

Figure 10–26b Starchy foods

Figure 10–26c High fiber foods

Starchy foods make up a major part of our diets. These foods come from plant roots, underground stems, and seeds. They include potatoes, rice, and products made from grain, such as breads, rolls, and pasta.

In addition to sugars and starches, there is a third kind of carbohydrate—**fiber.** It makes up the "tough" part of the plants we eat. Since plants do not have bones for support, they need some other method of staying upright. Fiber provides this support. Foods such as raw vegetables, celery, many fruits, and whole grain breads and cereals are high in fiber.

Figure 10–26 shows foods that contain sugar, starch, and fiber.

CARBOHYDRATES IN FOODS

Food	Total Carbohydrates	Sugars	Starch and Fiber
beans, 1 cup cooked	40 g	0 g	40 g
bread, 1 slice	12	1	11
broccoli, 1 cup	7	0	7
carrots, 1 cup	12	6	6
corn flakes, 1 oz.	24	2	22
pasta, ½ cup cooked	20	0	20
potatoes, 1 cup	30	6	24
rice, ½ cup cooked	20	0	20

As you can see on this chart, many foods contain more than one kind of carbohydrate. Carrots contain 12 grams of carbohydrate—6 of these grams are natural sugar and 6 are starch and fiber.

Most of our dietary carbohydrates come from plant sources. Foods from animal sources, such as meats and eggs, contain little, if any, carbohydrate. Dairy products, however, contain a type of sugar called lactose.

A Matter of Size

You might wonder why sugars, starches, and fibers are grouped together as carbohydrates. The answer involves the size of their molecules. Sugar is the smallest carbohydrate molecule. When many sugars are hooked together, they form either a starch or fiber molecule. We can digest starch, but we cannot digest fiber. The sugar molecules in fiber are hooked together in such a way that the body cannot break them apart.

All carbohydrates are made of carbon, hydrogen, and oxygen. That's all. They do not contain nitrogen, sulfur, copper, iron, or any other kind of atom. That is where the word *carbohydrate* comes from.

carbo = carbon
hydrate = water (hydrogen and oxygen)

Figure 10–27

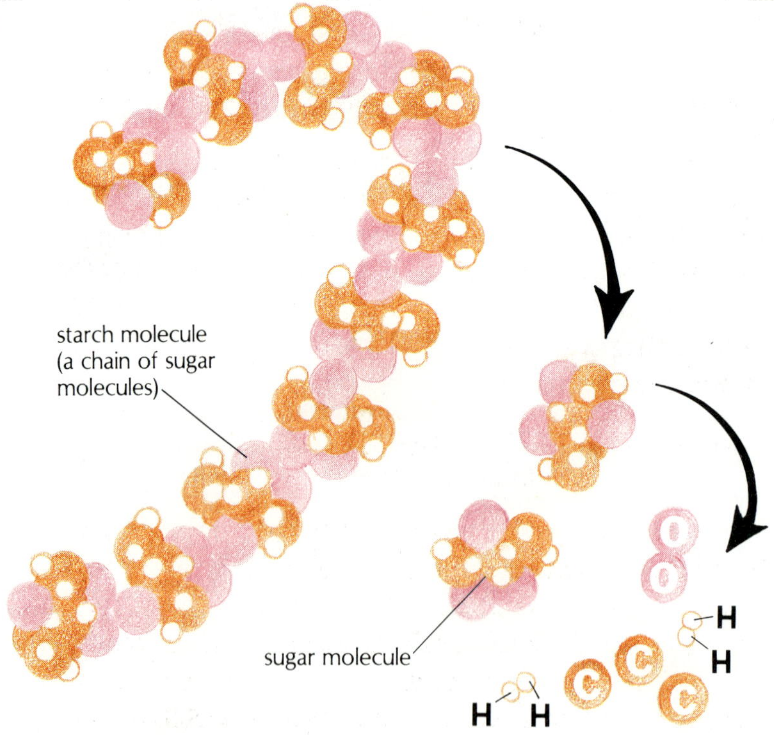

Carbohydrates in the Body

As with fats and proteins, our bodies use carbohydrates in a variety of ways, some of which are mentioned below.

Carbohydrates are an important *energy source*. Fats and protein can also be used as energy but not as easily. Cells have to "remake" fat and protein molecules before they can get the energy.

Fiber *helps foods move* through the digestive tract. People who eat enough fiber are not likely to have problems with constipation.

Foods that supply starches also *supply other nutrients* —vitamins, minerals, and some protein. Sugary foods usually are not rich in other nutrients.

Figure 10–28

Making Choices

Nutritionists recommend that at least half of our calories come from carbohydrates. Most of these carbohydrates should be in the form of starch or fiber. This means that you should be eating somewhere between 300 and 500 grams of carbohydrate each day. Instead of counting grams, it is much easier to remember these two eating guidelines.

EAT FOODS WITH ADEQUATE STARCH AND FIBER.
AVOID TOO MUCH SUGAR.

Hmmm . . . you may be thinking, "How does this help? What is an 'adequate' amount of starch? How much sugar is 'too much'?" In general, you will be eating adequate amounts of starch if you eat fruits and green vegetables each day, *and* if you include potatoes, rice, or beans with your meals. When possible, choose

Figure 10–29 By giving food selection some thought, it's usually possible to put together a nutritious and satisfying meal.

foods that provide both fiber and starch—whole grain breads instead of white bread, or a bran cereal instead of a sugary one.

As for sugar, some is okay, but "too much" sugar becomes a problem. The main problem with sugar is that it does not have any nutritional value. It is too easy to drink a soft drink and munch on a candy bar and then not be hungry for dinner. That is why it is important to eat foods that supply the necessary nutrients first and, then, if you are still hungry, to go ahead and eat something sweet. Read **Sweets** on pages 271 and 272 to learn more about the extra sugars.

At first, it may be hard for you to recognize the foods that contain carbohydrates. But, after you have checked the contents of quite a few foods in nutrient charts, you will begin to develop an "instinct" for what foods contain.

ANALYSIS QUESTIONS

1. What are the 3 types of carbohydrates?
2. Copy this list of foods onto your own paper. For each food, indicate if you think it is a good source of starch, sugar, or fiber. Remember, a food may contain more than one kind of carbohydrate.

mashed potatoes	celery	lima beans
saltine cracker	rice	sugar cookie
peppermint candy	corn	fresh pear
bran cereal	doughnut	white bread

3. What carbohydrate is made of the smallest molecule?
4. What is the difference between starch and fiber?
5. Give 3 reasons why we need to include carbohydrates in our diets.
6. Improve Mary's meal. The foods she ate are listed below. Write down the foods she could have eaten to make her meal more nutritious. You can leave out some foods, add foods, or substitute one food for another.

candy bar	pretzels	mashed potatoes
corn	Italian bread	sweet potatoes
cookies	fried chicken	

7. The nutrition information from a box of crackers is given in Figure 10–30.
   a. What is the main *nutrient* in these crackers—fat, protein, or carbohydrate?
   b. Look at the ingredients. Which ones are starches?
   c. Which ingredients are sugars?
   d. Could these crackers be labeled "high fiber"? Explain.
   e. Which fats are in these crackers?
   f. Are these fats "good" or "bad" for your health? Explain. (See pages 245 and 246 in **Fats and Oils** if you need to review kinds of fats.)

NUTRITION INFORMATION (per serving)

Serving Size	12 crackers (Approx. 1/2 oz.)
Servings per Container	15
Calories	70
Protein	1 gram
Carbohydrate	9 grams
Fat	3 grams
Polyunsaturated	1 gram
Saturated	less than 1 gram
Cholesterol	0 milligrams
Sodium	140 milligrams
Potassium	55 milligrams

PERCENTAGE OF THE U.S. RECOMMENDED DAILY ALLOWANCES (U.S. RDA)

Protein	2
Vitamin A	*
Vitamin C	*
Thiamine	2
Riboflavin	2
Niacin	2
Calcium	*
Iron	2

*Contains less than 2% of the U.S. RDA of this nutrient.

INGREDIENTS: ENRICHED WHEAT FLOUR CONTAINING NIACIN, REDUCED IRON, THIAMINE MONONITRATE (VITAMIN B_1) AND RIBOFLAVIN (VITAMIN B_2), VEGETABLE SHORTENING (PARTIALLY HYDROGENATED SOYBEAN OIL WITH TBHQ TO PRESERVE FRESHNESS), DEHYDRATED POTATOES, STEAMED CRUSHED WHEAT, SALT, SUGAR, DEFATTED WHEAT GERM, CORN SYRUP, MALT, ONION POWDER, MONOSODIUM GLUTAMATE, SOY LECITHIN, SODIUM BICARBONATE AND SPICE.

Figure 10–30

CHALLENGE

Read the ingredient lists on food package labels. Be a sugar detective! List as many kinds of sugar as you can find. Remember, the names of many sugars end with "ose."

8. Look again at the two dietary guidelines that apply to carbohydrates on page 267. Which one will be hardest for you to follow? Why?

EXTENSION ACTIVITIES

1. Bread can be a wholesome part of our diets. When you read the nutrition information on bread labels, you will see that some breads are more nutritious than others. Make a chart comparing the nutritional information for 4 different kinds of breads. Then answer these questions.

   What is an average serving size for bread?
   Which bread provided the *most* protein?
   Which was lowest in fat?
   Which bread was richest in other nutrients?
   Which bread is *most* nutritious? least?
   If you were buying bread for the school cafeteria, which one would you choose? Why?

2. Learn more about fiber. Dieticians can provide you with pamphlets about fiber. There have also been lots of magazine articles about fiber. Use the *Reader's Guide to Periodical Literature* to find several of the more recent ones. Find out which foods are high in fiber. Explain why are we supposed to include fiber in our diets. Keep a journal of the high fiber foods you eat each day for a week.

10–10 SWEETS

Americans are eating more sugary foods than ever before. Where does all of this sugar come from? Shouldn't we be cutting back on the sugar in our diets? This reading describes "extra" sugars—the syrups, table sugar, honey, and other sweets that we add to our foods.

Teaspoons of Sugar

Some people find it helpful to know the amount of sugar that is added to foods. This table shows how much sugar is needed to make some common foods.

Teaspoons of Sugar	Food (serving size)
6.0	sweetened drink (8 oz)
11.0	can of soft drink (12 oz)
0.0–3.0	cereal (1 oz)
1.5	bubble gum (1)
3.2	chocolate bar (1 oz)
4.5	popsicle (1)
2.5	pork and beans (½ cup)
1.5	cream corn (½ cup)
0.6	ketchup (1 tablespoon)

Empty Calories

One problem with sugar is that it provides "empty" calories. That is, you may get full on sugary foods without getting the protein, fiber, vitamins, calcium, and other minerals that your body needs. Bones and muscles do not grow on a diet of sugar. Blood cells need "real food" to do their jobs. Nerve cells cannot send messages on a diet of sweets. In fact, you do not need to eat any sugar. You could eliminate all sugar from your diet and still have healthy cells and a healthy body.

Tooth Decay

You know that sweets cause tooth decay. Do you know why? Bacteria love to eat sugar. They thrive on little bits of sweets that stick to your teeth. And, while they grow, bacteria produce acids. Acids ruin the surface of teeth. Eventually, the acids make a hole . . . a cavity!

Fatty Sweets

One problem with sweet foods is that they are often high in fat. Think of popular desserts—chocolate, ice cream, cakes, cheese cake—sweet rolls, and other pastries. They all are high in *fat* as well as being high in sugar. High fat diets, as you may remember, are associated with many health problems.

Sugar Rules

- Eat less of all sugars—white sugar, brown sugar, raw sugar, honey, and syrups.
- Choose snacks that are high in other nutrients.
- Brush your teeth after eating, especially after eating "sticky sweets."

10-11 OTHER NUTRIENTS

There is more to your food than fats, proteins, and carbohydrates. They also contain the other three nutrients—vitamins, minerals, and water.

Vitamins and minerals are the "helper nutrients." They are needed in very small amounts. They make it possible for us to use the other nutrients found in our foods. Each of the helper nutrients has its own role in the body. You can learn which vitamins and minerals are in a product by reading the nutrition information on its label. For instance, look at the nutrition information for instant pudding on the label in Figure 10–31. Which vitamins are listed? Which minerals?

NUTRITION INFORMATION PER SERVING

SERVING SIZE: ½ CUP SERVINGS PER PACKAGE: 6

	AS PACKAGED	PREPARED WITH WHOLE MILK
CALORIES	100	180†
PROTEIN	0	4 g
CARBOHYDRATE	25 g	30 g
FAT	0	4 g
SODIUM	460 mg	520 mg

PERCENTAGES OF U.S. RECOMMENDED DAILY ALLOWANCES (U.S. RDA)

	AS PACKAGED	PREPARED WITH WHOLE MILK
PROTEIN	*	10%
VITAMIN A	*	2%
VITAMIN C	*	*
THIAMINE	*	4%
RIBOFLAVIN	*	10%
NIACIN	*	*
CALCIUM	*	15%
IRON	*	*
PHOSPHORUS	25%	40%

†SAVE 30 CALORIES PER SERVING–USE SKIM MILK
*CONTAINS LESS THAN 2 PERCENT OF THE U.S. RDA OF THESE NUTRIENTS.

INGREDIENTS: SUGAR, CORN STARCH MODIFIED, COCOA PROCESSED WITH ALKALI, SODIUM PHOSPHATES (FOR PROPER SET), ARTIFICIAL FLAVOR, SALT, ARTIFICIAL COLOR (INCLUDING YELLOW 5), DI- AND MONO-GLYCERIDES (PREVENT FOAMING), HYDROGENATED SOYBEAN OIL WITH BHA (PRESERVATIVE), NONFAT MILK, HYDROXYLATED SOYBEAN LECITHIN (FOR UNIFORM DISPERSION OF COCOA), NATURAL FLAVOR. 170 g

GENERAL FOODS CORPORATION, WHITE PLAINS, NY 10625, U.S.A.

Figure 10–31

One nutrient that is not listed on labels is water. Even though it is not listed, it is extremely important. You could not survive for more than a few days without this nutrient. All body tissues contain large amounts of water. Water makes up about 75%, or three-fourths, of the brain and muscles, for instance. Even bone, which seems fairly dry, is 20%, or one-fifth, water. Overall, water makes up more than half of our body weight. Describing the many roles of water in the body could easily fill a book.

PROCEDURE

Here is a list of several nutrients. Choose one of them to study.

Water	Minerals	Vitamins
	iron	vitamin A
	calcium	vitamin B_1 (thiamin)
	sodium	vitamin B_2 (riboflavin)
	potassium	vitamin B_3 (niacin)
		vitamin C (ascorbic acid)

Write a report or make a poster that tells about the nutrient you chose. Answer these questions in your report or poster.

1. What foods are good sources of this nutrient?
2. Why do we need this nutrient? What does it do in the body?
3. What is the RDA (recommended daily allowance) for this nutrient?
4. What can happen if someone does not get enough of this nutrient or if they get too much?

EXTENSION ACTIVITY

Use your food diary to analyze your own diet for the nutrient you chose. Answer these questions.

- How much of the nutrient will you get at each meal? How much from your snacks?
- How does the amount you ate compare with the recommended amount?
- Should you have eaten more or less of this nutrient?

List at least 5 of your favorite foods that are good sources of this nutrient.

VITAMINS: Answers to Common Questions

Do I need to take vitamins?

Probably not. Very few people need to take vitamins. If you are healthy and eat a balanced diet, you do not need extra vitamins. There are a few exceptions. Some people have health problems and need vitamin supplements. Pregnant women and the elderly may also need to take vitamins.

I take a vitamin every morning. Is that bad?

No. If you are taking a multiple vitamin, it will not hurt you. However, it probably is not doing you any good either. Taking vitamins can be bad when people take large doses of one kind of vitamin or mineral.

Why would a person take lots of one kind of vitamin?

People take huge doses of one particular vitamin or mineral for a variety of reasons. They may have heard that it will make them "feel terrific," prevent cancer, turn their eyes green, or some other myth. Only people who have a few rare diseases can use these large doses.

What happens if you take large doses of a vitamin?

The body gets rid of many of the vitamins it cannot use. You flush extra vitamins down the toilet every day. Some vitamins, however, are not as easily gotten rid of because they are stored in body fat. Thus, they become dangerous. High doses of these fat soluble vitamins will make a person sick.

A final word

Your doctor will tell you if you should be taking something. Girls may want to ask about iron and calcium. Anyone who is not eating a balanced diet should discuss the possible need for vitamins with his or her doctor.

10–12 FAST FOOD CHOICES

A survey by the National Restaurant Association found that six out of ten Americans buy fast food more than twice a month. Therefore, you have probably eaten in a fast food restaurant at least once in the last month. How nutritious are the meals you choose when you eat at a fast food restaurant?

- Which nutrients are well supplied by a fast food meal?
- Which nutrients are undersupplied?
- What could you eat, or not eat, to make a fast food meal more nutritious?

Pick a fast food meal and analyze its nutritional value to answer these questions.

MATERIALS

fast food restaurant menu
fast food restaurant nutrition information
data page: ***Fast Food Choices*** (Figure 10–32)
colored pencils or crayons (orange, green, blue, and purple)

PROCEDURE

Work alone or with a partner.

Part A. Making Choices

1. Decide which fast food restaurant you would like to visit. Get the menu for your restaurant.
2. Pick a well-balanced, nutritious meal from the menu. Keep in mind what you have learned about protein, fat, carbohydrates, and other nutrients and remember the Dietary Guidelines. *And*, order a reasonable meal; you must "eat" everything you "order."
3. Write your food choices on the data table. Remember to include drinks!
4. When you have finished, turn in your menu.

FAST FOOD CHOICES

NAMES ____________

DATE ____________ PER. ____

Restaurant: ____________

Fast Food Item	MAIN NUTRIENTS Protein (grams)	Carbohydrates (grams)	Fats (grams)	Calories
TOTALS				

DAILY AMOUNT NEEDED				

Figure 10–32

Figure 10–33 If you would like nutrition information on your favorite fast foods, just ask the next time you order. Fast food restaurants have the information available for customers who would like it.

Part B. Meal Nutrients

1. Check out the nutrition information given by the restaurant you chose.
2. Look up each food item that you listed on your data table. Fill in the nutrition information for each food.
3. Total the number of grams of fat, protein, and carbohydrate. Then, add up the number of calories you would get with this meal.
4. Look at the chart. It summarizes the daily amount of each nutrient needed by most eleven- to fourteen-year-old students.

	Daily Amount Needed
protein	60–100 grams
carbohydrate	300–500 grams
fat	65–110 grams
calories	2000–3500

a. Decide about how much of these items you need. If you are small, you probably need the smaller amount. If you are in the middle of a growth spurt or are tall, you will need a larger amount.

b. Record the values *you* need at the bottom of your data page.

Examples

Mark loves to read. He spends all of his spare time reading mysteries. He is smaller than most of the other kids in his class.

Kendra is always on the go. She loves to be actively doing something—riding bikes, roller skating, swimming. She is not the tallest girl in her class but she has been growing steadily for the last year.

Here are the nutritional needs Mark and Kendra might choose for themselves.

DAILY AMOUNT NEEDED

	Mark	Kendra
protein	60 grams	90 grams
carbohydrate	330 grams	430 grams
fat	65 grams	100 grams
calories	2,100	3,000

Part C. Analyzing Your Meal

1. Imagine that you are going to eat the meal you selected.
   a. How much protein would you get?
   b. How much fat?
   c. How much carbohydrate?
2. Now compare the amounts of the nutrients you would get with the amounts your body needs.
   a. How much more protein will you need?
   b. How much more fat can you have?
   c. How much more carbohydrate will you need?
3. Use your green colored pencil to circle the foods that contain five or more grams of protein.
4. Look at the carbohydrates. Decide if the carbohydrates are fiber or sugars.
   a. Put a purple **SU** in the column by carbohydrates to show which ones contain sugar.
   b. Put a blue **FI** by the foods that you think provide fiber.
5. Look at the fats in your meal.
   a. If the nutrition information gives cholesterol amounts, use your orange colored pencil to write these amounts in the column by fats.
   b. If the information does not give cholesterol amounts, put an orange check mark in the column by fats to show the food items that you think contain cholesterol.
6. If you want to improve the nutritional value of this meal, how would you change your food choices?

CONCLUSION

Write a paragraph that answers the three focusing questions on page 276.

EXTENSION ACTIVITIES

1. Do a more complete analysis of your fast food choices by considering vitamins and minerals.
   How many milligrams (mg) of sodium are in this meal?
   Which item in your meal is highest in sodium?
   What is the total percentage for the vitamins and minerals?
   Did you get 100% of the RDA for any of the vitamins or minerals? If so, which ones?
   Which vitamins and minerals are not well supplied by this meal?
2. Use a computer to analyze your diet. There are several programs that will help you check the nutrients in your meals. MECC Elem. #13 disk will help you analyze many parts of your diet. Ask your teacher or the library media specialist if that program, or a similar one, is available for you to use.
3. Another way to analyze a meal is by thinking of the basic four food groups. Make a chart like the one shown below. List each of the food items from your Fast Food Choices on this chart to show in which group each belongs. Check your "balance." Do you have foods from each group? Ask the home economics teacher to review your analysis.

Food Item	Milk, Cheese	Meats	Fruit, Vegs.	Grains	Other
hot dog		✓			

"You Eat Like a Bird"

If you eat like a bird, you would soon weigh 200 or 300 pounds. Birds are voracious eaters. Birds' bodies "work faster" than ours do. The smaller the birds, the faster their bodies work. The tiny hummingbird has to eat the equivalent of its body weight each day to survive. (If you weigh 87 pounds, you would have to eat 87 pounds of food to eat like a hummingbird!)

CHAPTER 11

DIGESTION

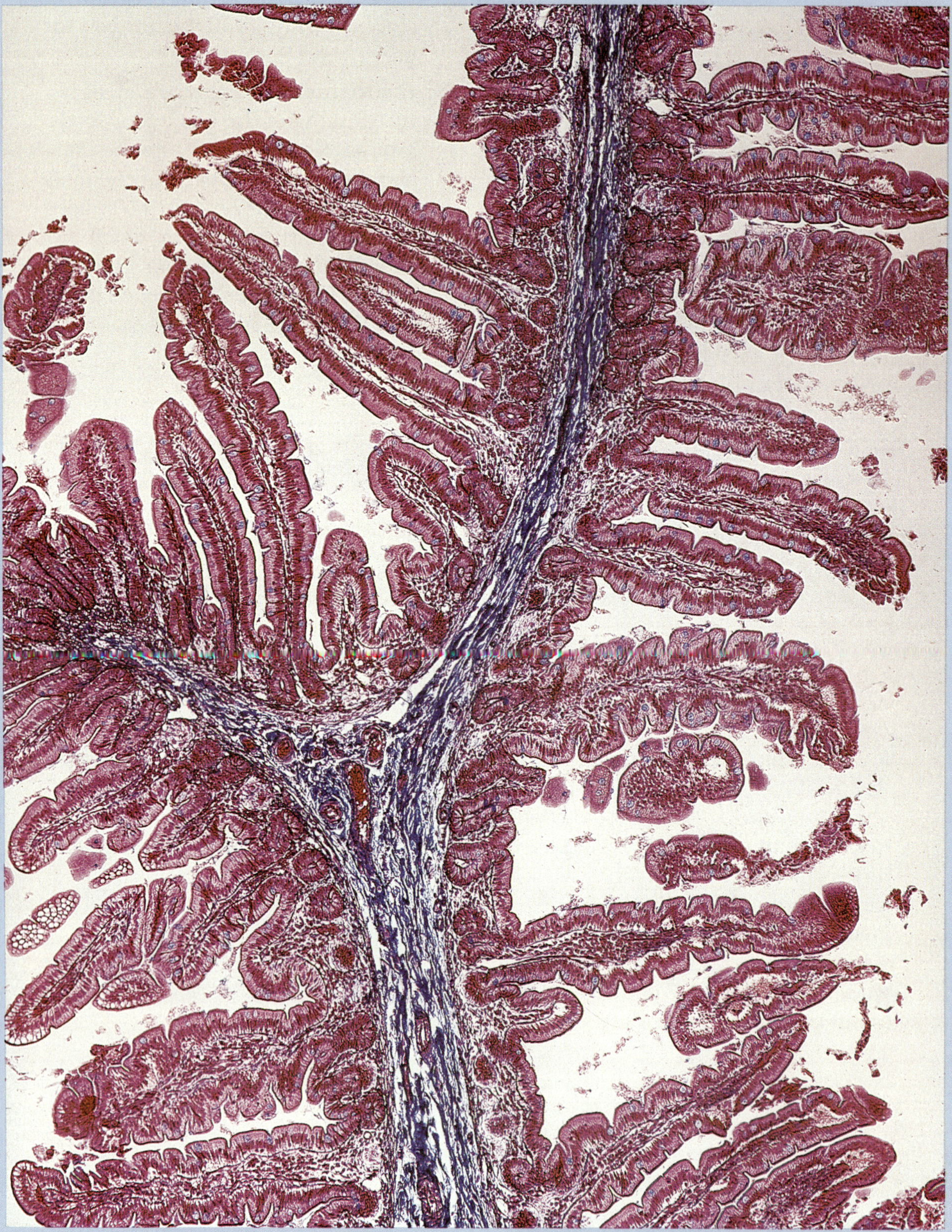

What part of the digestive systems is this?

11-1 THE MOUTH

Food cannot be used by the body unless it can get into the cells. Nutrients on the *outside* of a cell cannot be used for growth, energy, or any other cell function. Somehow, the food you eat has to be broken down into food you can use. This process is called digestion.

Digestion starts in the mouth and continues throughout the 30-foot-long digestive system. Let's start at the beginning of this system. Take a few minutes to think about your own mouth.

- How many mouth parts can you name and describe?

Use a mirror to inspect your own mouth and then reconsider your answer to this question.

MATERIALS

mirror
reading: ***Mouth Parts*** (page 282)

PROCEDURE

Part A. Mouth Inspection

1. Use a mirror to inspect your mouth. Keep your tongue *down*. Sketch what you see. Do not touch the mirror to any part of your mouth.
2. Now lift your tongue and make a second sketch. (You may want to color your sketches.)
3. Label as many mouth parts as you can on your two sketches. Try to label at least 8 things.
4. Look at some other people's sketches. Did they label anything that you did not put on your sketch? If so, add these structures to your drawings.

Part B. A Closer Look

Read ***Mouth Parts*** starting on page 282. While you read, use the mirror to find the parts you missed. Add these structures to your sketches and label them.

CONCLUSION

Describe one thing you learned from this activity.

Mouth Parts

What do you see when you inspect the inside of your mouth? Obviously, you see your teeth and tongue, but you can also see quite a few other structures. Almost every one of these is important in digestion.

The **tongue** is the most obvious mouth structure, probably because it takes up so much room and it moves. The tongue is almost all muscle. Try to make your tongue muscle contract and relax. Now swallow and notice what your tongue does. Can you swallow without moving your tongue?

Stick out your tongue and use a mirror to inspect its surface. You may see a whitish or grayish coating. This is normal. It is a combination of food and bacteria. This coating can be cleaned off by brushing your tongue when you brush your teeth. The tongue itself is not smooth; it is covered with thousands of tiny projections. Your **taste buds** are clustered in between these projections.

If you look underneath your tongue, you will see a membranous tissue that attaches your tongue to the bottom of your mouth. (See Figure 11–1.) The openings for one pair of **salivary glands** are located in this tissue but you probably cannot see them.

Another important pair of salivary glands empties next to your upper teeth towards the back of your mouth. Many small salivary glands are scattered throughout the mouth. They continuously secrete **saliva** to keep the mouth moist. A dry mouth would be miserable! Saliva also contains enzymes that help digest food. (You will be learning more about this in the next activity.)

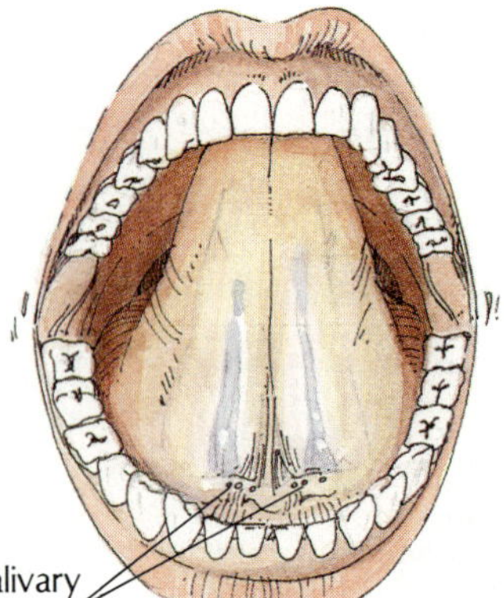

Figure 11–1 The mouth, as viewed with the tongue up

The insides of your **cheeks** form the walls of your mouth. Like your tongue, your cheeks are very muscular. Do your cheek muscles work when you chew or when you swallow? Try to look at the inside of one of your cheeks. Is it smooth or rough? pink or red? Do you see any blood vessels?

Feel your **palate**—the roof of your mouth—with your tongue or with your finger. The front of the palate feels hard because there is bone under the tissue. The back of the palate does not contain bone, so it feels soft. The soft palate ends in a flaplike structure called the **uvula.** Uvulas come in a variety of shapes and sizes. Some people are born without uvulas. Look for yours. Do you have one? If so, how would you describe it? Can you move it?

While you were looking for your uvula, you may have noticed your **tonsils.** They are the two bulges in the very back of your mouth. You can see where they are located in Figure 11–2. If your tonsils have been removed, someone else may let you look at theirs. Tonsils should be the same color as the rest of the mouth. If you see white spots on them, you might have an infection. (Tonsils are not part of the digestive system but this is a good time to look at them anyway.)

Next, take a look at your **teeth.** Hopefully, you brushed them this morning. When all of your teeth are in, you will have 32—16 in your upper jaw and 16 in your lower. How many teeth do you have now? Notice their different shapes and sizes. How do your front teeth compare with the molars in the back of your jaw? Figure 11–3 will help you locate the different kinds of teeth. When you chew, your teeth break down food into smaller pieces so it can be more easily digested. People do not eat well when their teeth hurt. Healthy teeth are important for a healthy body.

Your gums fit tightly around the teeth. Healthy gums are pale pink or brown and firm. If you have a gum infection, they become red, soft, shiny, and somewhat swollen. It is important to clean both your teeth and your gums in order to have a healthy mouth.

That completes the tour of your mouth. Inspect it again every now and then. If you see any sores that do not go away, ask your dentist or doctor about them. Keep your mouth healthy!

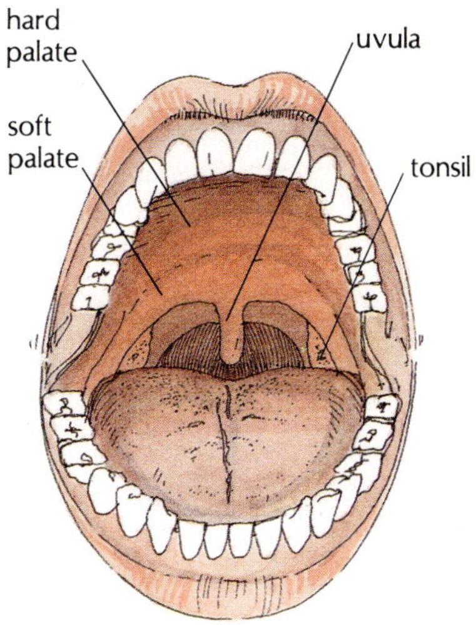

Figure 11–2 Did you see your tonsils when you looked in your mouth?

Figure 11–3 Different kinds of teeth have different functions. How does the shape of a tooth provide clues about its function?

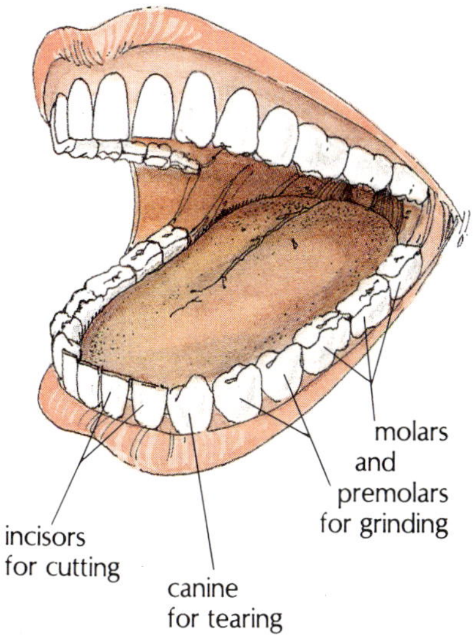

In Conclusion

Think about how the structures in your mouth might be involved in digestion. On the back of one of the sketches you made, list the following structures and explain how they function in digestion; tongue, salivary glands, cheeks, palate, teeth.

HOMEWORK

Choose one of the extension activities to do at home.

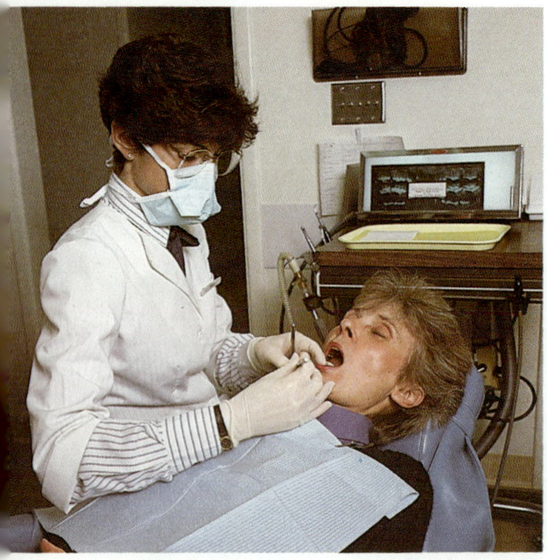

Figure 11–4 The dental hygienist can help you learn how to take care of your teeth and mouth.

EXTENSION ACTIVITIES

1. Interview someone who provides mouth care, such as a dentist, orthodontist, or dental hygienist. Write down at least 6 questions you want to ask. Include a question about the latest advancements in their field.
2. Many people have experienced cold sores. They are swollen, reddish areas that appear around the mouth. Find out more about them. Here are some questions for you to consider:
   What causes cold sores?
   Are they catching?
   What are the symptoms?
   How are cold sores treated?
   What is the difference between a *cold sore* and a *canker sore?*
3. Imagine that you have been asked to talk to a group of fourth graders about teeth. Do some research. Be sure to find out about types of tooth and gum diseases. Write out your talk. Explain what people should do to keep their teeth and gums healthy. Tell your audience what the biggest tooth problems are for people their age.
4. Write up a consumer report about toothpastes and dental flosses. Compare at least 5 brands of each. Read the labels, compare ingredients, and check prices. Summarize your data in table form. Make a recommendation if you think one product is better than the others.
5. There are lots of interesting lab activities that you can do to learn about your taste buds. Look in other science books until you find a taste bud lab that looks interesting to you. Do the lab and summarize your results.
6. It is possible to learn about an animal just by looking at its teeth. Do a research project about animals and their teeth. Here are some questions you could try to answer:
   Do grasshoppers have teeth?
   How do cow and wolf teeth compare?
   Are tusks really teeth?
   Can sharks bite?
   Since birds do not have teeth, how can they "chew" their food?

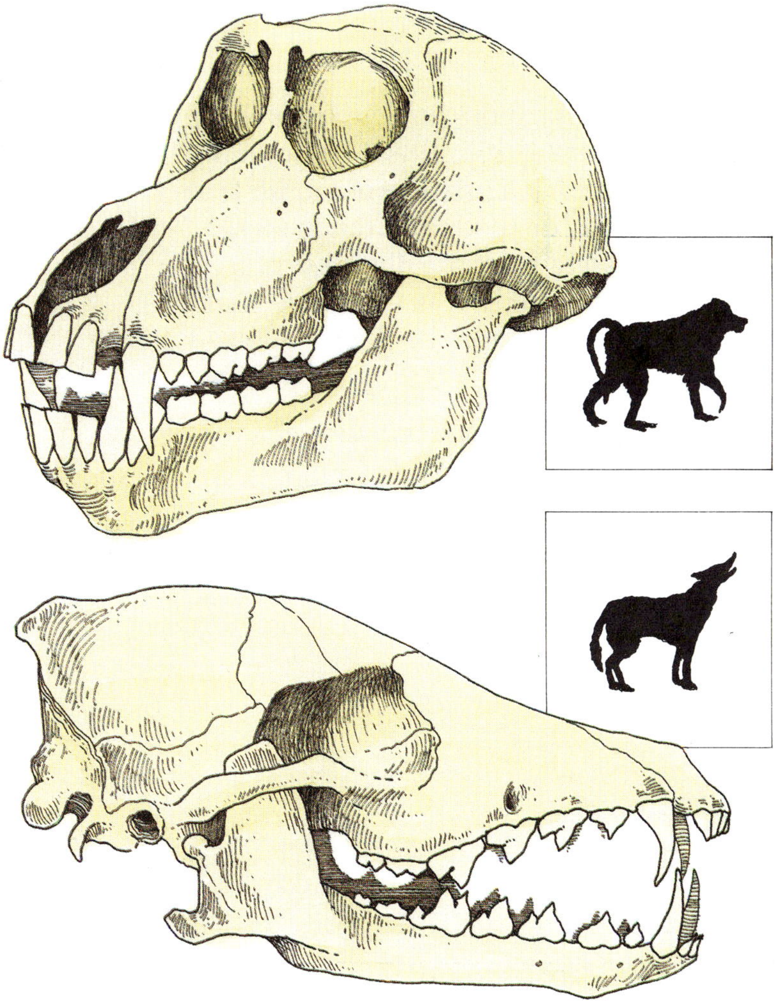

Figure 11–5 Skulls of a baboon and a coyote

You may also want to pick one of the two pictured in Figure 11–5 to research.

7. Many babies are born with a cleft lip or a cleft palate. Find out what causes these birth defects and how they are treated. You might want to pretend you are a baby who has one of these conditions. Write about your feelings, whether you are in pain, and if you have any problems eating. Describe possible surgeries for your condition.
8. Is there anything about the mouth or teeth that is of special interest to you? Talk to your teacher about making it into a special project.

11–2 SALIVA ACTION

During digestion, the three major nutrients have to be broken down so they can be used by the cells. Starch will be broken down into sugars, protein will become a mixture of amino acids, and fats will turn into fatty acids. The teeth and tongue tear, mangle, grind, and crush the food to start the digestive process. This action does not affect any molecules. It is the saliva that starts the chemical reactions which will break down the big molecules.

- What does saliva do to protein, fat, and carbohydrate?

In this activity, you will observe the effect of saliva on protein, fat, and carbohydrate in order to answer this question.

MATERIALS

6 test tubes
test tube rack
masking tape
starch (cornstarch)
iodine in dropper bottle
test tube brush
soap
bleach solution
Sugar Test station
  Benedict's solution in dropper bottle
  hot plate
  beaker
  test tube holder
  safety goggles

PROCEDURE

Work with a partner.

Part A. Getting Ready

1. Get 6 test tubes, a test tube rack, and masking tape. Label the test tubes as shown below. (See also Figure 11–6.)
   a. Label 2 test tubes *saliva.*
   b. Label 2 test tubes *starch.*
   c. Label the other 2 test tubes *starch and saliva.*

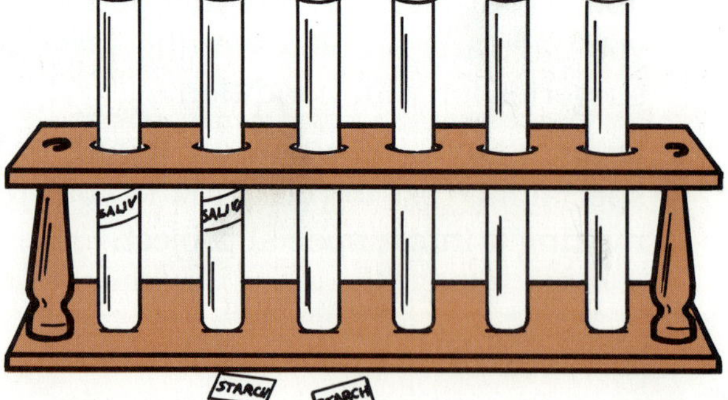

Figure 11–6

2. Rinse your mouth with water. Then collect about 3 ml of saliva in each of the 4 test tubes labeled *saliva.*
3. Place a pinch of cornstarch into each of the 4 test tubes labeled *starch.*
4. Gently swirl each tube to mix its contents.
5. Copy this data table onto a paper.

	Results of Starch Test	Results of Sugar Test
saliva		
starch		
saliva and starch		

6. Answer analysis questions 1 and 2.

Part B. Testing for Starch

1. Add a few drops of iodine to the test tubes marked *starch.* (A blue-black color change indicates that starch is present.) Record your results on the data table.
2. Test contents of the other two test tubes—saliva and starch/saliva—in the same way. Record your results.

Part C. Testing for Sugar

1. Use the remaining 3 test tubes to test for sugar.
2. Go to the Sugar Test station.
   a. **Put on your safety goggles.**
   b. Add 10 drops of Benedict's solution to each tube.
   c. Heat the test tubes in the water bath.
   d. Record your results. (See the Carbohydrate Test lab on page 260 if you need to review the procedure for this test.)
3. Clean your test tubes with soap and water, then rinse them in the bleach solution. Be careful with the bleach, it can ruin your clothes. Put your equipment away.
4. Answer analysis question 3.

Part D. Effect of Saliva on Fat and Protein

1. Pay close attention while your teacher explains the demonstration.
2. Observe the test tubes closely and then answer the rest of the analysis questions.

ANALYSIS QUESTIONS

1. What happens to each of the following substances during digestion?
   a. fat molecules
   b. protein molecules
   c. starch molecules
2. Think about what happened when you collected saliva.
   a. List 3 words to describe saliva.
   b. What do you think is the main "ingredient" in saliva?
3. Look at your results for the starch and sugar tests.
   a. In which test tubes did you get positive tests for starch?
   b. In which test tubes did you get positive tests for sugar?
   c. How can you explain these results?
4. What is the effect of saliva
   a. on fat?
   b. on protein?

5. Shawn forgot to do the sugar test on the cornstarch mixture. Why will it be hard for Shawn to tell if saliva digests starch?
6. Stephanie had eaten a candy bar right before she walked into science class. She forgot to rinse out her mouth before she collected her saliva.
   a. What will happen when she tests her saliva for starch?
   b. What will happen when she tests the starch mixed with saliva for starch?
   c. What will happen when she tests her saliva for sugar?
   d. What will happen when she tests the starch and her saliva for sugar?
   e. What will she be able to conclude from this experiment?
7. Which of these food are at least partly digested by saliva?

potato	cookie	cracker
tuna	hot dog	apple
celery	lettuce	butter

8. If you leave the test tubes out for two days, what do you think will happen to the
   a. fat in the saliva?
   b. protein in the saliva?
   c. starch in the saliva?

CONCLUSION

Write a paragraph that summarizes what you learned by doing this activity. Make sure that your paragraph includes an answer to the question on page 286.

EXTENSION ACTIVITIES

1. Test the action of saliva on the foods listed in analysis question 7. Which food is digested the most by saliva?
2. Choose some foods that you do not think are digested in saliva. Put each one in a small jar with saliva for a day or two. Observe any changes. How will you be able to tell if digestion has taken place?

11-3 DIGESTION BEGINS

The mouth is the entrance into the digestive system. Once food passes between the lips, it will begin a series of changes that will turn it into food for cells.

Mash and Mix

Bite, chew, and chomp. Digestion starts in the mouth where the teeth tear and grind large food particles into smaller ones. This also softens and moistens the chewed food. The lips and tongue are very muscular and keep moving the food to position it for chewing. The cheeks and lips keep the food from spilling back out of the mouth. Given enough time, chewing can turn food into fairly small pieces.

But, no matter how long you chew your food, you cannot break it down into small enough pieces for your body to use it. In order to be used, food has to be inside a cell. Even a tiny piece of food is not small enough to get through a cell membrane.

However, very small molecules can get into cells. In order to get such small molecules, chemical changes have to take place in the food. These start as soon as saliva mixes with the food.

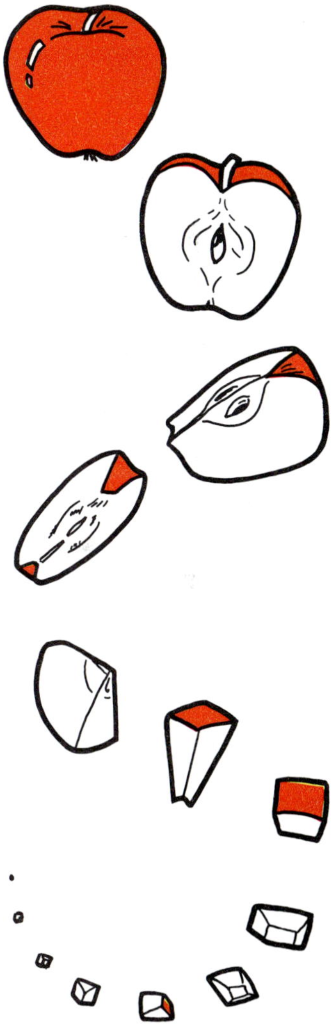

Figure 11-7 How small are the pieces of food when you swallow?

Saliva Action

Saliva contains enzymes that can break down starch into small sugar molecules. Starch molecules are too large to enter a cell but sugar molecules are small enough to go through cell membranes. Figure 11–8 on page 290 shows the relationship between a starch molecule, sugar molecules, and salivary enzymes.

There is not enough time for starch to be completely broken down when it is in the mouth. It will not be completely digested until farther along in the digestive tract.

Down the Tube

When you swallow, your tongue forces the chewed food to the back of your mouth. Muscles contract to pull the soft palate and uvula upward. This action blocks the

Figure 11–8 Enzymes in saliva can break the long starch chains into shorter lengths. Given enough time, the enzymes can completely break down starch into sugar.

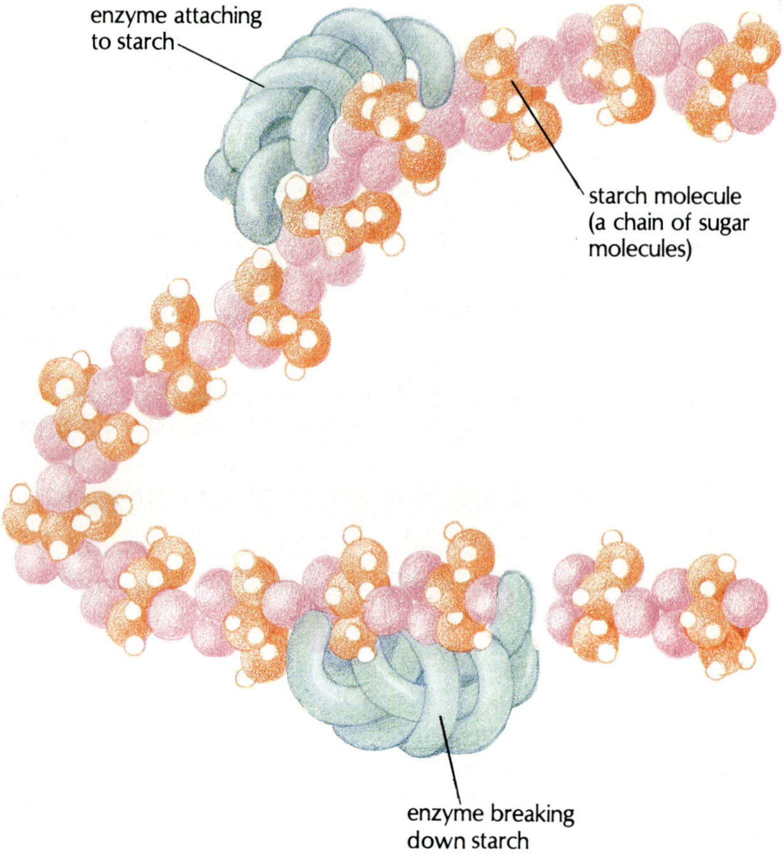

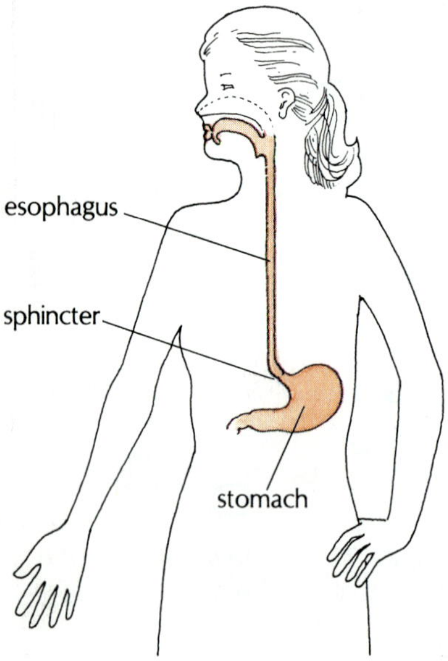

Figure 11–9 How long is your esophagus? Measure from the top of your throat to the bottom of your sternum to get an idea of its length.

opening to your nasal passages. (Food in the nose would not be pleasant.) A flap of tissue closes over the windpipe so nothing gets into your lungs. The food itself is moved into the **esophagus.**

The esophagus is a tube that extends from the back of your mouth to your stomach. (See Figure 11–9.) The walls of the esophagus are made up mainly of smooth muscles. These muscles circle the esophagus. As they contract and relax, they "squeeze" food down to the stomach. This keeps food moving in the right direction—toward the stomach. It even makes it possible for you to swallow if you are lying down or standing on your head.

Food spends only about ten seconds traveling through the esophagus. At the bottom of the esophagus is a muscle ring called a **sphincter.** The sphincter is normally closed. It keeps food from moving back into the esophagus once the food gets into the stomach.

When food passes through the sphincter, it is in the **stomach.** It is ready for the next step in digestion.

Enzymes

Chemical changes are always going on in your cells. Some molecules are being broken down while others are being assembled. These changes do not happen by themselves. A special group of proteins controls these reactions. These proteins are **enzymes.**

Each enzyme controls one chemical change. For instance, some enzymes work in the digestive system to help break down food. The enzymes in saliva can break down starch but they cannot break down fat or protein. There are other enzymes that digest fat; still others that digest protein. Which nutrients can be digested depends on which enzymes are present.

Not all enzymes are involved in digestion. Some enzymes make new cell membranes. Others control cell division. In fact, enzymes control every single body process.

ANALYSIS QUESTIONS

1. Digestion begins in the mouth.
   a. Describe how the appearance of food is changed in the mouth.
   b. What chemical change takes place in the mouth?
2. Why doesn't food get up into your nose when you swallow?
3. In your own words, explain what an enzyme is.
4. When you eat, how long do you think it takes food to get from your fork to your stomach?
5. Here is a lunch menu.
   ham on rye bread
   orange
   potato chips
   milk
   cookies
   a. Which of these foods will undergo some changes in form when they are in the mouth?
   b. Which ones will be chemically changed in the mouth?

11-4 ST. MARTIN'S STOMACH

Scientific discoveries come in many forms. We usually think of scientists as working in sterile laboratories, doing their investigations under carefully controlled conditions. This isn't always the case. Some very important experiments have been conducted under very unusual circumstances. For example, our understanding of the workings of the stomach was increased by a unique set of events.

To understand what happened, you need to transport yourself back to the early part of the nineteenth century, to an outpost near the present border between the United States and Canada. . . .

June 6, 1822

Fort Mackinac in the northern Michigan territory was a bustling trading post for the American Fur Company. Trappers returning from the wilderness brought pelts of fox and beaver that they had trapped during the winter. Besides receiving silver coins for some pelts,

Figure 11-10 Fur traders watching for otter

they traded others for Indian buckskins, liquor, tobacco, coffee, and other foods. Indians brought squash, wild rice, and game to exchange for American tobacco and wool blankets from England. A few American soldiers mingled with the crowd.

On the morning of June 6, 1822, an 18-year-old French Canadian voyageur*, Alexis St. Martin, was standing in the company's store. Suddenly a shotgun went off. St. Martin, who was standing only two feet from the gun when it accidentally fired, fell to the floor. Blood was pouring from his side. William Beaumont, a U.S. Army surgeon, made his way through the crowd to examine the wounded man.

*voyageur: a man who worked for a fur company transporting goods and men to and from remote areas

The shot had entered St. Martin's left side. His ribs were broken, a large piece of skin and muscle had been blown away, and his diaphragm was badly cut. Burned and bleeding parts of lung and stomach were sticking out through openings in the chest and abdomen. His stomach wall had been blown open and partly digested food was flowing out.

Dr. Beaumont removed some of the shot along with pieces of clothing that had entered the wound. St. Martin's injury was so severe that Dr. Beaumont did not expect him to live more than a day or two. However, St. Martin was young and healthy and, under Dr. Beaumont's care, he survived and continued to improve. Beaumont moved his patient into his own home, where he treated him. Day and night, Dr. Beaumont fed his patient, dressed his wounds, and provided constant care.

A Golden Opportunity

After three years, St. Martin was able to walk and feed himself. The wound had healed, but not the way Dr. Beaumont had hoped. The edge of the hole into the stomach had fused with the edge of the hole through the abdominal muscles and skin. This left a three-fourth inch wide opening in St. Martin's left side that led directly into his stomach!

Dr. Beaumont realized that he had a wonderful opportunity to study the way the stomach functions. He began observing and experimenting on St. Martin's exposed stomach in 1825 and continued to do so on and

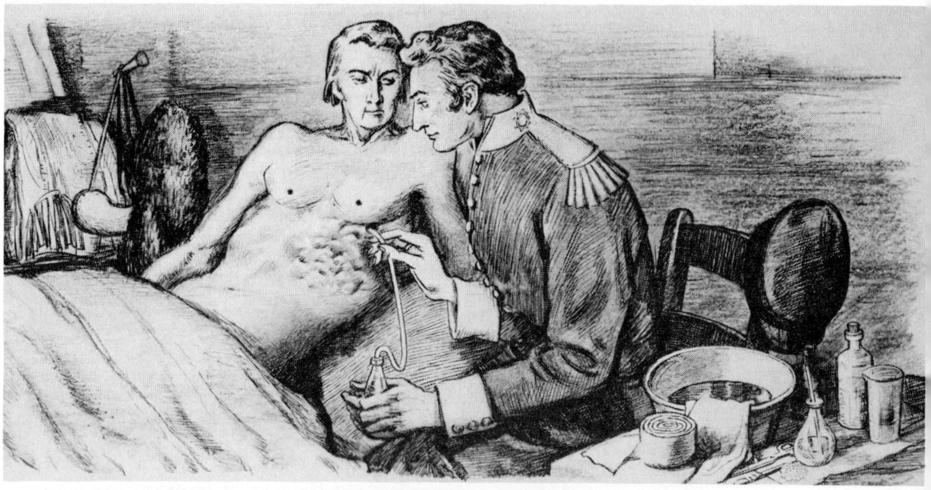

Figure 11–11 Dr. Beaumont's experiments on St. Martin provided considerable information about the function of the stomach in digestion.

off for eight years. (Figure 11–11 shows Dr. Beaumont in the midst of one of his experiments.)

Alexis St. Martin was not always a willing subject. He did not like being studied; he was often annoyed with the discomfort and inconvenience of the experiments. After the first set of experiments in 1825, he returned home. During his four-year stay in Canada, he worked as a voyageur for the Hudson Bay Company, married, and fathered two children.

Dr. Beaumont begged him to come back, offering money, appealing to the fact that he would be contributing to medical science, and reminding St. Martin that he, Beaumont, had saved his life. In 1829, St. Martin returned with his family for another series of experiments. Dr. Beaumont continued his experiments for the next four years. Afterwards, St. Martin returned to Canada where he lived until his death at the age of 83.

PROCEDURE

1. One of Beaumont's experiments on St. Martin is described on page 295. The description is based on Dr. Beaumont's records as originally published in 1833. Therefore, it is written from the point of view of Dr. Beaumont. As a class, read the experiment and discuss the following questions.
   - What specifically was Dr. Beaumont studying about the stomach?
   - How did he test his idea? What was his procedure?
   - What conclusion could he make as a result of this experiment?

- How do you think St. Martin felt during this experiment? How would you feel about having this experiment done on you?

2. With your group, read the experiment your teacher gives you and answer the same questions.
3. Help present your group's experiment to the class. Your group may use illustrations, skits, or another type of dramatization. Be creative!

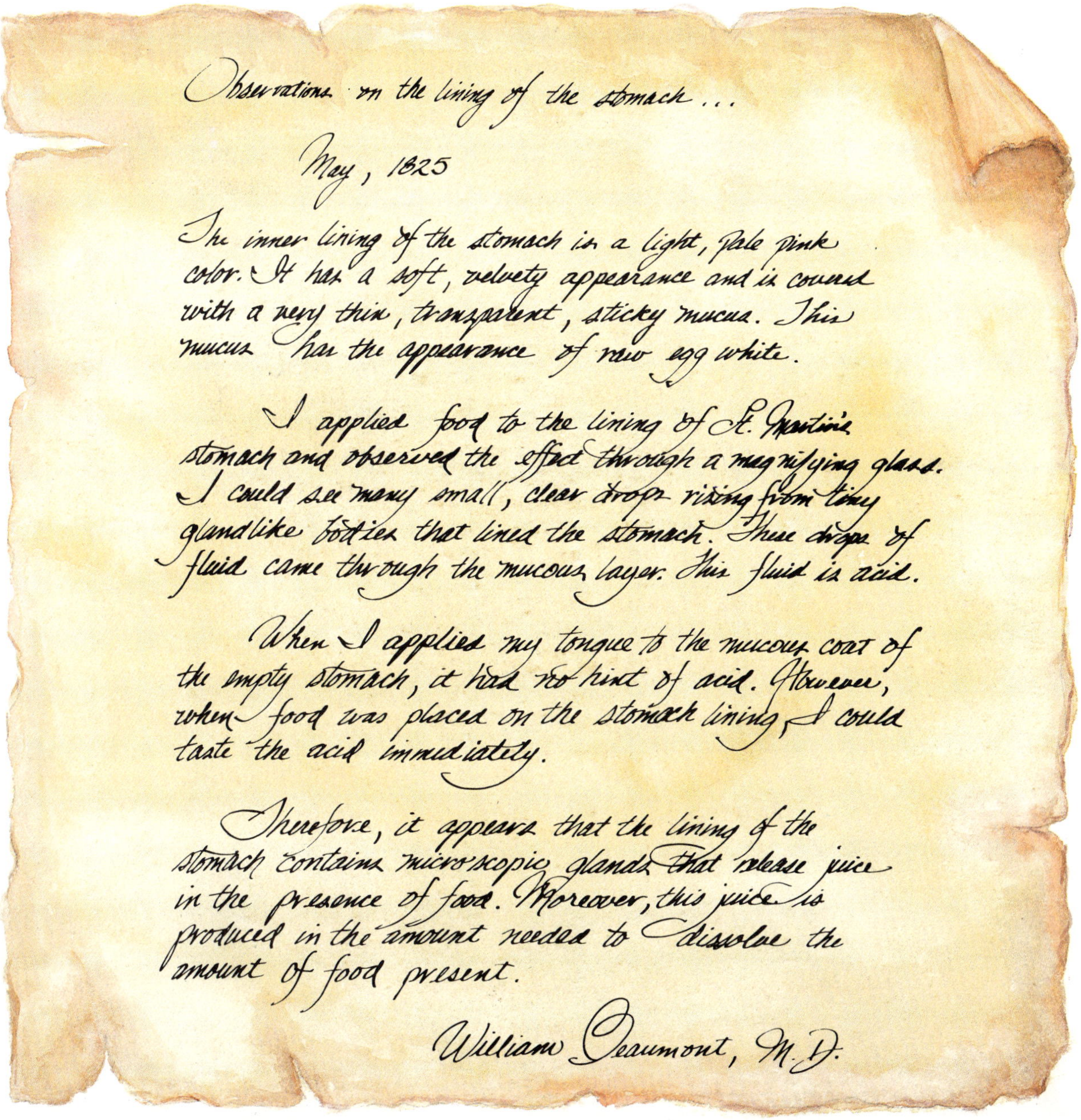

Observations on the lining of the stomach . . .

May, 1825

The inner lining of the stomach is a light, pale pink color. It has a soft, velvety appearance and is covered with a very thin, transparent, sticky mucus. This mucus has the appearance of raw egg white.

I applied food to the lining of St. Martin's stomach and observed the effect through a magnifying glass. I could see many small, clear drops rising from tiny glandlike bodies that lined the stomach. These drops of fluid came through the mucous layer. This fluid is acid.

When I applied my tongue to the mucous coat of the empty stomach, it had no hint of acid. However, when food was placed on the stomach lining, I could taste the acid immediately.

Therefore, it appears that the lining of the stomach contains microscopic glands that release juice in the presence of food. Moreover, this juice is produced in the amount needed to dissolve the amount of food present.

William Beaumont, M.D.

EXTENSION ACTIVITIES

1. Ask your teacher for a copy of some of the other experiments Beaumont did. Read them. Imagine that you are the doctor. What would you conclude from the results of each experiment?
2. Imagine that you are St. Martin. Write some diary entries starting with the day of the accident. Continue until your first trip home to Canada . . . or until your death at age 83. Record your thoughts, feelings, and observations. Of course, you do not have to fill in diary entries for every single day.
3. Do you want to know more about St. Martin and his well-known stomach? If so, read *Dr. Beaumont and the Man with the Hole in His Stomach* by Sam and Beryl Epstein. This short, but informative, book will give you a better idea as to the personalities of Beaumont and St. Martin.

11-5 THE STOMACH'S JOB

Down the esophagus, through the sphincter, plop! A wad of food lands in the stomach. This mass is a mixture of large chunks and tiny pieces. It contains protein, fat, and partially digested starch. Now what?

Stomach Action

The stomach stretches to hold all of the arriving food. It churns and grinds to mix the food with stomach juices. The stomach's digestive juice is made of water, acid, and enzymes. Your stomach makes about a half gallon of this juice each day.

Carbohydrates and fats are not digested in the stomach, but stomach enzymes start the digestion of protein. The enzymes break down proteins into amino acids.

$$\text{protein} \xrightarrow[\text{acid}]{\text{stomach enzymes}} \text{amino acids}$$

The acid in the stomach is important for digestion. It provides the right environment for the enzymes to work. It also kills some of the bacteria that are swallowed along with the food. Stomach acid is so strong that it would burn your skin. If this acid is so strong, why doesn't it dissolve the walls of the stomach? Sometimes it does!

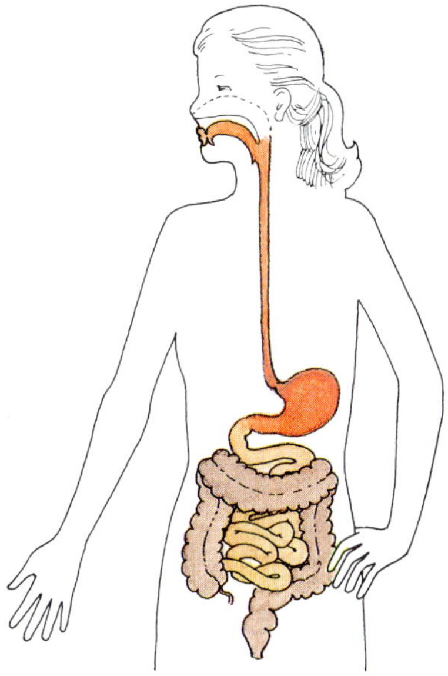

Figure 11–12 Do you know where your stomach is located? If you put your left hand over the bottom ribs on your left side with your fingers pointing towards your right hip, your hand will be resting over your stomach.

Stomach Structure

The stomach's lining looks like a series of folds when the stomach is empty. These folds are pictured in Figure 11–13 on page 298. When the stomach is full of food, the folds flatten out. A mucous coating protects the lining of the stomach. Sometimes acid gets through the mucus and causes continual stomach pain. A sore spot in the stomach lining is called an *ulcer.*

The stomach **walls** are made of smooth muscle cells. The muscle fibers allow the stomach to stretch when it is full and collapse when it is empty. When there is food in the stomach, the muscles contract. This

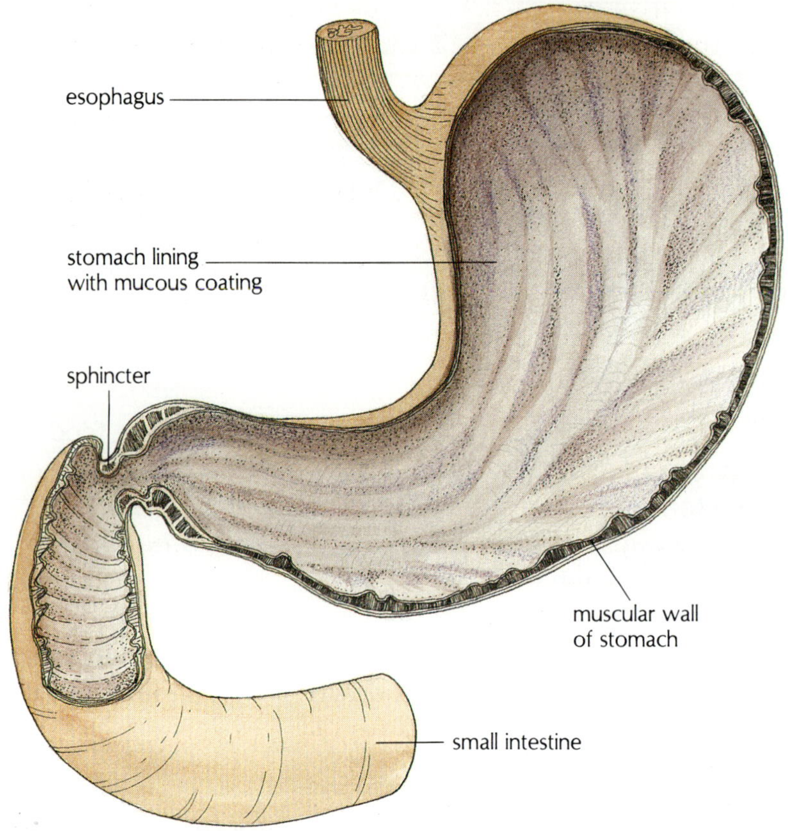

Figure 11–13 View of the inside of the stomach

causes the churning action that mixes the food and stomach juices. It also breaks up large pieces of food. If you eat too much, the muscle fibers are overstretched. What happens then? PAIN. Overeating is one cause of stomachaches.

Functions

The stomach has three important functions. You have already read about two of these functions. It churns and mixes the food and digests protein. The stomach is also a storage area. Food usually stays in the stomach for three to five hours before moving on through the digestive tract.

Food molecules are not absorbed through the walls of the stomach. So, even though you have eaten, your cells must wait awhile for the arrival of their "food."

Oops!

Sometimes the stomach muscles contract sharply and force the food *upwards.* This is what happens when a person vomits. This reaction can be triggered by some drugs, viruses, food poisons, or even some motions. You may have noticed a burning sensation when you vomited. That was from stomach acid.

We all swallow some air when we eat. If a person swallows too much air, gas builds up in the stomach. Lots of air in the stomach can be another cause of stomachaches. One way to get rid of this extra gas is by burping. Even though burping is a "natural action," not at the table, please! Our culture does not consider burping polite.

Moving On

Once food has been thoroughly churned and mixed in the stomach, it is ready to move on. It will pass through another sphincter to enter the small intestine.

ANALYSIS QUESTIONS

1. What 3 things are found in stomach juices?
2. Why doesn't stomach acid "eat" through the stomach walls?
3. What are the 3 main functions of the stomach?
4. You hear adults say they get "heartburn." They describe it as a burning feeling behind the sternum. What could cause this sensation?
5. It is not good to take too much antacid for stomach "relief." Why do you think this is so?
6. Which of these foods are digested, at least partly, in the stomach?

bread	cantaloupe	cottage cheese
noodles	lima beans	margarine
turkey	green beans	marshmallows

EXTENSION ACTIVITY

Demonstrate the importance of movement for digestion. Use 2 small pieces of hard candy and 2 small jars full of water. Put a piece of candy in each jar. Set one jar on the table. Shake the other jar until the candy dissolves. How long does it take for each piece to dissolve?

11-6 GUT REACTIONS—THE SMALL INTESTINE

In the stomach, food is slowly changed into a thick, soupy mixture. The muscular sphincter at the lower end of the stomach controls the flow of this mixture into the small intestine. Most of the real work of digestion takes place once the mixture leaves the stomach.

The Small Intestine—Function

The soupy mixture that once was food enters the small intestine in a series of spoon-sized squirts. Smooth muscles in the walls of the small intestine contract and relax, squeezing the food and moving it along. It will take one to two days for food to travel from one end of the small intestine to the other. During this time, two important digestive processes will take place:

1. The foods will be completely broken down.
2. Nutrients will be absorbed.

Both of these processes are very important. Food could not be used by the body if it did not go through the small intestine.

Very little of the food that arrives in the small intestine has been digested; now most digestion will take place. The rest of the starch will be broken down into sugars. Proteins will continue to be digested into amino acids and fats will be broken down. Burritos, pancakes, cupcakes . . . all foods will soon look the same in the small intestine. The complete breakdown of these foods involves many chemical reactions. Enzymes, again, are needed for these reactions to take place.

Some of these enzymes are produced in the lining of the small intestine. Others are produced in the pancreas and the liver. Figure 11–14 on page 301 shows the location of all of the structures involved in digestion.

Pancreas

The pancreas releases several different enzymes into the small intestine. Some of these enzymes digest

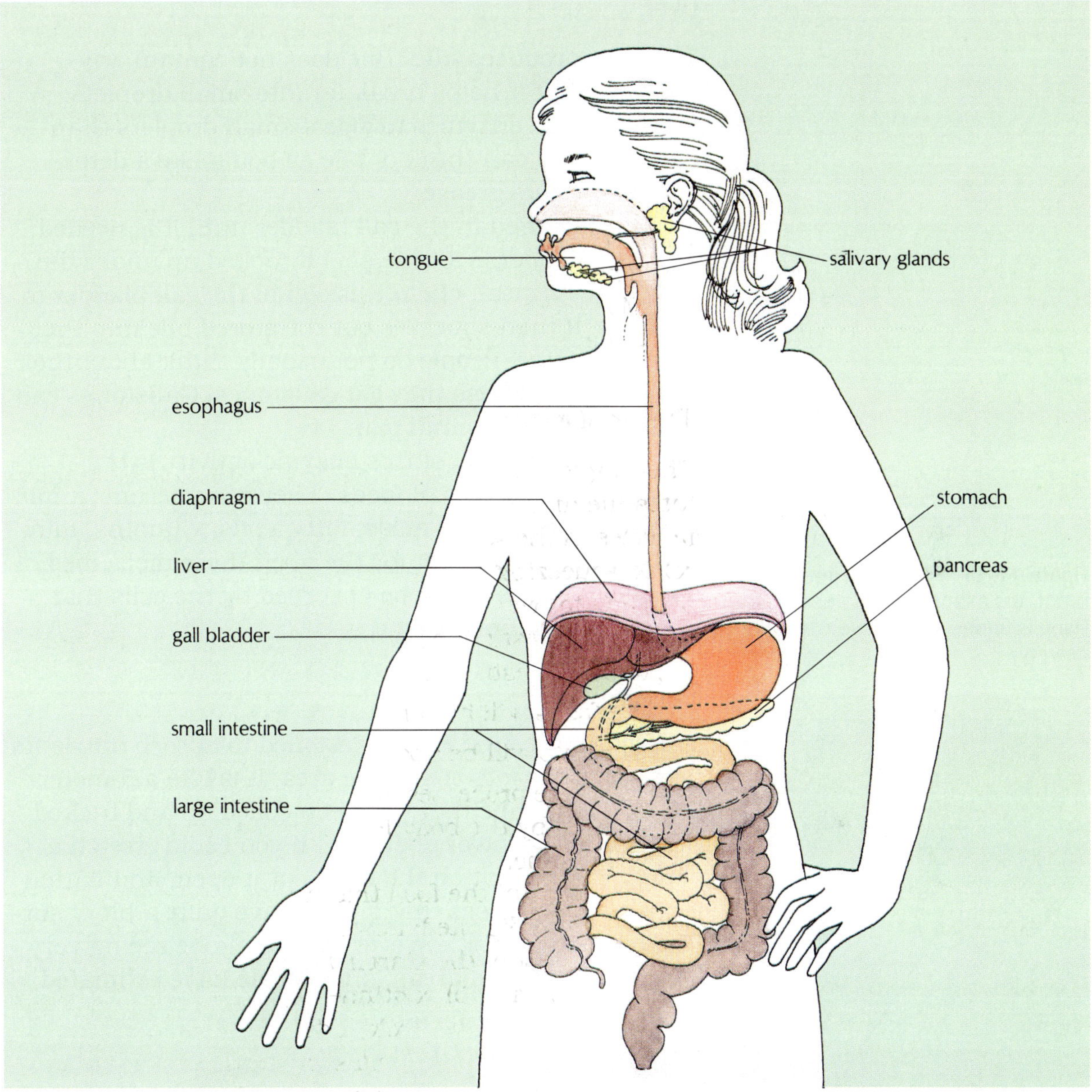

Figure 11–14 The digestive system. Through which of these structures does your food pass as it travels through the digestive system?

carbohydrates, others digest fats, and still others work on proteins. You probably didn't know that you give your pancreas a workout every time you eat!

In addition to containing enzymes, pancreatic juice neutralizes stomach acid that enters the small intestine. Otherwise, this acid would destroy the intestine's lining. The pancreas also produces hormones.

One of the hormones, insulin, is important for controlling blood sugar levels. If not enough insulin is produced, diabetes results. People who have diabetes know how important insulin is.

Liver

The liver produces **bile.** Bile does not contain any enzymes, but it helps break fat into small droplets. It is easier for enzymes to digest small droplets than big ones. You can think of bile as being like a detergent that "cuts grease."

Bile is stored in the gall bladder until it is needed. The gall bladder is a little pouch tucked up under the liver. After a meal, chemicals signal the gall bladder to contract. It then squeezes out the stored bile into the small intestine. People do not usually think about their gall bladders unless they get gallstones. Gallstones can cause sharp abdominal pain.

The result of all of this enzyme activity is the chemical breakdown of foods. They soon become a mixture of sugars, amino acids, fatty acids, vitamins, minerals, and fiber. Except for the fiber, these molecules are all small enough to be absorbed by the cells that line the small intestine.

Figure 11–15 The small intestine is not just a long, hollow tube. Its lining is folded and its surface looks velvety.

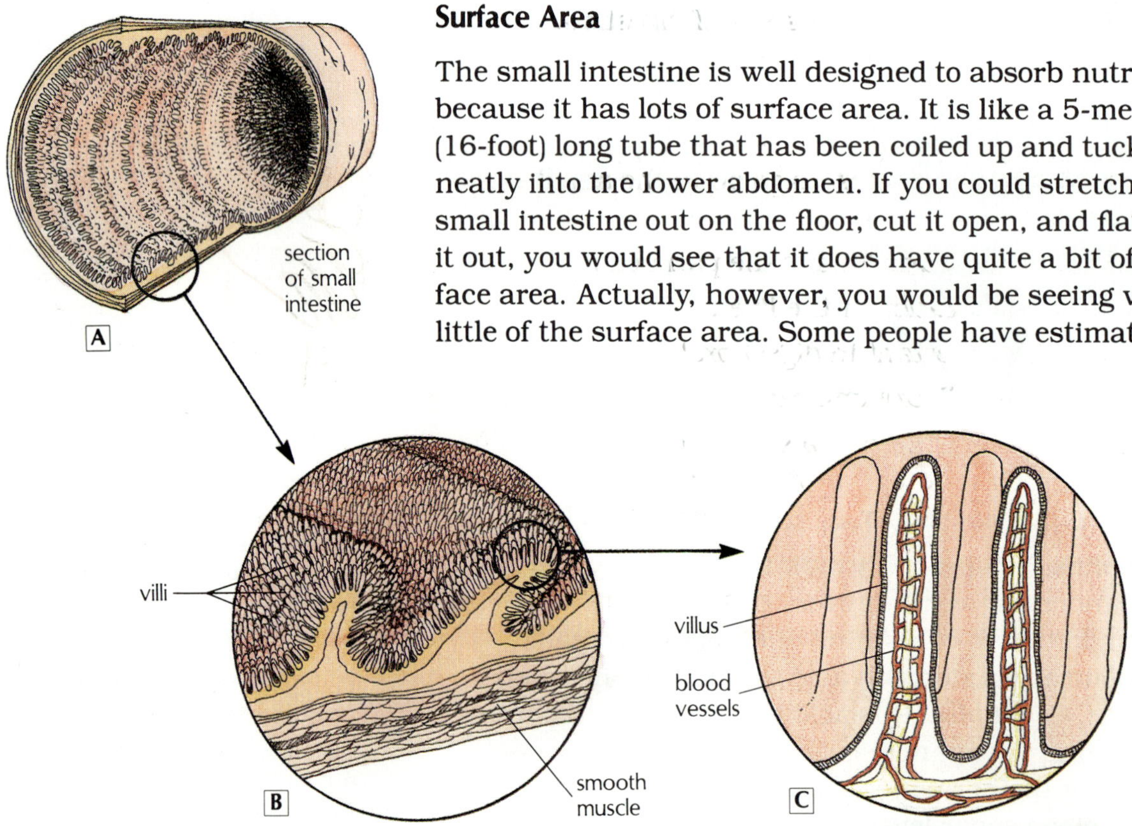

With magnification, you can see that the lining is covered with millions of tiny villi.

Each villus contains blood vessels that can carry away the absorbed nutrients.

Surface Area

The small intestine is well designed to absorb nutrients because it has lots of surface area. It is like a 5-meter- (16-foot) long tube that has been coiled up and tucked neatly into the lower abdomen. If you could stretch a small intestine out on the floor, cut it open, and flatten it out, you would see that it does have quite a bit of surface area. Actually, however, you would be seeing very little of the surface area. Some people have estimated that if an adult's small intestine were laid out flat, it would almost cover a football field!

The clue to all of this extra surface area is the lining of the small intestine. If you could look at the lining, you would see that it is not smooth. Instead, it looks velvety. Using a magnifying lens, you would see millions of fingerlike projections. These projections are **villi.** Each of the villi contain tiny blood vessels. As the nutrients flow through the small intestine, they are absorbed by the villi. They can then move into the tiny blood vessels. Once in the blood, the nutrients are circulated to every cell in the body. Food to the cells—finally! Study Figure 11–15 on page 302 to help you understand the structure of the small intestine.

The cells that line your intestines experience lots of wear and tear. They must constantly be replaced. In fact, you have a new cell lining in your intestines every three to six days!

Moving On

The liquid mixture that reaches the end of the small intestine no longer contains usable nutrients. The materials that remain are now ready to move through another sphincter and enter the large intestine. Here they will go through the final steps of digestion.

ANALYSIS QUESTIONS

1. What 2 digestive processes go on in the small intestine?
2. What are the 3 functions of the pancreas?
3. Describe the location of the liver.
4. How is bile important in digestion?
5. Explain why villi are important.
6. Figure 11–16 represents a section of the small intestine. Which arrow do you think shows the direction in which most absorption takes place—A, B, or C? Explain your answer.
7. Which structure do you think is most important—your stomach or your small intestine? Explain.
8. What do you think happens to food when it enters the large intestine?

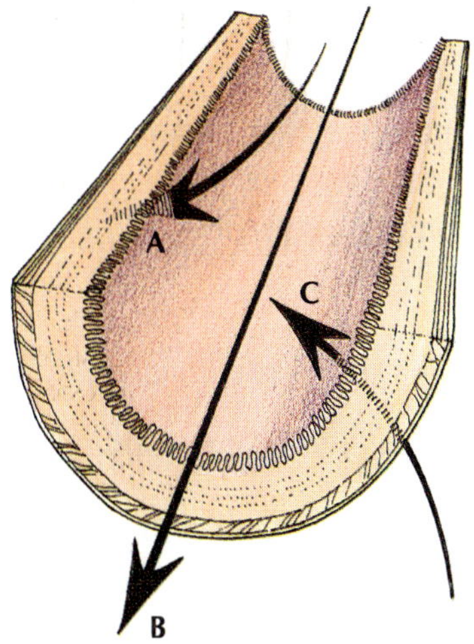

Figure 11–16

CHALLENGE

The villi increase the surface area of the small intestine. Think of a way to demonstrate this. You may want to make a model or draw a sketch to prove how this is possible.

11-7 CROSSING A MEMBRANE

The food in the small intestine no longer looks like food. It is a soupy mix of molecules. The nutrients in this mixture must get out of the intestine and into your body. This means they must pass into the cells that line the small intestine.

- Which nutrients are small enough to get through a cell membrane?
- How can you prove that molecules move when you cannot even see them?

Answer these questions after you have made a model that shows the movement of molecules.

MATERIALS

Day 1 *(Parts A & B)*
beaker
2 pieces dialysis tubing, 2″ each
2 rubber bands
plastic tube
corn syrup solution
cornstarch
graduated cylinder, 10 ml
stirring rod
masking tape
scissors

PROCEDURE

Day 1 Work with a partner.

Part A. A Cell Model

1. Prepare the dialysis tubing.
   a. Put about 150 ml water and 2 pieces of dialysis tubing in your beaker.
   b. Let the tubing soak 30 seconds.
   c. Open the tubing by rolling it between your thumb and finger.
   d. Cut along the length of each piece of tubing to make 2 squares.
2. Put your "cell" together.
   a. Get the 2 squares of dialysis tubing, the plastic tube, and the 2 rubber bands.
   b. Cover one end of the plastic tubing with a square of dialysis tubing. Attach it with a rubber band as shown in Figure 11–17.

Figure 11–17

Figure 11–18

3. Pour some water from your beaker into your "cell." Fill it about half full. Use the other square of dialysis tubing and other rubber band to seal the top of your cell. (See Figure 11–18.)

   Handle your cell carefully. You do not want it to leak!

Part B. Nutrient "Soup"

1. Use your graduated cylinder to add 5 ml of the corn syrup solution to the water in the beaker.
2. Add about ½ teaspoon cornstarch to the beaker. Gently stir the solution with a stirring rod.
3. Put your cell into the beaker of water. The level of liquid in the beaker should be slightly higher than the level in your cell. Add more water to the beaker if you need to. (See Figure 11–19.)
4. Use masking tape to put your names on your beaker. Put it where your teacher tells you. You will finish this experiment tomorrow.
5. Clean up your work area and answer analysis questions 1–5.

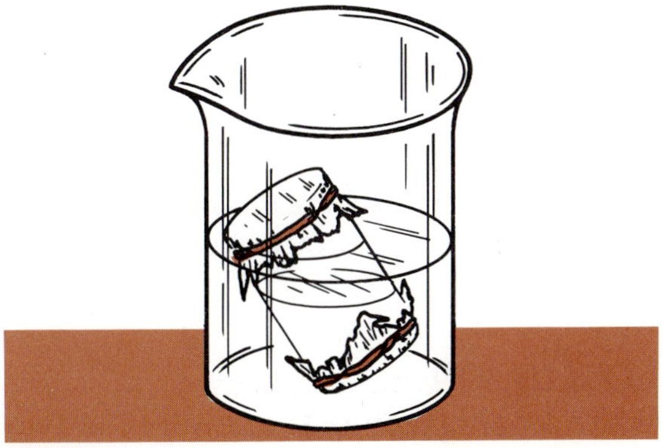

Figure 11–19

MATERIALS

Day 2 *(Part C)*

2 droppers
2 test tubes
masking tape
iodine solution in dropper bottle
Sugar Test station
- hot plate
- test tube holder
- safety goggles
- beaker
- Benedict's solution in dropper bottle

Part C. Sugar and Starch Tests

1. Design a data table to record your results. You will need to read the rest of this procedure to find out what data you need to record. Have your teacher check the data table you designed before you do the rest of this activity.
2. Prepare your materials.
   a. Label one test tube "inside cell" and the other "outside cell."
   b. Get 2 droppers. (You will use one dropper for liquid inside the beaker—outside the cell—and the other for inside the cell. See Figure 11–20 on page 306.) Do not mix up the droppers.

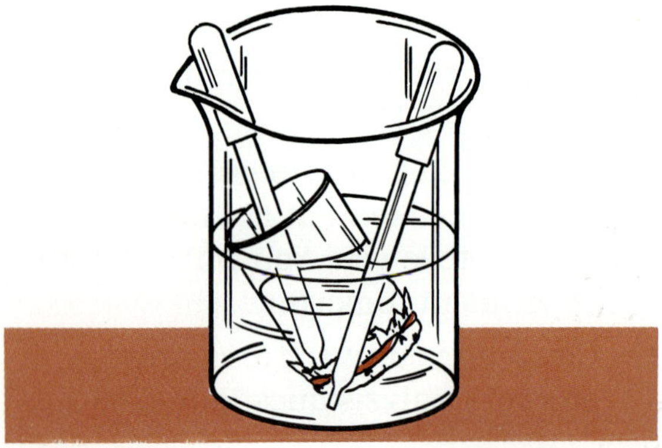

Figure 11–20

3. Carefully remove one end of your cell. Use a dropper to remove 15 drops of liquid from inside the cell. Put this "cell juice" into the test tube labeled "inside cell."
4. Use the other dropper to stir the solution in the beaker gently and to remove 15 drops. Put this liquid in the "outside cell" test tube.
5. Test *both* liquids for sugar. Go to the Sugar Test station.
   a. **Put on your safety goggles.**
   b. Add 10 drops of Benedict's solution to each test tube.
   c. Heat the test tubes in the water bath for a couple of minutes.
   d. Record your results.
   e. Rinse out your test tubes.
6. Test *both* liquids for starch.
   a. Add 12 drops of liquid from inside the cell to one clean test tube and 12 drops from the beaker to the other clean test tube. Make sure the right liquid goes in the right test tube! Gently stir each liquid with its dropper.
   b. Add 10 drops of iodine solution to each test tube.
   c. Record your results.
7. Clean up your equipment and work space.
8. Answer the remaining analysis questions.

ANALYSIS QUESTIONS

1. What is the purpose of doing this activity?
2. What did you put
   a. inside the cell?
   b. in the solution around the cell?
3. What would happen if you tested the inside of the cell on day 1
   a. for sugar?
   b. for starch?
4. What results would you get by testing the liquid outside of your cell on day 1
   a. for sugar?
   b. for starch?
5. On day 2 of this experiment, you will test the liquids in the cell and the beaker for both starch and sugar. What do you predict will be the results of these tests?
   a. inside the cell, test for starch
   b. inside the cell, test for sugar
   c. outside the cell, test for starch
   d. outside the cell, test for sugar
6. Look at your results.
   a. Where did you find sugar?
   b. Where did you find starch?
   c. How can you explain your results?
7. Compare your model cell with a cell in your intestine.
   a. List 2 ways they are alike.
   b. List 3 ways they are different.

CONCLUSION

Write a paragraph that answers the two focusing questions on page 304.

11–8 CELL "FOOD"

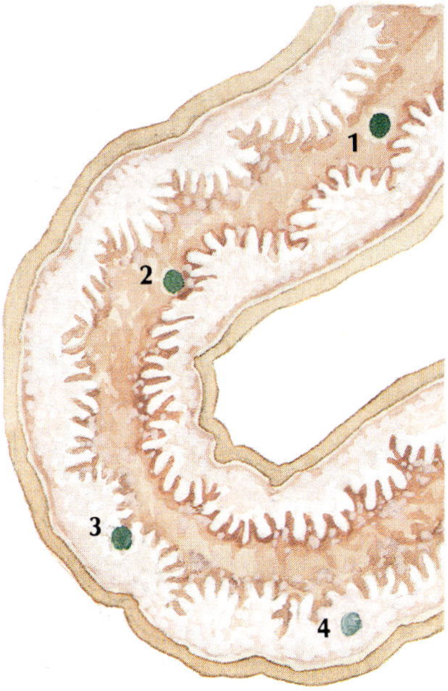

Figure 11–21 Nutrients are absorbed as they travel through the small intestine.

Food in the small intestine is a soupy mixture of molecules. It contains sugars, amino acids, fatty acids, vitamins, and minerals. These nutrients must get out of the small intestine and into the body. In order to do so, the molecules must pass through the cells that line the intestine.

Look at Figure 11–21. It shows one sugar molecule moving down the small intestine. It drifts near the wall of the intestine. It happens to touch the intestine lining. The next thing you know, the sugar molecule is *inside* the cell. It passed through the cell membrane.

Inside the Cell

Once the sugar molecule or any nutrient is in the cell, two things can occur. The cell may use the nutrient or it may pass it on to the next cell. The diagram in Figure 11–22 shows the two possibilities.

If the cell "needs" the nutrient, enzymes go to work. The nutrient may be broken down within the cell and used for energy. The nutrient could also be used to build new cell parts. It could become a part of the cell membrane, of the nucleus, or of any of the hundreds of microscopic structures in the cell.

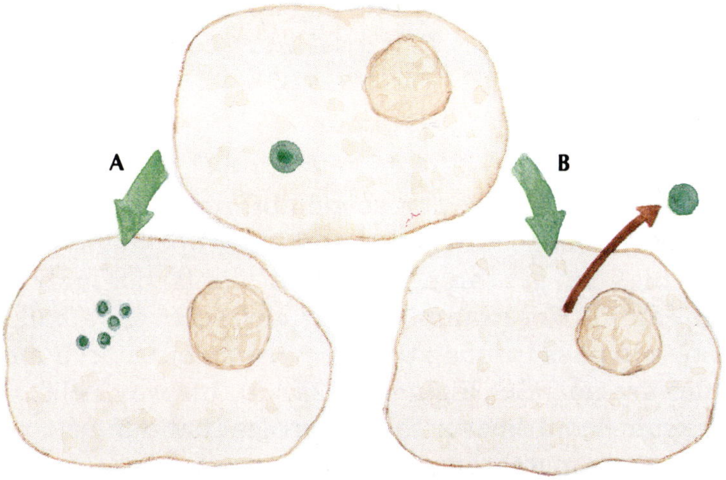

Figure 11–22 A sugar molecule that is inside a cell may be used for energy or it may leave the cell.

If the cell already has plenty of that particular nutrient, it can continue its travels through the body. It will cross the cell and pass through the membrane into the next cell. In this way, nutrients eventually reach the bloodstream. Once they are in the bloodstream, they can be carried to all parts of the body.

Diffusion

This movement of molecules is not as simple as it may sound. Sugar molecules cross cell membranes by a process known as **diffusion.** When molecules diffuse, they move from where there are more of them to where there are fewer. In other words, they move from a place of high concentration to one of low concentration. Molecules can diffuse into a cell or out of a cell. Diffusion stops when there are equal concentrations of molecules on both sides of the membrane. Diffusion is illustrated in Figure 11–23.

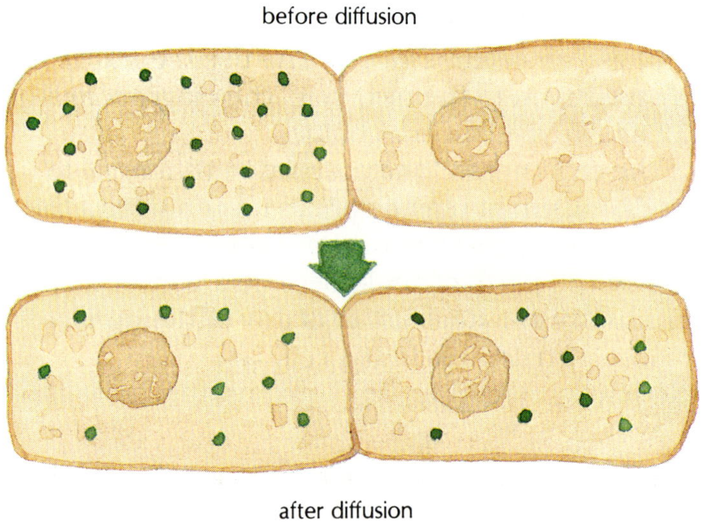

Figure 11–23 Diffusion results in equal concentrations of molecules on both sides of the membrane. Besides sugar, what kinds of molecules could the green dots in this drawing represent?

Membranes

Molecules can diffuse across membranes because membranes are not solid. They contain tiny openings. Small molecules can get through these openings, but large ones cannot. For example, sugar molecules are small enough to pass through the membrane, but starch molecules are not. (See Figure 11–24.) In this way, cells have some control over the molecules that diffuse across their membranes.

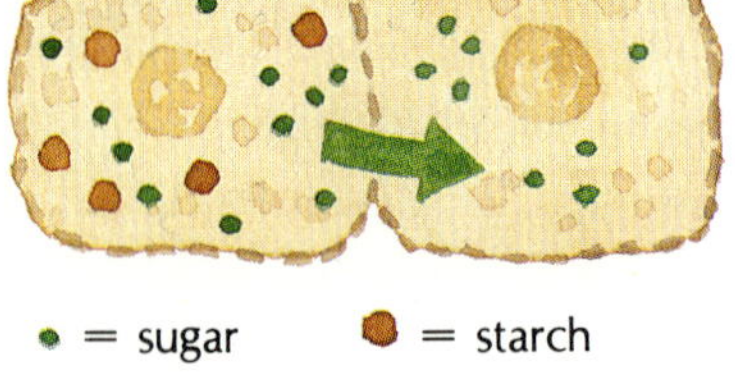

Figure 11–24 Cell membranes are not solid. They have very small openings that permit small molecules to enter and leave a cell.

Solutions and Concentrations

When you dissolve a substance in water, you make a solution. Solutions can be different concentrations. For example, you can dissolve 1 teaspoon of sugar in 100 ml of water to make one solution. Call it solution A. You could dissolve 3 teaspoons of sugar in another 100 ml beaker of water. Call it solution B. You would now have two different concentrations of sugar-water solutions.

Which solution is more concentrated?

Solution A

Solution B

ANALYSIS QUESTIONS

1. What are 2 ways that a cell might use a nutrient?
2. Define the word **diffusion.**
3. Choose one of these models and use it to explain diffusion. You can either write your explanation or illustrate it.
   a bottle of perfume
   freshly baked bread
   food coloring and water
   tea bag in hot water
4. Figure 11–25 shows three cells.
   a. Which cell has the highest concentration of sugar molecules?
   b. Which cell has the lowest concentration?
   c. Explain which way the sugar molecules will probably diffuse.

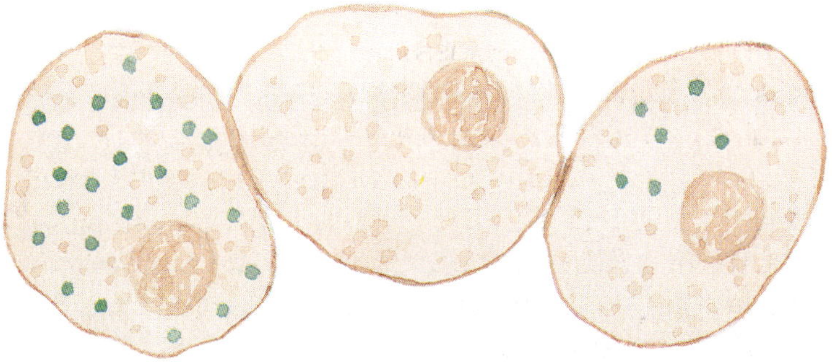

Figure 11–25

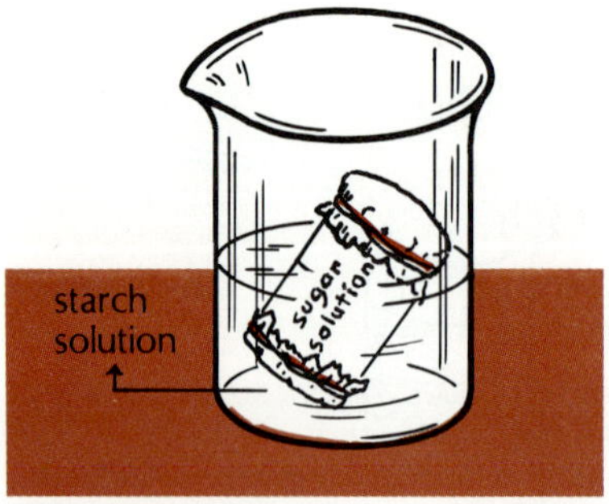

Figure 11–26

5. Why does food have to be digested before it can be used by the body?
6. Two students did another experiment with the dialysis tubing. (See Figure 11–26 for the set-up.) They tested the solutions on the second day. Predict the results of these four tests.
   - Sugar Test on the inside solution
   - Sugar Test on the outside solution
   - Iodine Test on the inside solution
   - Iodine Test on the outside solution

EXTENSION ACTIVITY

Set up the model cell with another set of conditions. Choose one of the conditions in Figure 11–27 to try. Predict what you think will happen. Explain your results.

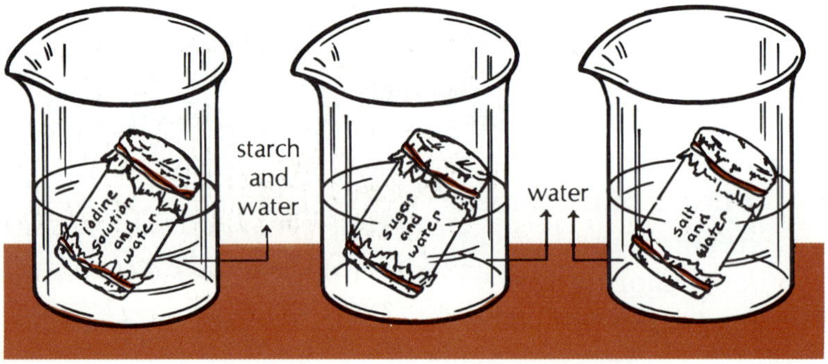

Figure 11–27

11-9 THE FINAL STEP— THE LARGE INTESTINE

The "food" that leaves the small intestine has been digested. It doesn't look at all like food anymore. Only the "leftovers" remain; the nutrients have all been absorbed. These leftovers will travel through the large intestine and, eventually, leave the body.

Structure

The large intestine is much shorter than the small intestine. It is only about 1.5 meters long (4 or 5 feet). It travels up your right side, crosses your abdomen just below your stomach, and makes an abrupt turn to go down your left side. Figure 11–28 shows the location of the large intestine.

Unlike the small intestine, the large intestine does not have villi; its lining is smooth. Like the other parts of the digestive system, it is coated with mucus. The walls of the large intestine also contain smooth muscle fibers. When these fibers contract, they move the undigested food along. Sometimes you can feel a "wave" of movement when the large intestine contracts.

The last part of the intestine is called the **rectum.** This is where the wastes are stored.

The opening from the rectum is the **anus.** There are two sphincter muscles in the anus. One is made of smooth muscle fibers and the other is skeletal muscle fibers. This makes it possible to control the opening and closing of the anus.

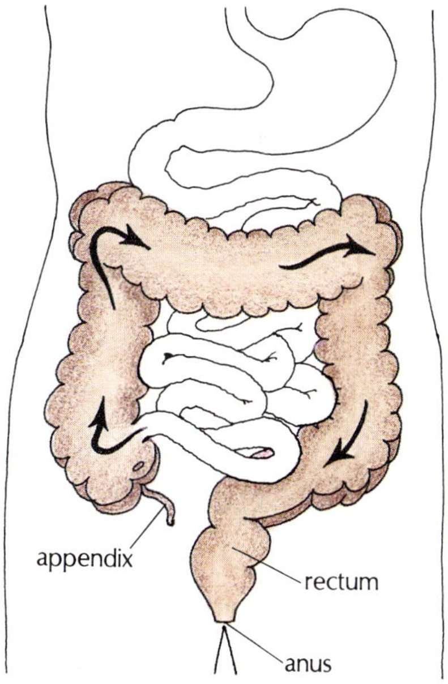

Figure 11–28 The final steps of digestion take place in the large intestine.

Functions

Two very important things happen in the large intestine. First, water is absorbed. This water comes from the liquids we drink, the foods we eat, and digestive juices. The large intestine rescues the water so it is not lost with the waste materials.

The large intestine also stores these leftovers. This allows us to choose the time when we get rid of the waste. The large intestine does not produce enzymes.

Figure 11–29 As the wastes travel through the large intestine, water is absorbed.

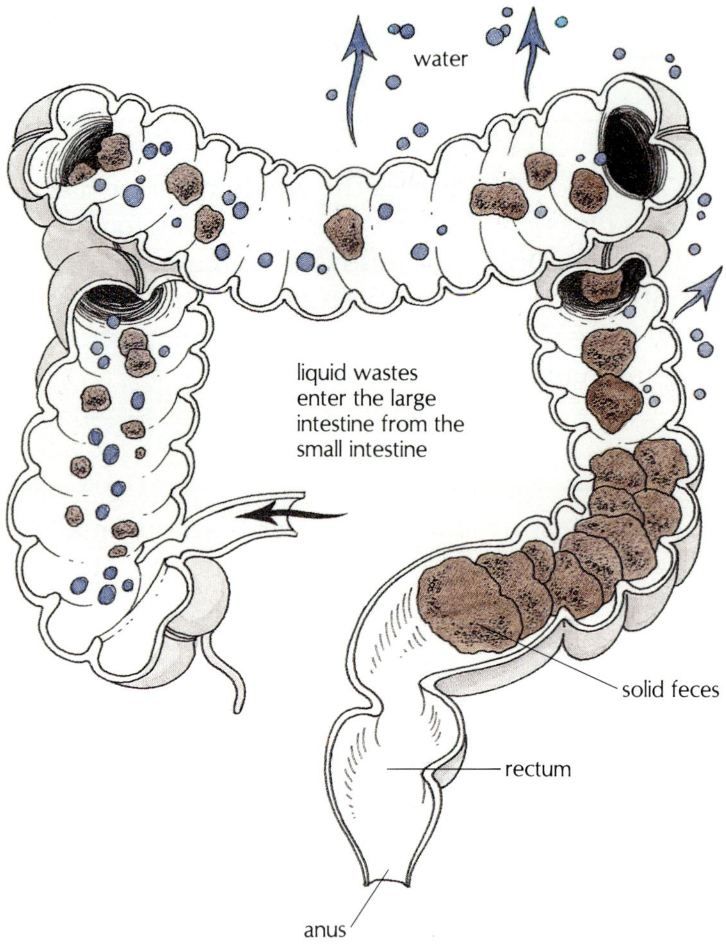

No digestion takes place here because there are no more nutrients to be digested.

The two functions are illustrated in Figure 11–29.

The Feces

The scientific word for the "leftovers" is **feces.** If you analyzed the content of feces, you would find that it is 75% water. The other 25% is bacteria, fiber, undigested food, mucus, and dead cells. Feces are brown because of the bile from the liver that was mixed into the food.

The odor of feces is due to the bacteria. When the bacteria feed on the leftovers, they produce gas. Some foods, such as beans and cabbage, contain molecules we cannot digest. These foods leave lots of leftovers for bacteria so they produce lots of gas. In large amounts, these gases feel uncomfortable and may be mistaken for a stomachache.

Figure 11–30 Many Americans feel they need to take laxatives for "regularity." Actually, eating more fiber and getting more exercise would be better for them and for their bowels.

Too Fast, Too Slow, or Just Right?

Bowel movements remove the feces from the large intestine. Many people think that a "once-a-day" bowel movement is essential. Actually, there is no rule for the right number of bowel movements. Some people are comfortable with a bowel movement once every two or three days.

If feces move through the large intestine too quickly, diarrhea results. The large intestine does not have enough time to absorb the water so the feces are too liquid. If a person, especially a young child, has diarrhea for very long, he or she will become dehydrated. Diarrhea can result from an infection, being nervous, or eating food that disagrees with you.

If feces movement is too slow, too much water is taken up by the large intestine. **Constipation** is the uncomfortable result. A high fiber diet can help prevent constipation. Fiber retains water to help keep the feces moving.

The Appendix

Another part of the digestive tract is the appendix. This 9 cm (3½ inch) pouch has no known function. Occasionally, however, the appendix swells and fills with pus. Pain is then felt in the lower right abdomen. Fever and nausea may follow. These are the symptoms of **appendicitis.** If the swollen appendix is not removed, it will burst. Bacteria are then released into the abdomen and a serious infection can follow. If you think you have appendicitis, call your doctor!

CHECK YOUR UNDERSTANDING

You have now studied both the small intestine and the large intestine. How many ways can you list in which these two structures are alike? How are they different?

Tract Records

It takes one to three days for food to travel the length of the human digestive system. Birds process their foods much more rapidly. A shrike can digest a mouse in 3 hours. Berries pass through the digestive tract of a thrush in just 30 minutes. Why do you think it is important for a bird to be able to digest its food quickly?

11-10 DIFFERENT FUNCTIONS, DIFFERENT STRUCTURES

Stomach, small intestine, and large intestine . . . these three body organs are important parts of the digestive system. Each of these organs has a different role in the digestive process. Each one also has its own specialized structure.

- Can you identify sections of small intestine, large intestine, and stomach?
- What clues will help you identify them?

Look at the three unlabelled specimens and use what you have learned about the digestive system to figure out what each one is. At the same time, you'll be answering the three focusing questions.

MATERIALS

preserved specimens A, B, and C
magnifying lens
dissecting tray
metric ruler
soap
paper towels
prepared microscope slides:
  small intestine
  large intestine
microscope
data page: ***Different Functions, Different Structures*** (Figure 11–31)

PROCEDURE

Work in a group of three.

Part A. Getting Ready

You will have 3 specimens to inspect. They are from the stomach, small intestine, and large intestine, but you will not know which is which. They are labeled A, B, and C. Your goal is to identify the 3 specimens.

1. Work in groups of three. Each of you will have a task. Decide who will do what.

   recorder—reads directions and fills in the worksheet while group works (R)
   handler—carefully handles specimens and holds them for recorder's and trotter's inspection (H)
   trotter—gets necessary materials and returns them neatly to where they belong (T)

2. Answer the first analysis question. If you are unsure of an answer, look back at ***The Stomach's Job*** on

DIFFERENT FUNCTIONS, DIFFERENT STRUCTURES

RECORDER ______

HANDLER ______

TROTTER ______

DATE ______ PER. ______

	A	B	C
1. Inspect the outside of the specimen. Each person should say *one* word that describes it. (All different words, please.)	R: H: T:	R: H: T:	R: H: T:
2. Do the same thing as you look at the inside of the specimen.	R: H: T:	R: H: T:	R: H: T:
3. Use a hand lens to look at the inside of the specimen. Think of additional descriptive words.	R: H: T:	R: H: T:	R: H: T:
4. Does this specimen contain villi?			
5. How thick are the walls of this specimen?			

Do not complete this section until you have looked at all three specimens.

This specimen is A. ______ B. ______ C. ______

Explain your decisions below.

Is your data table neat? Is it complete?

Figure 11–31

page 297, ***Gut Reactions*** on page 300, and ***The Final Step*** on page 311. The three of you should agree on the answers before you continue with Parts B and C.

Part B. Specimen Inspection

1. The trotter should get a dissecting tray, a metric ruler, a magnifying lens, and *one* specimen. It does not matter which specimen you inspect first.
2. The recorder should read the directions on the worksheet and fill in the appropriate column while the group works.
3. When you finish with a specimen, the trotter should return it and get a different one.
4. When you have looked at all 3 specimens, decide which specimen is which. Your worksheet should now be complete.
5. Clean up. The handler should wash any dirty equipment when washing his or her hands.
6. Answer analysis questions 2 and 3. Your teacher will tell you if you should answer these as a group or by yourself.

Part C. Close-Up Views

Two microscope slides have been set up for you to view. One shows the lining of the small intestine; the other the lining of the large intestine.

1. For each microscope slide, sketch and label what you see. Remember to record the magnification. Label the inside surface and outside surface of each one.
2. Answer analysis questions 4 and 5.

ANALYSIS QUESTIONS

1. Which structures—stomach, small intestine, large intestine—do the phrases below describe? (More than one answer may be needed.)
   a. its walls are made of muscle fibers
   b. it is shaped like a long tube
   c. it must be able to expand greatly
   d. its lining has fingerlike projections called villi
   e. its lining has folds
   f. its lining is smooth
2. Explain which structure is which. (Use complete sentences, please.)
   a. A is the ______ because . . .
   b. B is the ______ because . . .
   c. C is the ______ because . . .

3. What is the function of villi?
4. Describe the main difference between the small and large intestines when viewed through the microscope.
5. What do you think you will remember the longest about doing this activity?

CONCLUSION

Write a paragraph that answers the two focusing questions on page 315.

EXTENSION ACTIVITIES

1. The tongue is another part of the digestive tract that you can inspect. Buy a beef tongue—whole, not sliced—at a supermarket.
   a. Decide how it fits into the cow's mouth.
   b. Think of words that describe how the surface of the tongue feels.
   c. Use a good hand lens to take a closer look at the surface. What structures do you see now?
   d. Look for blood vessels. Where can you see them?
   e. Cut open the tongue. Does it look like muscle tissue—like steak?
   f. Decide how your tongue is like the one from the store and how it is different.
2. Hemorrhoids and polyps are two disorders of the intestine. Find out how common these two conditions are. Write a short report or design a pamphlet describing what these conditions are, their symptoms, and methods of treatment.

UNIT 4

BODY BASICS

CHAPTER 12

CIRCULATORY SYSTEM

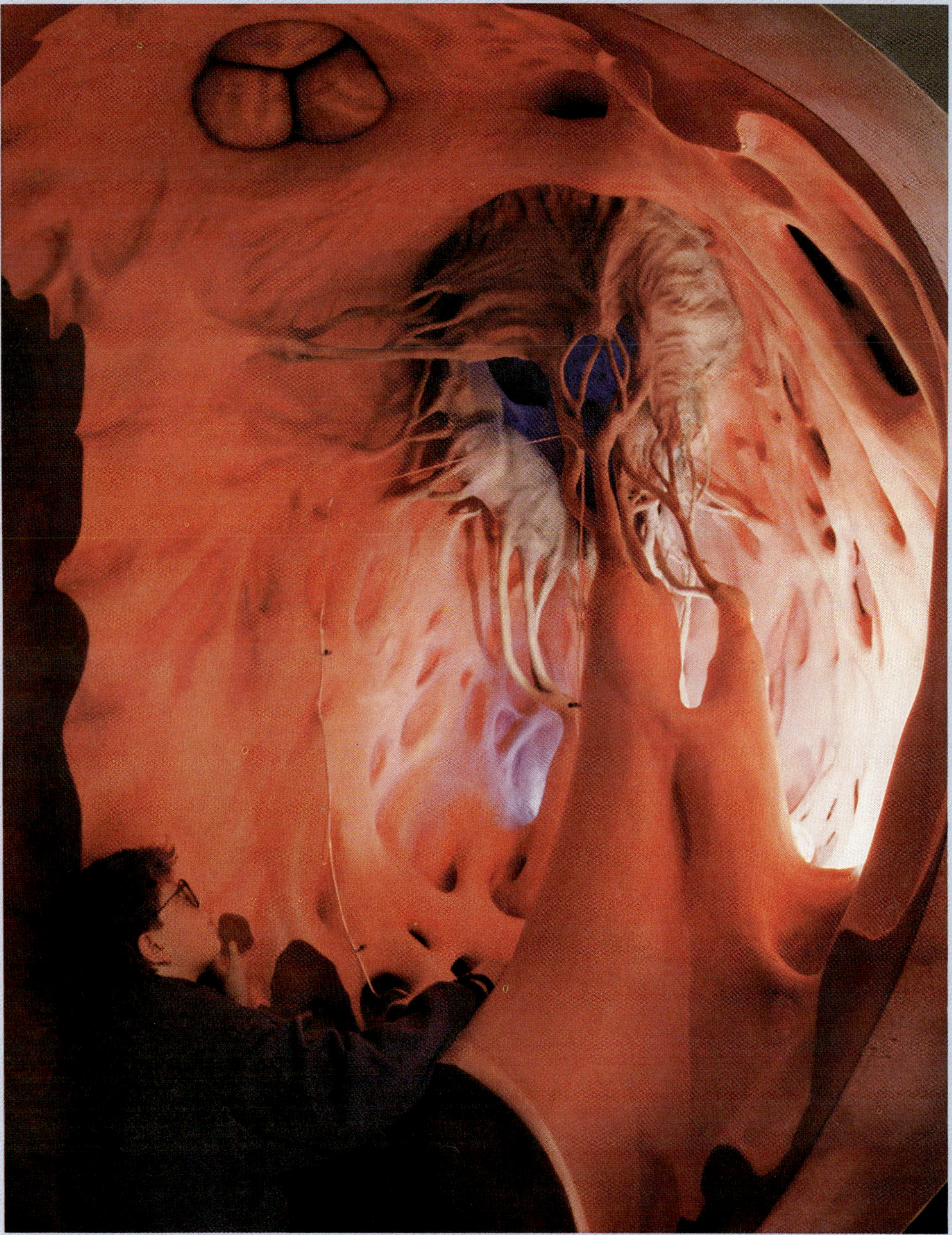

What do you think this boy will see when he walks through this 16-foot-tall model of the heart?

12–1 PHYSICAL FITNESS AND THE HEART

Running, swimming, and other exercises improve your muscle fitness; that is, they improve your strength, endurance, and flexibility. Exercise also strengthens bone and improves circulation to the skin. These benefits result from the increased blood supply bone and skin receive. In order to provide this "blood boost," your heart must work harder. So perhaps we should say that exercise improves both skeletal muscle *and* cardiac* muscle fitness.

- How can you use your pulse rate to check your heart's fitness?
- If a person gains weight, how can that affect his or her heart?

In order to answer these questions, you will measure your pulse rate when you exercise with and without extra weight.

MATERIALS

watch or clock with second hand
scale
6 kg of weight (such as textbooks or phone books)
data page: ***Physical Fitness and the Heart*** (Figure 12–2)
reading: ***Exercise for Your Heart*** (page 325)

PROCEDURE

Your teacher will announce times during this activity. Try to keep up!

Part A. Resting Pulse Rate

1. Find a pulse point on your wrist. (See Figure 12–1.)
2. Count your pulse for 15 seconds. Record the total on the data page.
3. Do this 2 more times.
4. Multiply each 15-second pulse count times 4 to find the number of pulse beats in 60 seconds. The number of pulse beats per minute is called **pulse rate.**

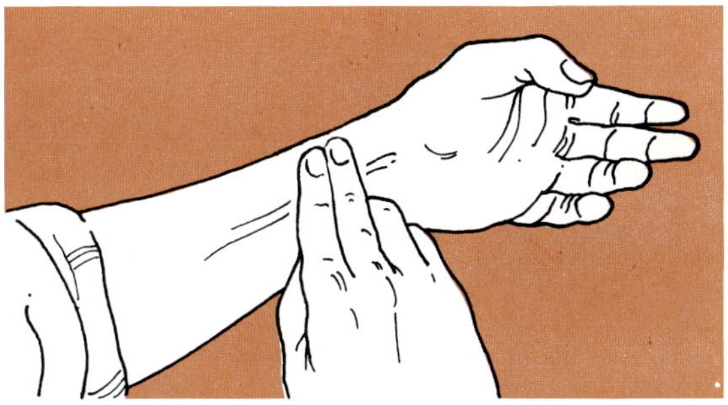

Figure 12–1

5. Calculate your average resting pulse rate.

Part B. Exercise and Pulse Rate

It is time to exercise and find out what happens to your pulse rate after you stop. Have your pencil and data page ready!

*The word *cardiac* always refers to the heart.

1. Repeat the following hop and jog exercise pattern for 3 minutes.
   hop to the right, to the left, to the front, to the back;
   jog in place 4 times
2. When your teacher calls "stop," sit down and find your pulse.
   a. Count your pulse for 15 seconds.
   b. Use the next 15 seconds to record your count and get ready to take your pulse again.
   c. Continue doing this for 6 minutes, remembering the timing: **count** (15 secs), **record** (15 secs), **count** (15 secs), **record** (15 secs).
3. Calculate your pulse rates and record them on the data page.

PHYSICAL FITNESS AND THE HEART

NAME ____________________

DATE ____________ PER ____

PART A. RESTING PULSE RATE

Trial Number	Pulses Per 15 Seconds	Pulse Rate Per Minute
1		
2		
3		

Average resting pulse rate (per minute): ________

PART B. EXERCISE AND PULSE RATE

Time	Pulses Per 15 Seconds	Pulse Rate (per minute)
0:00		
0:30		
1:00		
1:30		
2:00		
2:30		
3:00		
3:30		
4:00		
4:30		
5:00		
5:30		
6:00		

PART C. THE WEIGHT EFFECT

Time	Pulses Per 15 Seconds	Pulse Rate (per minute)
0:00		
0:30		
1:00		
1:30		
2:00		
2:30		
3:00		
3:30		
4:00		
4:30		
5:00		
5:30		
6:00		

Figure 12–2

Figure 12–3 How does exercising with additional weights affect your pulse rate?

Part C. The Weight Effect

1. Hold 6 kg (13 lbs) of weight and repeat steps 1–3 of Part B.
2. Read ***Exercise for Your Heart*** starting on page 325. Then answer the analysis questions.

ON YOUR OWN

Graph your data to show what happened to your pulse rate after you stopped exercising.

ANALYSIS QUESTIONS

1. What is your average resting pulse rate?
2. If Bill's pulse rate is 68, his heart is beating how fast?

a. 17 times a minute (17 × 4 = 68)
b. 68 times an hour
c. 68 times a minute
d. 100 times a minute
e. You can't tell.

3. Pulse rate is
   a. always the same as heart rate.
   b. sometimes faster than heart rate.
   c. 68 times a minute.
   d. 100 times a minute.
   e. You can't tell.
4. True or False? A pulse rate taken on the inside of the ankle is slower than the pulse rate taken at the wrist because the ankle is farther from the heart.
5. About how many times does your heart beat each day? Show your calculations.
6. Look at your data for exercising without holding the extra weight.
   a. What was your pulse rate when you first stopped exercising?
   b. How long did it take your pulse to return to its resting rate?
7. Why does pulse rate increase when you exercise?
8. The faster the pulse rate returns to its resting rate, the better it is for your heart. Why?
9. Look at your data for exercising with extra weight.
   a. What was your pulse rate when you first stopped exercising?
   b. How long did it take your pulse to return to its resting rate?
10. Compare your pulse rate data for exercising with and without extra weight.
    a. Which showed the highest pulse rate after exercising?
    b. Which took longer to return to its resting rate?
11. Explain one way that being overweight can affect a person's heart. (It may help to think about what happened when you exercised with 6 kg of weight.)

CONCLUSION

Write a brief paragraph to answer each of the focusing questions that appear on page 321.

EXTENSION ACTIVITIES

1. Analyze your pulse rate. Count your pulse at various times during the day. For example, take your pulse:
   before getting up in the morning
   when sitting or standing
   before or after eating
   after drinking something with caffeine
   when you are excited
   List the times when your pulse rate is fastest and slowest. Then decide what kinds of things cause an increase in rate.
2. How does age affect pulse rate? Try to answer this question by taking the pulses of babies, toddlers, children, young adults, and senior citizens.
3. Try to measure the heart rates for other organisms. Take the pulses of cats, gerbils, birds, different breeds of dogs, and any other animals that are handy. How did you measure their heart rates?

Heart Rate

Every animal has a characteristic heart rate. Here are some sample heart rates.

30	codfish
600	hummingbird
200	rabbit
70	human
400	mouse
25	elephant
800	shrew

Figure 12–4 A masked shrew

There is an association between animal size and heart rate. What do you think it is?

Exercise for Your Heart

Which of these exercises are best for your heart?

swimming	lifting weights
softball	biking
bowling	sprinting
jogging	volleyball

Choosing "Hearty" Exercises

Almost any exercise is better for improving heart fitness than doing nothing, but some exercises are better than others. Good heart exercises are said to be **aerobic**—that is, the muscles get plenty of oxygen while they are working. Here are three clues to help you choose aerobic activities.

- First, your pulse rate must go up. Of course, you do not want your heart to beat too fast. A good aerobic exercise would raise your heart rate to about 150 beats a minute, but no higher.
- Once the heart rate has increased, it should be kept at that level for 10 to 20 minutes.
- Finally, an aerobic activity is continuous; it does not involve a lot of starting and stopping.

In order to improve the heart's fitness, people should do aerobic exercises at least three times a week.

Figure 12–5 Activities like bicycling are not only fun, they can also be good for the heart.

Healthy Heart

Aerobic exercise helps your heart and blood circulation in several ways.

- It improves blood circulation throughout the body.
- It strengthens your cardiac muscle.
- It causes your resting heart rate to slow down. (A strong heart does not have to beat as often because it pumps more blood with each beat.)
- It develops arteries that go to the heart so if one of them becomes blocked, there will not be as much damage.

One way to check the health of your heart is to check your pulse rate before and after you exercise. As soon as you stop exercising, your heart gradually slows down. The time it takes for your pulse rate to return to normal indicates your heart's fitness. If your heart is in good shape, it will return to its normal pace fairly quickly.

Improving Heart Fitness

If you want to improve the health of your heart, participate in aerobic activities. Look at this list of exercises again. Which ones are aerobic exercises?

swimming	lifting weights
softball	biking
bowling	sprinting
jogging	volleyball

If you said that swimming, jogging, and biking are aerobic activities, you were right. These exercises are better for your heart than bowling, sprinting, lifting weights, or playing either softball or volleyball.

Are you exercising for your heart? If you do, you will feel better, look better, and possibly live longer.

Figure 12-6 Do you think this activity is aerobic?

12–2 HEARTBEATS AND STETHOSCOPES

One way to find out about the heart is to *feel* it beating. Another way to learn about the heart is to *listen* to it. Doctors and nurses use both of these methods. The **stethoscope** is the instrument that they use to listen to your heart. At some time before now, you may have listened to your own heart. If so, do you know what you heard?

- What causes heart sounds?
- How many different heart sounds can you hear?

Use a stethoscope to listen to your own heart and then you will be able to answer these questions.

MATERIALS

rubbing alcohol
stethoscope
cotton ball
worksheet: ***Heartbeats and Stethoscopes*** (Figure 12–8)

PROCEDURE

1. Work quietly today. The room must be very quiet or people will not be able to hear their heart sounds.
2. Clean the ear pieces on the stethoscope. Moisten a cotton ball with the rubbing alcohol and use it to wipe the ear pieces.

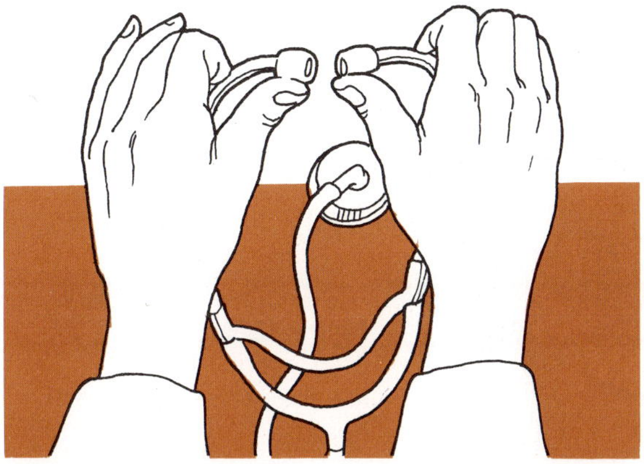

Figure 12–7 The earpieces of some stethoscopes can be tilted to make them more comfortable.

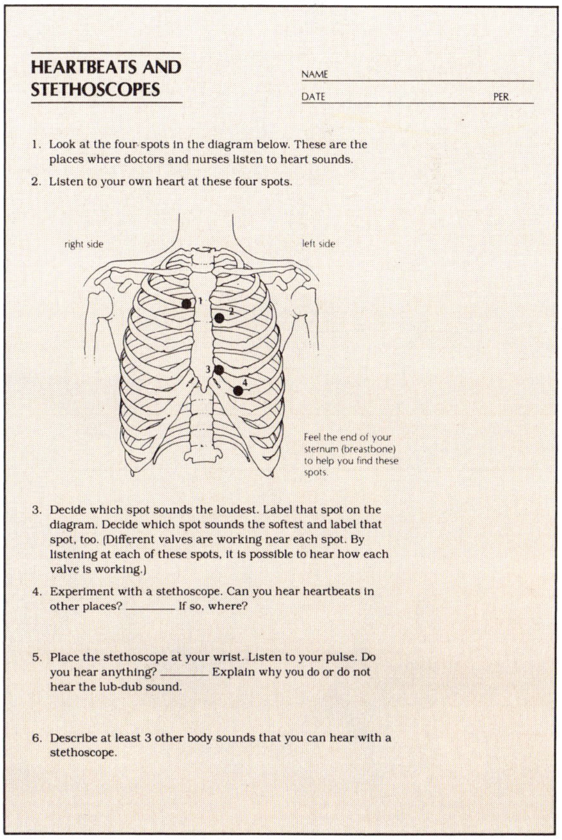

HEARTBEATS AND STETHOSCOPES

NAME ______ DATE ______ PER. ______

1. Look at the four spots in the diagram below. These are the places where doctors and nurses listen to heart sounds.
2. Listen to your own heart at these four spots.

3. Decide which spot sounds the loudest. Label that spot on the diagram. Decide which spot sounds the softest and label that spot, too. (Different valves are working near each spot. By listening at each of these spots, it is possible to hear how each valve is working.)
4. Experiment with a stethoscope. Can you hear heartbeats in other places? ______ If so, where?
5. Place the stethoscope at your wrist. Listen to your pulse. Do you hear anything? ______ Explain why you do or do not hear the lub-dub sound.
6. Describe at least 3 other body sounds that you can hear with a stethoscope.

Figure 12–8

3. Listen to your own heart with the stethoscope. The sounds you hear are described as ''lub-dub'' sounds. They are made by the closing of valves in the heart.
4. Read ''Lub-Dub'' below to find out why the heart makes these sounds.
5. Continue to listen to heart sounds while you complete the worksheet ***Heartbeats and Stethoscopes.*** (See Figure 12–8 on page 327.)

CONCLUSION

Answer the focusing questions on page 327. When you are done, reread your answers. Do they say what you want them to?

Lub-Dub

When blood moves from one part of the heart to the next, it passes through a **valve.** The valve snaps shut to keep blood from flowing backwards. There are several different valves in the heart. Each of these valves makes a distinctive, lub-dub sound every time it closes. These are the sounds we call the heartbeat.

Heart murmurs are often caused by valves that do not close completely. The ''murmur'' is the sound of blood squeezing through the opening. Many people have murmurs. Some murmurs are normal and do not cause any problems; others are more serious. They may indicate that the heart is not functioning properly. People with serious valve problems may have surgery to replace their faulty valves with plastic ones.

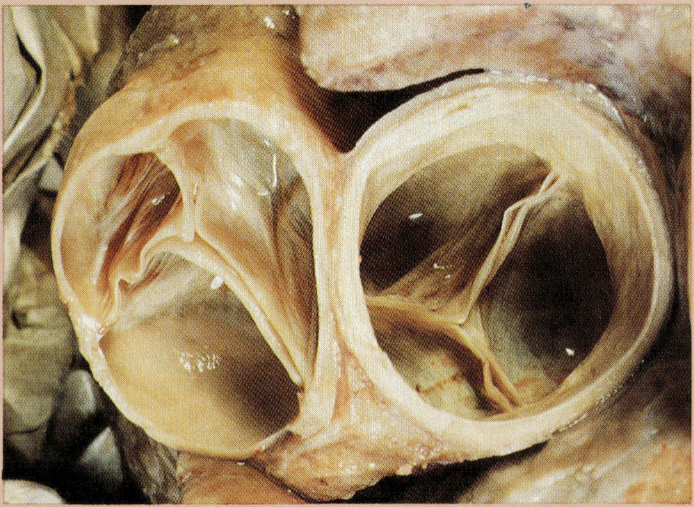

Figure 12–9 In this photograph, you can see the valves in a human heart.

EXTENSION ACTIVITY

Talk to your school nurse, physician, or other health professional about heart murmurs. Ask questions such as these:

What causes a heart murmur?
How does a heart murmur sound?
Which valve(s) "leak" when a person has a murmur?

Early Stethoscopes

The original stethoscopes were short wooden tubes from 7- to 12-inches long. The physician would press one end of the tube to the chest or back to listen to a person's lungs and heart. The modern stethoscope has two ear tubes so the health care provider can listen with both ears.

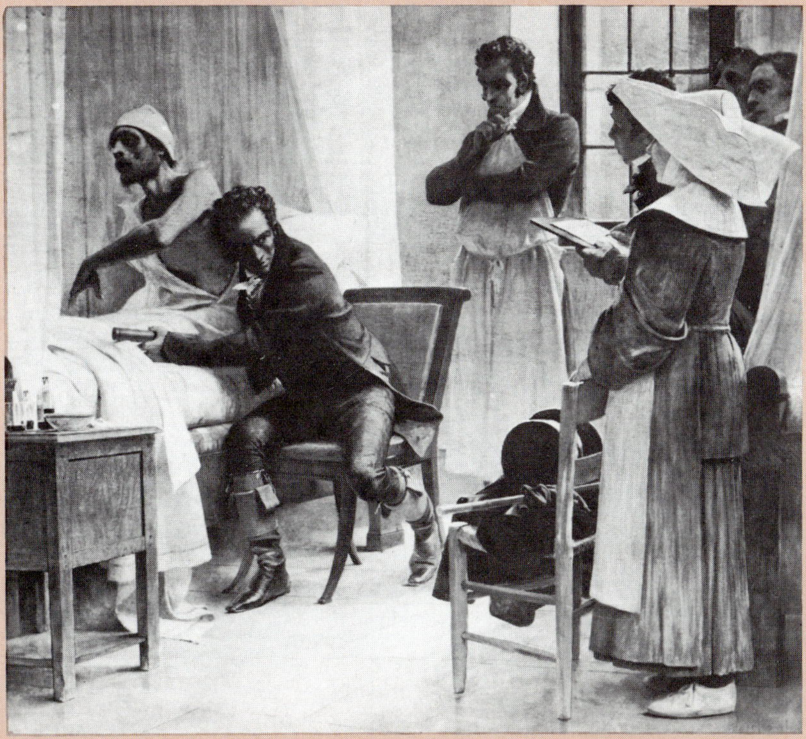

12-3 HEART PARTS

True, you can learn some things about the heart by feeling it beat or by listening to the "lub-dub" sounds, but you can learn much more about the heart by examining one. Maybe you have seen, touched, or even eaten a heart before. But, really, how much do you know about the parts of a heart?

- Which parts of a heart can you locate and name?
- Which part of the heart is the largest?
- How would you describe a preserved heart to a person who has never seen one?

In this activity you will inspect a preserved heart to answer these questions.

MATERIALS

preserved heart (pig, sheep, or cow)
dissecting pan
soap
paper towels
scissors
21 pins
checklist: ***Heart Parts*** (Figure 12–10) one per group
diagram: ***Heart Parts*** (Figure 12–11) one per person

PROCEDURE

Work in groups of four.

1. Each person in your group should choose one of the tasks below.
   - reader—reads checklist and keeps group on task
   - cutter—cuts apart list of heart parts as needed
   - pinner—pins name to correct spot on preserved heart
   - handler—carefully handles and positions heart so everyone can see it

NAMES ____________________
____
____
____

HEART PARTS DATE __________ PER. ____

PART A. OUTSIDE OF HEART

Put the two halves of the heart together. Check off the steps as you complete them.

_____ 1. Hold up the heart. Decide how you think it would be positioned in your body.

_____ 2. Find the *fat*. It is cream-colored; it does not look like muscle tissue. Label it.

_____ 3. The *large blood vessels* at the top of the heart were cut when this heart was removed. How many can you find? _____
Gently put your finger in one. How does it feel? Write *rough* or *smooth*. __________
Label one of these vessels.

_____ 4. Find the *ventricles*. They make up most of the heart muscle. There are two ventricles—a right one and a left one. From the outside they look like one structure. Label them.

_____ 5. Find each *atrium*. The atria look like two flaps and are located on top of the heart. They are much smaller than the ventricles. Label each one.

_____ 6. A netlike system of blood vessels serves the heart muscle. You can see some of these vessels on the outside of the ventricles. They are called *coronary vessels*. Label one of them. (They may look like lines.)

_____ 7. Have your teacher check the labels and initial this step.

PART B. INSIDE OF HEART

Separate the two halves and lay them on your pan.

_____ 1. Find the ventricles. Each ventricle is a chamber with a muscular wall. The left one has thicker walls than the right. In fact, the *left ventricle* is the biggest structure of the heart. Label it on each half.

_____ 2. The *right ventricle* is smaller than the left one. Label it on each half.

HEART PARTS

Cut out these words as you need them.

PART A.
fat
large vessels
ventricles (from outside)
atrium (from outside)
atrium (from outside)
coronary vessels

PART B.
left ventricle
left ventricle
right ventricle
right ventricle
left atrium
left atrium
right atrium
right atrium
right AV valve
right AV valve
left AV valve
left AV valve
bottom
front
back

Figure 12–10 (Checklist, page 1)

2. Follow the directions that the reader reads from the checklist.
   a. When it says to label a heart part, have the cutter cut that word off the list of words.
   b. Once everyone agrees on the location of the part, have the pinner pin the word in place.
3. When your group has finished labeling the preserved heart, label your ***Heart Parts*** diagram. Compare the real heart with the drawing as you label the structures.
4. Remove the pins. Return the heart to the plastic bag and clean up.

EXTENSION ACTIVITIES

1. Ask a butcher, hunter, or packing house manager for a fresh heart. Dissect it and try to find the same structures that you found on the preserved heart. How are the two hearts alike? How are they different?
2. Do worms have hearts? Obtain a night crawler or a large earthworm. Place it in alcohol for about ten minutes and then dissect it. Look for its heart and blood vessels. How is this heart *like* the mammal heart? How is it different? Look up worms in a science textbook to learn more about their hearts.

NAMES ______________

HEART PARTS
(PAGE 2)

____ 3. The *left atrium* is above the left ventricle. Label it on each half.

____ 4. The *right atrium* is above the right ventricle. Label it on each half.

____ 5. Blood flows from the right atrium into the right ventricle. The *right AV valve** prevents the blood from going back into the atrium. Valves are flaps of tissue. These flaps are connected to the ventricles by threadlike tendons. Label the *right AV valve* on each half of the heart.

____ 6. Find the *left AV valve* on each half of the heart. Label them.

____ 7. Carefully hold up both halves of the heart. The large vessels are at the top of the heart. Label the *bottom* of the heart on one of the halves.

____ 8. Imagine that this heart is in your chest. Notice where the left and right sides are. Label the *front* of the heart.

____ 9. Label the *back* of the heart.

____ 10. Have your teacher check the labels and initial this step.

____ 11. **LABEL THE HEART PARTS DIAGRAM** (Figure 12–12).

____ 12. Remove all labels and throw them away.

____ 13. Put the heart back in its container. Clean up your work space.

CONCLUSION

Write a sentence to answer each of the three focusing questions on page 330 in your text.

*The AV valve is the valve between the atrium and ventricle.

Figure 12–10 (Checklist, page 2)

NAMES ______________

HEART PARTS

DATE ______ PER. ____

Use this diagram to help you remember the parts of the heart. You should be able to label all of the parts that you located on your preserved heart. Label these heart parts on this diagram.

left atrium	left ventricle	left AV valve
right atrium	right ventricle	right AV valve
large vessels		

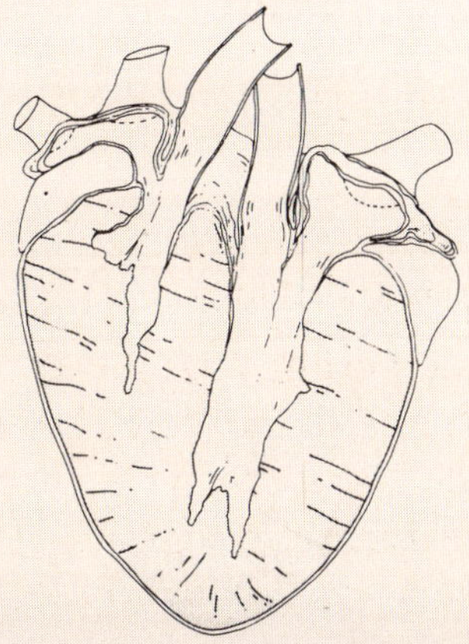

Which half of the heart is shown here, the front half or the back half?

Figure 12–11

12-4 THE HEART: A DOUBLE PUMP

A powerful, fist-sized muscle known as the heart is located inside the chest behind the sternum. The sternum, together with the ribs, forms a bony cage that protects the heart. The heart is not actually in the exact center of the chest; it is a little more to the left of the sternum than to the right. (See Figure 12–12.)

About once a second—sometimes more often, sometimes less—the muscles in the wall of the heart contract. This repeated contraction and the relaxation that follows is what we call the heartbeat. If you could watch the heart beat, it would look like a pulsing, writhing mass of muscle. Since you have seen a preserved heart, you may be able to imagine what one looks like when it is alive and working.

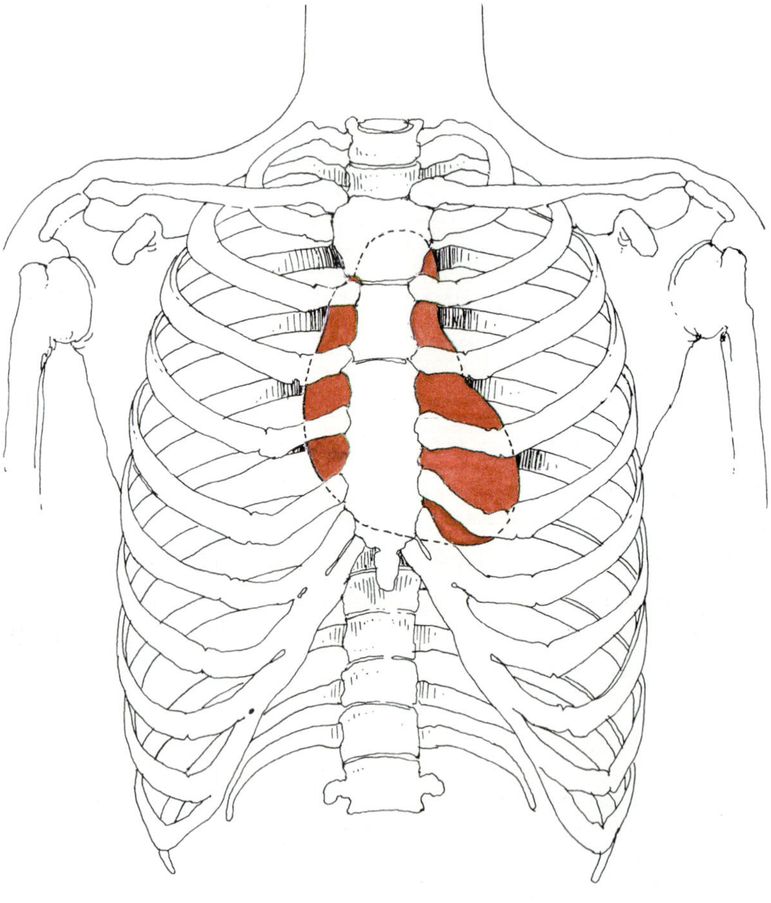

Figure 12–12 The heart is protected by the rib cage.

Power of the Pump

At the rate of 60 or 70 beats a minute and with two or three ounces of blood being pumped with each beat, the heart's pumping system moves about five quarts of blood in a minute. In one day, the heart could pump enough liquid to fill a 2,000-gallon tanker.

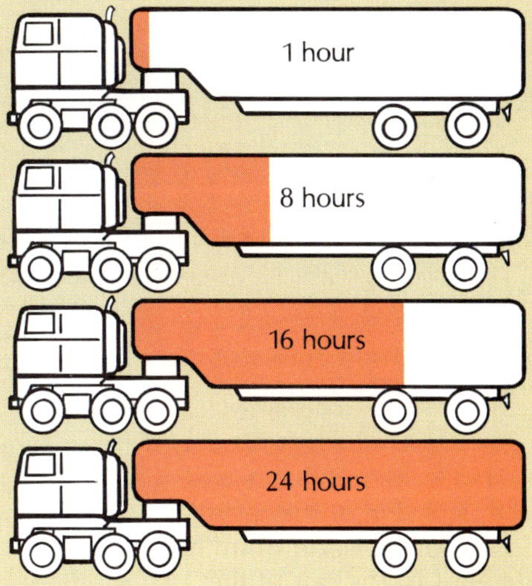

Inside the Heart

Try to imagine what is going on inside the heart when it is beating. The chambers of the heart are being squeezed closed and then are reopening. When the chambers open, they fill with blood; when they close, the blood is squeezed out. Open and fill, close and empty; open and fill, close and empty. This pumplike action is repeated over, and over, and over again. On the average, your heart pumps blood about 60 times a minute. And, with each pump, two or three ounces of blood spurts from the heart. After the blood has circulated through the body, it returns to the heart, only to be pumped back out again.

If you think about it, you may wonder why the heart needs *four* chambers. It might seem that *one* chamber would be enough. Blood could flow into this single chamber and then be squeezed out. While this design might work as a pump, it would not work as a heart. That is because a human heart actually works like two pumps; there is a pump on the left and a pump on the right.

A Double Pump

The action of the two pumps is shown in Figure 12–13 on page 334. When the ventricles contract, that pumping action moves the blood. The pump on the

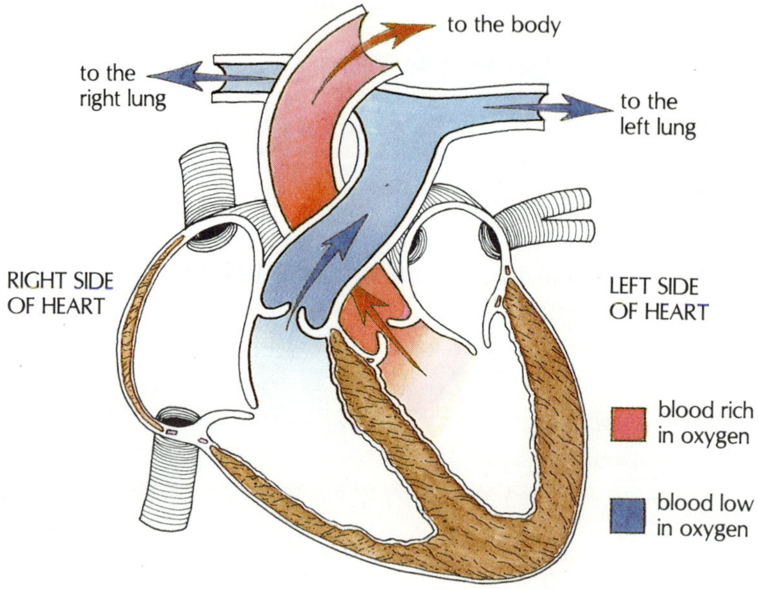

Figure 12–13 Blood that leaves from the right side of the heart goes to the lungs. Blood from the left side goes to the rest of the body.

right sends blood to the lungs. The one on the left sends blood to the entire body. This explains why the left ventricle is so much more muscular than the right. The trip to the toes, fingers, and nose is much longer than the one to the lungs!

Not only must blood be pumped *from* the heart, it must return *to* the heart. On its return, blood enters the atria; but which atrium? Think about it for a minute. The blood that has circulated through the body is low in oxygen. It returns to the right side of the heart so it can be sent on to the lungs. This route is shown in Figure 12–14. The blood that is returning from the lungs to the heart is rich in oxygen. It is ready to go to the rest of the body so it is returned to the left atrium.

Figure 12–14 Blood that has circulated through the body returns to the right side of the heart. Blood that has circulated through the lungs returns to the left side of the heart.

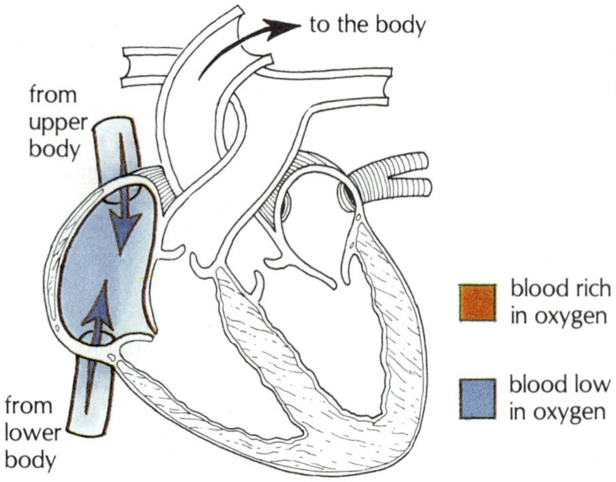

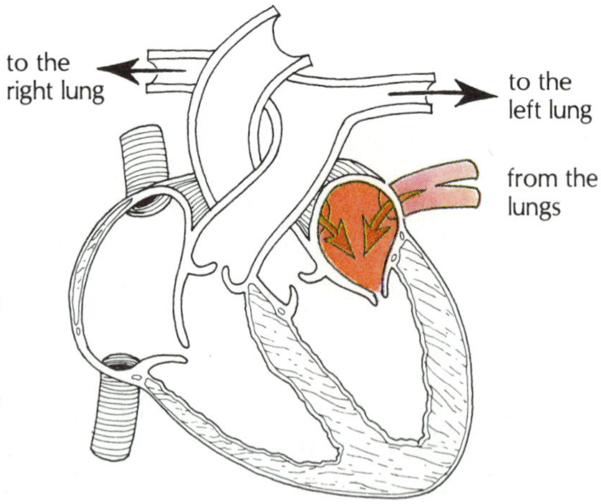

Your Turn

Put your finger on the left ventricle. Trace the route you would take if you were moving along with the blood. Think about what is happening to the blood as it travels. "The blood that leaves the left ventricle is rich in oxygen. As it flows through the. . . ."

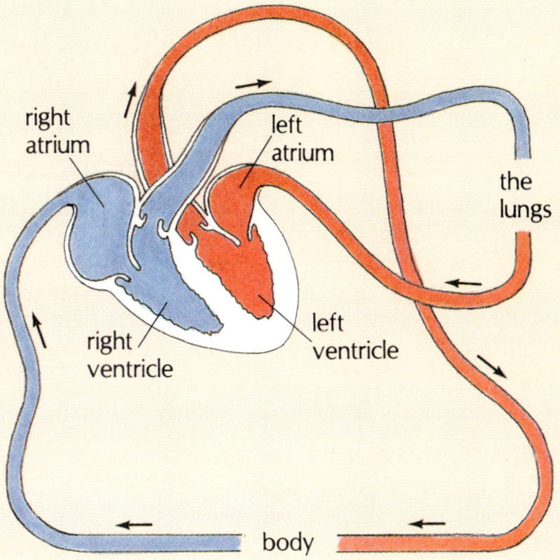

In Action

Even though the left and right ventricles are like two separate pumps, they pump at exactly the same time. That is because cardiac muscle encloses both ventricles. When it contracts, it sends blood out of both sides of the heart at the same time. Likewise, the ventricles relax at the same time. When the ventricles are relaxed, the atria contract. This pushes the blood through the AV valves to fill the ventricles. When the ventricles are full, they contract again. This action closes the AV valves and sends the blood out of the heart and on to the lungs and the rest of the body. Study the drawings in Figure 12–15. They summarize the actions.

Figure 12–15 When the heart is beating, the atria and ventricles take turns contracting. When the atria contract, which valves close? Which are open? Why?

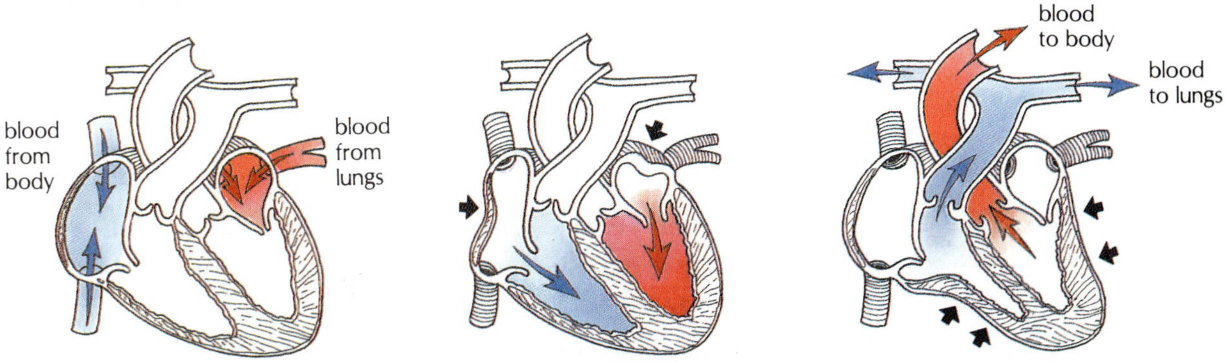

1. Rest

2. Atria contract (blood is forced into the ventricles)

3. Ventricles contract (blood is forced out of the heart)

Contract. Relax. Contract. Relax. Contract. . . . The heart will beat more than two and a half billion times in a 75-year lifetime. We cannot make a pump as reliable. When surgeons replace heart valves, the manufactured materials become battered after only a few years. Yet the living tissues of a healthy heart will stand up under the beating of a lifetime.

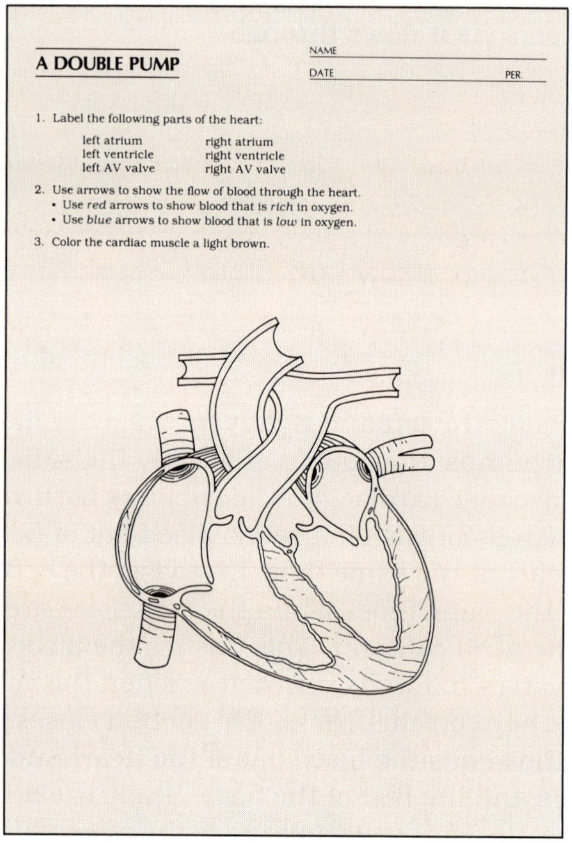
A DOUBLE PUMP

NAME
DATE PER.

1. Label the following parts of the heart:

left atrium	right atrium
left ventricle	right ventricle
left AV valve	right AV valve

2. Use arrows to show the flow of blood through the heart.
   - Use *red* arrows to show blood that is *rich* in oxygen.
   - Use *blue* arrows to show blood that is *low* in oxygen.
3. Color the cardiac muscle a light brown.

Figure 12–16

ANALYSIS QUESTIONS

Before you answer these questions, label and color your copy of the worksheet entitled **A Double Pump.** (See Figure 12–16.)

1. Which ventricle has thicker walls? Why?
2. How is the blood in the right ventricle different from the blood in the left ventricle?
3. How many times a minute does one of *your* AV valves open and close? (Think about this one; it's easy.) Explain your answer.

4. Imagine that your left AV valve is not closing all the way. What will happen when your left ventricle contracts?
5. Even though your heart is located almost in the center of your chest, you can feel it beat much better on your left side than on your right. Why do you think this is so?
6. Why is a four-chambered heart a better design for humans than a one-chambered heart would be?

EXTENSION ACTIVITY

Find articles about one of the artificial hearts and read them. Look for answers to these questions:

How is it *like* a real one? How is it *different?*

How does blood flow through this heart?

What does it looks like?

If you were going to have a heart transplant, what would you want to ask the cardiologist—heart specialist? List at least 5 questions you would want to ask.

"Why Does the Heart Beat?"

The heart beats because the cardiac muscle contracts. Why does it contract? You may remember that nerves trigger muscle contraction. Unlike skeletal muscles, there is not a single nerve that "tells" the heart to beat. Instead, there is an area of specialized nerve tissue built into the right atrium. Note its position. It sets the pace of muscle contraction, and so is called the **pacemaker.**

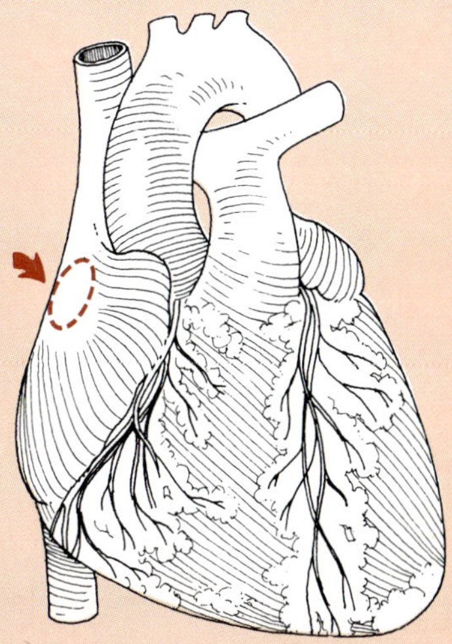

The pacemaker triggers all of the muscle cells in the two atria to contract at the same time. (Imagine what would happen if the fibers contracted at different times.) One-tenth of a second later, the ventricles contract.

An artificial pacemaker can be used if a person's own pacemaker does not work properly. It sends electrical impulses to make the cardiac muscle contract.

12-5 BLOOD FLOW

While you read about the flow of blood through the circulatory system, complete the ***Blood Flow*** *worksheet (Figure 12–17). In this reading, the parts of the circulatory system are in bold type. Label these parts on the worksheet. The numbers in the reading indicate the order in which blood flows through each of the body parts. Number the parts on your worksheet to show this order. Use a pencil in case you want to change an answer.*

Imagine that you are a red blood cell in the left ventricle of the heart. You are about to take a 60-second trip through the circulatory system. On your way, you will travel through a series of blood vessels and pass through the

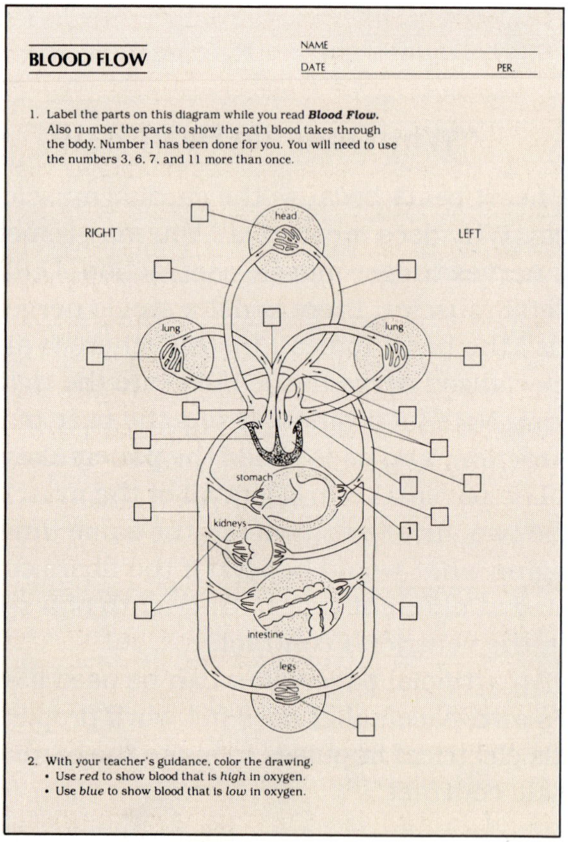
BLOOD FLOW

NAME

DATE PER.

1. Label the parts on this diagram while you read ***Blood Flow.*** Also number the parts to show the path blood takes through the body. Number 1 has been done for you. You will need to use the numbers 3, 6, 7, and 11 more than once.

2. With your teacher's guidance, color the drawing.
   - Use *red* to show blood that is *high* in oxygen.
   - Use *blue* to show blood that is *low* in oxygen.

Figure 12–17

heart again. At the end of 60 seconds, you will be back in the left ventricle. What parts of the circulatory system will you "see" on this trip?

To the Body

When the heart contracts, blood spurts out of the **left ventricle** (1) through the **aorta** (2). This blood is rich in oxygen. On its "journey," it will deliver this oxygen to all of the tissues in the body. Almost immediately, the aorta branches, giving off **arteries** (3) that carry blood to the upper part of the body. (When you feel the pulse in your neck, you are feeling the blood in one of these arteries as it spurts toward your brain and the other parts of your head.)

The main branch of the aorta (3) continues down, carrying blood to the lower parts of your body. This artery keeps branching to deliver oxygen-rich blood to various tissues and organs. For example, some arteries branch to supply blood to the walls of the stomach (4). Others branch to the intestines (5), while yet others go to the kidneys, bladder, and other body organs.

As the arteries continue to branch, they form smaller and smaller vessels. The very tiniest vessels are called **capillaries** (6). When blood reaches the capillaries, it is as far from the heart as it is going to get.

From the capillaries, the blood will start the return trip to the heart through many **veins.** Tiny veins (7) drain the oxygen-poor blood from the capillary beds. (See Figure 12–18.) The small veins empty into somewhat larger ones. These veins come together to form even larger vessels. The veins from the upper part of the body drain into the **upper vena cava** (8). The veins from the lower part of the body drain into the **lower vena cava** (8). The upper and lower venae cavae are the largest veins of all; they return blood to the heart.

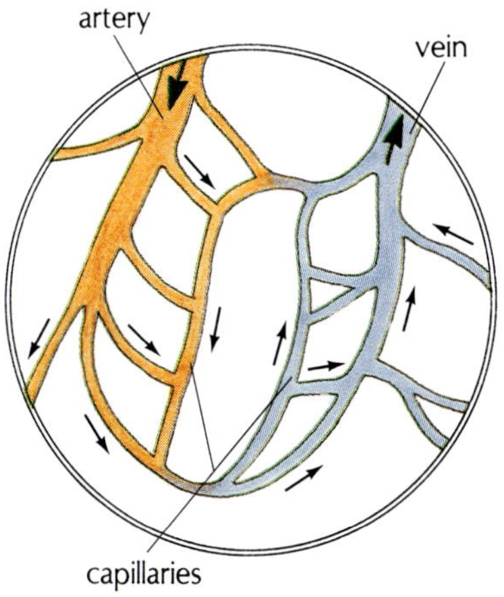

Figure 12–18 Capillaries link arteries and veins.

To the Lungs

The blood in the venae cavae enters the **right atrium** (9) of the heart. At this point, the blood contains lots of carbon dioxide. It must get rid of the carbon dioxide and pick up more oxygen. In order to do this, the blood needs to go to the lungs.

The blood flows from the right atrium into the **right ventricle** (10). When the heart contracts, the blood is forced out into the **pulmonary artery** (11). The pulmonary artery carries the blood to the lungs. Once in the lungs, it branches to form smaller and smaller vessels. The smallest of these arteries branch to form a network of tiny **capillaries** (12) in the lungs. The blood in these capillaries picks up oxygen and gets rid of the extra carbon dioxide. The blood is now ready to return to the heart.

It helps to remember that
- *cardiac* refers to the heart,
- *pulmonary* refers to the lungs.

Small veins drain the blood from the capillary beds. As always, small veins empty into larger veins. These veins eventually join to form the **pulmonary veins** (13). The pulmonary veins return blood to the *left* side of the heart—into the **left atrium** (14). The newly oxygenated blood now flows into the left ventricle . . . and will leave the heart through the aorta again!

Every minute, of every day, throughout your life, blood flows—following this circulatory pathway. Its route never changes directions. It is always the same. *From the left ventricle to the body. Through the body and back to the right atrium. From the right ventricle to. . . .*

ANALYSIS QUESTIONS

1. Which vessel carries blood
   a. to the lungs from the right ventricle?
   b. from the left ventricle to the entire body?
   c. from the lungs to the left atrium?
   d. from the body to the right atrium?
2. Imagine that you are doing a study on blood. For your study, you need to use blood that is rich in oxygen. Which of these structures carry the oxygen-rich blood that you need?

right atrium	left ventricle	aorta
right ventricle	pulmonary artery	coronary artery
left atrium	pulmonary vein	tibial vein

3. Why do doctors worry when fat starts to build up *inside* the aorta?
4. The coronary arteries supply the heart with blood. You may have seen these if you dissected a preserved heart. (See also Figure 12–19 on page 341.)
   a. How do you think blood gets into a coronary artery? Where does it come from?
   b. Where does the blood in a coronary artery go?
   c. Why do you think we need coronary arteries?

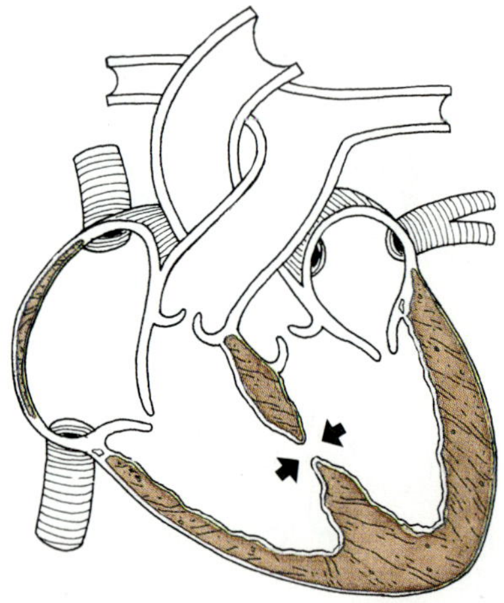

CHALLENGE

Sometimes a baby is born with an opening between the right and left ventricles, such as is shown above. How do you think this affects the baby?

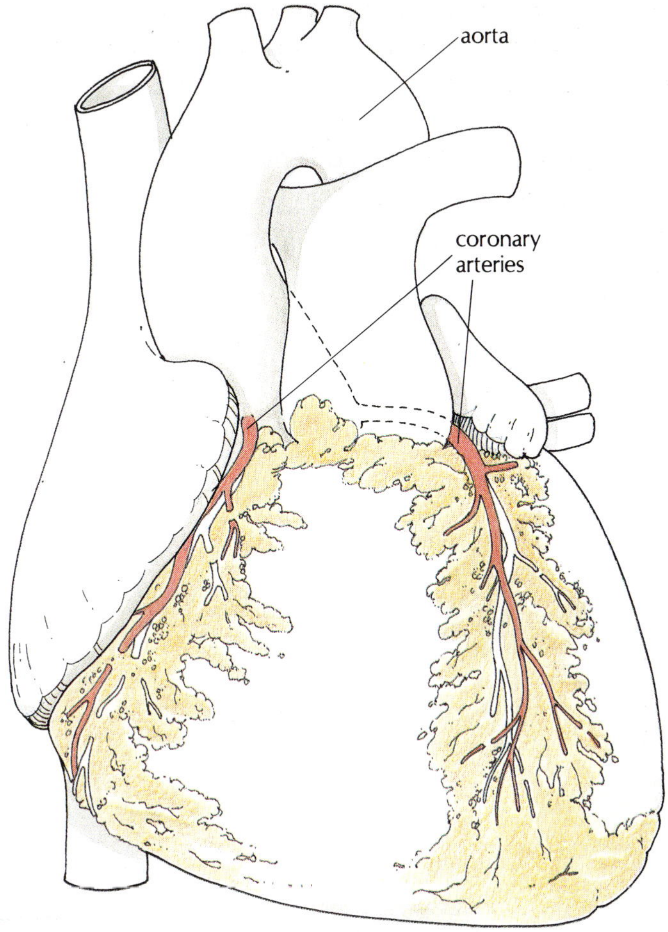

Figure 12–19

EXTENSION ACTIVITIES

1. Use a preserved heart to trace the flow of blood through the vessels and heart. (It may be difficult to identify the larger vessels since they are often damaged when the heart is removed.)
   a. First find the atria and ventricles on the preserved heart. Figure out which is the left side.
   b. The vessel that leaves from the left ventricle is the *aorta.* Find it. (Remember that the blood in the aorta goes to the entire body.)
   c. Blood returns to the heart through the *venae cavae.* Find where these vessels enter the right atrium.
   d. When blood leaves the right ventricle, it goes to the lungs. The *pulmonary artery* takes it there. Probe gently from the inside of the ventricle to find this vessel.
   e. *Pulmonary veins* return blood to the heart. Find these vessels. They enter the left atrium.

f. Look for the valve that blood passes through to get into the aorta. Can you find valves at the bases of the other large vessels?
g. Find the coronary vessels. Follow them up to the large vessels. Where do they attach?

2. Amphibians, such as frogs and salamanders, and some reptiles have three-chambered hearts. Do some research and then compare the structure of this kind of heart with a four-chambered heart. How is the pattern of blood flow similar? How is it different? What are the advantages or disadvantages of these two kinds of heart?

12-6 BLOOD VESSELS

The heart pumps to move blood through the circulatory system. This circulatory pathway is made up of blood vessels. The vessels—arteries, veins, and capillaries—are not all alike. Each kind has its own characteristics.

- How can you tell the difference between an artery and a vein when you look at them through a microscope?
- Why are capillaries so important?

This activity is made up of two parts—this reading and some microscope work. It does not matter which part you do first. Your teacher has set up the microscopes with slides of arteries and veins. Record your microscope observations on the worksheet that is like the one in Figure 12–20.

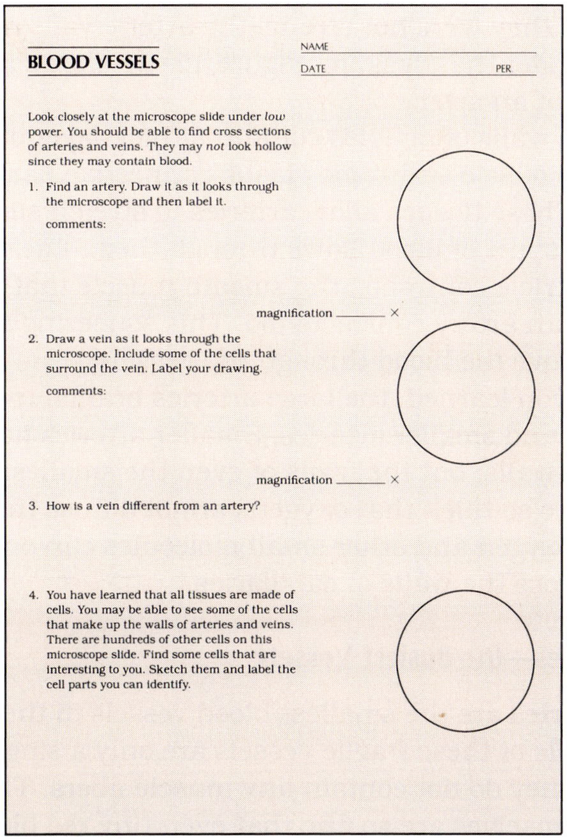

BLOOD VESSELS

NAME
DATE PER.

Look closely at the microscope slide under *low* power. You should be able to find cross sections of arteries and veins. They may *not* look hollow since they may contain blood.

1. Find an artery. Draw it as it looks through the microscope and then label it.

   comments:

   magnification ______ ×

2. Draw a vein as it looks through the microscope. Include some of the cells that surround the vein. Label your drawing.

   comments:

   magnification ______ ×

3. How is a vein different from an artery?

4. You have learned that all tissues are made of cells. You may be able to see some of the cells that make up the walls of arteries and veins. There are hundreds of other cells on this microscope slide. Find some cells that are interesting to you. Sketch them and label the cell parts you can identify.

Figure 12–20

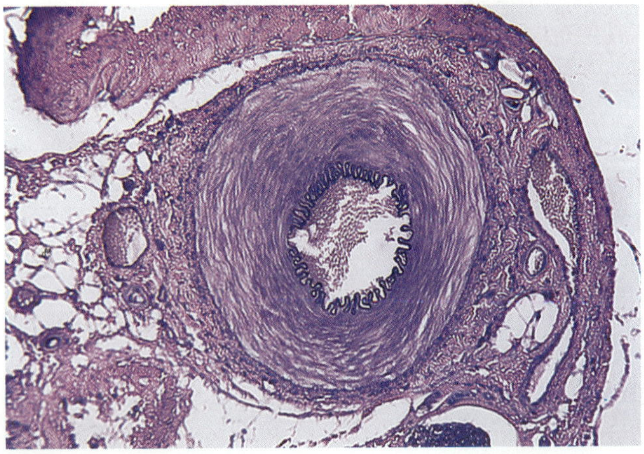

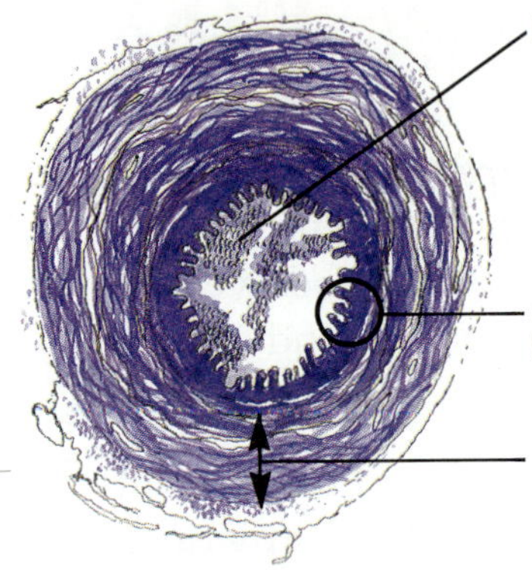

Figure 12–21 Cross section through an artery. In the photograph, try to identify the parts labeled in the drawing. What do you think is in the center of the artery?

Arteries

Every time the heart beats, blood surges through two large arteries—the pulmonary artery and the aorta. Like all arteries, these two carry blood away from the heart. If you could look at a human aorta, you would find it to be about as big around as a garden hose. You would also note that it has very sturdy walls. Artery walls have to be strong. Each time the heart beats, blood surges into the arteries and pushes against the walls. If they were not strong, the artery walls might burst under the pressure. Figure 12–21 shows a cross section of an artery.

The walls of arteries contain a thick layer of smooth muscle and some elastic connective tissue fibers. These tissues allow arteries to stretch slightly when a spurt of blood flows through them. Each time the ventricle contracts, the smooth muscle that wraps around an artery contracts too. This squeezing action helps move the blood through the vessels.

As you learned, the large arteries branch to form smaller and smaller arteries. Smaller arteries have thinner walls; but the walls of even the smallest arteries are so thick that oxygen cannot diffuse through them. Oxygen and other small molecules can only diffuse across the walls of capillaries.

Capillaries—the Busiest Vessels

Capillaries are the smallest blood vessels in the body. The walls of these fragile vessels are only a single cell thick. They do not contain any muscle fibers. The vessels themselves are so fine that even tiny red blood

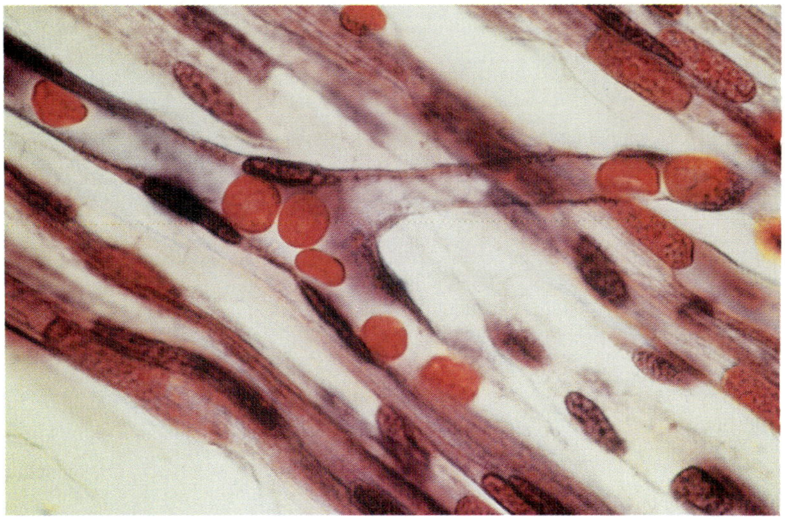

Figure 12–22 This photograph of a capillary was taken through a microscope. Notice that only one blood cell can flow through a capillary at a time.

cells must bend and twist to squeeze, in single file, through their narrow openings. (See Figure 12–22.)

It is impossible to sketch a "real" capillary pathway. There are so many of these microscopic vessels that the sketch would look like a tangled mass of spider webs. Most tissues are so laced with capillaries that just about every cell is within *one-millionth of an inch* of a capillary. This is important because oxygen and food molecules must be able to diffuse out of the capillaries to supply the cells. In addition, carbon dioxide and other waste molecules must be able to diffuse out of the cells and into the capillaries. No other blood vessels have walls that are thin enough for diffusion to take place. As you can see, these vessels are where the important functions of circulation occur. (See Figure 12–23.)

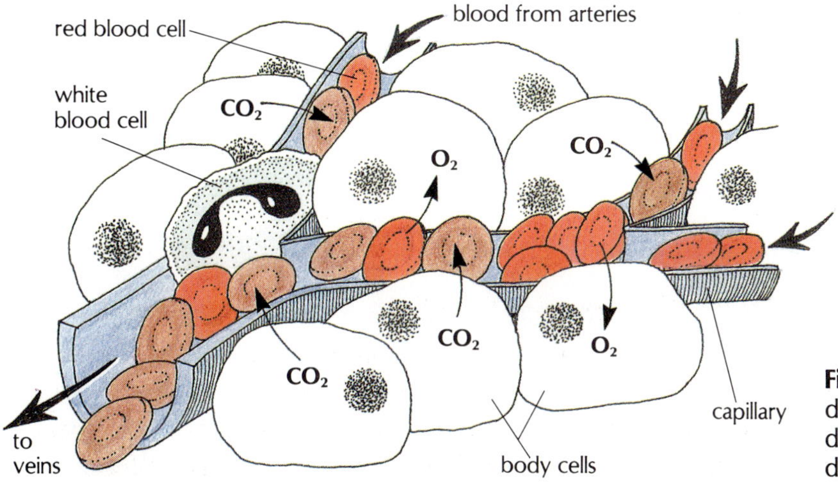

Figure 12–23 Small molecules can diffuse across capillary walls. What diffuses into capillaries? What diffuses out?

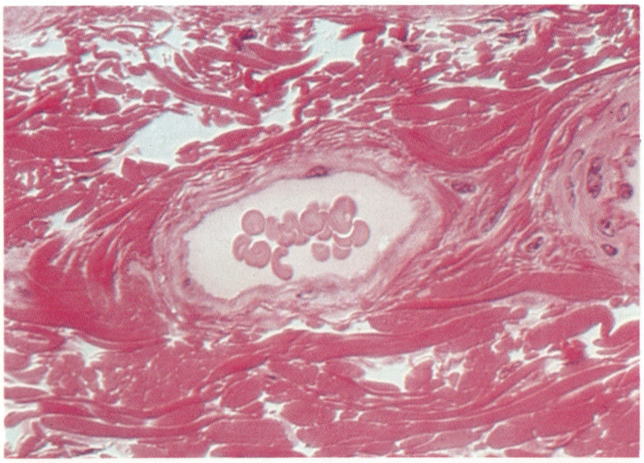

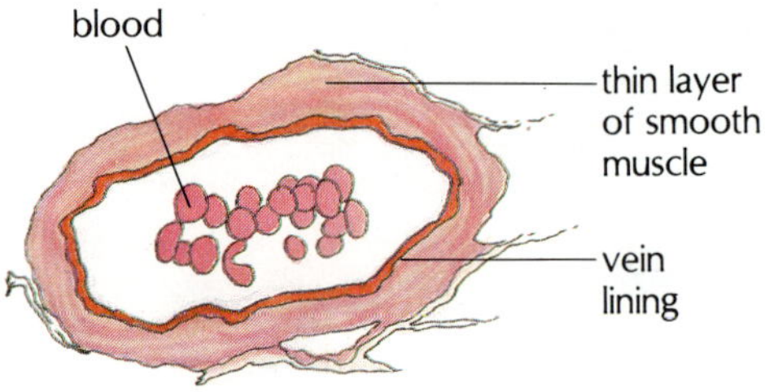

Figure 12–24 Cross section through a vein. In the photograph, try to identify the parts labeled in the drawing. How does the vein compare with the artery in Figure 12–21?

Veins

Blood is drained from the capillary beds by very small veins. Then, from these small veins, the blood flows into larger and larger veins on its way to the heart. The walls of veins are thinner than artery walls. They contain less smooth muscle and elastic tissue. Because their walls are thinner, veins may be collapsed when you look at them under a microscope, as shown in Figure 12–24.

In the veins, blood flows fairly slowly. That's because there is not much pressure on the blood. In fact, there is so little pressure that the blood may flow backwards. In order to prevent this, most veins contain valves. These one-way valves open when blood is flowing toward the heart; they close when blood starts to flow the wrong way. Veins eventually combine to form the large veins that return blood to the heart—the venae cavae.

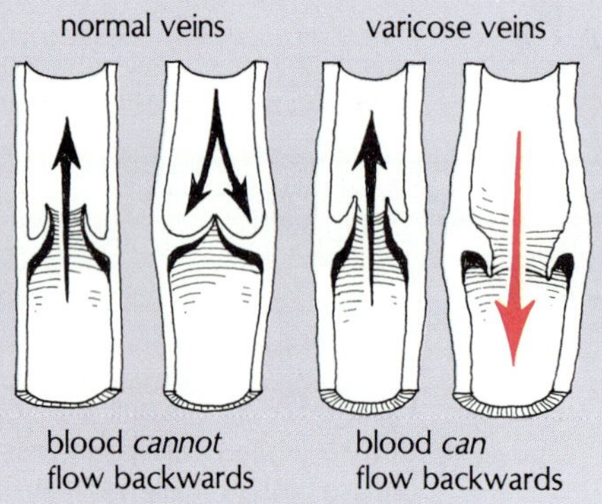

Faulty Valves

When the valves do not close properly, the blood backs up and causes the veins to swell. You may have seen swollen veins. They look like raised, bluish lines on a person's legs. These are called **varicose veins.**

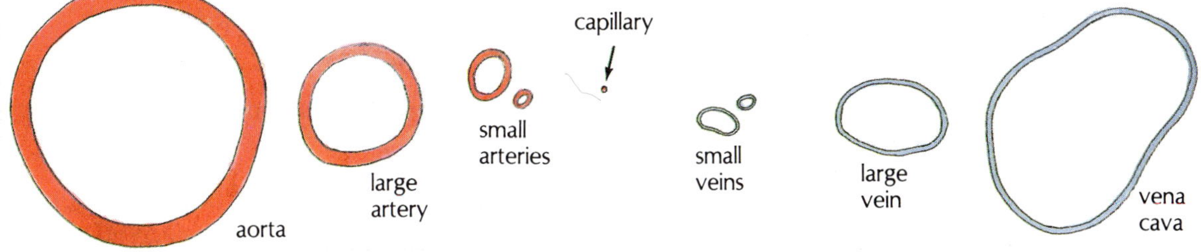

Figure 12–25 Cross sections of blood vessels

Sizes

If you could compare the vessels as they would look in cross section, you would notice quite a difference in their sizes. The largest artery—the aorta—and the two largest veins—the venae cavae—are each about 2.5 cm (1 in.) wide. A capillary, however, is much smaller than the dot at the end of this sentence. Note the sizes of the vessels shown in Figure 12–25.

ANALYSIS QUESTIONS

1. Copy this table onto your own paper and then fill it in.

SUMMARY TABLE—ARTERY, VEIN, AND CAPILLARY

	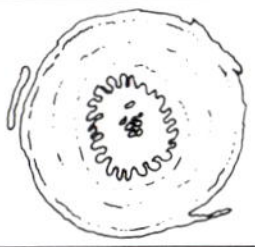	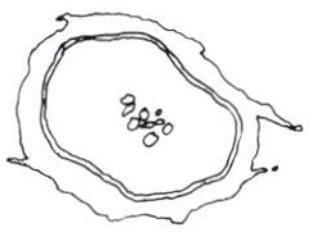	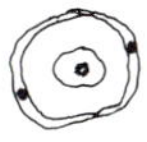
Describe its walls. (thick, medium, or thin?)			
What are its walls made of?			
This vessel is a/an _____ .			
Does blood flow toward or away from the heart in this vessel?			

Decide whether statements 2–9 are true or false. If a statement is false, explain why it is false.

2. T F All arteries carry blood away from the heart.
3. T F All veins carry oxygen-poor blood.
4. T F A cut vein would probably bleed more than a cut artery.
5. T F Arteries and veins look alike.

6. T F Gases and other small molecules are exchanged between body cells and the blood in arteries.
7. T F The upper vena cava contains oxygen-poor blood.
8. T F Blood cells never get out of blood vessels.
9. T F Pulmonary arteries take blood to the lungs.
10. You have a deep cut on your arm. Blood is gushing from it in spurts. Do you think that you have cut an artery or a vein? Explain your answer.
11. A stroke occurs when a blood clot blocks an artery to the brain. How will this affect the brain?

CONCLUSION

Write a few sentences to answer the focusing questions on page 343.

EXTENSION ACTIVITY

Figure 12–26

The tail of a live goldfish is a good place to observe the movement of blood through capillaries. Soak a piece of cotton with water. Wrap a live goldfish in the cotton and place it in a shallow dish. (See Figure 12–26.) Do not let the goldfish get dry! Work quickly. Place the dish on the stage of a dissecting microscope. Focus on the tail. Describe the size of the capillaries and the movement of the red blood cells.

A Whale of a Vessel

Many years ago a scientist was dissecting a whale's heart when his foot slipped. He slid down into the heart and ended up with both feet in the aorta. Luckily, he was rescued by a bystander. To show how narrow his escape was, he sliced a ring of tissue from the aorta. It could pass over his head and down to his feet without even stretching! What would have happened if he had slipped a little farther?

12-7 A SHOCKING EXPERIENCE

Marcie Larkin and her friends could barely sit still. They had won the basketball game in the last 10 seconds! Mrs. Larkin was laughing with the kids when she started through the green light. No one noticed the oncoming car.

Marcie didn't remember hitting her head on the windshield. Dazed, she reached up to touch her head. She felt something warm and wet. Was it blood? Someone was telling her to lie down on the sidewalk. She felt numb, and hot, and weak. Where was everyone? Were they okay?

Mr. Walsh had taken a first aid class but he had not expected to need it. He looked carefully at Marcie. Her head was bleeding and she looked pale. She did not seem to have any other injuries or broken bones.

While Mr. Walsh wrapped his handkerchief over the cut, he talked calmly to Marcie, "Everything's going to be all right. Your mother and friends are okay. They're being helped now. Here, let me put this coat under your feet so that your legs will be higher than your heart." As he continued to talk, he covered her with his jacket. He did not want Marcie to get cold.

In the Emergency Room

A few hours later, it all seemed like a bad dream. The doctor was saying that Marcie and Mrs. Larkin could go home. She explained that Marcie had been in shock but was otherwise unharmed. Marcie was not reassured and said, "I don't remember much about the accident except that a man was trying to help me. I felt sick to my stomach, hot, cold, and sweaty all at the same time. Are you sure I'm okay?"

"Yes, those are all normal symptoms of shock," said the doctor. "Mr. Walsh, the man who helped you, has been calling to find out how you're doing. He did a good job of handling the situation. You were weak and pale because your capillaries had all opened. There wasn't enough blood in your larger vessels to keep your circulation going."

"Oh," Marcie interrupted, "that's why he told me to lie down and why he made me keep my legs up."

"That's right," the doctor agreed. "He wanted to make sure your blood was flowing to your heart. You had enough blood, it just wasn't circulating; so you were going into shock. If you had gone into shock because of blood loss, you would have needed a transfusion. A little rhyme that will help you remember what to do goes like this:

If the face is pale, raise the tail.
If the face is red, raise the head.

Makes sense doesn't it?"

SHOCK . . . What Is It?

Figure 12–27

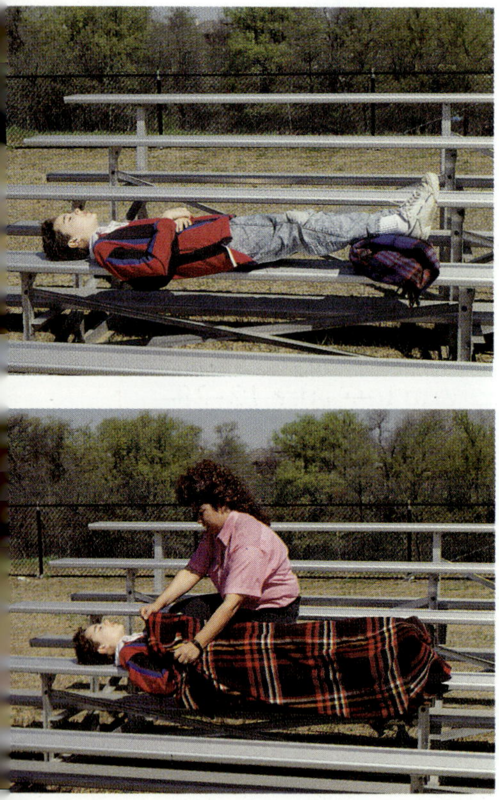

Marcie had experienced one of the body's reactions to a particularly stressful situation. When a person goes into shock, many capillary beds deep within the body open up. The blood stays in these capillaries and does not circulate. As a result, the body organs do not get enough oxygen and may stop functioning.

Some symptoms of shock include the following:

- sweating, faintness, nausea
- panting
- pale, cool, and moist skin
- dropping blood pressure and an increase in pulse rate
- drowsiness, confusion, and loss of consciousness

If you think a person may be in shock, take these first aid steps. (See also Figure 12–27.)

1. Have the person lie down. Raise the legs about 30 cm (1 foot).
2. Keep the person warm. She or he may feel hot at first but a chill will follow.
3. Do not give the victim anything to eat or drink.
4. Be reassuring.
5. **Get medical help!** If untreated, shock can lead to death.

12-8 PRESSURE ON YOUR ARTERIES

When the heart contracts, blood spurts out of the ventricles. This surge of blood puts pressure on the walls of the arteries. Between beats, the pressure drops. A blood pressure reading is a measurement of both of these pressures.

- Why do you need a stethoscope to measure blood pressure?
- What do the numbers in a blood pressure reading mean?
- Why is high blood pressure dangerous?

In this activity, you will learn to take someone's blood pressure. After doing that, you should be able to answer these questions.

MATERIALS

blood pressure cuff
stethoscope
reading: ***Blood Pressure and Health*** (page 354)

PROCEDURE

Part A. Finding the Artery

1. Sit comfortably and quietly.
2. Roll up your left sleeve.
3. Rest your left arm on the table in front of you. Your elbow should be at about the same level as your heart.
4. An artery crosses the inside of the elbow. Use two fingers to feel the pulse of the artery, as shown in Figure 12–28. Remember where this pulse point is. You will need to find it again later.

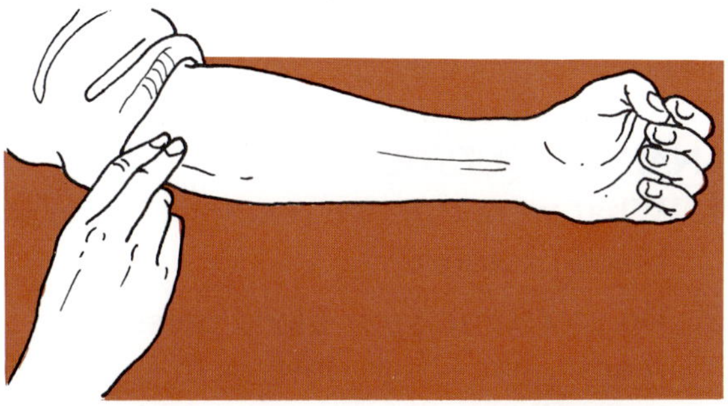

Figure 12–28

Part B. Inspecting the Equipment

From now on, work with a partner. Choose a role: pressure taker—takes your partner's blood pressure, *or* subject—lets your partner take your pressure.

You will need a stethoscope and a blood pressure cuff to take a blood pressure. Use the diagram in Figure 12–29 to help you identify the parts of the cuff. **Be careful with this equipment. It is very sensitive.**

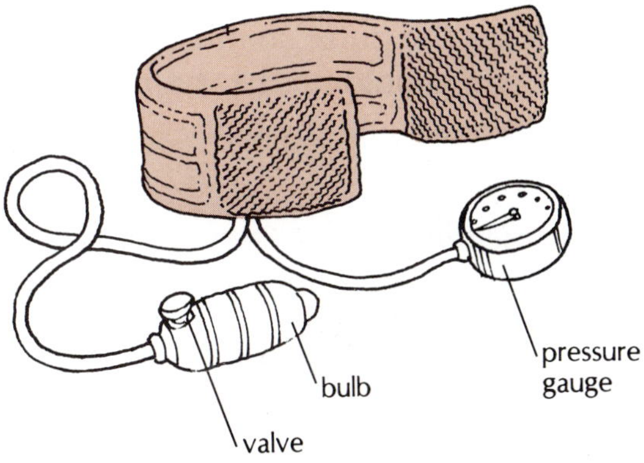

Figure 12–29 Blood pressure cuff

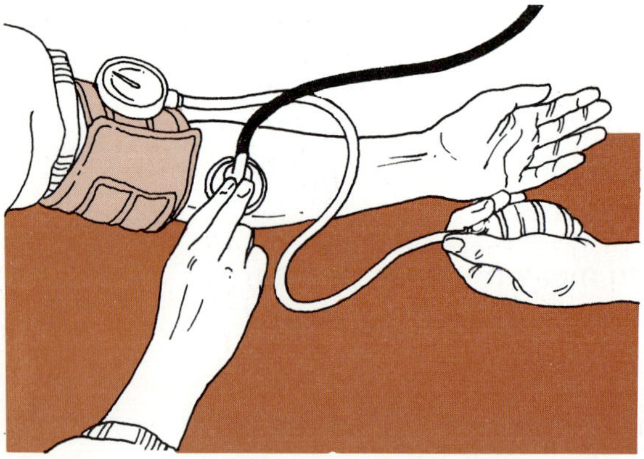

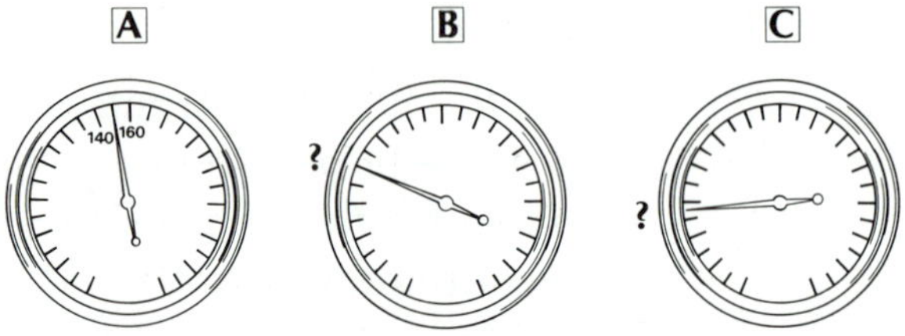

Figure 12–30

Part C. Getting Ready to Measure Blood Pressure

1. Put the cuff around the subject's left arm. It should be between the elbow and the shoulder.
2. The tubes on the cuff should come over the artery. Wrap the cuff so it is snug but not too tight. Fasten it with the Velcro.
3. Close the valve and squeeze the bulb a couple of times. Watch the pressure gauge reading. It should go up until the pressure reaches 60.
4. Adjust the valve so the air slowly escapes. It should take 5 to 10 seconds to return to zero.
5. Have the subject hold the end of the stethoscope over the artery.
6. If you are the pressure taker, put on the stethoscope and be ready to listen for the pulse sounds. Listen now; you should not hear anything. Continue to hold the bulb.

Part D. Measuring Blood Pressure

Read this procedure *before* you take a blood pressure reading!!

1. Close the valve and pump the bulb until the gauge reads 150, as shown in Figure 12–30A.
2. Loosen the valve so air escapes. Look at the number on the gauge as soon as you hear pulse sounds, as shown in Figure 12–30B. Remember this number.
3. Keep listening while the air is being released. When you cannot hear pulse sounds any more, look at the gauge again. Remember the number.
4. Record both numbers.
5. Remove the cuff from the subject's arm.

Now switch roles. Then switch again. Each person should have two chances to take a blood pressure.

Part E. Finishing Up

Read ***Blood Pressure and Health*** starting on page 354. Then answer the analysis questions.

ANALYSIS QUESTIONS

1. What were the two readings for your blood pressure?
2. Do you think your blood pressure reading was high, low, or in the normal range?

3. List 3 things that could cause a blood pressure reading to be wrong.
4. What do the two numbers that make up a blood pressure reading mean?
5. What would a blood pressure reading of 0/0 mean? Explain your answer.
6. In which vessels is blood pressure the *lowest*—veins, capillaries, or arteries? Explain.
7. Your dad's blood pressure reading is 152/105. He doesn't take his high blood pressure medication because he feels fine. What is one problem he might experience from his high blood pressure?
8. Ms. Roth has high blood pressure. If you could measure the pressure throughout her body, where would you get high readings? Choose the places from the list below where the blood pressure would be high.

upper left arm	lungs	eyeball
upper right arm	brain	left foot
stomach wall	aorta	kidney

9. Jason and Wes are both 13 years old. Jason's blood pressure measured 95/65. Wes' blood pressure reading was 128/87. The nurse took both measurements under the same conditions. Which of these statements explains the results?
   a. Something is probably wrong. Boys the same age should have the same blood pressure.
   b. Something is probably wrong. Jason's blood pressure is too low for a 13-year-old boy.
   c. The measurements are reasonable. Blood pressure varies from person to person.
   d. The measurements are reasonable. Something is wrong only if the difference is more than 40.
10. What are several things *you* can do to lower your chances of developing hypertension?

CONCLUSION

Write a paragraph that answers the three focusing questions on page 351.

EXTENSION ACTIVITIES

1. Find out what affects blood pressure. Take your blood pressure at various times of the day. Take it before and after eating, before and after exercising, or when you are excited, nervous, or depressed. Take it standing up, sitting down, and with your arm over your head. When is your blood pressure the lowest? When is it the highest?
2. All animals have a blood pressure. Call a veterinarian. Find out if vets take blood pressures of animals. Then ask how they can evaluate the circulatory system of an animal such as a dog or if pets can get heart disease.
3. Have someone else check your blood pressure. Ask someone who knows how to use the equipment, such as the school nurse, a parent, or a neighbor. *Always* ask what your blood pressure is when it is taken.

Blood Pressure and Health

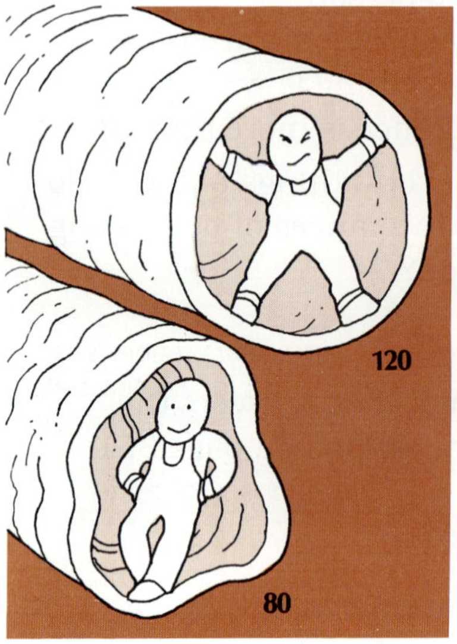

Figure 12–31 Blood pressure is a measure of the pressure that blood puts on your arteries.

When blood is pumped out of the ventricles, it pushes against the walls of the arteries. This pushing, or pressure, is what keeps blood circulating through the body. A "blood pressure reading" is a measurement of how hard the blood "pushes" against the walls of the vessels.

A blood pressure reading consists of two numbers. Let's say, for example, that a health care provider tells you your blood pressure is 120 over 80. What does this mean? The larger number measures the pressure your blood puts on your arteries when it spurts out of your heart. The smaller number tells how much pressure is on the arteries between beats. The cartoon in Figure 12–31 will help you understand a blood pressure reading.

Pressure Changes

Normal blood pressure varies from person to person. It even varies throughout the day for the same person. For instance, your blood pressure changes when you shift your position. It will differ depending upon whether you are lying down, sitting, or standing. Blood pressure is also affected by such things as bladder fullness, amount of exercise, nicotine usage, and amount of food in the stomach. Blood pressure changes to meet the body's needs.

Blood pressure also changes with age. It increases slowly as we get older. A six-month-old baby might have a reading of 90 over 60. An average reading for someone your age is 115/65.

High Blood Pressure

By checking blood pressure, doctors can learn about the condition of the heart and blood vessels. If the pressure is too high, the heart is working too hard to pump blood through the body. This strain is bad for the heart. High blood pressure is not good for the arteries either. Very high blood pressure could cause a vessel to burst.

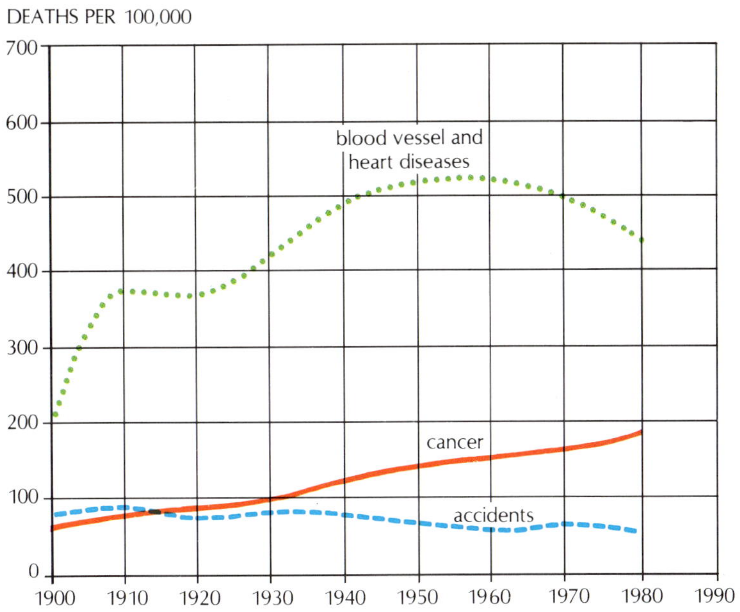

Figure 12–32 The number one cause of death in the United States is diseases of the heart and blood vessels. High blood pressure contributes to many of these diseases.

If this happens in the brain, it can cause a stroke. A pressure above 140 over 90 (140/90) is usually considered high.

While high blood pressure is rare in children, it affects about one out of every twenty adult Americans. Will you be one of them?

If one of your parents has high blood pressure, your chances of developing it are higher. People who are overweight are also more likely to have high blood pressure. Emotions such as anger and pain or stressful situations can increase blood pressure. Some studies have linked hypertension—high blood pressure—with high salt diets.

People with high blood pressure rarely show any symptoms and may not even be aware they have it. This doesn't mean that high blood pressure can be ignored. Even mild high blood pressure may lower life expectancy. In most cases, it cannot be cured; but it can be controlled by diet, exercise, and medication.

Low Blood Pressure

It is usually healthier to have a *low* blood pressure than a *high* one. However, if a person's blood pressure is too low, the blood will not circulate. When this happens, there may not be enough blood flowing to the head. The person will feel faint or dizzy.

12-9 YOU AND HEART DISEASE

High blood pressure and high fat diets are not good for the heart. They both can contribute to heart disease, but they are not the only factors that can lead to heart disease. Scientists have identified other factors that put a person at risk.

You may wonder if you are at risk for heart disease. Take this short self-inventory to find out. Read the statements below. Keep track of the ones that are true for you.

- I get at least 30 minutes of exercise three times a week.
- I am not overweight.
- I try not to eat fatty foods such as fried foods, fatty meat, and butter.
- I use very little salt.
- I do not smoke.
- Neither one of my parents has had a heart attack.
- Neither one of my parents has high blood pressure.
- I am usually a calm person.

How many items apply to you? The more that do, the healthier is your heart. You also probably have thought of some questions. Usually, when students read this list, they ask questions such as the ones that follow.

What is heart disease anyway?
Heart disease is any medical problem that involves the heart or blood vessels. Two common examples of heart disease are heart attacks and hardening of the arteries.

According to the list, exercise is good for the heart. Why would this be true?
When you exercise, you build up your cardiac muscle. A strong heart can pump more blood so it doesn't have to pump as often. In the long run, a strong heart will not "wear out" as quickly. Exercise also improves circulation and helps with weight control.

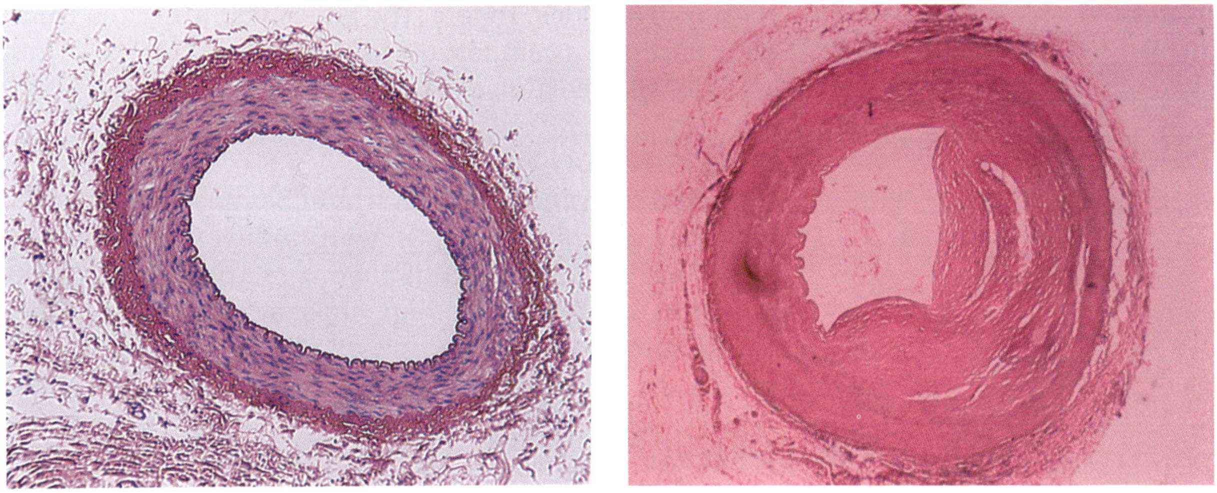

Figure 12–33 These are cross sections through two arteries. The one on the left is healthy; the one on the right is partly clogged by fat.

The list stresses not being overweight. Is fat bad for the heart?
People who are overweight are more likely to have other problems that can lead to heart disease. For instance, they may develop diabetes or high blood pressure. Both of these conditions are hard on the heart. Heavier people also have more fat circulating in their blood.

Fat in the blood doesn't sound good; but how can that hurt the heart?
Some fats, such as cholesterol, can form deposits on the walls of the arteries. As fat builds up, the insides of the arteries become narrower. (See Figure 12–33.) Less blood can flow through the arteries. The parts of the body that receive blood from these arteries will not get enough blood.

My parents are careful about what they cook because they both have "high triglycerides." What does that mean?
Because triglycerides are a kind of fat, your parents have probably had their blood tested for fat levels. They learned that they have too much fat circulating in their blood. They know that cutting down on certain fats in the diet may lower the level of blood fat. You're lucky! Your folks' cooking is good for *your* heart, too!

Me? I don't have to worry about heart disease!
Oh, yes, you do! Studies have shown that fat can start accumulating in blood vessels of children. You may be at risk for heart disease since both of your parents have

high triglycerides. In fact, the genes you inherited from your parents are important indicators of whether or not you will get heart disease.

Can I check my genes?
No, but a heart gene screen would be nice. You should be aware of what goes on in your family. Listen and ask questions about the health of your relatives. High blood pressure, heart attacks, high cholesterol or triglyceride levels, strokes . . . these are all clues to family heart problems.

Heart disease kills more people in the United States each year than all forms of cancer combined.

National Institute of Health Consensus Report, December 1984

What can I do to prevent heart disease?
Look over the self-inventory list on page 356 again. Can you change your lifestyle in order to lower your risk of heart disease?

It's your turn to answer a question.

What can you do to have a healthy heart?

12-10 BLOOD—WHAT IS IT?

"You've been sick for a couple of weeks now. I recommend that we do a blood count. It's safe, simple, and almost painless. This test, which requires about two teaspoons of blood, will tell us more about what's going on than any other lab test. By doing this, we can check your hemoglobin concentration, red blood cell count, and white blood cell count."

Your Doctor

- How do you prepare a blood smear?
- What does blood look like through a microscope?

You will prepare a microscope slide of your own blood to answer these questions.

MATERIALS

2 clean microscope slides
soap
cotton ball
alcohol
sterile, disposable lancet
Wright's stain in dropper bottle
2 beakers (already set up for you)
microscope
bleach solution
clock or watch with second hand
prepared slides: leukemia
anemia

PROCEDURE

Part A. Preparing a Microscope Slide

1. Wash and dry your two glass microscope slides.
2. Your hands must be very clean so wash them with soap and water.

CAUTION
Since blood can carry infectious diseases, this activity describes safe procedures for handling blood. Follow the directions carefully.

3. Put rubbing alcohol on a cotton ball. Use it to clean the finger you are going to prick. After you have cleaned your finger, do not let anything touch it.
4. Take the lancet out of its wrapper. **Since it is sterile, do not let the sharp end of the lancet touch anything.**
5. Quickly prick the top of your clean finger with the lancet. Put the used lancet into the beaker labeled "Used Lancets." **It must *not* be reused.**

Figure 12-34

6. Squeeze 1 drop of blood onto the end of the microscope slide, as shown in Figure 12–34.
7. Use another slide to smear the drop of blood.
   a. Set the edge of the slide near the drop. Then slowly move it towards the drop so the blood fills the angle between the two slides, as shown in Figure 12–35.

Figure 12-35

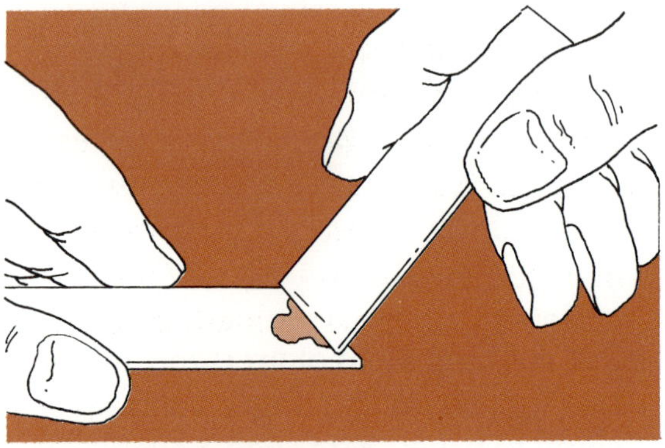

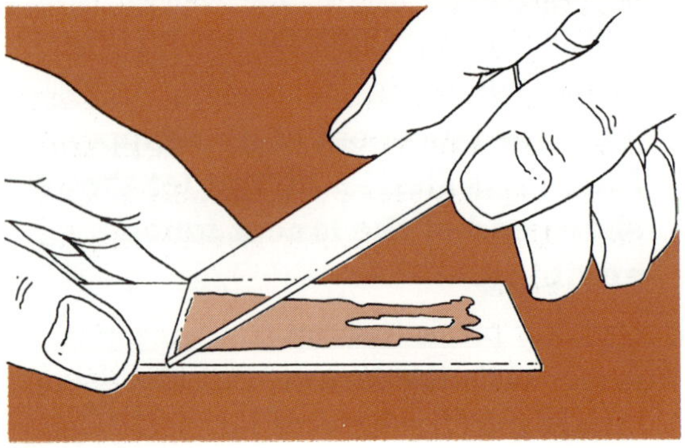

Figure 12-36

   b. Slowly and smoothly push the second slide away from the drop, as shown in Figure 12–36. (The blood will form an even smear.)
   c. Set second slide aside.
8. Look at the slide through the microscope. Use low power first, then high power. Describe what your blood looks like.

Part B. Staining Your Microscope Slide

1. Drop 1 or 2 drops of Wright's stain onto the blood smear. Leave on for one or two minutes.

CAUTION

Wright's stain will discolor almost anything it touches including books, clothes, desks, and fingers. Handle it carefully.

---

2. Dip the slide into the beaker of water labeled "First Wash," as shown in Figure 12–37. Leave it there for 30 seconds.
3. Dip the slide in and out of the beaker labeled "Second Wash."
4. Let the slide air dry.

Figure 12-37

Part C. Looking at Your Blood

1. Hold the slide up to the light. Can you see any blood cells?
2. Examine your slide with the microscope. Use low power first, then high power.
3. While looking with high power, draw what you see. Label red blood cells, white blood cells, and platelets.

> *Red blood cells* are pale pink and round. They do not have nuclei.
> *White blood cells* are larger and darker. They have dark purple nuclei.
> *Platelets* look like blue specks.

4. Clean your microscope slides and your desk top with the bleach solution. (Bleach should kill any infectious microorganisms that are present.)
5. Answer the analysis questions.

ANALYSIS QUESTIONS

1. Can you see blood cells without using the microscope? If so, what do they look like?
2. Which blood cells were most numerous in your blood smear—red blood cells, white blood cells, or platelets?
3. List 3 differences you noticed between red and white blood cells.
4. Why did the procedure say to put Wright's stain on the blood smear?
5. Several students could not see anything on their microscope slides. List at least 3 possible reasons for their problem.

CHALLENGE

Your teacher has two prepared microscope slides. They are labeled A and B. One is a blood smear from a person with leukemia. The other is from a person with anemia. Decide which is which, then answer the questions below. You will also need to read ***Blood and Blood Tests*** starting on page 362 before you can answer the questions.

- Which microscope slide is the blood smear from the person with leukemia, A or B? Explain your decision.
- How does the leukemia blood smear compare with your blood smear?
- How does the anemia blood smear compare with your blood smear?

CONCLUSION

Answer the first focusing question on page 359.

EXTENSION ACTIVITIES

1. Think of a method for counting red and white blood cells using a microscope. Use your blood smear to test your method. Have someone else try your technique. How do their numbers compare with yours? If you are curious, find out how medical technologists count blood cells.
2. Imagine that you just found out that you have leukemia. Write a story about the first few weeks after you learned about this diagnosis. In your story, include what leukemia is and what treatments are suggested. Consider these questions when you plan your story:

   What causes leukemia?
   Could it have been prevented?
   What worries you most about having this condition?

12-11 BLOOD AND BLOOD TESTS

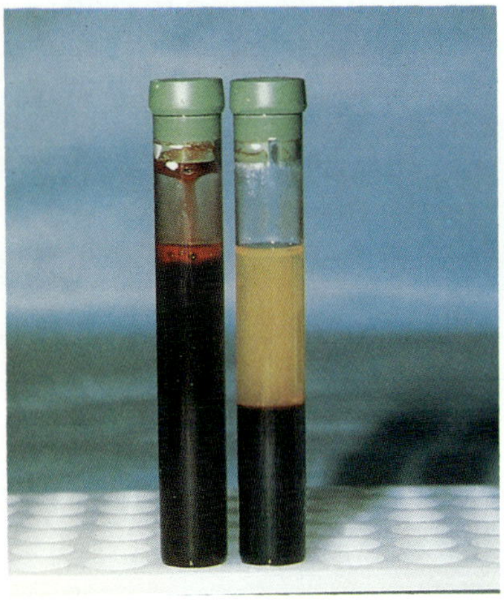

Figure 12–38 The cells can be separated from the plasma in a sample of blood. The blood in the test tube on the left has just been drawn. The test tube on the right shows what blood looks like when the cells have settled to the bottom of the tube.

So far, you have studied the heart and blood vessels. These are both important parts of the circulatory system. But, in order to understand the purpose of the circulatory system, you also need to know about blood.

When you look at blood under a microscope, you begin to realize that it is more than just a red liquid. It can be separated into two basic parts—fluid, or plasma, and cells. The cell portion contains red blood cells, white blood cells, and platelets. The makeup of blood is illustrated in Figure 12–38.

Plasma

Plasma is yellowish in color. It makes up a little more than half of the blood, about 55%. Plasma is mostly water but it also contains thousands of different substances. These include vitamins, glucose, fats, enzymes, hormones, and a variety of proteins. Two groups of proteins are especially important. One group helps blood clot. The other proteins are antibodies for fighting germs.

Sometimes a physician will suggest a blood test to check substances in your plasma. For instance, you may have your blood drawn in order to have your sugar level checked. Blood fats are also commonly analyzed from a blood sample.

Red Blood Cells

Most of the cells in blood are **red blood cells.** They are small, disk-shaped cells with depressions on both sides. (See Figure 12–40 at the bottom of page 363.) We need lots of these cells because they carry oxygen and carbon dioxide. These gases do not float around in the blood cells. Instead, they bind to large molecules in the cells called **hemoglobin.** These cells could not transport as much oxygen as they do if they did not contain hemoglobin.

A red blood cell is a package of hemoglobin molecules. If you could "unzip" the red cell membrane, you would see thousands of hemoglobin molecules and not much else. They do not even have nuclei. For this reason, a red blood cell is a good example of a highly specialized cell.

Blood is pumped out of the heart with such force that it takes only about a minute for a blood cell to travel throughout the body and return to the heart. This puts a lot of wear and tear on a cell. Because it does not have a nucleus, a red blood cell cannot repair any damage that might occur. Thus, the average red blood cell lives only about 120 days.

Several blood tests can be used to check red blood cells. A *red blood cell count* is exactly what it sounds like. It involves counting the number of red cells in one cubic millimeter of blood. There should be about 5 million red cells in this small amount of blood. (See Figure 12–39.)

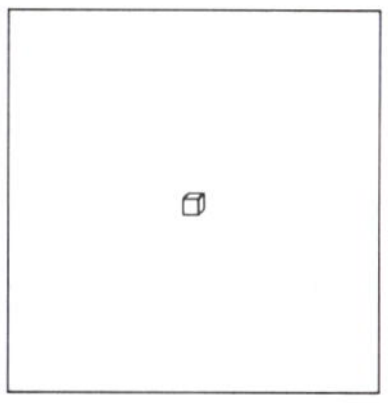

Figure 12–39 Five million red blood cells can fit in one cubic millimeter of blood. (What size box would it take to hold something larger, such as five million marbles?)

A second test measures the *amount of hemoglobin* in each cell. Sometimes a person can have the right number of red blood cells but not enough hemoglobin. Why would this be a problem?

Both of these tests are ways to check for anemia. This condition results when blood cannot carry enough oxygen. When this happens, a person may feel tired and weak and look pale.

White Blood Cells

While **white blood cells** are called blood cells, they actually are found throughout the body. These cells can squeeze out through the little spaces in the capillary walls. Once they are in the tissues, they can do their

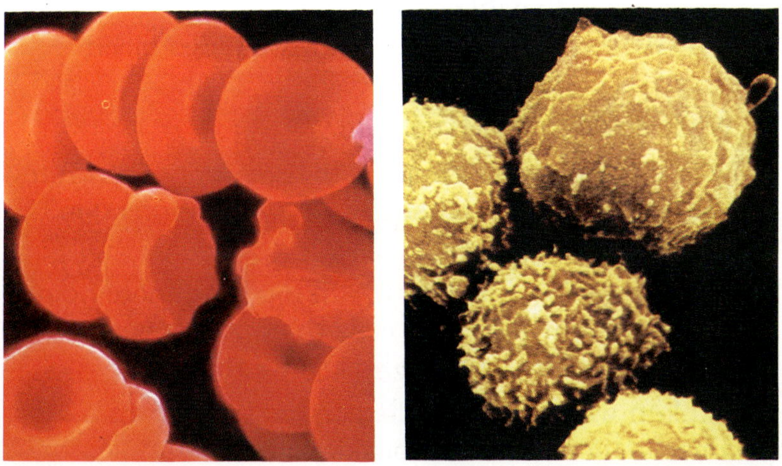

Figure 12–40 The red blood cells on the left have been magnified 1,300 times. How do they compare with the ones you saw through your microscope? The white blood cells on the right have been magnified 1,650 times. Compare them with the red blood cells. In what ways do these two cell types differ?

jobs. There are several different kinds of white blood cells, each one having a specific job. White blood cells fight infection; some do so by producing antibodies. Others engulf bacteria and destroy them. Still other white cells destroy harmful substances and break down fragments of old cells. (See Figure 12–40 at the bottom of page 363.)

One helpful medical test is a *white blood cell count.* There are usually about 8,000 white blood cells in a cubic millimeter of blood. If there are more, it could mean that the person has an infection.

Extremely high numbers of white blood cells can be a sign of leukemia. Leukemia is a cancer of the white blood cells. It results when too many white blood cells are being produced. These extra cells interfere with the functioning of other parts of the body.

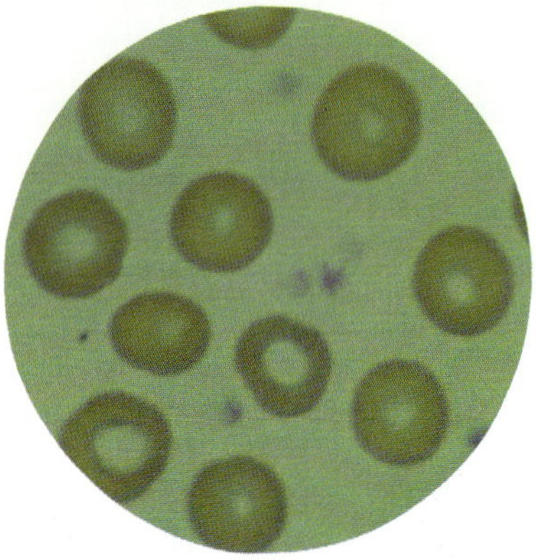

Figure 12–41 These platelets have been magnified 1000 times. Platelets are much smaller than red and white cells.

Platelets

The third type of "cell" is the **platelet.** Platelets are not really cells; they are actually just pieces of cell membrane. They help stop bleeding when a blood vessel is injured by plugging the hole. If the injury is serious, platelets are important in causing a blood clot to form. A person with a low platelet count could have a bleeding problem. (See Figure 12–41.)

ANALYSIS QUESTIONS

1. Compare and contrast red and white blood cells.
   a. How are they alike?
   b. How are they different?
2. What are platelets?
3. Why do you need hemoglobin in your red blood cells?
4. What do white blood cells do?
5. What do you think happens to the number of white blood cells when a person is fighting an infection? Explain your answer.
6. People who live on a mountain top have more hemoglobin in their red blood cells than people who live by the seashore. Why do you think this is so?
7. Explain why a single-celled organism, like an amoeba, does not have blood.

In one cubic millimeter of blood, there are about . . .
5,000,000 red cells
8,000 white cells
300,000 platelets

"Blood Red"

Ask any child the color of blood and you will immediately be told, "It's red." However, if *you* look at blood under a microscope, it doesn't look red at all. It looks red only when you see masses of cells. The red color is due to the hemoglobin in red blood cells. The shade of red varies according to the amount of oxygen that is bound to the hemoglobin. Oxygen-rich blood looks bright red; oxygen-poor blood looks bluish-purple. You probably have never seen blood when it is low in oxygen. Do you know why?

EXTENSION ACTIVITIES

1. Call the laboratory at a hospital and talk to the medical technologists who do blood tests. Find out what tests they do and why. You might also want to ask them about what is the most interesting part of their jobs and what is the most boring. Find out what training one needs to become a medical technologist.
2. For hundreds of years, people believed that bleeding a sick person was a way to cure just about every ailment. Sometimes they would simply cut open a blood vessel to drain some blood; other times they would apply leeches to suck the blood from underlying tissues. These "cures" often weakened the patients even more, sometimes killing them. Today, doctors are finding that leeches can be useful following some types of surgery. The leeches remove blood that would otherwise clot and interfere with tissue repairs. Learn more about leeches. Do research to find answers to these questions:
   What do leeches look like?
   Where are leeches found?
   How do leeches use blood?
   What do leeches normally live on?

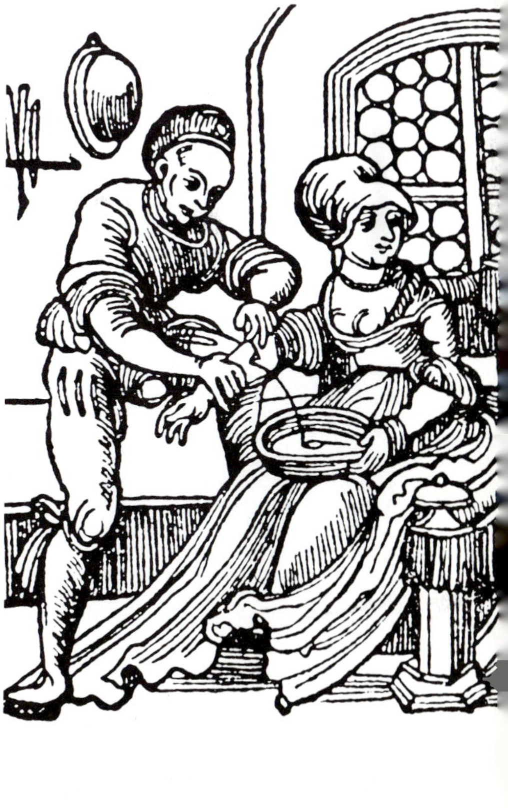

CHAPTER 13

URINARY SYSTEM

Why is it so important to drink plenty of water?

13-1 URINE TESTS

It's time for your physical. The nurse checks your pulse and blood pressure, listens to your heart and lungs, and measures your height and weight. Then you are given a cup and asked to collect some urine. Have you ever asked what is done with it?

- What are five characteristics of normal urine?
- How can you test normal urine for sugar and protein?

After doing several tests on a "urine" sample, you will be able to answer these questions.

MATERIALS

6 test tubes
test tube rack
graduated cylinder, 10 ml
stirring rod
masking tape
safety goggles
"normal urine"
"mystery urine"
red and blue litmus paper
Sugar Testing Station
- hot plate
- test tube holder
- safety goggles
- beaker
- Benedict's solution in dropper bottle

Protein Test Station
- nitric acid in dropper bottle

data page: ***Urine Tests*** (Figure 13–1)

PROCEDURE

Work with a partner.

Part A. Getting Ready

1. Before you begin the tests, make some predictions about the appearance and smell of normal urine. Decide also if you think it is acidic, contains sugar, and/or contains protein. Record your predictions on the data table.

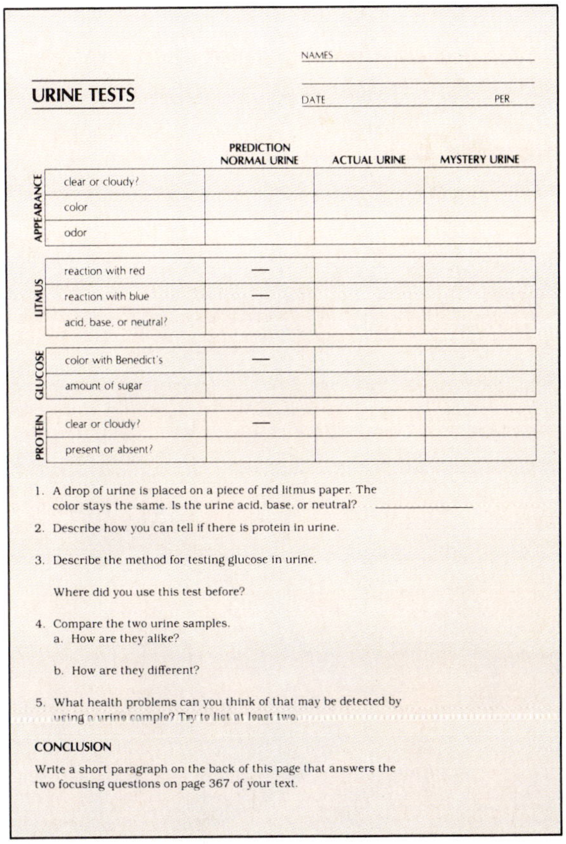

NAMES ______

URINE TESTS

DATE ______ PER ______

		PREDICTION NORMAL URINE	ACTUAL URINE	MYSTERY URINE
APPEARANCE	clear or cloudy?			
	color			
	odor			
LITMUS	reaction with red	—		
	reaction with blue	—		
	acid, base, or neutral?			
GLUCOSE	color with Benedict's	—		
	amount of sugar			
PROTEIN	clear or cloudy?	—		
	present or absent?			

1. A drop of urine is placed on a piece of red litmus paper. The color stays the same. Is the urine acid, base, or neutral? ______
2. Describe how you can tell if there is protein in urine.
3. Describe the method for testing glucose in urine.
   Where did you use this test before?
4. Compare the two urine samples.
   a. How are they alike?
   b. How are they different?
5. What health problems can you think of that may be detected by using a urine sample? Try to list at least two.

CONCLUSION

Write a short paragraph on the back of this page that answers the two focusing questions on page 367 of your text.

Figure 13–1

2. Using masking tape, label the 6 test tubes as follows:
   normal
   normal—glucose
   normal—protein
   mystery
   mystery—glucose
   mystery—protein

Part B. Testing Normal Urine

1. **Appearance**
   a. Put about 5 ml of "normal urine" in each of the three test tubes labeled "normal."
   b. Describe its appearance. Does it look clear or cloudy? What color is it?
   c. Describe its odor.
   d. Record your observations.

CAUTION

Never put your nose over an open container. Ask your teacher to show you how to "waft" the odor toward your nose. (See Figure 13–2.)

Figure 13–2 "Wafting"

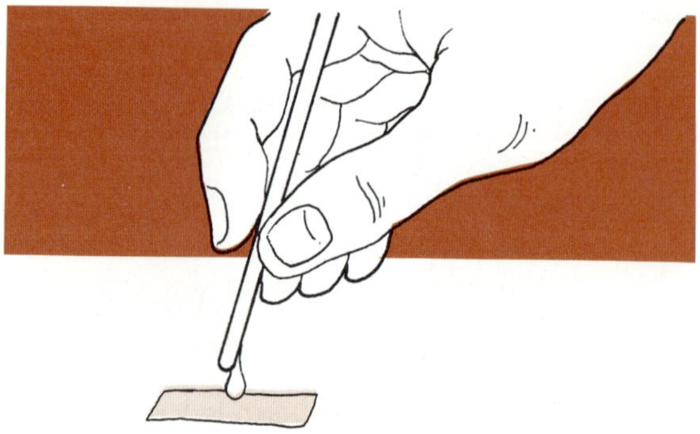

Figure 13–3

2. **Litmus Test**
   (Litmus paper is used to test whether a substance is an acid or a base.)
   a. Use the stirring rod to get a drop of urine from the test tube.
   b. Touch the drop to a piece of red litmus. (See Figure 13–3.)
   c. Repeat with a piece of blue litmus.
   d. Analyze your results. The meaning of the color changes is explained below:
      blue litmus turns red = acid
      red litmus turns blue = base
      no color change = neutral—neither acid nor base
   e. Record your results on the data table.

3. **Glucose—a simple sugar**
   a. Add 10 drops of Benedict's solution to the urine in the test tube labeled "normal—glucose." Swirl the test tube gently to mix the liquids.
   b. **Put on your safety goggles.** Heat test tube in a water bath for 2 or 3 minutes.
   c. Record the color and amount of sugar on the data table, using the chart on the next page.

Color	Amount of Sugar
blue	0 (none)
green	+ (a little)
green-yellow	++ (some)
yellow	+++ (medium)
orange-red	++++ (a lot)

d. Keep this test tube in the rack so you can compare its contents with the "mystery urine's" later.

4. **Protein**

a. Add 10 drops of nitric acid to the urine in the test tube labeled "normal—protein."

CAUTION

Acid burns. Wear safety glasses. If you get any acid on your skin, rinse with cold water and tell your teacher immediately!

b. Analyze the results, using this information:

yellow "cloud" in urine = protein present

no "cloud" forms = protein not present

c. Record your results.

d. Keep this test tube in the rack so you can compare its contents with the "mystery urine's" later.

Part C. Testing Mystery Urine

1. Your teacher has set up several samples of "mystery urine." Repeat the urine tests using one of these samples. It may help to compare the test tubes of "normal urine" and "mystery urine" while you work. Record your results on the data table.
2. Clean the test tubes with soapy water.
3. Answer the questions on the data page.

EXTENSION ACTIVITIES

1. Ask your family doctor or the school nurse what they look for when they do a urine test. See if you can have a copy of your last urinalysis results.
2. Some companies require their employees to submit urine samples for testing. The tests are used to find out whether the employees have used certain drugs. People are divided on this issue. What are your feelings? Should all people be required to have their urine tested? Should the testing be limited to people whose jobs affect the lives and health of others, such as air traffic controllers, train engineers, pilots, police officers, doctors? How would you feel if all students were required to have their urine tested?

13–2 URINALYSIS

A urinalysis is a series of tests done to find out what substances are present in urine. It is a routine part of a physical examination that provides important information about a person's health. Normally, urine is about 95% water. The other 5% is salts and other dissolved substances. However, certain illnesses can change the characteristics of urine.

Figure 13–4 Urine is 95% water.

Appearance

Fresh urine is clear, but becomes cloudy after standing a few hours. If fresh urine is cloudy, it may contain pus, blood, or bacteria from an infection in the urinary system.

Urine is normally yellow to yellow-orange in color. A lighter color means there is more water in the urine. A brown-to-red color may mean there is blood or hemoglobin in the urine. Blood in urine may indicate an injury somewhere in the urinary system.

Certain foods and dyes can also affect urine color. For instance, after eating beets, some people may be surprised to notice that their urine is red.

Odor

Fresh urine has a characteristic odor. This odor can be affected by urine concentration, food, its age, and disease.

Concentration—The more concentrated the urine, the stronger it smells. Drinking more fluids makes urine more dilute.

Food—Eating cabbage and asparagus can result in an unpleasant odor in the urine.

Age—When urine stands for a while, it begins to smell like ammonia. You may have noticed this if you have changed a baby's diaper.

Disease—Bacteria from urinary tract infections can produce a foul smell. Diabetes can cause a "fruity" odor.

Acid, Base, or Neutral

Urine is usually slightly acid. However, some foods, medicines, and diseases can cause urine to become more acidic or even somewhat basic. For instance, eating large amounts of citrus fruits, such as oranges, lemons, and grapefruit, can make urine more acid. After you take an antacid such as Tums, urine is more basic. Also, as urine gets older, it becomes more basic.

Protein

Normally there is very little protein in urine. If the protein level increases, the kidneys may not be functioning properly. However, strenuous exercise, fever, and stress can also cause protein to be present in the urine.

Glucose

Normally, glucose is present in urine in such small quantities that it cannot be measured using the standard tests. If sugar is detected, the amount of sugar circulating through the body is not being controlled properly. Sometimes this happens after eating a big meal or when a person is under emotional stress. But if the glucose level is high under normal circumstances, it may indicate diabetes. (See Figure 13–5.) People who have diabetes may have high levels of glucose in their blood. When this happens, their kidneys cannot "reclaim" all the sugar and some of it is lost in their urine.

Figure 13–5 People with diabetes can easily check their urine for glucose by using specially treated papers. After they dip the paper strip in urine, they can compare it with a color chart to determine how much sugar is present.

Other Tests

The tests just mentioned are not the only tests done on urine. Lab technicians also measure urine concentration. In addition, they look through a microscope at samples of urine to see if it contains blood cells. The presence of white blood cells or bacteria in the urine often means a urinary tract infection.

But wait! If normal urine does not contain protein, glucose, or blood cells, what does it contain? Remember, 95% of urine is water. The other 5% includes small amounts of ammonia, sulfur, sodium, calcium, other mineral salts, and urea. **Urea** is a waste product formed when proteins are used by the body. It makes up half of the dissolved substances in urine.

Did you realize how much this waste product can reveal about your health?

ANALYSIS QUESTIONS

1. What is a urinalysis?
2. Describe 3 ways urine can be affected by the foods a person eats.
3. What might cause protein to be present in urine?
4. How could a urinalysis reveal if your kidneys have been injured?
5. Copy the table below. It lists some of the substances found in urine and blood. Write "yes" if you would expect to find the substance in normal urine or blood or "no" if you would not.

	Water	Glucose	Protein	Urea
Urine				
Blood				

   a. Do you think more glucose is normally found in blood or urine? Why?
   b. Do you think more urea is normally found in the blood or urine? Why?

6. You are baby-sitting your 9-month-old brother. When you change his diaper, you notice that his urine is a strange color. What could you do?

CHALLENGE
Look at the results of the "mystery urine" you tested. What explanations can you give for these results?

13–3 URINARY SYSTEM

The body has several ways of getting rid of waste products. When breathing, we exhale carbon dioxide. In addition, some salt is lost through the sweat glands in the skin. Most waste products, however, are lost in the urine. How do these wastes end up in urine?

From Circulatory to Urinary System

Waste molecules are produced in the cells. If these wastes were allowed to build up, they would poison the cells; but they do not build up. The wastes are transferred to the blood as it flows through the capillaries. Then, fortunately, the body has a built-in "filtration system." The kidneys filter the blood, removing waste molecules and producing urine.

Kidneys to Urethra

The kidneys are two bean-shaped organs; each one is about 11 cm (a little over 4 inches) long. The kidneys are located just above the waist on either side of the spine. They are protected by the ribs and a cushion of fat.

A pair of arteries delivers blood to the kidneys. Once the blood has been filtered, it leaves by way of a pair of veins. Waste molecules and some water are left behind. The fluid is now called **urine.** Urine is continuously trickling out of the kidneys through two narrow tubes called **ureters.** These carry urine to the bladder. (Locate the ureters in Figure 13–6. Refer back to the figure to locate other parts of the urinary system as you read about them.)

The **bladder** stores urine. Its wall is made of smooth muscle fibers. The lining is folded and is elastic so it can stretch as it fills. An adult's bladder can stretch to hold about 500 ml (2 cups) of urine. How much do you think your bladder can store? How could you test this?

A ringlike **sphincter** muscle prevents urine from leaking out of the bladder. This muscles is important for bladder control. When the bladder is full, nerves

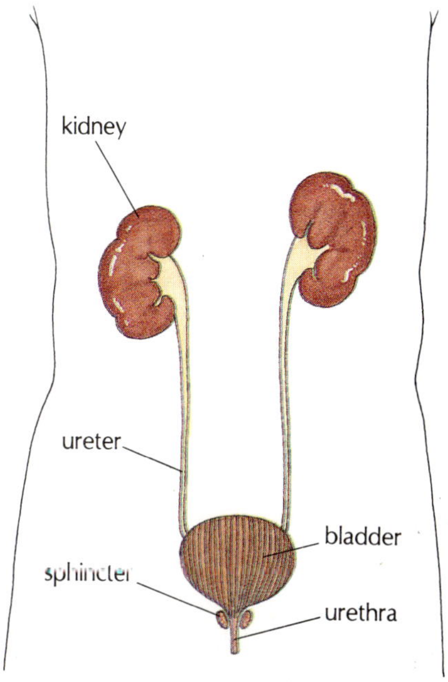

Figure 13–6 The urinary system

sends messages to the brain that the bladder needs to be emptied. When the time is right, the brain signals the sphincter to relax so urine can be released.

Very young children cannot control their sphincter muscles. That is why they may have "accidents." As they mature, they develop better control of this muscle. Urine leaves the bladder through a muscular tube called the **urethra.**

See how well you understand the structure of the urinary system by completing a copy of the worksheet. (See Figure 13–7.) Look back over this reading to make sure you label the structures correctly.

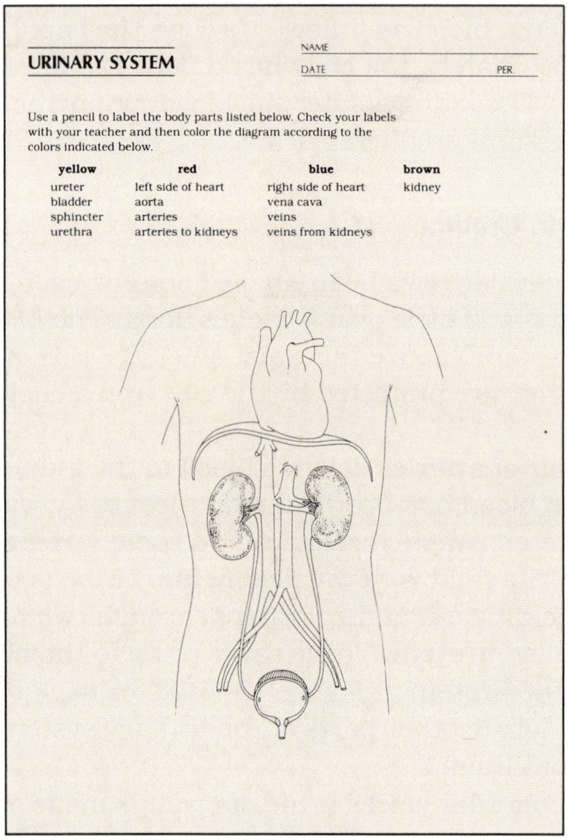

Figure 13–7

13-4 KIDNEYS—HOW DO THEY WORK?

Close your eyes and picture a kidney. Now imagine that the kidney is being cut in half. What do you think you will see?

If you are like most people, you probably do not have any idea what a kidney looks like. And yet, like the heart, kidneys are vital organs. You could not live without them. They keep the amounts of water and other substances in the blood in balance. Every time you eat, drink, or exercise you change this balance. It then becomes the job of the kidneys to remove the right amounts of water and dissolved substances to keep your body functioning normally.

Kidney Structure

Even though a kidney feels fairly firm, it is actually a mass of microscopic tubes and blood vessels. Urine is formed in the tiniest tubes called **tubules.** They empty

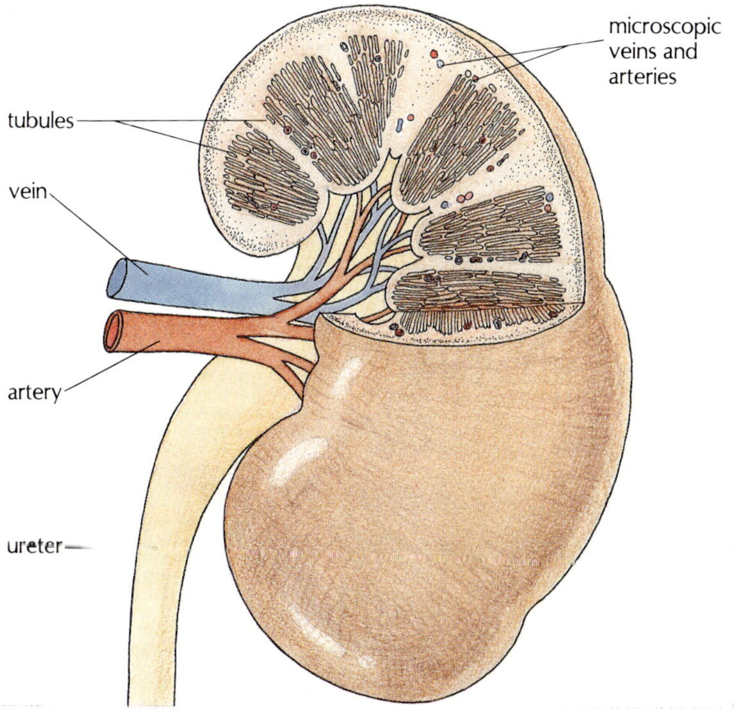

Figure 13–8 The kidney is a mass of microscopic veins, arteries, and tubules.

into slightly larger tubes and these, in turn, empty into even larger tubes. Eventually, these tubes empty into a space within the kidney. Urine is drained from this space by the ureters. (See Figure 13–8 on page 375.)

Urine does not "just appear" in the tubules; it is made by filtering blood. Everyday, large arteries deliver about 170 liters (42 gallons) of blood to the kidneys. Once in the kidneys, the large artery immediately divides to form smaller arteries. These arteries divide again and again to form smaller and smaller vessels until finally they become a thick mass of capillaries and tubules. Capillaries and tubules—these are the two structures in the kidneys needed to form urine.

Urine Formation

As blood travels through the kidneys, water, urea, oxygen, and other small molecules diffuse out of the capillaries into the tubules. Blood cells and proteins are

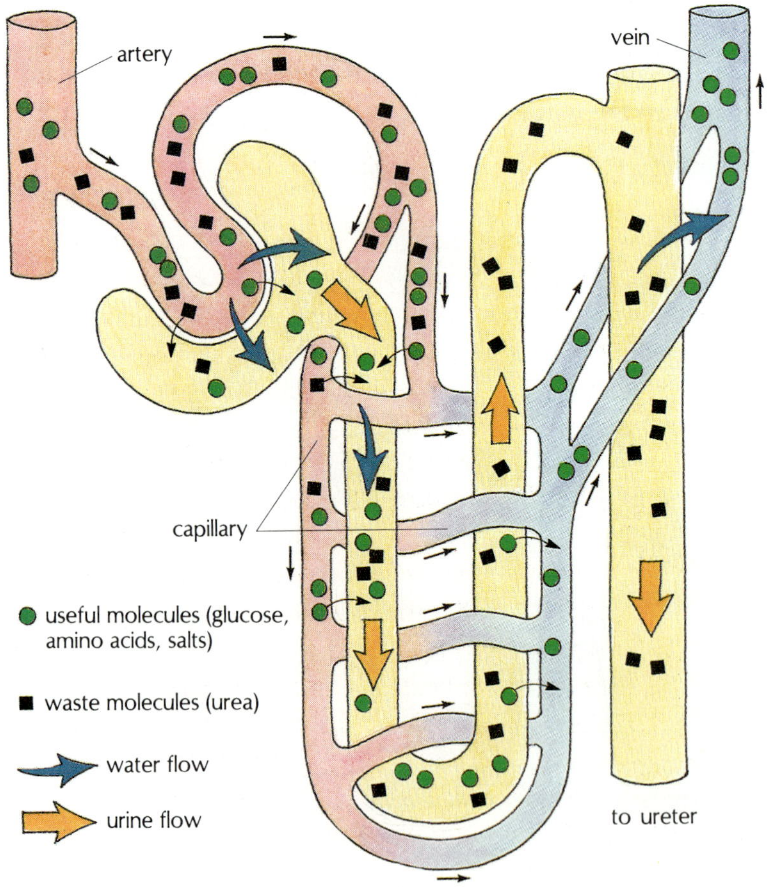

Figure 13–9 Trace the path of useful molecules. Where do they end up? Where do the waste molecules go?

too big to pass out of the capillaries so they stay in the blood. This filtering process is the first step in the formation of urine. The fluid in the tubules contains water and waste molecules as well as molecules the body needs. Somehow the useful substances have to be saved.

As the fluid flows through the tubules, the useful molecules, such as glucose, amino acids, water, and salts, diffuse back into the blood. (See Figure 13–9 on page 376.) Small veins carry this filtered blood to larger veins, which eventually form one large vein. This large vein carries the filtered blood from the kidney to the vena cava. The vena cava returns the blood to the heart.

Meanwhile, water, some salts, and urea are left in the tubules. This waste fluid, now called urine, continues down the tubules and out into the ureter. From the ureter, the urine enters the bladder. This process goes on every minute of every day.

Kidney Failure

Occasionally both kidneys stop functioning. When this happens, the person looks swollen or puffy because water is being stored in his or her system. If the problem is not treated, wastes build up and the person will die.

A kidney machine can be used to filter blood. This filtering process is called **dialysis;** it is shown in Figure 13–10. When a person undergoes this treatment, his or her blood flows out of an artery and into the dialysis machine through a tube. A thin membrane in the machine filters the waste molecules from the blood. The filtered blood is returned through another tube into one of the patient's veins. This procedure takes about six hours and is repeated two or three times a week.

Sometimes a person's diseased kidney is replaced with a healthy one. The ideal donor, the person who gives the kidney, is an identical twin. If there is not an identical twin, a close relative can be the donor. However, transplants involving unrelated donors are becoming increasingly successful. A donor can afford to give up a kidney because the human body can function fine with one kidney. Isn't it nice to know you have a spare?

Figure 13–10 Patient undergoing dialysis

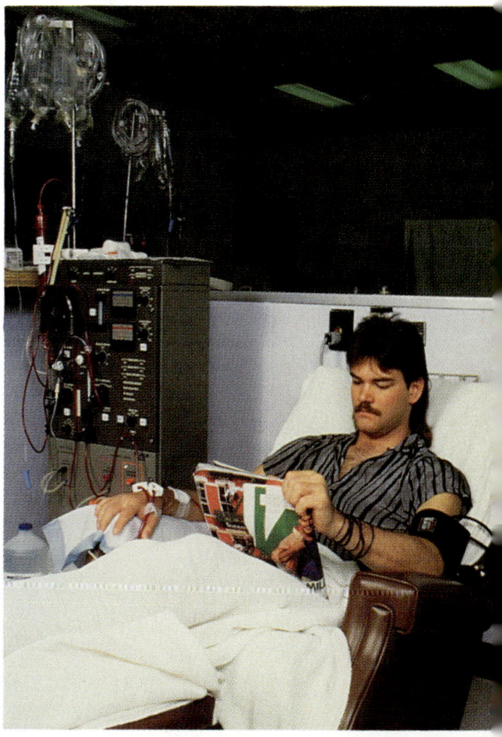

Healthy Kidneys

Two simple guidelines will help you keep your kidneys healthy.

- Drink at least three glasses of water each day. This helps flush waste substances out of the kidneys.
- Have your blood pressure checked each year. High blood pressure damages capillaries in the kidneys.

Sometimes kidney damage can be prevented if problems are taken care of right away. Notify your doctor if you experience any of these warning signs:

swollen ankles or hands
back pain below the rib cage
pain or burning during urination

TEST YOURSELF

A molecule of glucose is in an artery heading toward the kidney. Where will it end up, in the blood or urine? Decide the fate of each type of molecule listed below and give reasons for each answer.

Molecule	Blood or Urine?	Why?
glucose		
urea		
water		
amino acids		
salts		

EXTENSION ACTIVITIES

1. Several organizations provide information about kidney-related conditions. The National Kidney Foundation, the American Diabetes Association, and the Juvenile Diabetes Foundation have pamphlets and brochures that you can request.
2. Do you know someone who uses a dialysis machine? If so, ask permission to interview him or her. Decide what questions you would like to ask. Share your impressions with your class.
3. In some states, people who wish to donate their organs sign a permission form on their driver's licenses. If they die, their kidneys can be removed and donated to people whose kidneys have failed. Find out if your state has an Organ Donor program and, if so, learn which organs can be donated.

CHAPTER 14 IMMUNE SYSTEM

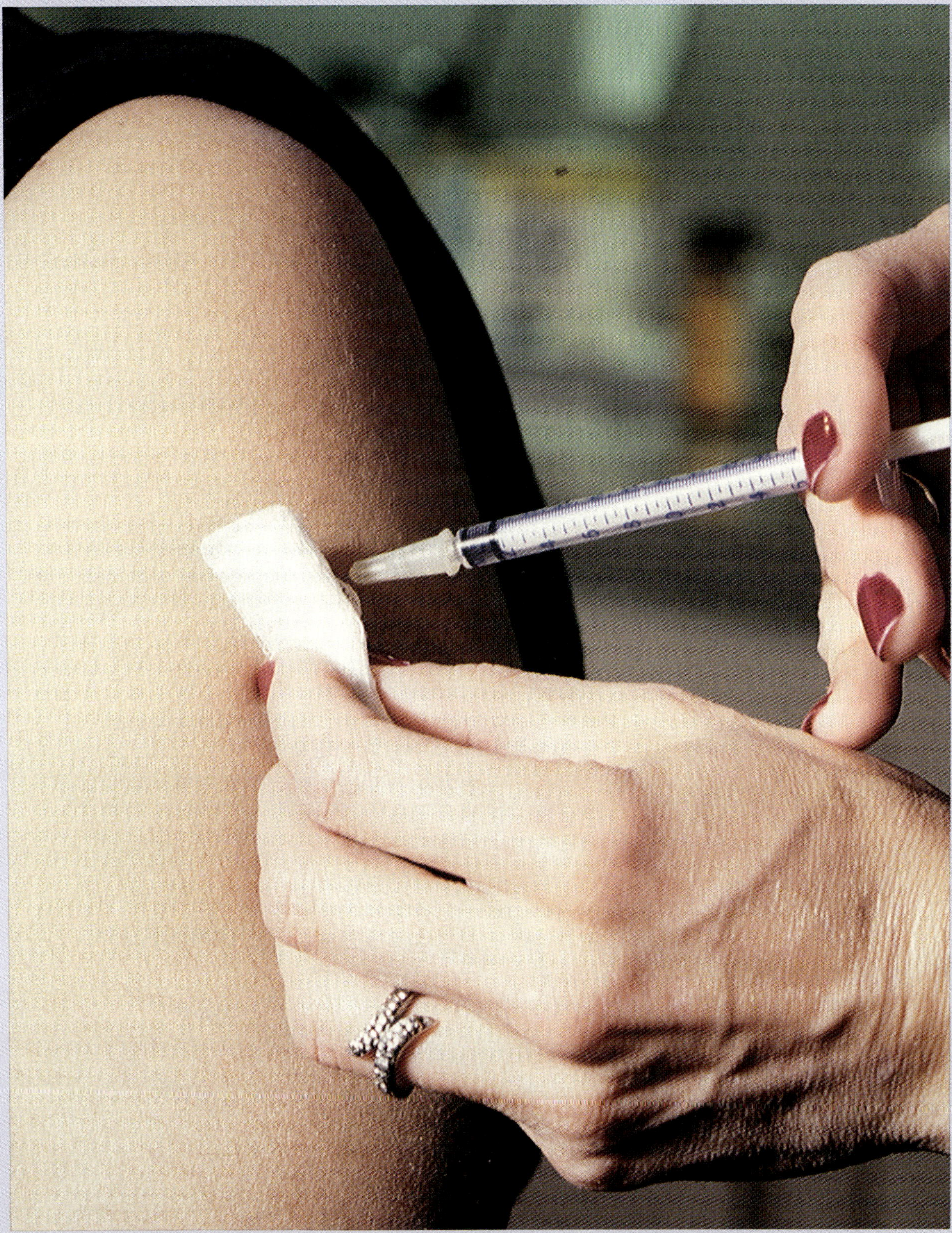

Why do you think this girl needs a measles shot? Do you need one?

14–1 LIVER, SPLEEN, AND TONSILS

Tonsils, liver, and spleen . . . these structures are all involved in some way with your blood. You can survive without your tonsils, but you would die if a surgeon removed your liver. What about your spleen? Can you live without it?

Liver

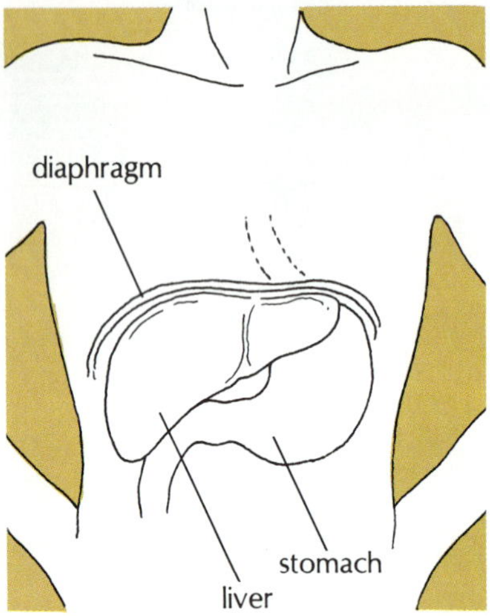

Figure 14–1 Location of the liver

The liver is one of the most vital body organs. Like your heart, kidneys, brain, and lungs, the liver is an organ you cannot live without. Put your hand over your bottom ribs on your right side. Your liver is behind these ribs, just below your diaphragm. It is a fairly large structure and weighs a couple of pounds. (See Figure 14–1.)

You may have seen an animal's liver. Its characteristic dark reddish-brown color is due to the many blood vessels that it contains. As blood flows through these vessels, it is "filtered." Harmful or waste substances are removed and essential molecules are replaced. For instance, the liver helps regulate the level of sugar in the blood. If there is too much blood sugar, liver cells can remove some of it and store it for later use. Liver cells can also break down the fatty acids that circulate in the blood after they have left the digestive tract.

In addition to regulating sugars and fatty acids, the liver acts on amino acids. Liver cells contain enzymes that can break down these acids or rebuild them to

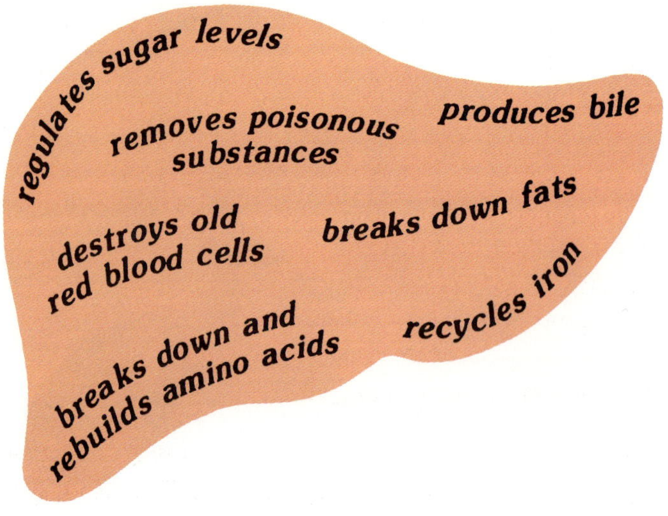

Figure 14–2 Functions of the liver

make other amino acids. When amino acids are broken down, ammonia is formed. Ammonia is very poisonous to humans. The liver removes it from the blood and quickly changes it into urea. The urea is then returned to the blood and will be removed from the body by the kidneys.

The removal of ammonia from blood is one example of another important liver function—it clears poisonous substances from the blood. Liver cells also break down drugs such as acetaminophen (a nonaspirin pain reliever) and alcohol. Cancer-causing chemicals are also broken down in the liver.

Another liver function is to "clean up" after old red blood cells. When these cells get old, they break apart. The membrane fragments that remain are broken down by liver cells. The liver also breaks down the hemoglobin that was inside the red cells. It recycles the iron to make more hemoglobin and uses the rest of the molecule to make bile. In this way, the liver functions in digestion.

Some of the functions of the liver are summarized in Figure 14–2 on page 380.

Spleen

Next to the liver but on the left side of the body is the spleen. It, too, is below the diaphragm and behind the ribs so you cannot feel it. (See Figure 14–3.) The spleen contains many blood vessels. If all of the capillaries in the spleen open up, they can hold 1 liter (.3 gallons) of blood. Some people think of the spleen as a blood reservoir since it can store blood. It, too, helps destroy old red blood cells. The white blood cells in the spleen help fight infections.

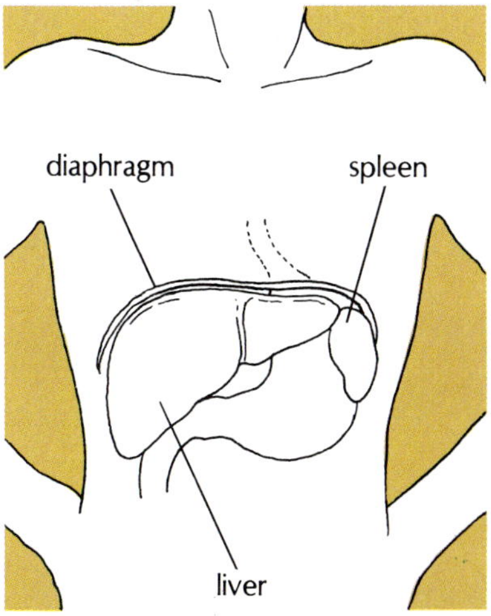

Figure 14–3 Location of the spleen

Even though the spleen has several important functions, we *can* live without it. For instance, you may hear that a victim of an automobile accident had his or her ruptured spleen removed. (This was necessary because the spleen cannot be stitched up—it is too spongy and its covering is too thin.) The spleen is removed to prevent internal bleeding.

Tonsils and Other Lymph Nodes

"It looks like a bean!" That's what you would probably say if you could look at a **lymph node.** You have lymph nodes throughout your body. These nodes contain white blood cells that engulf and destroy disease-causing microorganisms. They are infection fighters. They may become enlarged and tender when they are active. Some lymph nodes have names. The lymph nodes in the back of your throat are called **tonsils.** They are in a good position to trap the infectious organisms that enter your mouth. (See Figure 14–4.)

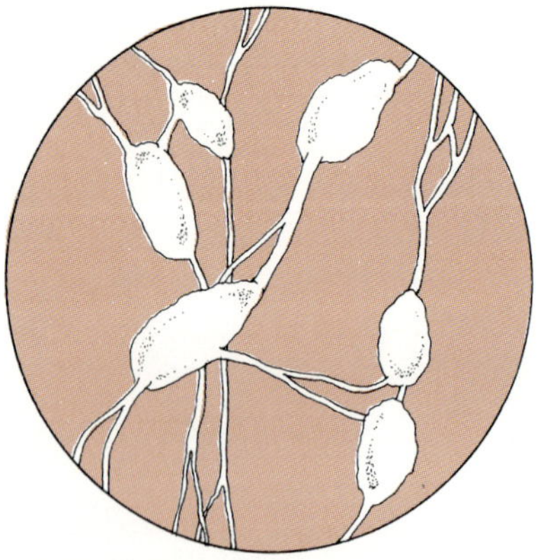
Figure 14–4 Lymph nodes

Liver, spleen, and tonsils . . . they are all important structures. Why don't we hear more about them?

ANALYSIS QUESTIONS

1. Copy this table onto your own paper. It will help you summarize what you read. Fill in the functions for each structure.

Structure	Functions
liver	1. 2. 3. 4. 5.
spleen	1. 2. 3.
tonsils and other nodes	

2. Put an asterisk on the table by each function that involves blood or blood cells.
3. When you go to the doctor for a physical exam, you lie down on the examining table. You are told to take a deep breath, then the doctor gently pushes on your abdomen to feel your liver. Why do you have to take a deep breath for this part of the exam?
4. Decide which statement is false. *Explain why you decided it is false.*
   a. Tonsils can swell if there is an infection.
   b. Tonsils are lymph nodes.
   c. Tonsils contain many red blood cells.
5. *Cirrhosis* of the liver means dead liver cells. Why do you think alcohol can cause cirrhosis?

EXTENSION ACTIVITIES

1. Buy some liver at the supermarket. Inspect it. How does it compare with other tissues?
2. Practice looking at tonsils. Use a flashlight and something to hold down your tongue, such as a spoon handle. Look at your tonsils in a mirror. Find out what they look like when you are *healthy*. Then remember to look at them when you have a sore throat. If you see red and swollen tonsils, you could have an infection; white spots could mean you have strep throat. If you see any of these signs, call your doctor.

14–2 PROTECTION FROM GERMS*

*The word *germs* refers to all disease-causing microorganisms. It is not a scientific term.

Complete the study sheet (Figure 14–5) while you read. It will help you remember the important ideas in this reading.

White blood cells are part of the body's defense system. They produce antibodies and destroy germs. Since we have white blood cells to fight infections, why do we need shots and pills? In order to answer this question, you need to know more about the body's protection system.

Natural Protection—White Blood Cells

Some white blood cells are always on the lookout for microorganisms that could make you sick. They destroy

PROTECTION FROM GERMS

NAME ______________________

DATE ______________ PER. ______

NATURAL PROTECTION—WHITE BLOOD CELLS

1. List 2 ways that white blood cells defend your body.

2. What kinds of antibodies have your white blood cells produced? Use checks to show which antibodies your body has made.

   ____ antimeasles ____ antiflu ____ anti"strep throat"

   ____ antichicken pox ____ antirabies ____ antiringworm

   ____ antimononucleosis ____ antimumps ____ anti"pinkeye"

3. Describe in your own words what the word *immune* means. (You can ask someone else or look it up in the dictionary, but put the answer in *your own words*.)

EXTRA HELP—VACCINATIONS

4. What is in a vaccine?

5. Complete this table.

Age	Vaccinated Against	How Given
0–12 months	DTP (3 times)	shots
0–12 months	(2 times)	by mouth
15 months	measles, mumps, rubella	shot
18 months		
18 months		
5 years		
5 years		
14–16 years		
every 10 years		

Figure 14–5 (Study sheet, page 1)

PROTECTION FROM GERMS
(PAGE 2)

NAME ______________________

6. A 10-year-old probably has had many of each of these vaccinations?

   ____ DTP ____ polio ____ measles, mumps, rubella

7. What vaccination is given throughout life?

8. What is an antibiotic?

9. What 2 things can happen if you take too much antibiotic?

10. Fill out this table about several diseases. You may need to use a dictionary or encyclopedia to find out what causes one or more of them.

Disease	Cause (virus, bacteria, other)	Can it be helped by an antibiotic?
strep throat		
flu		
tetanus		
diabetes		
pneumonia		
measles		
a cold		
hemophilia		
staph infection		

CHALLENGE

You have had 3 colds this year. You have probably had 30 or 40 colds since you were born. How is this possible? Shouldn't you have plenty of "anticold" antibodies?

Figure 14–5 (Study sheet, page 2)

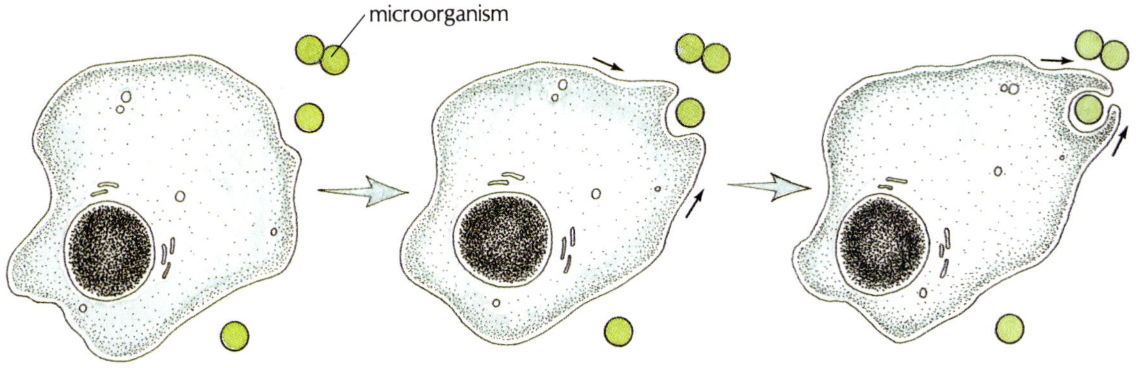

Figure 14–6 Some white blood cells can engulf microorganisms.

these germs by "eating" them. That is, they engulf and destroy the microorganisms. (See Figure 14–6.)

Some white blood cells make specialized protein called **antibodies.** These proteins leave the white cells and move with the blood throughout the body. Antibodies are attracted to substances that are "foreign" to your system. When they bump into a foreign cell, such as a germ, they attach to it. This "labels" the germ for destruction—they are then engulfed and destroyed by white blood cells. (See Figure 14–7.)

When you catch a disease, your body makes antibodies against that one kind of microorganism. For example, if you catch measles, you make measles antibodies. Measles antibodies can only attack the measles virus. They cannot fight the mumps virus, a flu virus, or any other kind of virus. There are thousands of different antibodies. Each antibody recognizes only one kind of microorganism. Later, if you "catch" the same microorganism again, you already have some of the needed antibody *and* your body can quickly make more. You are already *immune.* That is why you only get some diseases, like measles, one time.

Wouldn't it be nice if you were born with a supply of all the different antibodies? You would never get sick from infectious diseases!

Figure 14–7 Some white blood cells produce antibodies. Others engulf microorganisms that have been labeled by the antibodies.

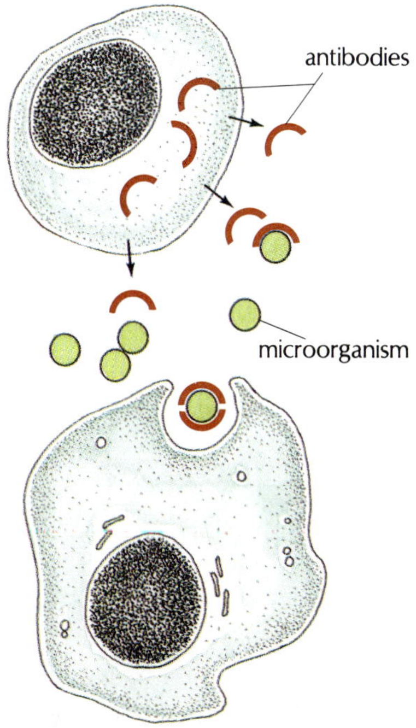

Extra Help—Vaccinations

You have probably never had polio, diphtheria, or tetanus. But even though you have never had these diseases, you do have antibodies to protect you from getting them. How did this happen? How did your body know how to make these antibodies?

*Another name for *vaccination* is *immunization*. You may hear both words being used.

You have some antibodies already because of **vaccinations*** you have received. A vaccination contains killed forms of a particular microorganism. You probably had three DTP vaccinations before you were a year old. This vaccine contains the microorganisms that cause Diphtheria, Tetanus, and Pertussis. (Pertussis is another name for whooping cough.) Your body made antibodies against these dead germs. In this way, you became immune and will not catch these diseases.

Some vaccines last a lifetime. In other words, the antibodies stay in your body forever. The vaccine for measles, mumps, and rubella (MMR) is an example of a lifetime vaccine. It usually is given when a child is about 15 months old.

Your body has to be "reminded" how to make other antibodies. That is why the DTP shot is given a second time when a baby is 18 months old and then again when the child is around 5 years old. These are called **boosters.** A Td shot is needed between 14 and 16 years of age. (Td is for tetanus and diphtheria. The pertussis vaccine is not given because it makes older people sick.) You will need to get Td shots about every ten years throughout your life.

There is one more vaccination that is routinely given—it is to prevent polio. This vaccine comes in drops that are placed in the mouth. A baby gets two polio vaccinations before he or she is a year old. Booster vaccines are given when the baby is 18 months old and again at 5 years.

Your family doctor keeps a record of when your shots were given; but it is a good idea to make your own record in case you change doctors or need your records quickly.

ROUTINE VACCINATIONS

Diphtheria, Tetanus, Pertussis (DTP)
0-12 months – 3 times

Measles, Mumps, Rubella (MMR)
15 months

Boosters

DTP	18 months
	5 years
Td	15 years
	25 years
	35 years
	.
	.
	.
	105 years

Polio
0-12 months – 2 times
18 months
5 years

More Help—Antibiotics

Sometimes disease-causing microorganisms reproduce faster than the white blood cells can destroy them. Thousands of them may invade your cells and make you sick. When this happens, your body may need help to fight the infection.

Luckily, we have drugs to help in this battle against certain diseases. They may stop germs from growing or they may kill them. These powerful drugs are called **antibiotics.** Unfortunately, antibiotics only work for bacterial infections; they do not affect viruses. Some frequently used antibiotics and their effects on bacteria are listed in the following table:

Antibiotic	Action
Penicillin Bacitracin	prevent cell walls from forming around bacteria
Streptomycin Erythromycin Tetracycline	prevent bacteria from making proteins
Sulfa drugs	prevent bacteria from making DNA for chromosomes

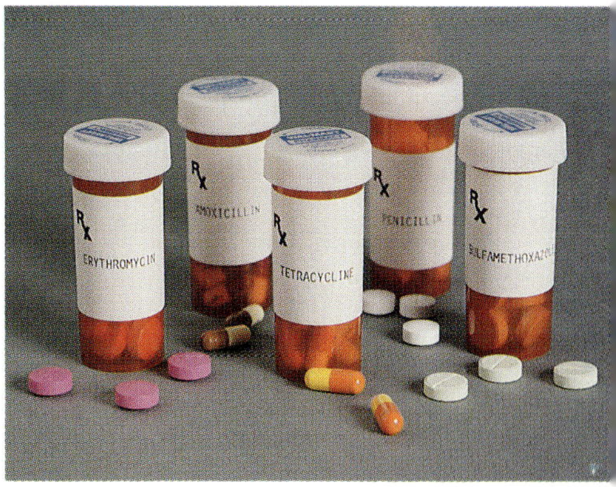

Figure 14–8 Have any of these antibiotics ever been prescribed for you?

Antibiotics have to be used carefully so you need a doctor's prescription in order to buy them. If used too often, antibiotics can kill the *good* microorganisms that help keep you healthy. If you take too much antibiotic, you could kill your own cells.

EXTENSION ACTIVITY

Did you ever hear about David, the "Bubble Boy"? His immune system did not work so he had to live in a germ-free environment. Find out about David and his medical problem. Imagine that you have the same condition. Describe what your life is like and what treatments you are receiving. Explain your chances of being cured.

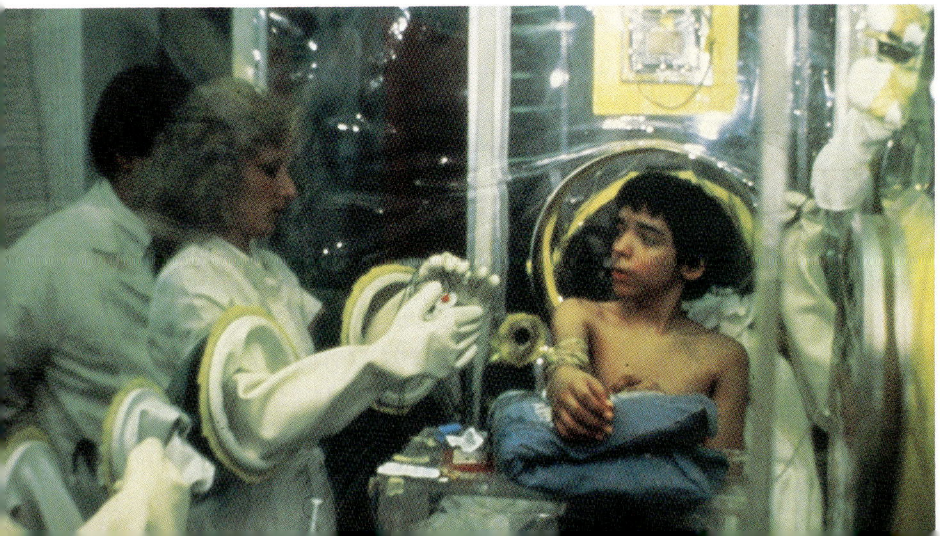

Figure 14–9 How would David's life be different from yours?

14-3 IMMUNIZATION RECORDS

It is impossible to remember when you had which vaccinations. For this reason, it is important to keep a record of all your shots. You should record the date, which vaccination you got, and who gave it to you. It is also a good idea to keep track of any reaction you might experience such as swelling or a fever.

VACCINE vaccine		DATE GIVEN dada en la fecha	DOCTOR OR CLINIC doctor o clinica	MEDICAL NOTES notas del medico
Oral Polio Vaccine	1.	3-5-2010	Flynn Health Dept.	
	2.	5-10-2010	"	
	3.	6-2-2011	"	
	4.	3-22-2015	"	
	5.			
DTP Diphtheria, Tetanus, and Pertussis	1.	3-5-2010	Flynn Health Dept.	
	2.	5-10-2010	"	
	3.	7-17-2010	"	
	4.	6-2-2011	"	
	5.	3-22-2015	"	
Td Tetanus and Diphtheria				
MMR				
Other		4-15-2018	Suerto Clinic	smallpox

The immunization record for Meghan Futura is shown above. Meghan was born on January 1, 2010. Imagine that today is January 2, 2023. Is Meghan Futura up-to-date with her shots? Look on page 386 of ***Protection from Germs*** for the suggested times to be vaccinated.

1. At what ages did Meghan get polio vaccines (OPV)?
2. Does Meghan need another polio vaccine (OPV)?
3. When will Meghan need another DTP or Td shot?
4. Meghan never had the combined measles, mumps, and rubella vaccine (MMR). She had measles when she was four. She thinks she had mumps when she was seven. Do you think she should get this combined vaccine?
5. Smallpox vaccine is not routinely given to people in the United States. Why do you think Meghan got the vaccination?
6. Which disease sounds the most unpleasant to you—polio, diphtheria, or tetanus? Why?

THINK ABOUT IT

It is especially important that females be immunized against rubella. In fact, many employers test to see if women have had the rubella vaccine when they are hired. If a woman does not have the rubella antibody, she may be asked to get the vaccination. Why do you think this is so?

HOMEWORK

Complete your own immunization record. Ask your parents, your school nurse, or your doctor to help you fill it in. Hopefully, you will be able to find all the information you need to complete your own vaccine record. Sometimes it seems like you need to be a detective to track down medical records. That is why you should be keeping your own records. Look over your records and answer the following questions.

1. When do you need your next polio vaccine (OPV)?
2. When do you need your next Td shot?
3. Do you need an MMR vaccination?

EXTENSION ACTIVITY

Each state has its own laws about vaccinations. Call your local health department. Ask them to send a copy of the immunization requirements for your state. Also ask them what you would need to do to get a Td shot. Immunization and booster shots are often very inexpensive, maybe even free, at the local, county, or state health department office in your area.

CHAPTER 15

RESPIRATORY SYSTEM

What do you think holds more air — your lungs or these bubbles?

15-1 X-RAYS AND LUNGS

Infections can develop anywhere in the body. If they develop on the skin or in the mouth, they are easy to diagnose. Infections that develop inside of the body can be harder to detect.

The lungs are a common site of infection. Since it is impossible to look directly at the lungs, x-rays can be used to help detect an infection.

- What structures can you see when you look at a lung x-ray?
- What happens to the lungs when someone has pneumonia?

By looking at lung x-rays and reading about pneumonia, you will be able to figure out the answers to these questions.

There are three parts to this activity—reading, inspecting x-rays, and looking at microscope slides. Your teacher will tell you which part to start with; it does not matter which part you do first. Answer the questions and do the sketches on your own paper while you work.

MATERIALS

x-ray packet: normal lungs
lungs with pneumonia
normal lungs with contrast medium
microscope
dictionary
reading: ***Pneumonia*** (page 393)
microscope slide: *Streptococcus pneumoniae*

PROCEDURE

Part A. X-Rays

When you look at the x-rays, imagine that you are using x-ray vision to look through a person's back.

1. X-rays are black and white. Look at the x-ray of normal lungs.
   a. What tissues look white on the x-ray?
   b. What tissues look black?
   c. Sketch the shape of a lung.
   d. Is a lung hollow? Use the x-ray to explain your answer.
   e. How do you think the size of these lungs compares with the size of your own?
   f. Name at least 4 body structures, other than lungs, that you can see in this x-ray. Did you find the heart?
2. Look at the x-ray of the lungs with pneumonia.
   a. What is the main difference between the lungs in this x-ray and the normal lungs?
   b. What do you think is the cause of this difference? (If you have

not read ***Pneumonia*** starting on page 393 yet, you may want to come back and answer this after you do so.)

c. Do you think this person is a baby, child, teen, or adult? Explain.

3. Look at the x-ray of normal lungs with contrast medium.
   a. What can you see in these lungs that you could not see in the other x-rays? (Describe what you see. Making a sketch might also help.)
   b. Why might a doctor request this type of x-ray?
4. How many liters of air do you think these lungs could hold? Look at the liter bottle your teacher has set out to help you make your estimate.)

Part B. Microscope Work

Look at the microscope slide of *Streptococcus pneumoniae*. This kind of bacteria is often the cause of pneumonia. Remember, bacteria are very small. Be patient as you search for them on the microscope slide.

5. What magnification did you use?
6. Sketch what these bacteria look like through the microscope.
7. How do you think *Streptococcus pneumoniae* gets into a person's lungs?

Part C. Doing the Reading

Read ***Pneumonia*** on pages 393 and 394.

8. What can cause pneumonia?
9. What happens to the lungs when a person has pneumonia?
10. Would antibiotics be useful to treat pneumonia? Explain.

CHALLENGE

Many other health problems can affect the lungs.

a. List as many of these as you can. Try to think of at least 4.
b. Look up 3 of the conditions in a dictionary. Write down their definitions.

CONCLUSION

Write brief answers to the two focusing questions on page 391.

EXTENSION ACTIVITIES

1. People who have pneumonia may be admitted to a hospital. An inhalation therapist works with them during their stay. Find out what inhalation therapists do. Maybe you can visit one and see the types of equipment being used. Imagine that you are an inhalation therapist. Write a "hospital report" about what you did during the course of a day. Include information about your patients, their problems, and your work with them. Also comment on what you *like* or *don't* like about your job.
2. Cystic fibrosis is a genetic disorder that involves the lungs. People with cystic fibrosis often get pneumonia. Look up *cystic fibrosis* and find answers to these questions:
   What are symptoms of cystic fibrosis?
   How does it affect the lungs?
   Why do people with cystic fibrosis often get pneumonia?

Pneumonia

"You missed more than a week of school. You must have been really sick."

"Yes, I had pneumonia. I felt awful."

"Didn't the doctor give you antibiotics? I had pneumonia last year. As soon as I started on antibiotics, I felt much better."

"The doctor said antibiotics wouldn't do any good because pneumonia is caused by a virus. You know, *viral pneumonia.*"

"Are you sure you had pneumonia? Antibiotics sure helped me."

Who is right? Is pneumonia caused by a virus? Will antibiotics help someone who has pneumonia?

Causes

Actually, pneumonia is not one single disease. It is a term given to several different kinds of lung infections. These can be caused by bacteria or viruses. Most of the time pneumonia is caused by bacteria.

Bacteria like to grow in warm, moist environments. The lungs are perfect habitats. Luckily white blood cells usually destroy the dangerous microbes before they "settle in." However, when people have been sick or if their bodies are not in good shape, their defense systems do not work as well as they should. This is when the microbes take over. (See Figure 15–1.)

As the numbers of bacteria grow, they produce more and more waste products. These irritate the lung cells. As a result, the lung produces fluids to help "heal" the irritated cells. These fluids are also good for bacterial growth so the infection spreads.

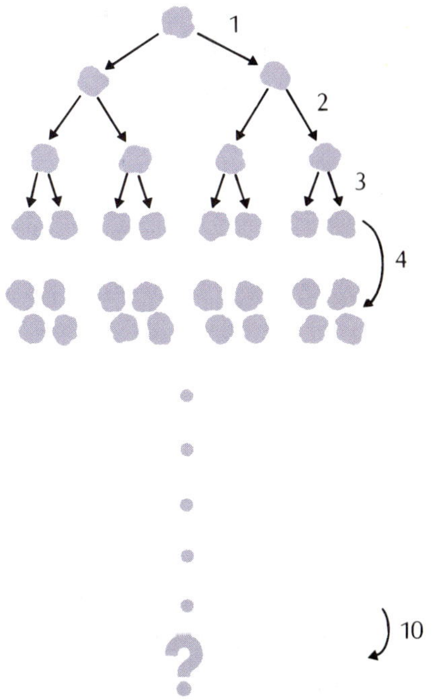

Figure 15–1 If you start with one bacterium, how many bacteria would you have after ten rounds of cell division?

Symptoms

A cough is a common symptom of pneumonia. Coughing can clear the fluid out of a person's infected lungs. It may hurt to breathe, and a fever with chills is likely to develop. The exact symptoms depend on the type of microorganism that is causing the disease.

Diagnosis and Treatment

When pneumonia is suspected, the doctor will use a stethoscope to listen for fluid in the lungs. If it sounds like there is fluid, x-rays can show which parts of the lungs are affected. Next, it is important to decide what is causing the pneumonia. Laboratory tests on mucus or blood will determine if it is a bacterial or viral infection.

Antibiotics are extremely helpful in treating bacterial pneumonia. Since viruses are not affected by antibiotics, viral pneumonia is more difficult to treat. No matter which type of pneumonia a person has, it is important to see a doctor. If the condition is not treated, lung damage may result. Before we had antibiotics, pneumonia was a major cause of death. It is still a serious illness for the sick and elderly.

15–2 BREATHING PARTS

Ask a young child to take a deep breath and to tell where the air goes. He or she may say that the stomach fills with air. You probably would have said the same thing when you were little. However, when we are older, we realize that the **lungs** and not the stomach hold air. But how about the other parts of the respiratory system? Do you know what they are? How would you describe their functions?

Getting In—The Nose and Mouth

Room air might destroy our lungs. It is usually too dry, too dirty, and too hot or too cold. That is why the **nose** is important. The nose is the main entrance to the respiratory system. It filters, moistens, and warms or cools the air as it enters. (See Figure 15–2.)

Short hairs at the front of the nose act as a filter. They remove pollen, dust, and other small particles from the air. A **mucous membrane** lines the nose. It moistens the air to prevent the lungs from drying and can trap small particles. The mucous membrane also contains many capillaries. The blood circulating in these microscopic vessels warms the air.

When the nose passage is blocked, air enters through the **mouth.** The mouth does not prepare the air for entering the lungs.

Figure 15–2 How would you describe the air you breathe?

- Take 10 deep breaths through your mouth. Describe how your throat feels.
- Take 10 deep breaths through your nose. Describe how your throat and nose feel.

Moving On—The Trachea and Bronchi

After air passes through the nose or mouth, it enters the **trachea.** The trachea, or windpipe, is in front of the esophagus. It is important that food goes down the esophagus and air goes down the trachea. You can see that the openings to these two tubes are close together.

So, in order to prevent a mix-up, we have a built-in safety device, the **epiglottis.** It is a flap of tissue that can cover the opening to the trachea. The epiglottis folds down automatically when you swallow and lifts when you breathe. This is shown in Figure 15–3 on page 396.

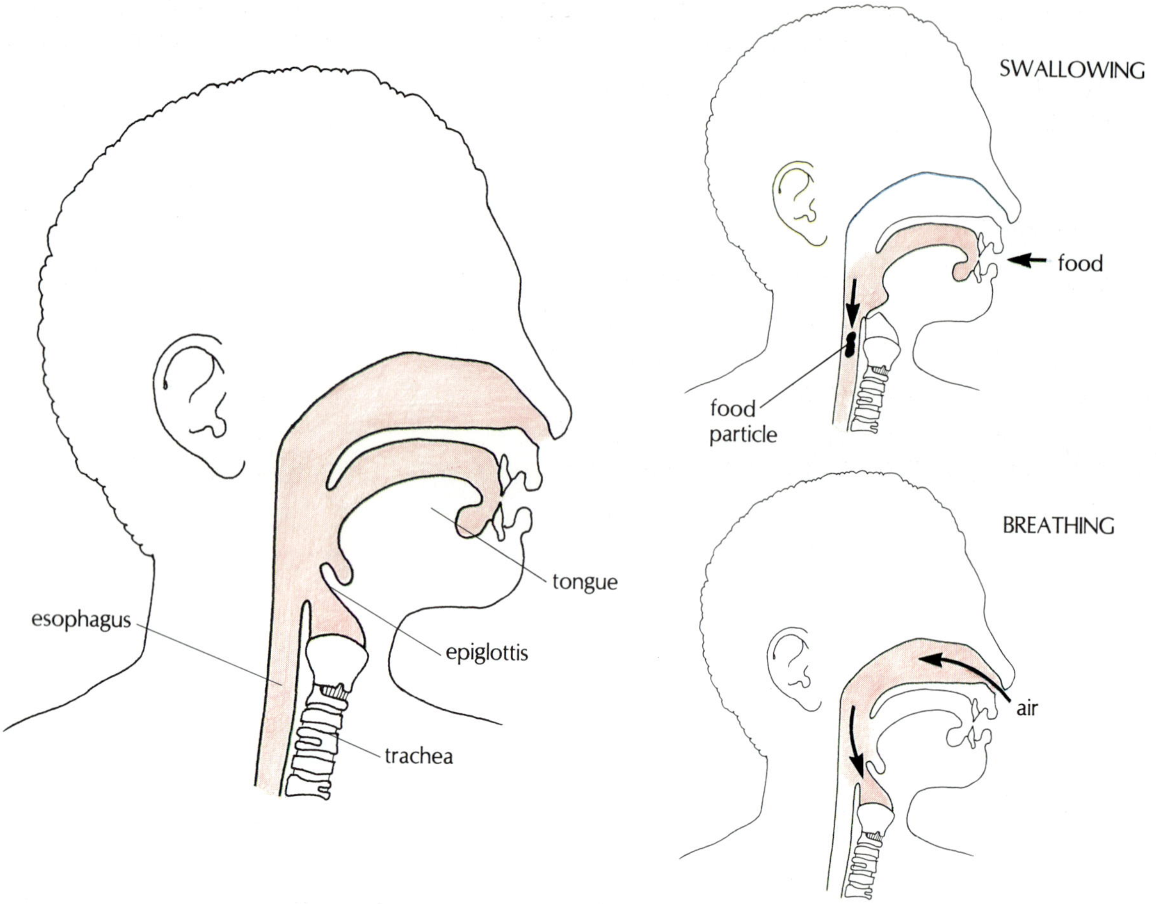

Figure 15–3 Look at the position of the epiglottis in each of these drawings. Notice also that the trachea is in front of the esophagus.

It can be dangerous if food gets into the trachea by mistake. Food in the trachea causes a reflex reaction—choking. A cough can force the food back up and out of the trachea.

The trachea continues down along the front of the throat. Find your trachea; it is the firm tube you feel at the base of your throat. The trachea feels firm because it is made of cartilage. Most other structures are made of smooth muscle and connective tissue. (See Figure 15–4.)

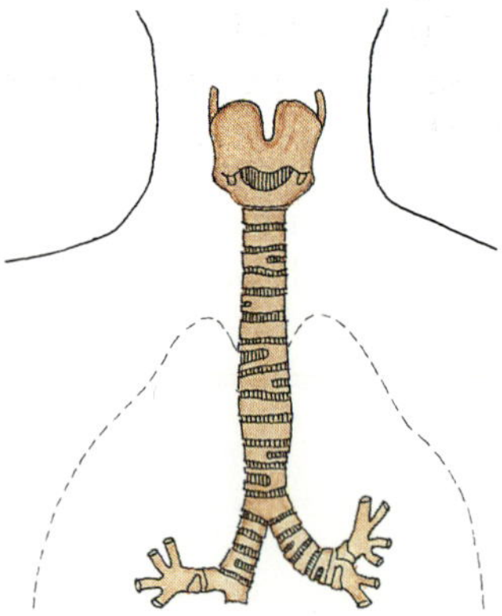

Figure 15–4 Can you feel the rings of cartilage? What do you think would happen if your trachea were made of smooth muscle?

Once in your chest, the trachea branches to form two **bronchi.** One bronchus goes into each lung. The bronchi divide again and again into smaller and smaller tubes.

The bronchi and trachea are lined with **cilia.** Cilia are fine hairlike structures on the outside of some cells. They wave back and forth to move mucus up the tubes. The mucus cleans dust and pollutants out of the tubes as it moves. (See Figure 15–5 on page 397.)

Sometimes microorganisms such as a cold virus may get into the bronchi. When the linings of the bronchial tubes are infected, the condition is called **bronchitis.**

Figure 15–5 Cells with cilia line the air passages of the respiratory tract.

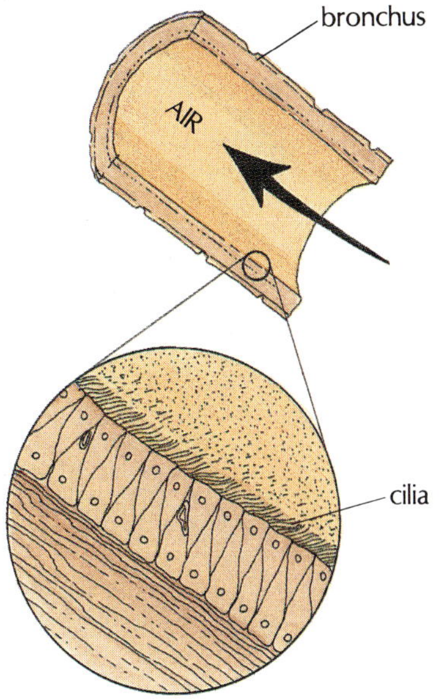

The Destination

At the ends of the smallest tubes are microscopic **air sacs.** They are like tiny balloons. (See Figure 15–6.) Each lung is made up of millions of these little air sacs. These sacs do not contain any muscles or bones. Because of all of the air spaces, lung tissue is light. Normally, a piece of lung tissue will float in a glass of water.

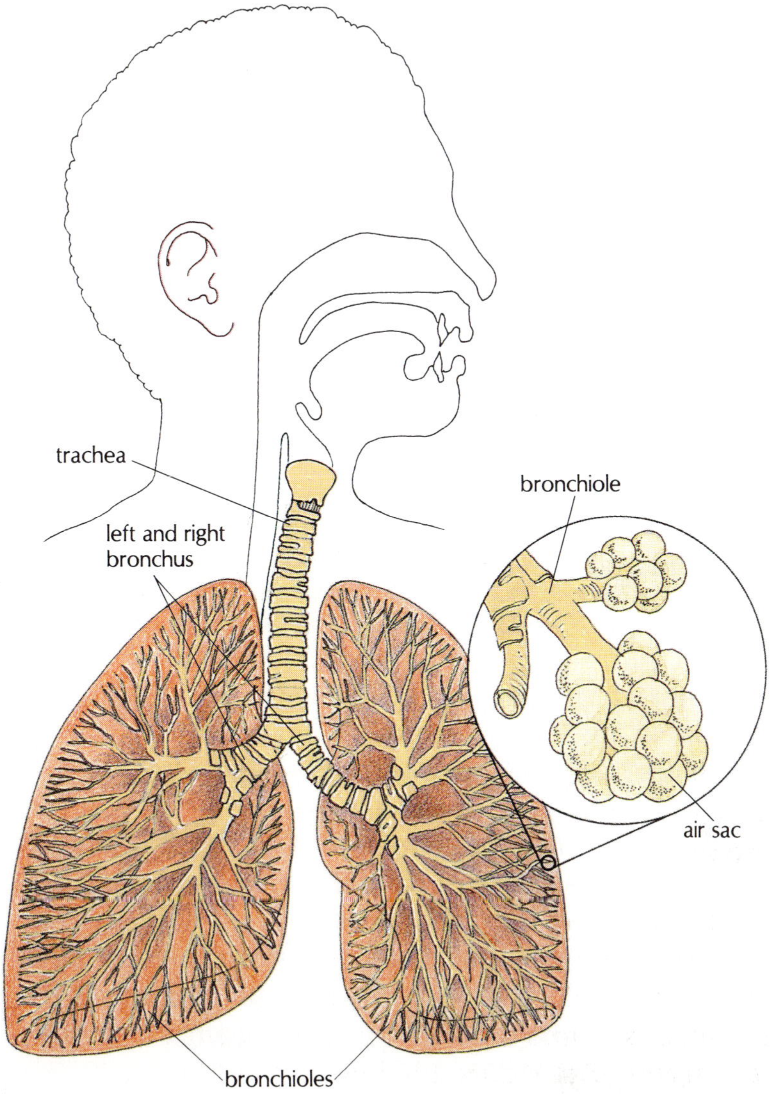

If you could open all of the air sacs in your lungs and spread them out, you'd need a space at least 60 square meters in size. That is probably more floor space than you have in your classroom!

Figure 15–6 At the ends of each of the bronchioles are microscopic air sacs.

Figure 15–7 How would you describe the position of the diaphragm when a person is inhaling and when a person is exhaling?

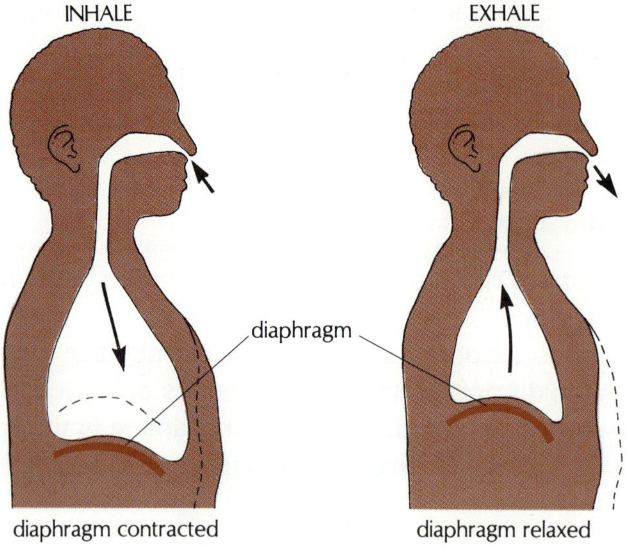

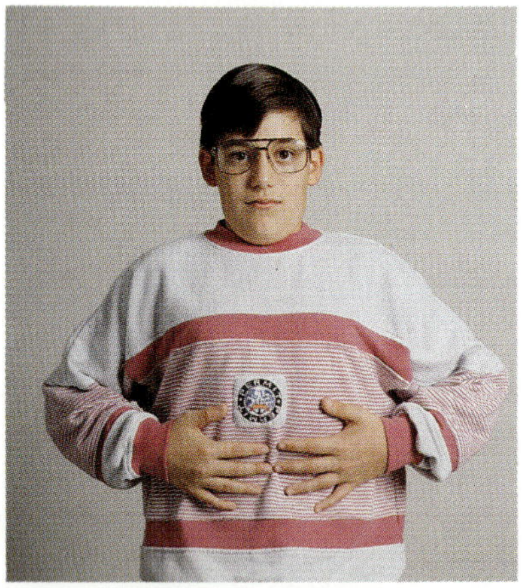

Figure 15–8 Try to breathe without moving your ribs.

Moving the Air—Ribs and Diaphragm

Since lungs are not made of muscle, they cannot move by themselves. You need to use your diaphragm and ribs to expand and contract your lungs. These actions are illustrated in Figure 15–7.

The **diaphragm** is a sheet of muscle just below the lungs. When it contracts, it flattens and air enters the lungs. When it relaxes, it pushes up on the lungs and forces air out. If your diaphragm contracts more vigorously, you get the hiccups. When you hiccup, you inhale in short gasps that sound very different from normal breaths.

Ribs form your rib cage. Muscles attach to the ribs. When these muscles contract and relax, the rib cage moves in and out. This action helps move the air.

Open your mouth. Try to say "How are you?" without changing position of your jaw. Now try to say it without moving your lips. Can you say it without moving your tongue?

Try humming when you breath in. Try again when you breathe out.

One Final Word

You may want to know about your **larynx.** Sometimes it is called a voice box or Adam's apple. You may be able to feel it at the top of your trachea. Look back at Figure 15–4 on page 396. How does the location of the larynx help you to speak?

Two bands of tissue cross the inside of the larynx. These bands are the **vocal cords.** Air passes between the vocal cords to make sounds. A strong blast of air will produce a loud sound; very little air is needed to whisper. You turn the sound of moving air into words by using your jaw, lips, and tongue.

ANALYSIS QUESTIONS

1. List 3 ways that the nose prepares air to enter the body.
2. Copy the chart below onto your own paper and then fill it in.

	Trachea	Esophagus
Which system is it part of?		
What is its function?		
What kind of tissue is it made of?		

3. Remember what you learned about cilia.
   a. Where are they found?
   b. What do they do?
4. Why is it better for a lung to be made of millions of air sacs instead of being one big sac like a balloon?
5. Explain what happens to your ribs, diaphragm, and lungs when you inhale.
6. When you talk, is air entering or leaving your lungs? Explain.

CHALLENGE

Look up the definitions for the following four words:
tracheotomy
bronchoscope
laryngitis
pneumonectomy
Write a two- or three-sentence story that uses at least three of these words.

EXTENSION ACTIVITIES

1. Call a meat packing plant—slaughter house—and ask for an animal's lung or a piece of one. Inspect the tissue. Comment on its color and texture. Try to find as many structures as possible. Start with the largest tube you can find and follow it down. How many times does it branch? How big is the *smallest* tube you can find? Look at a piece of the tissue under a dissecting microscope.
2. Find out about respiratory system infections. Look up colds, laryngitis, emphysema, and asthma. Answer these questions about each condition.
   What arc the symptoms?
   What causes it?
   What part of the respiratory system does it affect?
   How serious is it?
   What is the treatment for this condition?

15-3 GAS EXCHANGE

A person inhales about 2,000 liters (525 gallons) of air each day. Two thousand liters . . . that's a lot of air! What is air anyway? What happens to air after it is inhaled?

The Air We Inhale

Some people think air is all oxygen. Actually, air contains more nitrogen than anything else. Air is about 79% nitrogen, 20% oxygen, and 1% other gases. Water vapor is also in air. The amount of water varies from place to place and from day to day. For instance, if you live by an ocean, there is more moisture in the air there than you would find if you live in the desert.

Air also contains pollutants. Some examples of solid pollutants are tiny particles of sand and salt from the seas; other pollutants are types of gases. Some pollutants are dangerous; they can damage the environment and cause health problems. When we breathe, we cannot make molecule choices. We inhale it all—the normal gases along with the pollutants.

Figure 15-9 How does the content of the air you exhale differ from the air you inhale?

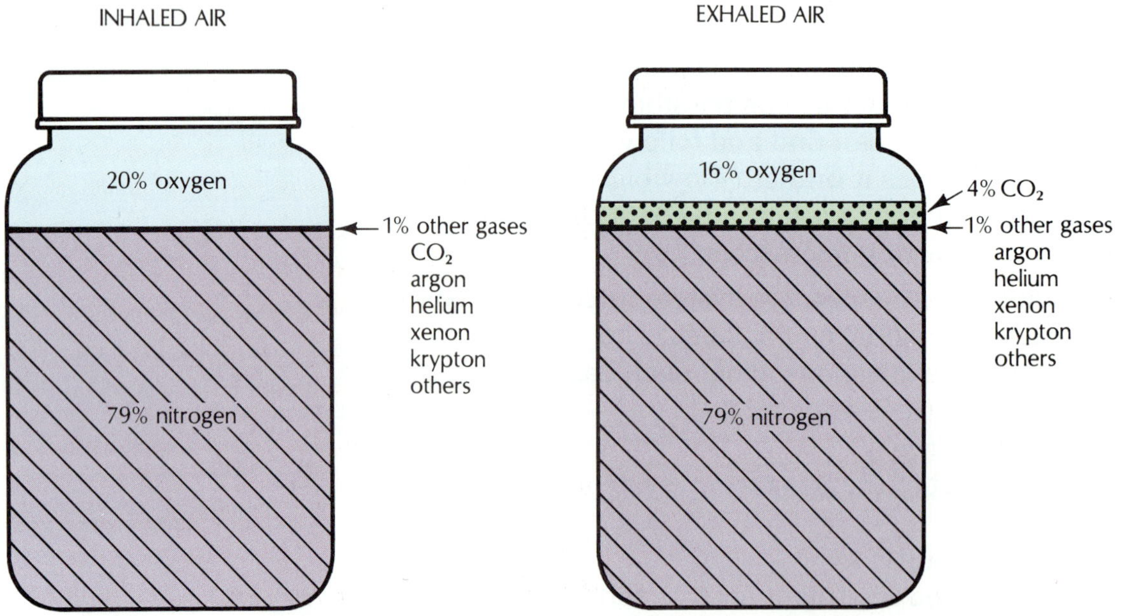

The Air We Exhale

The air you breathe *out* is not the same as the air you breathe *in*. Breathe on your hand for two or three breaths. The air feels warm, doesn't it? It also contains more moisture and a different concentration of gases than the air you inhaled. Exhaled air is still 79% nitrogen and 1% other gases; but now it contains 4% carbon dioxide and only 16% oxygen.

Figure 15–9 on page 400 shows information about the air we inhale and exhale.

Something happens to the air while it is inside your body; a gas exchange goes on. You keep some oxygen and get rid of some carbon dioxide. How does this switch take place?

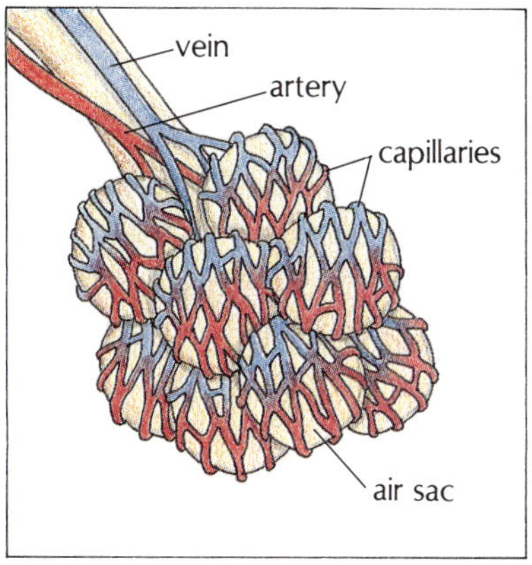

Figure 15–10 The air sacs are enclosed by a network of capillaries.

In the Lungs

When air enters the lungs, it goes through the bronchial tubes and into the air sacs. Each air sac is surrounded by a network of capillaries. (See Figure 15–10.) When blood flows through these capillaries, oxygen diffuses out of the air sacs. It "hitches a ride" in the blood. It can then circulate throughout the body. At the same time, carbon dioxide diffuses out of the blood and

Figure 15–11 The arrows in these drawings show the directions that gas molecules move in air sacs and capillaries. How many differences can you find between the two drawings?

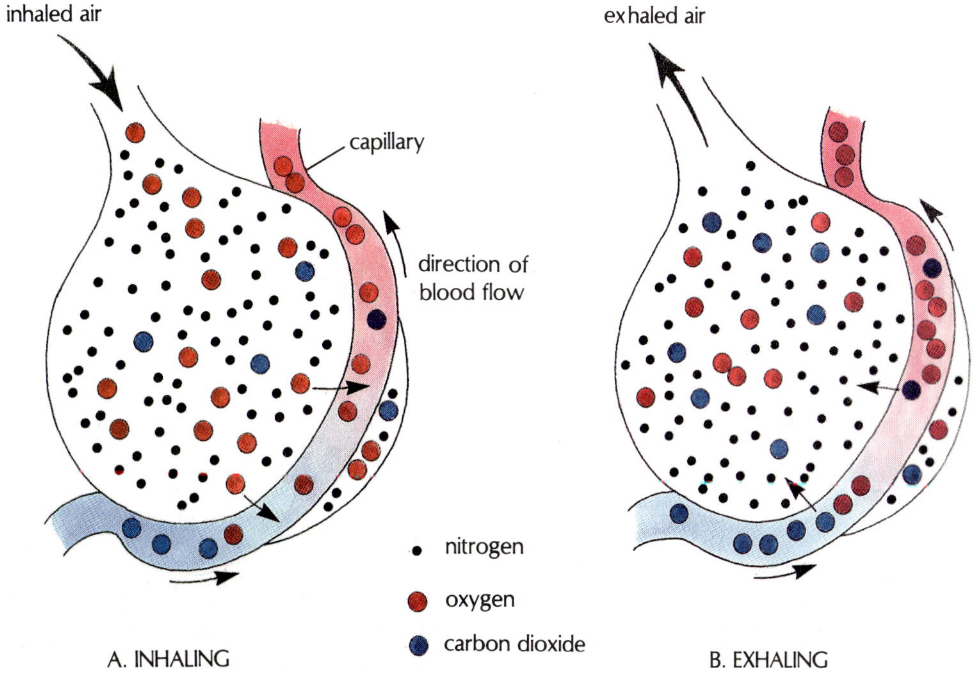

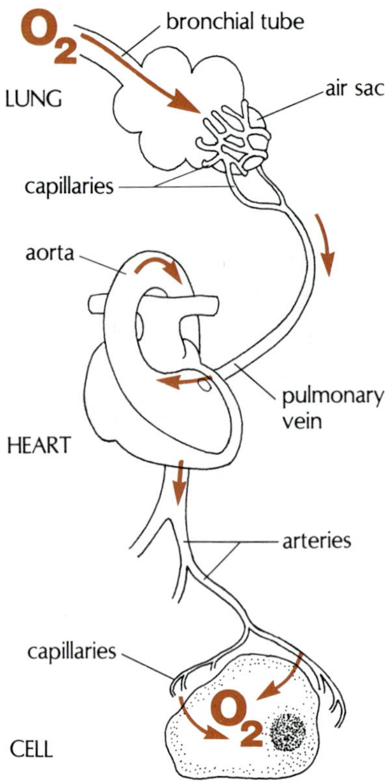

Figure 15–12 An oxygen molecule on its journey to a body cell

Figure 15–13 With the help of enzymes, glucose and oxygen are changed to energy in the cell.

into the air sacs. The exchange is done! This process is summarized in Figure 15–11 on page 401.

Stop! Ask yourself **why?** Why does this gas exchange take place? Why does the blood that enters the lungs have extra CO_2? What did your body do with the oxygen?

In the Cells

To understand what happens to the oxygen, imagine that *you* are an oxygen molecule. You have just been inhaled. You are moving through a series of bronchial tubes. It's dark in here! Soon, you reach a dead end. You are in an air sac. Since the wall of the air sac is only a single cell thick, you diffuse across the membranes and swoosh! You are swept up by a red blood cell.

Now you can cruise along. You make a quick trip through the heart and find yourself moving out through the aorta. There is no time to think; it is like being in a river. You flow with the current through a series of smaller and smaller arteries until you find yourself in a bed of capillaries. Suddenly, you feel a force pulling you out of the blood cell. You can't resist it. You diffuse out of the blood cell, through the capillary wall, and into a body cell. (This trip is illustrated in Figure 15–12.)

Oops! Now what? That cell *needs* you. It cannot break down food until you—oxygen—are present. An enzyme clings to you. Here comes a glucose molecule. **Wham!** Energy is released. You are gone. The food is gone. You have just been used in a chemical reaction. You may now find yourself as part of a water molecule (H_2O) or as part of a carbon dioxide molecule (CO_2).

Take another look at what happened to the oxygen when it was inside the cell. The cell needed the oxygen to break down food molecules, such as glucose, so a chemical reaction took place. It released energy, carbon dioxide, and water. It can be summarized like this:

food + oxygen → carbon dioxide + water + energy
(O_2) (CO_2) (H_2O)

This process is essential. Cells would die without oxygen; and if your cells died, you would die too. This process of combining food and oxygen to produce energy is called **respiration.**

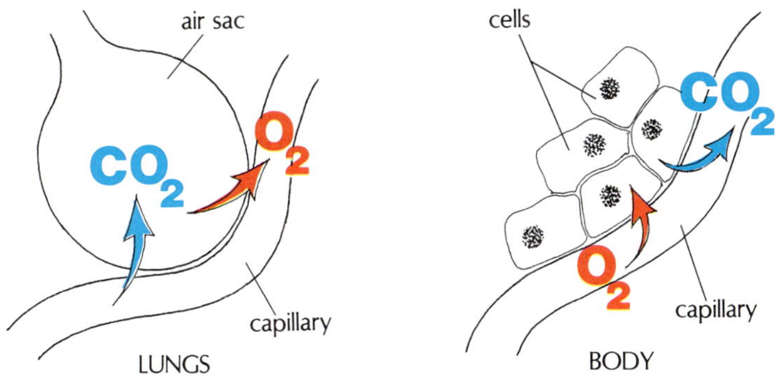

Figure 15–14 Gas exchange in the lungs and in body cells

Gas Exchange

By now you should be aware that a gas exchange occurs in two places—in the lungs and in the cells. In both of these places gases diffuse across membranes. If it were not for these gas exchanges, our cells would not get the oxygen they needed and we would not be able to get rid of carbon dioxide. (See Figure 15–14.)

CHECK YOUR UNDERSTANDING

Now it is your turn. Imagine that you are a carbon dioxide molecule. You are in a muscle cell in someone's leg. How are you going to get out of that person's body?

15-4 CIGARETTE SMOKE

A regular smoker probably shortens his or her life by about 5½ minutes for each cigarette smoked. The average smoker smokes 15–20 cigarettes a day.

American Medical Association
Family Medical Guide, 1987

- How is a lung model like a real lung?
- How does a new lung model differ from one that has "smoked"?

By using a lung model to "smoke" a cigarette, you will learn the answers to these questions.

MATERIALS

lung model
  clear plastic 10- to 16-oz bottle with small hole punched near the top
  10–15 cotton balls
  modeling clay
  rubber tubing, 8 cm long
cigarette
matches or alcohol burner
can with sand for cigarette stubs

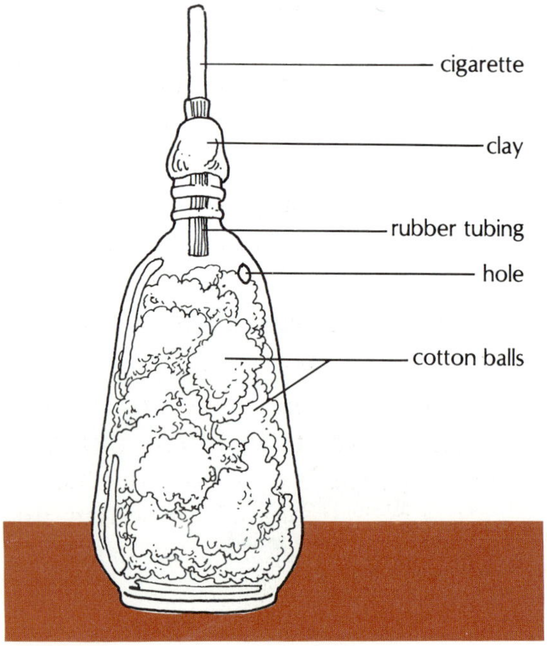

Figure 15–15 Lung model

PROCEDURE

Work in groups of two or three to do this activity.

Part A. Making a Lung Model

1. Get the materials you need to make a lung model.
2. Pull the cotton balls slightly apart in order to fluff them.
3. Fill the bottle with cotton balls. Do not pack them in.
4. Use clay to hold the tubing in the neck of the bottle. About 3 cm of tubing should stick above the top of the bottle. (See Figure 15–15.)

Part B. Practicing Breathing

Each person in your group should "practice breathing" as described below before the group continues. (See also Figure 15–16.)

1. *Exhale*—Squeeze the air out of the bottle. Leave the small hole uncovered. Watch how the air gets out of the bottle.
2. *Inhale*—Cover the small hole with your finger. Slowly release the bottle. Watch how air enters the bottle.
3. Repeat the process. Exhale. Inhale. Exhale. Inhale. Breathing should be slow and steady.

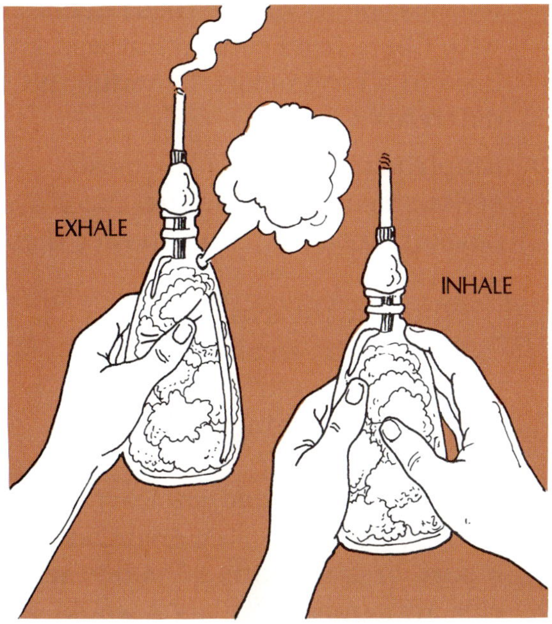

Figure 15–16 "Breathing" with a lung model

Part C. "Smoking" the Cigarette

1. Have one person take the lung model to your teacher. Your teacher will position a cigarette in the rubber tubing.
2. Have one person light the cigarette while another person holds the lung model. Do so by following these steps:
   a. Squeeze bottle (exhale).
   b. Light cigarette.
   c. Cover hole and release the bottle—air will enter through the cigarette (inhale).
3. Continue to "inhale" and "exhale" with the model until the cigarette is burned. Remember to "breathe" slowly!
4. Put the burnt cigarette stub in the disposal can.

Part D. Finishing Up

1. Closely observe the cotton. Record any changes.
2. Compare your lung model with others. Make sure you see the model that has not "smoked" any cigarettes.
3. Clean up and put away all of your equipment.

ANALYSIS QUESTIONS

1. List at least 2 ways that this model is like a lung.
2. Why did you cover the hole when you released your grip on the squeeze bottle?
3. What was the purpose of using one lung model without a cigarette?
4. List 3 differences between a lung model that smoked a cigarette and one that did not.
5. Imagine that you are looking at the cotton in two of the different lung models made in this activity. The cotton balls in one of the bottles are very brown. The cotton balls in the other one are a light yellow. Give at least 2 reasons why they could look so different.

Before you answer the rest of the questions, read the first paragraph on page 404 again.

6. About how many cigarettes does the average smoker smoke
   a. each day?
   b. every week? (Use the larger number and show your work.)
   c. every year?
7. If the American Medical Association is right, calculate how much a smoker's life is shortened by smoking 20 cigarettes a day
   a. for one day.
   b. for one year.
   c. for 20 years.
8. How can the lung model help us understand the effects of smoking on real lungs?
9. Your friend reads you a sentence from an article. It says "There is no definite proof that cigarette smoking is bad for a person's health."
   a. List at least 3 questions you could ask your friend about this article.
   b. Explain to your friend why you agree or disagree with that sentence.
10. Many states have laws that make it illegal to sell cigarettes to anyone under 18 years of age. Do you think these laws are necessary? Explain your answer.

CONCLUSION

Write brief answers to the focusing questions on page 404.

EXTENSION ACTIVITIES

1. Make a lung model at home and use it to "try" a cigar and an unfiltered cigarette. Are there any differences in the way the cotton balls looked? (Ask a parent to help you if you decide to try this experiment.)
2. Look at cotton fibers from clean cotton under the low power of the microscope. Compare these fibers with fibers from cotton that was used in a lung model. Describe the differences.
3. Emphysema is one of the most serious health threats that faces a smoker. Look it up and answer these questions.
   What are the symptoms?
   What is the treatment?
   Can it be cured?
   What other things can cause emphysema?

   Tell your class what you learned. Do so by making a poster, giving an oral report, or presenting a "public service announcement" as it would appear on radio or television.

15-5 INTERVIEWING A SMOKER

You can read about cigarettes, you can listen to news documentaries about smoking, and you can study the research findings about the effects of smoking, but there will still be something missing from your understanding of cigarettes and smoking. That "something" is hearing the smoker's viewpoint.

- What do most people who smoke have to say about smoking and the hazards of smoking?

Interview someone who smokes cigarettes and then share what you have learned with your classmates. Use the information to answer the focusing question.

MATERIALS

interview form: ***Interviewing a Smoker*** (Figure 15–17)

PROCEDURE

Part A. Getting Ready

1. Look over the interview form so you will be familiar with the questions.
   a. If there are any that do not seem to make sense, ask your teacher about them.
   b. Answer analysis questions 1–3.
2. Decide which adult you would like to interview. Choose someone who you think will feel comfortable talking about smoking.
3. Think about what you are going to say when you contact the person you want to interview. Remember to explain that this is a class assignment. Tell them also that it is all right if they choose not to answer a question.
4. As you interview, record the answers on the form. Remember to thank the person when you are done.

Part B. Summarizing the Interviews

1. Trade interview forms with someone in your class. Read each other's

INTERVIEWING A SMOKER

NAME ______

DATE ______ PER. ______

Talk to an adult you know who smokes cigarettes. Explain that you need some information about smoking for one of your classes. Remember to thank the person when you are done.

1. How old were you when you smoked your first cigarette?
2. What was it like? Was it fun?
3. Why did you begin to smoke?
4. About how many cigarettes did you smoke each day when you started?
5. How many cigarettes do you smoke each day now?
6. How much money do you usually spend on cigarettes in a month?
7. Have you ever tried to quit smoking? (If the answer is "no," go to question 8.)
   a. Why did you decide you wanted to quit?
   b. How long did you go without smoking?
   c. Why did you start smoking again?
   d. How many times have you tried to quit?
8. Do you think you have ever had any health problems because of smoking? Please explain.
9. What advice would you give to someone my age who was thinking about starting to smoke?

Information about the person who was interviewed.

sex: ____ male ____ female

age: ____ 20–29 ____ 30–39 ____ 40–49 ____ 50–59 ____ 60 or older

Figure 15–17

notes. Talk about your interviewing experiences and then answer analysis question 4.

2. After the class discussion, answer the rest of the analysis questions.

ANALYSIS QUESTIONS

1. How do you predict that adults will answer
   a. question 3?
   b. question 9?
2. Which question do you think looks the most interesting? Why?
3. Imagine that you just asked Mrs. Harris if you could interview her. She seems angry and says, "No, I'd rather you didn't."
   a. Why do you think she might be angry?
   b. What will you say to her now?
4. No two people answer questions exactly the same.
   a. For which question did the people's answers differ the most?
   b. What was one question to which most people gave similar responses?
5. How did most people who were interviewed answer
   a. question 3?
   b. question 9?
6. What was the most interesting thing you learned by doing this activity?

CONCLUSION

Think about the interview you did and the class discussion of the interview responses. Then write a summarizing paragraph to answer the focusing question on page 407.

15-6 SMOKING AND THE HUMAN BODY

There are two parts to this activity, a reading and microscope work. It does not matter which part you do first. Analysis questions 1–9 go with the reading. Answer analysis questions 10–12 when you look at the microscope slides. When you have completed both the reading and the microscope work, answer question 13.

Everybody knows that smoking cigarettes is a health hazard. It's been reported over and over again that smoking increases a person's chances of getting cancer and emphysema. Actually, lung cancer and emphysema are not the only health hazards associated with cigarettes. Smoking has widespread effects on the entire body.

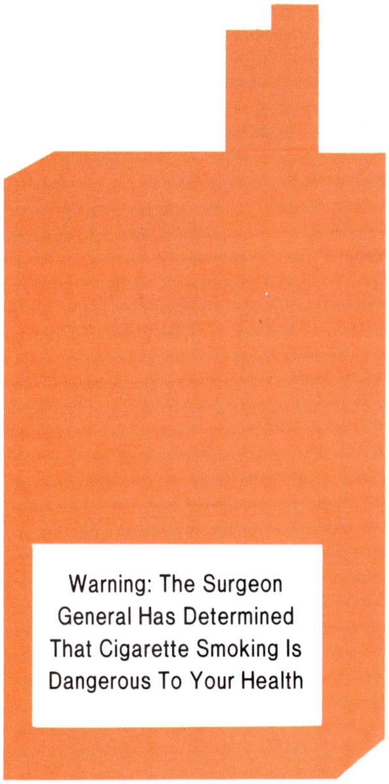

One Cigarette

The damage to a smoker's body starts with just one cigarette. After years of smoking, the average smoker is 20 times more likely to die from cancer than a nonsmoker. Smokers also double their chances of dying from a heart attack. The longer people smoke, the more damage they do to their bodies.

Smoke

The health problems are caused by the cigarette smoke. This smoke contains hundreds of chemicals. Three of the most dangerous ones are tars, nicotine, and carbon monoxide.

If you used the lung model to smoke a cigarette, you saw the **tars** build up on the cotton. Tars also build up in human lungs. (See Figure 15–18 on page 410.) Tars are brown, sticky substances that irritate the cells in the respiratory system. They also interfere with the gas exchange that takes place in the air sacs.

Nicotine is the addictive chemical in cigarettes. It acts on the body's nervous system. Some people crave

JUST 1 CIGARETTE . . .

- speeds up the heart rate
- increases blood pressure
- interferes with gas exchange in the lungs
- drops temperature of fingers and toes

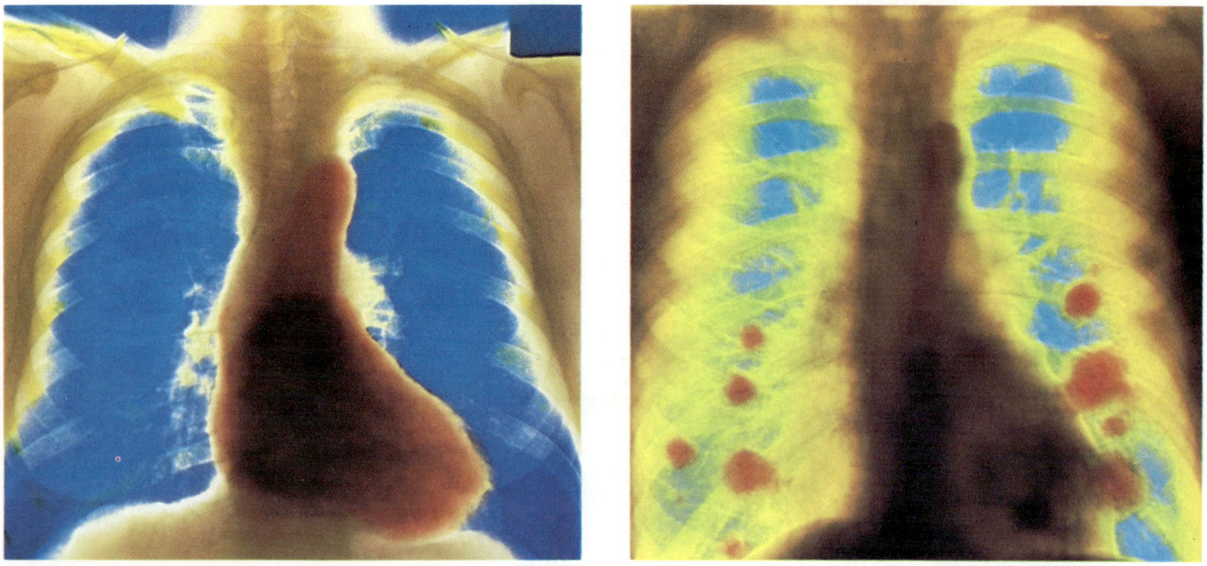

Figure 15–18 The x-ray on the left shows what a healthy lung looks like. The x-ray on the right was taken of someone who has lung cancer.

the feeling they get from nicotine. This craving can make it very hard for people to stop smoking once they have started.

Carbon monoxide (CO) is one of the gases in the smoke. It interferes with the gas exchange in the lungs. The hemoglobin in the red blood cells picks up carbon monoxide instead of oxygen. The body will not get the oxygen it needs if too much carbon monoxide is present.

Effects of Smoking

Doctors say that one of the best things you can do for your health is not smoke. Everyday they see people suffering from health problems. Figure 15–19 on the following page illustrates some of the most common health problems as well as other things that are associated with smoking.

Special Situations

Recent studies are showing that *nonsmokers* can get sick from other people's smoke. Sometimes the level of carbon monoxide in a smoke-filled room gets so high that it violates air pollution standards. How many times have you found yourself breathing someone else's cigarette smoke?

This secondhand smoke affects your body whether or not you realize it. Your heart rate increases and your

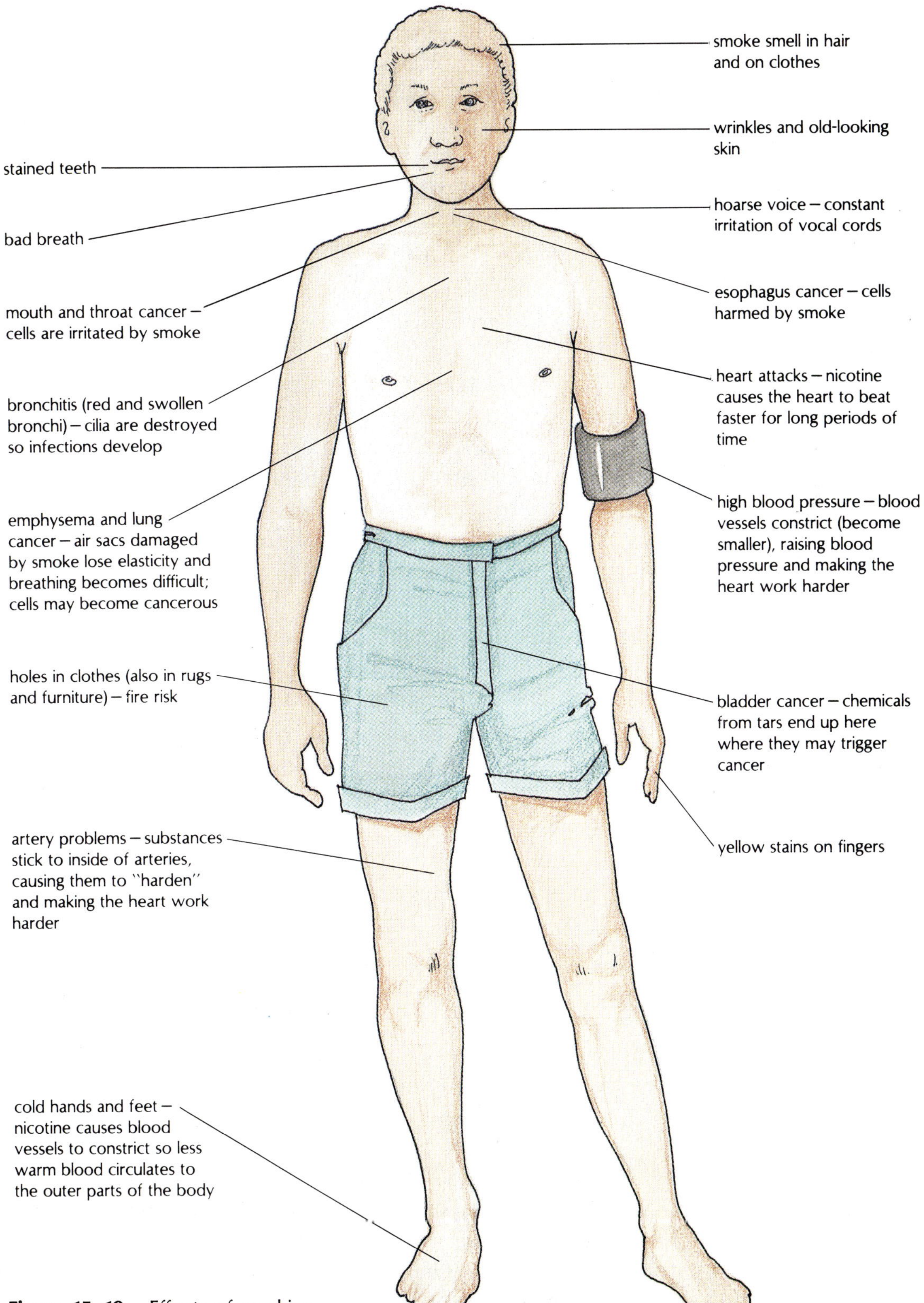

Figure 15–19 Effects of smoking cigarettes

Figure 15–20 Pregnant women should not smoke.

blood pressure goes up. Many people also complain of sore eyes, headaches, itchy throats, and stuffy noses.

Pregnant women need to be especially aware of the problems associated with smoking. Nicotine and carbon monoxide can slow their babies' growth. Children born to women who smoke are often smaller than average at birth. This is a concern because small babies are more likely to have health problems in infancy. (See Figure 15–20.)

Quitting

It does pay to quit. Almost immediately, the ex-smoker will notice that food tastes better, clothes smell fresher, and coughing stops. People who stop smoking for 15 years have about the same chances of dying from heart disease or cancer as does a nonsmoker. Quitting is not easy. Yet, this is one time when being a quitter is something to boast about!

Chewing Tobacco

You may have wondered if it is any safer to chew tobacco than it is to smoke it. Actually, tobacco chewers also suffer a variety of problems. Some of these include loss of teeth, bad breath, nicotine addiction, mouth sores, gum disease, and even mouth cancer.

Chewers can also be social rejects. Many people find chewing and spitting to be disgusting habits. So . . . is chewing tobacco "safe"? Definitely not!

Marijuana

During any discussion of the dangers of cigarette smoke, someone will probably ask about marijuana. The smoke from marijuana contains even more cancer-causing substances than cigarette smoke; it also contains more tars. People who use marijuana often inhale it very deeply and smoke the joints to the very end—where the tar concentrations are the highest. Thus, these people may have an even higher risk of developing cancer than people who smoke tobacco. They also have a higher risk of getting other respiratory problems such as bronchitis and asthma.

Of course, some of the greatest dangers of smoking marijuana involves the nervous system as well as the respiratory system. *Any* substance that affects the mind must be handled cautiously or not taken at all.

ANALYSIS QUESTIONS

Decide if these four statements are true or false. If a statement is false, explain why it is false.

1. T F Smoking one cigarette does not affect a person.
2. T F A person who smokes heavily will die from lung cancer.
3. T F Carbon dioxide is a dangerous gas in cigarette smoke.
4. T F Smoking cigarettes can make a person's feet feel cold.
5. Name 3 of the most damaging chemicals found in cigarette smoke.

Use the diagram that shows the effects of smoking on the human body in Figure 15–19 on page 411 to answer questions 6, 7, and 8.

6. If someone has just started to smoke, which of the problems shown on the diagram might he or she notice first?
7. List at least 3 health problems that involve the respiratory system.
8. List at least 3 health problems that involve the circulatory system.
9. What do you think is the biggest health problem that can result from
   a. smoking cigarettes?
   b. smoking marijuana?
   c. chewing tobacco?
10. Sketch what a nonsmoker's lung looks like when viewed through a microscope.
11. Sketch the smoker's lung section too.
12. Compare the two microscope slides.
    a. List several ways that the lung sections are alike.
    b. List ways that the lung sections are different.

13. Your friend Katie has just started to smoke. You try to talk her out of it. She tells you she is not worried about health problems because her uncle has smoked for 25 years and he is fine. What would you say to her next?

EXTENSION ACTIVITIES

1. A law was passed that requires cigarette packages and advertisements to carry stern warnings about the health hazards of smoking. Four different warnings were approved by the surgeon general. Look on cigarette wrappers and ads to find the four warnings. Write them down. Which one would worry you the most? Which one might worry your parents?
2. There are many reasons people choose to smoke even though they know it is dangerous. Advertising may have something to do with their decisions. Cut out at least 10 different cigarette advertisements. Answer these questions while you look at each ad.
   - What does the ad tell you about people who smoke?
   - How old are the people in the ad?
   - What dangers of smoking do the ads mention?
   - What age group are the advertisers trying to attract?
   - If you wanted to design an ad to discourage smoking, what would you include?

15-7 CLEAN AIR, DIRTY AIR

Even when the air smells absolutely terrible, you breathe. You cannot help it. Oxygen is so important to the body that you breathe even when you know it is dangerous to do so. The healthiest thing you can do for your lungs is to avoid dirty air. So, what is dirty air?

Dirty Air

"Dirty air" is another way to say **air pollution.** It can be caused by thousands of different chemicals and gases in the air. Some pollutants are gases that we *normally* breathe; but, in high concentrations, even they can be dangerous. For example, when you look at the list below, you can see that carbon monoxide is normally found in air. It only becomes dangerous at high concentrations.

GASES IN THE AIR*

nitrogen	oxygen	nitrogen dioxide	butane
argon	propane	carbon dioxide	neon
ethylene	ethane	carbon monoxide	methane
acetylene	krypton	formaldehyde	hydrogen
helium	ozone	nitrous oxide	isobutane
ammonia	xenon	sulfur dioxide	

*This is not a complete list.

Air pollutants come from many sources. Cigarette smoke is one example of a pollutant. Dust, sand, and salt from ocean spray are pollutants carried by the wind. The burning of fuels and wastes are major sources of air pollution. Factory and automobile emissions add to the problem. Even wood stoves and fireplaces contribute to dirty air.

The Air You Breathe

Some cities have serious air pollution problems. For days on end, their air may contain high levels of pollutants. Sometimes these pollutants cause health problems. Cities have a system for determining when the air quality is unhealthful. They use an air quality rating system to make their decisions. The table below explains this rating system.

AIR QUALITY RATING SYSTEM

Classification	Rating	Air Quality
Good	0–49	This is the best air to breathe.
Acceptable	50–99	Ok, but could be better.
Poor	100–199	Unhealthful. People may get sore eyes and headaches.
Extremely Poor	200–299	Very unhealthful. Healthy people complain of sore eyes and headaches. People with heart or lung problems should stay inside.
Dangerous	300–399	Hazardous. Triggers asthma, bronchitis, and heart attacks. Everyone should stay inside.
Extremely Dangerous	400–500	Hazardous. Causes death in the ill and elderly. Healthy people feel sick.

In order to come up with a rating, the levels of five dangerous substances are measured. These five substances are sulfur dioxide, nitrogen dioxide, carbon monoxide, ozone, and solid matter that is floating in the air. The table that follows gives some information about each of the pollutants. Air pollution is a problem that is suddenly not going to go away. Be alert to the condition of the air in your own town.

AIR POLLUTANTS

Pollutant	Description
Carbon monoxide (CO)	An invisible, odorless gas. It combines with hemoglobin in the blood. When this happens, blood cannot carry enough oxygen. Large amounts of carbon monoxide cause death. Small amounts cause dizziness, headaches, and tiredness.
Sulfur dioxide (SO_2)	Sulfur dioxide irritates the eyes, nose, and throat. It causes breathing problems and can damage the lungs. It kills plants.
Nitrogen dioxide (NO_2)	Nitrogen dioxide irritates the eyes and nose. It can cause respiratory problems such as bronchitis and pneumonia.
Ozone (O_3)	Ozone is formed when nitrogen oxides are exposed to the energy from sunlight. Because sunlight is needed to produce ozone, ozone levels increase during the day and decrease at night. This pollutant irritates the lungs and causes eyes to "burn." It damages plants and materials such as rubber.
Solid suspended particles	These include dust, smoke, soot, ash, and pollen. Some dust particles can be radioactive, poisonous, and/or cancer-causing. These particles may stay suspended in the air or settle to the ground. They darken the air and cause objects to become dirty. They interfere with normal breathing and may injure the lungs.

ANALYSIS QUESTIONS

1. Look at the list of gases in the air on page 415.
   a. List 3 gases that you have heard of before.
   b. List 3 gases that you have never heard of.
2. List at least 5 places or things in your community that add pollutants to the air you breathe.
3. What is the air quality rating on an unhealthful day?
4. What symptoms might you experience if the rating were 230?
5. Which of the pollutants listed in the table on page 417 can cause respiratory problems?
6. Which pollutant sounds most dangerous? Why?

EXTENSION ACTIVITIES

1. Cut out newspaper articles on air pollution for one week. Make a list of the pollutants that are described. Explain why each of these substances is a concern. According to your articles, where is air pollution a problem? What is causing it? Use your articles to make an air pollution poster.
2. Go through your shelves and cupboards at home. Check the labels on cleaners, spray cans, herbicides, and other poisons. Which ones warn you not to breathe the contents of the containers? Which ones say to use in a well-ventilated area? Find out why these chemicals are dangerous to breathe and how they can harm you. Determine where the closest poison control center is to you. See if they can help you find some of the answers.
3. See if your local newspaper gives air quality readings for your area. These may be on the page with the weather. If you find them, record the values for one week. You can make this into a bigger project by graphing the air quality for one month or more. On your graph, also record the major weather condition for each day—hot and sunny, rainy, snowy, windy, and so on. You may be able to see how weather affects air quality.

Clean Your Air! Buy a Spider Plant!

Outdoor air pollution used to be our only worry. Now we hear that indoor air pollution can also be a health hazard. According to the National Academy of Sciences, thousands of people may be dying every year as a result of air pollutants in homes and offices. Millions of other people may suffer from allergies, bronchitis, and other health problems that result from these pollutants.

A gas stove is an example of an indoor air polluter. In less than an hour it can fill a small kitchen with more carbon monoxide and nitrogen dioxide than national standards allow for city air.

Bill Wolverton, an environmental scientist for NASA, may have a solution. He thinks spider plants can help fight indoor air pollution. For some unknown reason, spider plants remove gaseous pollutants at a tremendous rate. Perhaps, it is because their "babies" get their food from the air. Even the soil in the pot helps out. Soil microorganisms break down formaldehyde, carbon monoxide, and other chemicals. An average-size home would need at least fifteen spider plants to keep the air clean.

In spacecraft, soil microorganisms and plants could help purify the air; but, right now, NASA's plant man is helping solve air pollution problems right here on Earth!

UNIT 5

BODY CONTROLS

CHAPTER 16

SENSES

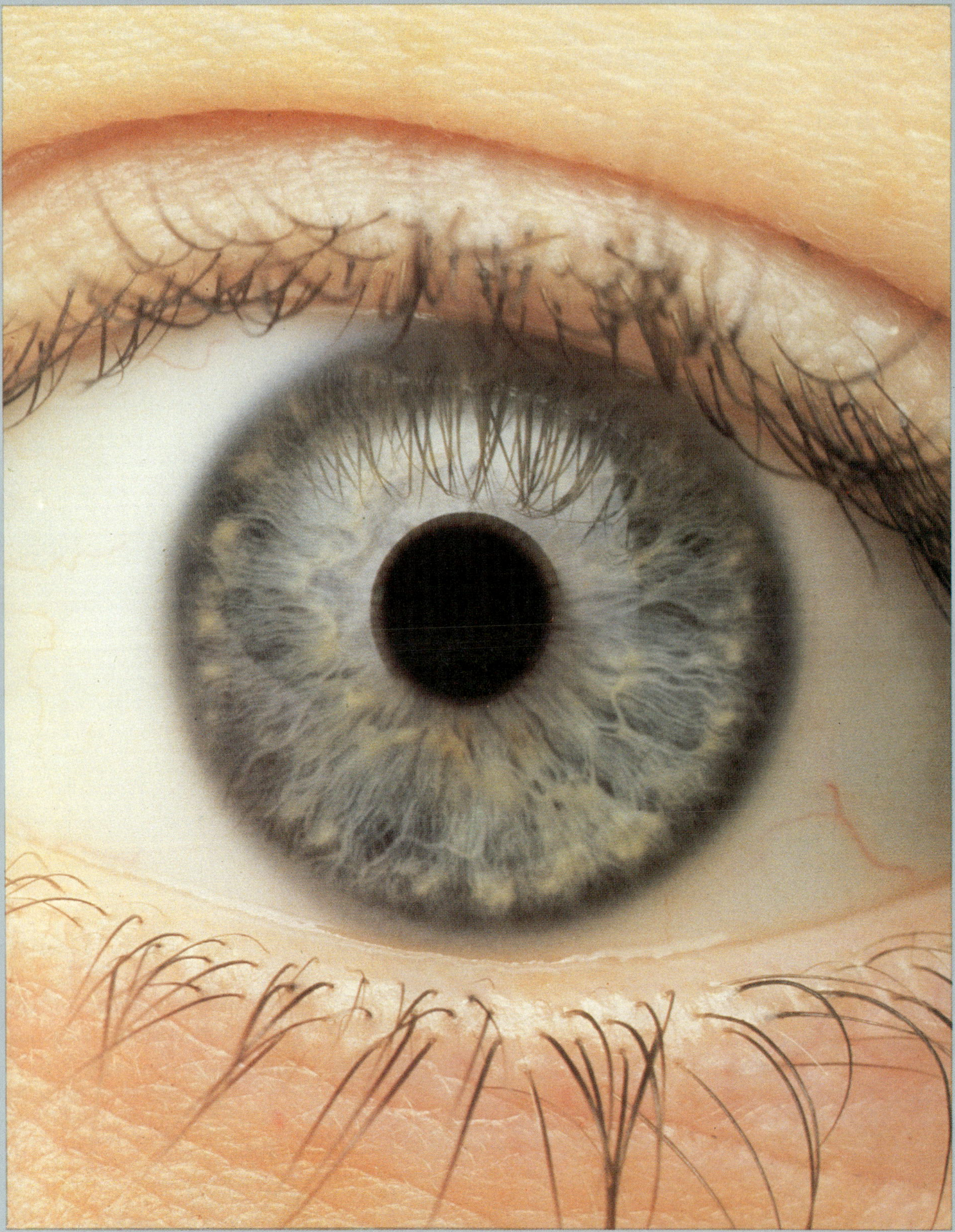

How does the iris—the colored portion of the eye—help a person see? What do you think the iris is made of?

16-1 HEARING TEST

A world without sound is hard for most of us to imagine, but some degree of hearing loss is actually quite common. About one out of every thirty children has a noticeable hearing loss. As people grow older, having a hearing loss becomes more common. One out of every four people over age 65 has lost some hearing. Most hearing problems are difficult to recognize; a young child who has never heard does not know anything is missing and older persons usually lose their hearing so gradually they do not realize that it is happening. Although many people have hearing problems, few people are totally deaf.

- How do words sound when someone has a hearing loss?
- What sounds may be hard to hear for people who have a hearing loss?

You will be able to answer these questions after you experience what it would be like to have a hearing loss.

MATERIALS

recording of the hearing test
pencil and paper

PROCEDURE

1. Make a data table like the one below on a piece of paper.

	A	B	C
1			
2			
3			
4			
5			
6			
7			
8			
9			
10			

2. Listen to the recording of the hearing test.
3. Take the hearing test when the recording says to do so. Record the words on your data table.
4. Answer the analysis questions.

ANALYSIS QUESTIONS

1. How many words did you hear correctly
   a. the first time?
   b. the second time?
   c. the third time?
2. What was done during the recording of each list to make it sound different from the others?
3. How did you feel when you could not understand the words?

4. If someone has a hearing loss, what could you do to help him or her understand you when you talk? (Think about what would have helped you understand the words on the recording.)
5. Do any of your family members have a hearing loss? How old were they when the loss was noticed?
6. What are some situations—besides spelling tests—when good hearing is especially important?
7. What do you think can cause a hearing loss? List as many causes as you can.

EXTENSION ACTIVITY

Visit a clinic that fits hearing aids. Find out about different kinds of hearing aids, how much they cost, and how the different kinds of hearing loss are helped by the use of a hearing aid. Try on some hearing aids. How do they affect your hearing?

"What?" Asked the Insect

Even if an insect could talk, it would not do much good. Most insects cannot hear. Only locusts, cicadas, crickets, grasshoppers, and most moths have ears. They have simple ears that allow them to detect the sound of a possible mate or maybe an enemy.

Figure 16–1 Cricket

16–2 NOISE POLLUTION

Some sounds can be painful. You try to protect your sense of hearing from these painful sounds without even thinking about it. That is an important reflex action because loud sounds can damage your hearing. Some other sounds, even if they are not loud, can be dangerous, too.

- How can sound be dangerous?
- What sounds in your environment can damage your hearing?

To answer these questions, you will analyze familiar sounds and read about noise pollution.

MATERIALS

reading: ***Noise Hurts*** (page 428)

PROCEDURE

1. Use 2 pieces of paper to make two data tables like the one below. One table will be for inside sounds and the other for outside sounds.

SOUND INVENTORY

Location: ____________________

Sounds	Decibels

2. Use one of the data tables to take a "sound inventory" in one room of your home. List all of the sounds you hear in 3 minutes.
3. Choose an area outside and take another 3-minute sound inventory. Remember to record the sounds you hear on your second data table.
4. Use the information given in Figure 16–2 on page 426 to find the decibel level for each item on your two data tables. Record the appropriate decibel level for each item you listed. If you listed an item that is not on the chart, choose something that is about the same loudness and record that number.
5. Read ***Noise Hurts*** starting on page 428. Then answer the analysis questions.

ANALYSIS QUESTIONS

1. What were the 3 loudest sounds you heard inside your home?
2. What were the 3 loudest sounds you heard outside?
3. What is the lowest decibel level that can cause hearing loss?
4. Which of the loud sounds that you listed can cause permanent hearing loss?
5. Music at some rock concerts has been measured at 120 decibels. How long could you listen to the music without damaging your hearing?

Decibels

A sound may be as soft as a whisper or as loud as an explosion. The loudness of the sound is measured in decibels.

DECIBEL LEVELS OF COMMON SOUNDS

Decibels	Sounds
0	hearing begins
10	leaves rustling • normal breathing
30	soft whisper • library
40	quiet office or room • refrigerator
50	light auto traffic – 100 feet away
60	**normal conversation**
70	dishwasher • vacuum cleaner • noisy restaurant
80	shower • alarm clock – 2 feet away • motorcycle
90	large heavy truck
100	garbage truck • lawn mower – 3 feet away • rapid transit rail noise
110	rock music
125	loud portable head stereo
140	jet engine – 75 feet away

You are familiar with other measures. Centimeters measure length and seconds measure time. You can add centimeters to get total length—3 cm + 4 cm = 7 cm—and you can add seconds to get total time—5 secs + 40 secs = 45 secs. However, you cannot add decibels to get total loudness. That is, 60 decibels plus 60 decibels does not equal 120 decibels. It equals 63 decibels. The total loudness of all the sounds around you is about the same decibel level as the loudest sound by itself.

Figure 16–2

6. What are 2 things you do that are restful to your hearing? How do you know they are restful?
7. Read Figure 16–2 on page 426.
   a. About what would be the total decibel level for your outside location?
   b. How long could you safely stay there without damaging your hearing?
8. List at least 4 things that noise can do to your body besides cause a hearing loss.
9. How do you do the "walkaway" test?
10. Describe 3 things you could do to lower the decibel level in one room of your home.

CONCLUSION

Answer the questions on page 425 now that you have taken your sound inventories and completed the reading.

EXTENSION ACTIVITIES

1. Take sound inventories at 3 other locations, such as a street corner, supermarket, or woodworking shop. Which "soundmaker" has the highest decibel level at each location? How could the noise level be lowered at each of these locations?
2. Call your police department or another office at city hall to find out if your community has noise level laws. If so, ask someone to explain them to you. Then ask if you can have copies of the laws. Find out how these laws are enforced.
3. Ask the band director or school nurse if there is a sound meter that you could use. Measure the sound level in various parts of your school. Where is it the quietest? the noisiest? Are any areas in your school that are dangerous to your hearing?
4. Talk to an audiologist to find out what his or her job involves. Audiologists work at places such as hospitals, hearing clinics, and health departments. Ask questions such as the following:

   What does a hearing test involve?

   What are some causes of hearing loss?

   What do you think about the effects of loud music on hearing?

Figure 16–3 An audiologist giving a hearing test

Noise Hurts

Noise has become a serious concern to many people because our environment is becoming noisier all of the time. Motor vehicles, small engines, voices . . . just about everything contributes to environmental noise. Noise is a concern for at least three reasons.

1. Noise can cause hearing loss.
2. Noise can harm a person's health.
3. Noise can make it hard to work and study.

Hearing Loss

When you listen to music, where do you set the volume? Do you have it turned up quite loud? If so, think about this. A recent study found that one out of every three students who were entering college had a significant hearing loss. It seemed that this loss was from listening to loud music. This kind of hearing loss is permanent. Loud noises destroy cells that detect sounds. Once the cells are destroyed, hearing is gone forever.

Figure 16–4 The lower she keeps the volume, the less her chances of damaging her hearing

Some teenagers listen to loud music even after they have been told that they are hurting their hearing. Why? How about you? Do you have any hearing loss? You may not know it if you do because a hearing loss sneaks up on you. When you realize your hearing has gone, it is too late.

Health Effects

Hearing loss is not the only problem caused by noise. Recent medical research has shown that noise affects our bodies in many ways. Some of the things noise can cause are as follows:

- high blood pressure
- stomachaches and ulcers
- feelings of nervousness
- an increase in heart rate
- muscle tension

Most people do not realize that noise causes health problems. However, people who live in noisy areas should be aware that they are at risk for developing these noise-related problems.

Figure 16–5 One cause of noise in a big city.

Constant Noise

Airports are noisy places. People who live by airports may claim that they adjust to the noise. They often explain that they "tune out" the sounds of the planes. As proof, they may say that they sleep well at night in spite of air traffic. However, studies have shown that they are wrong. Even though the people thought they slept well, their heart rates jumped each time there was a loud sound. Most of them also had higher blood pressure readings than average and more muscle tension. In addition, many people were moody the day after a noisy sleep. Their bodies really had not adjusted to the noise.

One study found that noise also made it harder for children to learn. The researchers compared schools by the Los Angeles airport with schools in quiet neighborhoods. One of the things they found was that children in the quiet schools did better solving puzzles and math problems.

SOUND EFFECTS

Decibels	Description of Effect	Permanent hearing loss can result if it continues for longer than this amount of time.*
0	barely noticeable as a sound	–
20	softest sound most people hear	–
60	**normal speaking levels**	–
80	annoying	20 hours
90	poor conditions for working	8 hours
100	sounds very loud	2 hours 15 minutes
110	louder yet	35 minutes
120	uncomfortably loud	7 minutes
130	strong "tingling" in ear	1 minute
140	sharp pain in ear	

*These levels were established by OSHA (Occupational Safety and Health Administration—a government agency).

Dangerous Sounds

There have not been enough studies done yet to know which sounds can cause each of the health problems mentioned earlier; however, there is strong evidence showing which sound levels cause hearing loss. In general, sounds below 60 decibels (a normal speaking level) are not harmful. Any sound above 80 decibels can cause permanent hearing loss if it continues for a long enough time. The table at the top of the page will give you some idea of "dangerous times" for different sound levels.

Make It Safer

If you are concerned about your health and hearing, you should be concerned about noise levels. If an area seems noisy to you, check the sound level. Since you aren't likely to have a decibel chart or a sound meter with you, do the "walkaway" test described below.

Ask a friend to read aloud or talk in a normal voice. You should be able to walk away and still understand the words when you are 25 feet (8 meters) away. If you cannot hear the words, the noise level is high enough to stress your body and possibly cause hearing damage.

Luckily, some noises can be avoided or at least turned down. Think about your own environment. Which areas are particularly noisy? What noises can *you* turn down?

16-3 EARS AND HEARING

Hearing is a complex process. If you could inspect the parts of an ear, it would help you understand how the ear works. The best way to do this would be to dissect one. However, this is very difficult to do since most of the ear is encased in bone. When you remove the bone, you ruin the ear.

Since it is not possible to dissect an ear, you can learn about its parts by reading. As you read this section, label the diagram on your worksheet. (See Figure 16–6.)

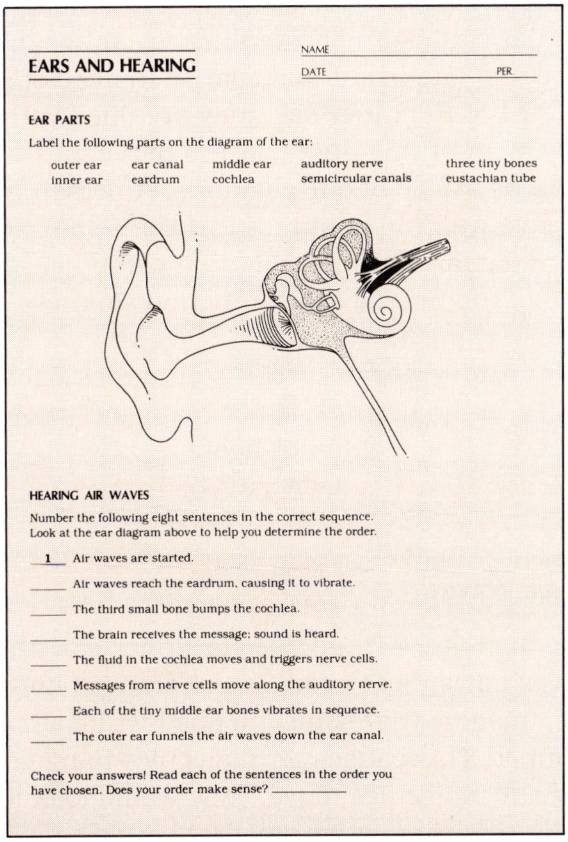

EARS AND HEARING

NAME ____________

DATE ____________ PER. ____

EAR PARTS

Label the following parts on the diagram of the ear:

outer ear	ear canal	middle ear	auditory nerve	three tiny bones
inner ear	eardrum	cochlea	semicircular canals	eustachian tube

HEARING AIR WAVES

Number the following eight sentences in the correct sequence. Look at the ear diagram above to help you determine the order.

__1__ Air waves are started.

_____ Air waves reach the eardrum, causing it to vibrate.

_____ The third small bone bumps the cochlea.

_____ The brain receives the message: sound is heard.

_____ The fluid in the cochlea moves and triggers nerve cells.

_____ Messages from nerve cells move along the auditory nerve.

_____ Each of the tiny middle ear bones vibrates in sequence.

_____ The outer ear funnels the air waves down the ear canal.

Check your answers! Read each of the sentences in the order you have chosen. Does your order make sense? ________

Figure 16–6

Ear Parts

There are three areas of the ear: outer ear, middle ear, and inner ear. You may already know the parts of the **outer ear.** These are the parts of the ear you see. The

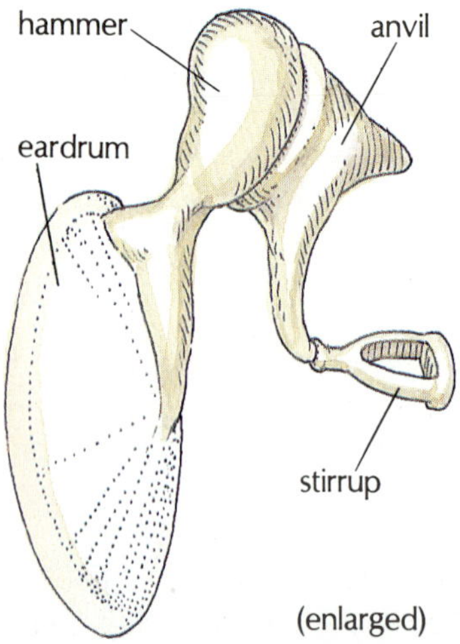

Figure 16–7 These three bones are the tiniest ones in the body. Their shapes give you clues to their names — hammer, anvil, and stirrup.

most obvious one is the cartilage structure that we call the **ear.** It funnels sound into the **ear canal.** If you could look down the ear canal, you would see that it leads to the **eardrum.** The eardrum separates the outer ear from the middle ear.

Three tiny bones make up the **middle ear.** These bones are shown in Figure 16–7. They bridge the space between the eardrum and part of the inner ear—the cochlea.

The **inner ear** is a delicate structure; it is enclosed by the bones of the skull. While both the outer and middle parts of the ear are filled with air, the inner ear is filled with fluid. The easiest part of the inner ear to recognize is the **cochlea** because it is shaped like a snail. The cells that line the cochlea are very sensitive. They can be damaged by loud sounds. Three **semicircular canals** attach to the cochlea. Even though these semicircular canals are located in the ear, we think of them as being important for balance instead of hearing. Like the rest of the inner ear, they are filled with fluid. The movement of this fluid triggers nerve messages that make us aware of our positions. A person may have trouble walking or standing if the semicircular canals are damaged. If the fluid in a person's canals moves all of the time, a type of motion sickness may result.

The **auditory nerve** carries sound messages from the cochlea to the brain. It is not really a part of the ear but, without it, you could not hear.

Hearing Air Waves

Together, the outer ear, middle ear, inner ear, and auditory nerves allow us to hear. To understand how this happens, imagine that someone has just blown a note on a trumpet. This causes air molecules to move. If you could see the molecules move, they would look like waves. For this reason, we say that sound travels in sound waves. The outer ear collects these waves and directs them to the middle ear. This causes the bones in the middle ear to vibrate. In the inner ear, the vibrations are changed into a nerve message. Your brain then "interprets" the message so you can hear the trumpet.

Take a break from reading and see if you can decide how your ear works. Put the sentences at the bottom of your copy of Figure 16–6 in the correct order.

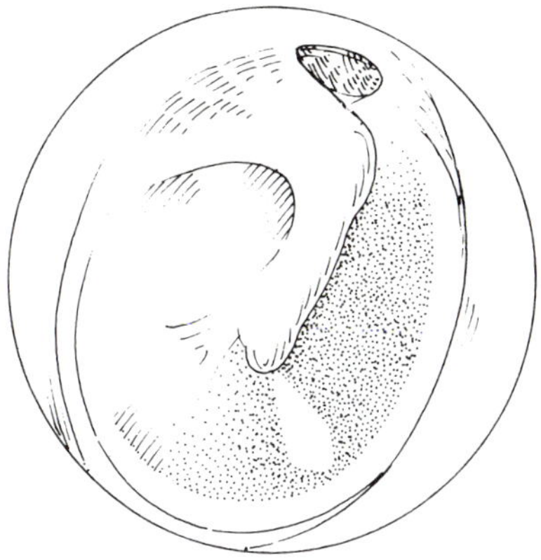

Figure 16–8 This is an eardrum as it appears when viewed through an otoscope.

Eardrums

Even though all parts of the ear are important for hearing, people talk most about eardrums. Figure 16–8 shows what an eardrum looks like when you look down the ear canal at it with an otoscope. It is a thin membrane that separates the outer ear from the middle ear. You can picture the eardrum as being like a piece of a balloon stretched over a ring. Like this balloon, a real eardrum can move in and out. If you close your mouth, plug your nose, and try to breathe, you can feel your eardrums move. When air pressure changes, the eardrums move. This is illustrated in Figure 16–9.

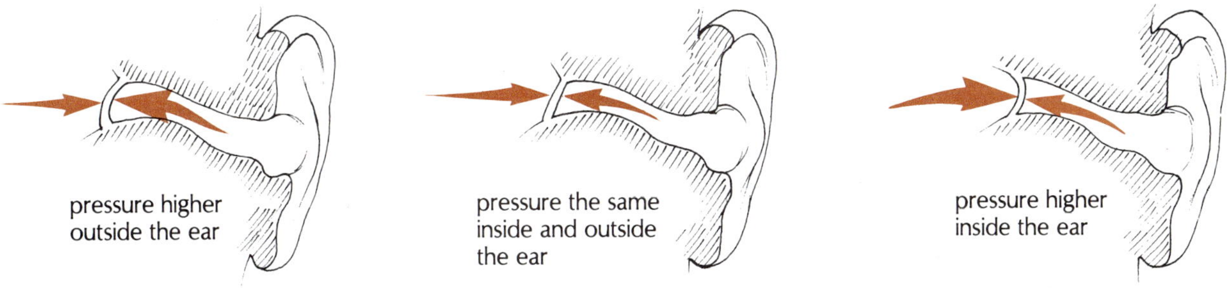

Figure 16–9 The eardrum moves when air pressure changes.

When an eardrum stretches too much, it tears. You protect your eardrums from tearing without realizing it. When your ears start to feel uncomfortable, you try to "make them pop" by yawning, chewing gum, or swallowing. By doing this, you are moving the air in your **eustachian tubes** to make the air pressure equal on both sides of your eardrum. The eustachian tubes are openings from the middle ear into the throat. *(Remember to label the eustachian tubes on the ear diagram on your worksheet.)*

Infections

Sometimes, bacteria or viruses travel up a eustachian tube from the throat into the middle ear. When this happens, a middle ear infection can develop. If, however, bacteria settle in the ear canal, an outer ear infection results. Thus, the treatment for an ear infection depends on which part of the ear is involved.

CHECK YOUR UNDERSTANDING

1. Imagine that you have a cold and your ear starts to ache. Are you more likely to have an infection in your ear canal or middle ear? Explain.
2. "Don't put cotton swabs, hairpins, fingernails, or anything else into your ears." When people say this, what are they afraid you could do to your ear?
3. What kind of infection do you think could be treated with ear drops—an infection in the ear canal or middle ear? Explain.
4. Suppose you know someone who has a hearing loss in one ear. Look at the ear diagram on your worksheet. Think of the things that happen in the ear to allow you to hear. Explain at least 2 things that could go wrong with the structure of the ear that could explain your friend's hearing problem.

EXTENSION ACTIVITIES

1. Ask the school nurse if you can look at an eardrum with his or her otoscope. Find out how you can tell if someone has an infection by inspecting an eardrum.
2. The ear is important for balance. Demonstrate this by walking a log three times. The first time just walk the log. The second time use headphones to listen to music while you walk the log. The third time adjust the headphones to play music into one ear only. How does the music affect your ability to walk the log?

Different Pitches

Different animals can hear different pitches. Frogs can hear sounds much lower than we can hear. We hear low sounds better than dogs, cats, and birds; they, however, can detect high sounds that we never hear.

16–4 READING BY TOUCH

Do you know anyone who does not have a sense of touch? Probably not. Very rarely are people born without touch receptors in their skin. Even though our sense of touch is very important, we seldom give it much thought unless we have to rely on it for some reason.

People who have lost one of their other senses may develop a very sensitive touch. For instance, blind people are very aware of touch because they use this sense to read.

- How can reading Braille be used to test your sense of touch?
- What problems do you think you would have if you were learning to read Braille?

In this activity you will use your sense of touch as well as your sense of sight to answer these questions.

MATERIALS

Braille alphabet card
2 mystery word cards

PROCEDURE

Work with a partner.

Part A. The Braille Alphabet

You will need a Braille alphabet card to do this part of the activity.

1. People who have lost their sense of sight use their sense of touch to read Braille. Braille is a language of raised dots. Each letter is a particular combination of from one to six dots. The six dots in their specific locations are called a *full cell.* This is shown in Figure 16–10. Letters, numbers, and punctuation are made from different patterns of these six dots.
   a. Find the full cell on the Braille card.
   b. Figure out how to tell if a symbol is a letter or a number.

1 o o 4
2 o o 5
3 o o 6

Figure 16–10
A full cell

2. Use your fingertips to feel the dots. Can you tell how many dots make up each symbol? Try using different fingertips to find out if one is more sensitive than another.
3. Answer analysis questions 1–4.

Part B. A Mystery Word

You and your partner will take turns being "blind" in this part of the activity. The blind person will try to read a "mystery word" in Braille. This word is printed in two places on a card. The letters in the middle of the card have extra space between them to make the word easier to read. A sample mystery card is shown in Figure 16–11 on page 436.

1. If you are the blind person, close your eyes so you will not see the mystery word. You will soon find out how well you can use your sense of touch.

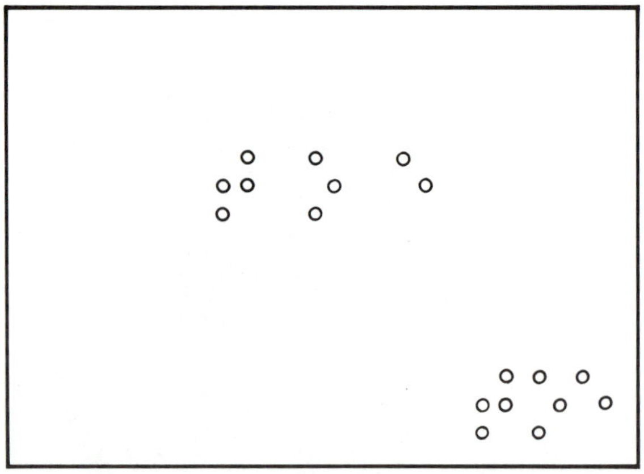

Figure 16–11 Mystery word card

2. Your partner should get one mystery word card, record the identifying letter of the card on your paper, and hand the card to you.
3. Describe the arrangement of dots that form the letters on your mystery word to your partner. (Remember to keep your eyes closed.) Using your paper, your partner should record the pattern of dots you describe. Later you can compare this pattern with the dots on the mystery word card.
4. Switch roles with your partner and get a new mystery word for him or her to "read."
5. Answer the rest of the analysis questions.

ANALYSIS QUESTIONS

1. Write your name in Braille.
2. Several words are printed on the last four lines of the Braille alphabet card. What is the last word?
3. How can you tell the letter *a* from the symbol for a comma (,)?
4. Which of your fingertips is most sensitive to the dots?
5. What was the dot pattern on your mystery card? What is the word?
6. Imagine that you have lost your sense of sight. Which of your other senses do you think would help you the most—your sense of touch, taste, hearing, or smell? Explain.
7. List at least 3 things that could cause blindness in a person your age.
8. Imagine that you are showing your classroom to a three-year-old girl who is blind. Write a short paragraph explaining how you would help her "see" the room so she can *safely* find her way around.

CONCLUSION

Write a paragraph that answers the two focusing questions on page 435.

EXTENSION ACTIVITIES

1. Find out more about reading Braille. If someone in your school is blind, invite him or her to visit your class and demonstrate how to read Braille. What does a Braille book look like? How fast can he or she read?

2. Read a book about Helen Keller. She was both blind and deaf. She started to learn to talk when she was ten years old. This passage was taken from her autobiography *The Story of My Life.*

   "In reading my teacher's lips I was wholly dependent on my fingers: I had to use the sense of touch in catching the vibrations of the throat, the movements of the mouth and the expression of the face; and often this sense was at fault. In such cases I was forced to repeat the words or sentences, sometimes for hours, until I felt the proper ring in my own voice. My work was practice, practice, practice."

   While you read, keep these questions in mind:

   How did Helen Keller lose her sight and hearing?

   How did Helen Keller learn to communicate?

   What did she enjoy doing in her spare time?

   How did she use her senses of touch, smell, and taste to learn about the world around her?

16-5 EYEBALLS

The eye is one of the most complex organs of the body. One way to help you understand an eye is to take one apart. Since other animals have eyes that are very much like ours, you can look at an animal eye to learn more about your own.

- Which parts of an eye can you identify?
- What is inside an eyeball?

By dissecting a preserved eye, you will find the answers to these questions.

MATERIALS

preserved eyeball
dissecting pan
scissors
paper towels
data page: ***Eyeballs*** (Figure 16–12)

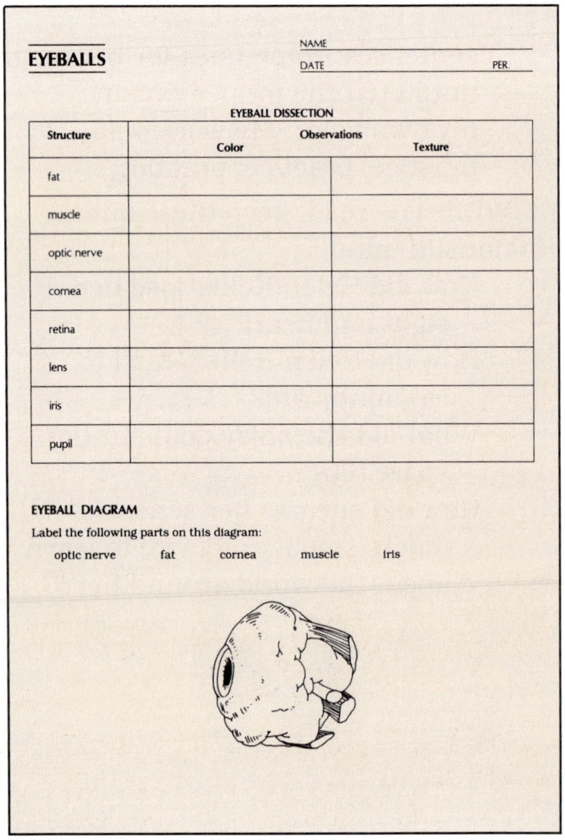

EYEBALLS

NAME
DATE PER.

EYEBALL DISSECTION

Structure	Observations: Color	Texture
fat		
muscle		
optic nerve		
cornea		
retina		
lens		
iris		
pupil		

EYEBALL DIAGRAM
Label the following parts on this diagram:
optic nerve fat cornea muscle iris

Figure 16–12

PROCEDURE

Record your observations on the data page as you find each structure printed in bold type.

Part A. The External Eye

1. Look at the eye. You will find **fat** and **muscles** attached to the back and sides of the eyeball. (Remember to record your observations for fat and muscle.)
2. Locate the *top* of the eye. Try to determine how the eye was attached to the skull.

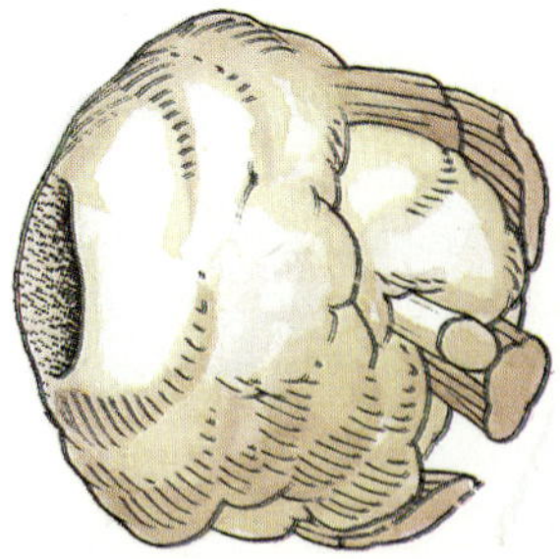

3. If your eye seems to have a lot of fat around it, carefully trim off some of the fat.
4. The structure coming out of the back of the eye is the **optic nerve.** It may be hidden in the other tissues. If yours is hard to find, check for it on someone else's preserved eye.
5. Locate the tough outer layer covering the front of the eye. This layer is called the **cornea.**
6. Before you continue, make sure you have recorded your observations for the first four structures on the data table.

Part B. The Internal Eye

1. Prepare to cut open the eye.
   a. Choose a spot about halfway between the front and back of the eyeball. (See Figure 16–13.) Carefully snip at this spot with the scissors to form a short slit. Be sure not to plunge the point in too deeply.
   b. Continue cutting about three-fourths of the way around the eye.
2. Gently open the eyeball. A jellylike material may be inside. If it is, carefully remove this substance. (In life, this substance would be more liquid; it helps give the eyeball its shape.)
3. Look inside the eyeball. Try to locate the optic nerve from the inside. On the back surface you can see a whitish tissue—the **retina.**
4. Carefully lift out the retina. You will find a layer of dark tissue. It contains blood vessels. (This layer of tissue is not found in human eyes.) A tough outer tissue encloses the eyeball.
5. Look at the front of the eye from the inside. Find the **lens.** Gently lift the lens to see how it is held in place. Remove the lens.
6. Rinse the lens in water. Try to look through the lens and read. Note how the letters appear.
7. Now you can see the entire **iris**—the colored part of the eye. The iris surrounds the opening called the **pupil.** You can also see the inside of the cornea. (You saw the cornea from the outside in step 5 of Part A.)

You just inspected an eyeball. Isn't it amazing that this structure allows an animal to see!

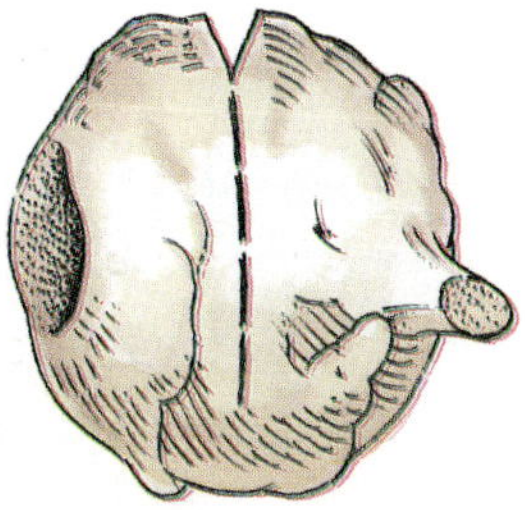

Figure 16–13

Part C. Finishing Up

1. Clean up your equipment and work area.
2. Label the diagram on your data page and answer the analysis questions on your own paper.

ANALYSIS QUESTIONS

1. What structures are probably transparent in a living eye? Why?
2. What function do you think the fat serves?
3. Where does the optic nerve go when it leaves the eyeball?

4. List the following eye parts in the order that light passes through them: retina, lens, cornea, pupil, fluid.
5. Light causes the pupil to change size. How could you test this?
6. Cataracts are due to "clouding" of the lens. How would this affect vision?

CONCLUSION

Write a summary of this activity that includes the answers to the two focusing questions on page 438.

EXTENSION ACTIVITIES

1. Compare your eyes with your pet's eyes. List as many ways as you can that your eyes and your pet's eyes are similar. List the differences. See if you can get your pet to look at something off to one side without turning its head.
2. Use a microscope to look at the eyes of several different insects, such as a fly, grasshopper, mealworm, and ant. Sketch what you see. How many eyes does a spider have? Use a magnifying glass and count carefully!

An Internal "Mirror"

The eyes of many animals seem to glow at night. This glow is simply the light reflected from a streetlight, headlight, flashlight, or other source of light. The eyes glow because they have a **tapetum.** The tapetum is a layer of tissue located at the back of the eyeball; it acts like a mirror. The tapetum increases the amount of light available in the eye so the animal can see better at night.

Figure 16–14
Bobcat

16-6 EYES AND VISION

Everyday your eyes go through a muscular workout. They move continually to allow you to "read the world." Eye muscles pull your eyes in the directions you want to look. But, even if you are facing the right way, you cannot see unless there is light. Why do you need light to see? How does an eye work?

Inside an Eyeball

In order to understand vision, you need to know something about the structure of an eye. If you dissected a preserved eyeball, the diagram in Figure 16–15 should look familiar. It illustrates the parts of an eye as they look in a cross section.

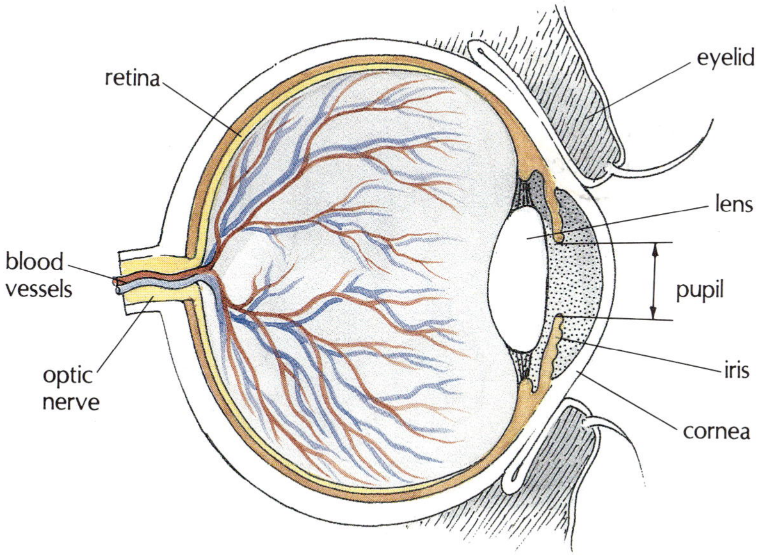

Figure 16–15 Eyeball structure. This drawing shows what a human eyeball looks like in cross section.

The **cornea** is the tough outer layer that protects the front of the eye. Under the cornea is the colored part of the eye—the **iris.** It is made up of connective tissue and smooth muscle fibers. These muscle fibers contract and relax to control the size of the **pupil.** Directly behind the pupil is the **lens.** It is a small disk made of clear, living cells. The lens allows us to focus on the objects we want to see. The **retina** lines the eye. It is made up of highly specialized cells that change a light image into a nerve message. This message can then travel along the **optic nerve** to the brain. Together, all of these structures allow us to see.

The Vision Pathway

Imagine that you are looking at a flower. What must happen in order for you to see the flower? Read the "vision pathway" below. The six steps describe what happens when you see an object.

A Six-Step Vision Pathway

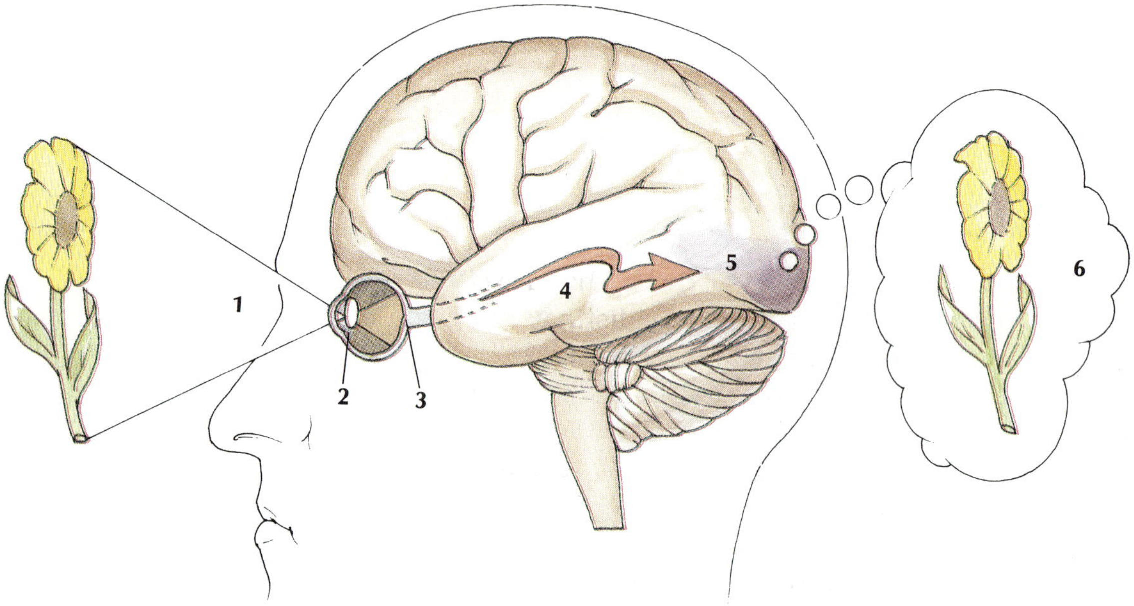

1 Light reflects off the flower and into the eye through the *pupil.*

2 The light is focused as it passes through the *lens.*

3 It then passes on through the *eyeball* and stimulates cells in the *retina.* Here, the light image is changed into a nerve message.

4 This nerve message travels to the brain by way of the *optic nerve.*

5 The optic nerve delivers the message to the *vision center* of the *brain.*

6 The brain interprets the message so you can see the flower.

Distance Vision

Some people can easily focus on objects that are close to their eyes. Others can more clearly distinguish objects that are far away. Your ability to see objects that are near or far or both depends on your distance vision. Distance vision is measured with an eye chart.

To measure distance vision, stand 20 feet away from an eye chart, cover one eye, and read as many of the lines of letters as you can. Then record the numbers printed by the smallest line of letters that you can read.

If you have normal vision, you can read the line labeled "20/20." Suppose the smallest letters you can read are on the line labeled "20/70." This means that you must stand 20 feet from the chart in order to read the line that a person with normal vision can read at 70 feet. A person is legally blind if his or her vision is 20/200 *with glasses.*

What Else?

As well as being able to focus on objects, your eyes can do many other things. Because of the way your eyes are made, you can distinguish colors and you can see more than one thing at a time. You can see in strong light or in very weak light. You can see while standing on your head or with both feet on the ground and while sitting in a moving car or while standing on a cement platform. Your eyes and brain have a lot to do in order for you to see under all of these conditions.

ANALYSIS QUESTIONS

1. These structures are part of the vision pathway: brain, lens, cornea, optic nerve, pupil, retina. List them in the order they are used when you look at an object.
2. Why is light necessary in order for you to see?
3. Do you think drawing A or drawing B in Figure 16–16 shows the way a pupil would look in bright light? Explain your answer.

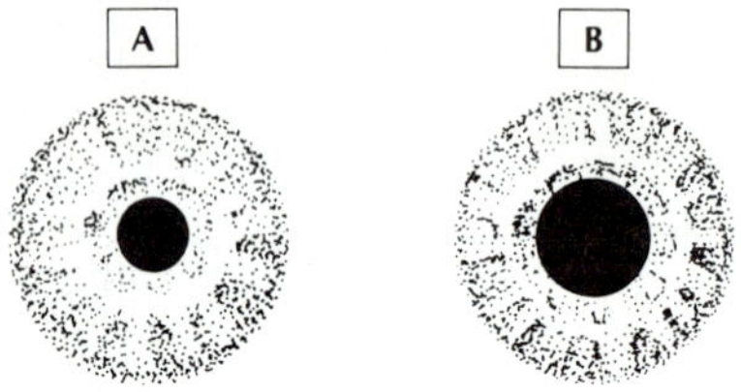

Figure 16–16

4. What does it mean if a person has
   a. 20/20 vision?
   b. 20/100 vision?
5. Is a person with 20/70 vision nearsighted or farsighted? Explain.
6. Which of these vision measurements might an eye doctor find as a result of doing an eye exam?
   a. right eye: 20/20 left eye: 20/20
   b. right eye: 20/20 left eye: 20/100
   c. right eye: 20/100 left eye: 20/100
   d. right eye: 20/100 left eye: 20/200
7. Some people are born with damaged retinas. Why is it impossible for them to see?

EXTENSION ACTIVITIES

1. Find out more about the lens of the eye. Look in a high school biology text to find out how the lens works. Read on to find out what causes nearsightedness. If you feel inspired, use a glass lens to show what you have learned to the rest of your class.
2. There are many kinds of vision loss. Some people are born blind; others lose their vision as they get older. Explain some causes of vision loss. Pamphlets from eye clinics and organizations concerned with vision loss are good sources of information.

CHAPTER 17

NERVOUS SYSTEM

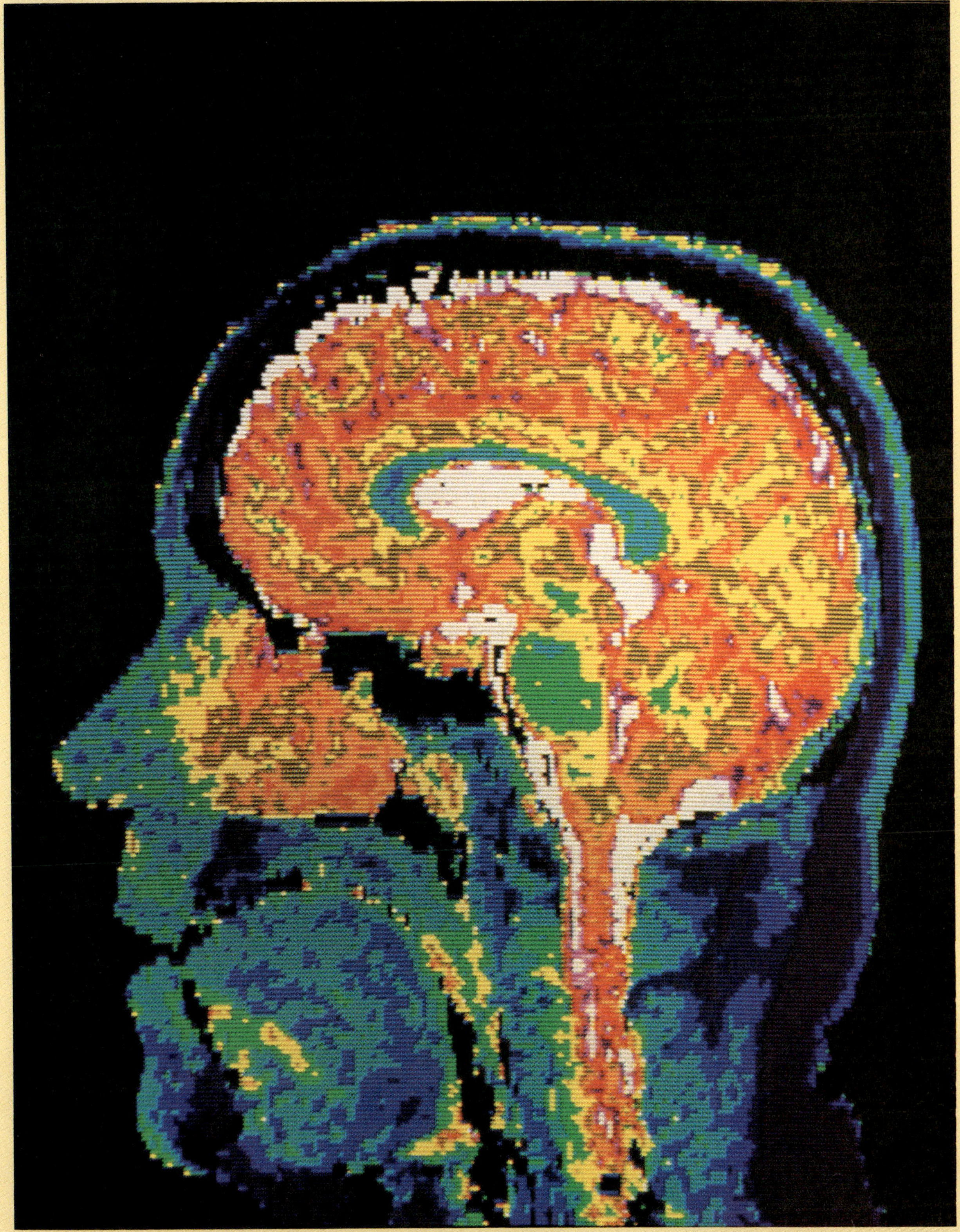

What clues help you recognize that this is a picture of a brain?

17-1 NERVES AND NERVE CELLS

When you studied the sense organs, you heard about several highly specialized nerves. The *optic nerve* carries messages from the retina of the eye to the brain. These messages allow you to see. The *auditory nerve* picks up messages from the cochlea and carries them to the brain. These enable you to hear.

Other specialized nerves sense other conditions in the environment. Their nerve endings may detect odor, heat, cold, pressure, or texture. Their messages, too, are "translated" in the brain.

Nerves

Nerves, along with the brain and spinal cord, make up the **nervous system.** Some nerves carry messages *to* the spinal cord and brain. They help you sense the environment. Other nerves carry messages *away* from the spinal cord and brain. They allow you to respond to the environment. Where a nerve attaches to the spinal cord, it is fairly large. It branches again and again until it forms a network of tiny nerves. These small nerves reach all parts of the body.

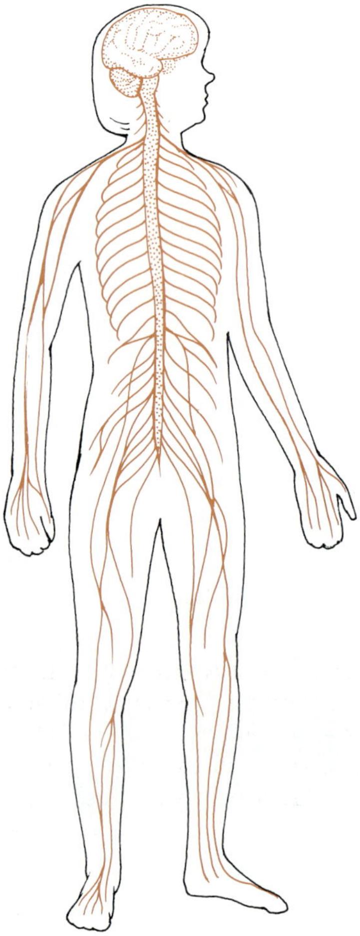

Figure 17-1 A network of nerves reaches all parts of the body.

Responses: Decision or Reflex?

Most of the time, you respond to the environment by planning what you want to do. For instance, if it is cool outside, you may decide to put on a sweater. In order for you to do so, thousands of messages will travel from your brain along nerves to the muscles in your shoulders, chest, back, arms, and hands. These messages allow you to position the sweater so you can put your arms in the sleeves and your head through the neck.

However, not all muscle messages come from the brain. When you need to react quickly, it would take too long to think about what you needed to do. For example, if your hand touches a hot iron, you quickly pull your hand away. You do not have to think about doing this. Nerves from your hand carry messages to the spinal cord. In the spinal cord the message is immediately relayed to another nerve that tells your muscles to contract. The message does not need to go to the brain in order for you to react. Quick **reflexes** like these are important for survival.

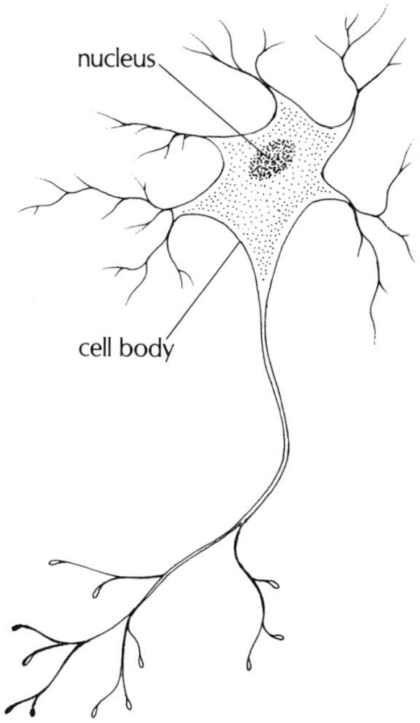

Figure 17–2 Messages travel from one end of a nerve cell to the other.

Nerve Cells

The brain, spinal cord, and nerves are made up of billions of nerve cells. Nerve cells are specialized to carry messages from one part of the body to another. As you might expect, these cells do not look like other cells. They are shaped to carry messages great distances. (See Figure 17–2.)

A message starts at one end of the nerve cell, moves through the cell body, and continues on down the other end. When it reaches the end of the cell, the message may be passed on to another nerve cell. Eventually the message reaches its destination and causes an action to occur.

ANALYSIS QUESTIONS

1. What is meant by the term **reflex?**
2. List at least 3 examples of reflexes.
3. What is the difference between a nerve and a nerve cell?
4. How is a nerve cell
   a. like other body cells?
   b. different from other body cells?
5. Mark was in a car accident. He cut a nerve in his upper arm. Now he has lost all sensation in his fingers. Explain how this could happen.
6. Where in your body would you *not* expect to find nerve cells? Explain your answer.
7. Consider what you have learned about the circulatory system and the nervous system.
   a. List at least 2 ways that these systems are similar to each other.
   b. List at least 3 ways that they differ from each other.

CHALLENGE

Look at nerve cells through a microscope. Do they look the way you expected them to? Explain your answer. Draw and label *one* nerve cell.

17-2 THE HUMAN BRAIN

Suppose it were possible to transplant Kim's brain into Pat's body. After the operation, who is the person—Kim or Pat?

While such transplants are not likely to ever take place, this question does point out the difference between the brain and other body organs. We do not just depend on our brains, we *are* our brains.

The brain is the single most important organ in the body. Although it looks quite simple, it is incredibly complex. Some people doubt if we will ever totally understand what goes on inside the brain. For instance, think about these questions:

How does a nerve message turn into an image we can see?

How does our memory work?

What happens in the brain when we solve a math problem?

We do not know the answers to these and many other questions, but we do know quite a bit about the structure of the brain. We also know something about what each part of the brain controls.

In this reading you will learn about the major parts of the brain. While you read, color and label a diagram of the human brain.

A Protective Case

Because the brain is so important, it is well protected. The bony **cranium** provides a sturdy case for the brain, but this case is not enough. Every time you bump your head, your brain would be jarred if it weren't for the **fluid** that surrounds it. A tough **membrane** keeps this fluid from leaking out.

Label the membrane and color it yellow. Shade the space that contains the fluid a pale yellow.

The Learning Center

The largest, most obvious, structure of the brain is the **cerebrum.** (Find it on Figure 17–3 on page 449.) This is

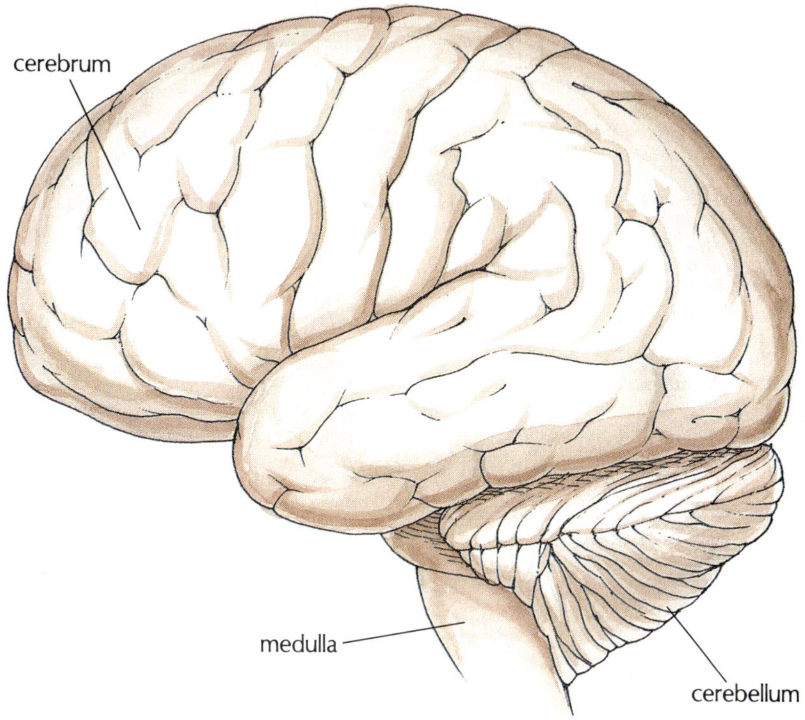

Figure 17–3 The main parts of the human brain

the structure most people think of when they hear the word *brain*. The thin outer surface of the cerebrum contains billions of nerve cells. Even though this layer is only a few millimeters thick, it is incredibly important. The various areas of this layer have different functions. For instance, the very back of the cerebrum receives messages from the optic nerve and is responsible for vision. Injury to this part of the brain could cause blindness.

Other areas of the cerebrum are responsible for hearing, taste, and touch sensations. The cerebrum is also the "brainy" part of the brain. It contains areas for intelligence, speech, memory, and even personality.

Label the cerebrum and color it blue.

Muscle Coordination

The **cerebellum** is located under the cerebrum near the base of the brain. (Find it on Figure 17–3.) When you decide to move, your cerebrum starts the action but the cerebellum quickly takes over. Learning to ski, playing a piccolo, or riding a unicycle requires intense concentration—that's the cerebrum's job. Once the motions are learned, they become more automatic and can be regulated by the cerebellum. As you might have realized from reading this, the cerebellum coordinates skeletal muscle movement.

Injury to the cerebellum can cause coordination problems. When this happens, muscle control becomes difficult. For instance, a person might stagger instead of walking smoothly if he or she had cerebellar damage.

Label the cerebellum and color it green.

Basic Body Control

Right below the cerebellum is the **medulla.** It connects the brain and spinal cord and looks like a slightly enlarged part of the spinal cord. (Find it on Figure 17–3 on page 449.) Even though the medulla is only a few centimeters long, it is extremely important. It monitors such things as heart rate, blood pressure, and body temperature. Reflex actions like coughing, swallowing, blinking, and vomiting are also controlled by the medulla.

The medulla controls all of the automatic functions that keep us alive. When the medulla is injured, death may follow.

Label the medulla and color it red.

A Bundle of Nerves

The medulla narrows to form the **spinal cord.** It is made of nerves that carry messages to or away from the brain. Like the brain, the spinal cord is protected by bone. These bones make up the vertebral column.

Every year, spinal cord injuries paralyze thousands of people. For example, when riders are thrown from their bikes, their backs are jarred when they slam against the pavement. X-rays may reveal spinal cord damage. If the spinal cord is totally cut, paralysis from that point down results. That's because the brain is no longer able to communicate with the nerves below the injury. No amount of willpower can move muscles when nerves are severely damaged.

The training and equipment used in many sports prevent injuries to a player's head, neck, and back. Helmets, pads, and back supports are designed to reduce the possibility of damage to the nervous system.

Label the spinal cord and color it purple.

ANALYSIS QUESTIONS

1. Write these brain parts in order of size, from largest to smallest: cerebellum, cerebrum, medulla.
2. Sometimes people say that the skull is like an eggshell. What do they mean when they say that?
3. Make a chart like the one below.

Cerebrum	Cerebellum	Medulla

   a. Decide which part of the brain is most important for controlling each of the following activities or body functions. Then write each activity or body function in the appropriate column on the chart.

thinking	swimming
seeing	solving a problem
skiing	making decisions
swallowing	learning
sneezing	feeling sad
breathing	pulse rate

   b. Think of 1 additional activity or body function to add to each column.

4. By looking at a brain, do you think it is possible to figure out the following information?
   a. What kind of an animal it is from
   b. How smart the animal was
   c. How old the animal was
5. Football teams are penalized if one player grabs another's face mask. Baseball players wear batting helmets when it is their turn at bat. What kinds of nervous system protection are used in other activities?

EXTENSION ACTIVITIES

1. Many scientists are studying the way the brain works. One area of research is about left and right brain functions. What do people mean when they

talk about having a dominant right brain? Write a one-page summary of their ideas. Do you think your right or left brain is dominant?

2. People with spinal cord injuries undergo physical therapy to improve their muscle tone and to learn new ways to carry out everyday activities. If there is a physical rehabilitation center near you, ask to speak to a physical therapist. Decide in advance what questions you will ask. You want to include the following ones:

   What exercises are important for someone with spinal cord injuries?

   How does a physical therapist help someone adjust psychologically to the loss of sensation and movement?

   How do most spinal cord injuries occur?

   How can they be prevented?

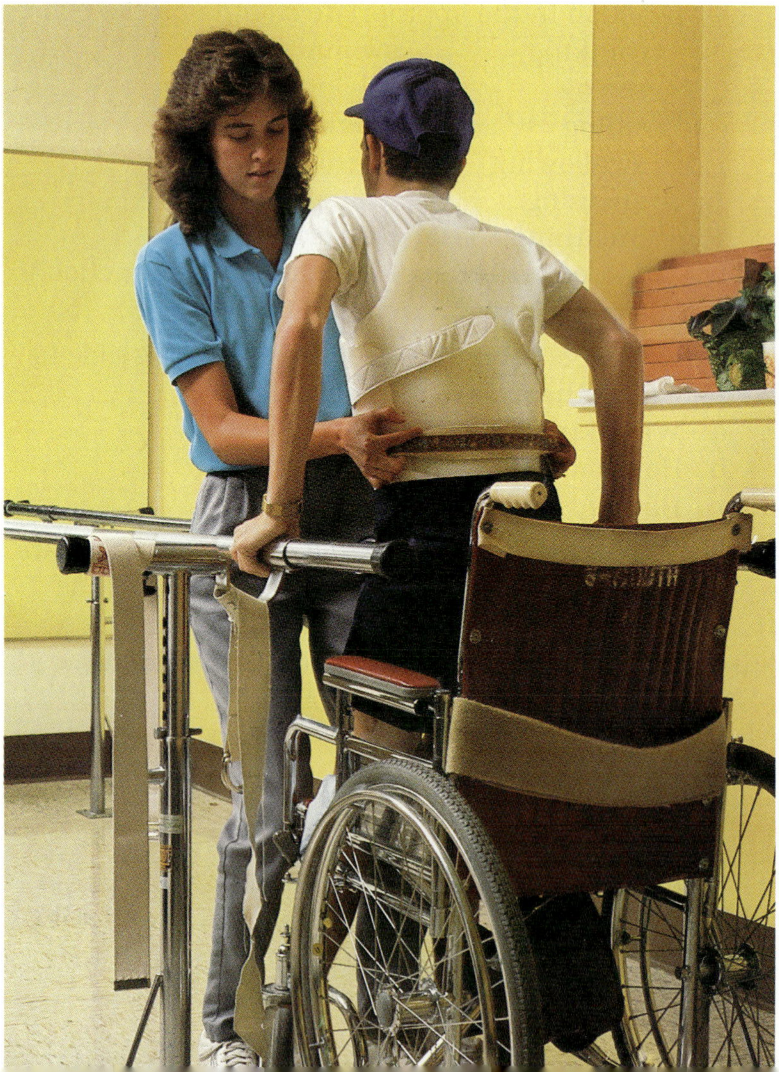

Figure 17–4 The physical therapist is working with a student whose spinal cord has been injured.

17-3 THE WEST TURKANA BOY

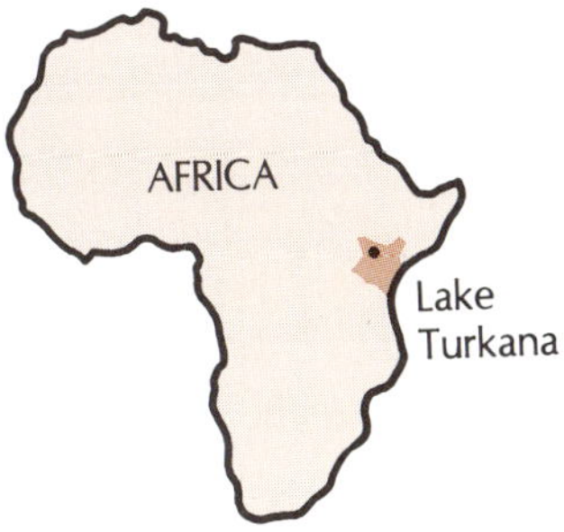

In 1984 Kamoya Kimeu was working on an archaeological dig near the shore of Lake Turkana in Kenya. After months of sorting through sand, rocks, and rubble, he found a small piece of human skull. With growing excitement, he and his co-workers kept digging. They next found most of the skull, the jaw, and several teeth. Eventually they found enough bones to assemble much of the skeleton. All they were missing were some arm, hand, and feet bones.

What happened on the shore of this lake? Was there an accident? a death from a deadly disease? a murder? Did the body wash in from somewhere else? We do not know why this person died and probably never will. However, we can determine the age of the bones. Studies have now shown that this skeleton, known as KNM-WT 15000, is 1.6 million years old.

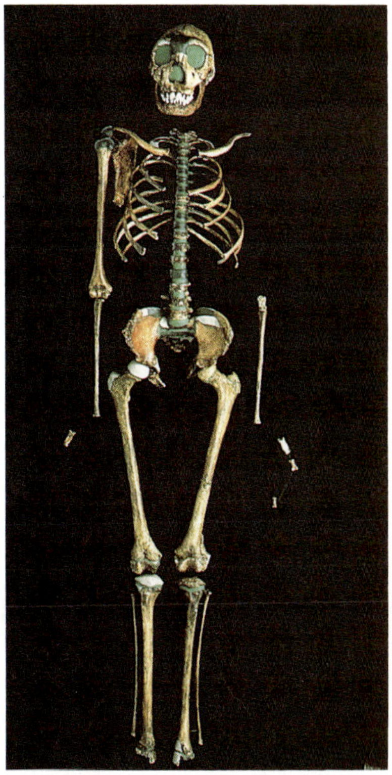

Figure 17-5 West Turkana boy's skeleton. (The brown pieces are actual fossils; gray areas were filled in by mirror imagery, and green and yellow were missing parts filled in here for completeness.)

A Young Boy

Even though these bones are ancient, paleontologists* can tell us something about the person from whom they came. For instance, the shape of the pelvis reveals that KNM-WT 15000 is a male. Females have shallow, broad pelvises; this pelvis is narrow and more upright. (See Figure 17–6 on page 454.) Next, the teeth reveal that this male was young. His wisdom teeth had not erupted—they were still in his jaw.

There are other signs of his youth. The joints in his skull had not hardened; this usually occurs somewhere between 12 and 15 years of age. His long bones also indicate that he was still growing at the time of his death; the ends of his bones had not fused. If he had lived, he probably would have grown to a height of 6 feet. Researchers have used all of these clues to estimate that this boy was about 12 years old when he died.

Scientists are continuing to study the bone fragments to learn more about this boy. At the same time, they will discover more about the story of human

*A paleontologist is someone who studies fossils to learn about ancient life forms.

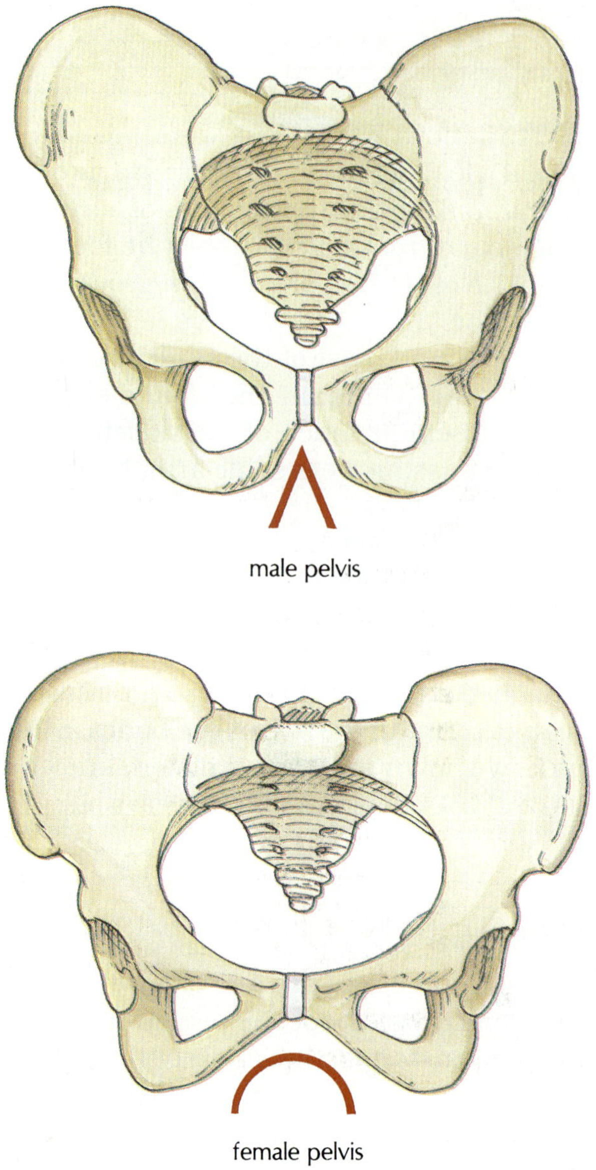

Figure 17–6 An expert can tell the difference between a male pelvis and a female pelvis. Do you think you could?

history. Here are some of the questions they are considering:

Was KNM-WT 15000 healthy?
What did he eat?
How large was his brain?
Was he capable of abstract thinking?
How well-developed was his speech?

Brain and Spinal Cord

Scientists have started to answer some of these questions already. For instance, by looking at the shape and size of a skull, it's possible to estimate brain size. Examination of KNM-WT 15000's skull reveals that his brain would have reached a volume of about 880 ml. In comparison, the average volume of a brain in today's human is 1350 ml. (See Figure 17–7.) This means that this boy was as big as a modern American but he had much less brain with which to control his body. We can't help but wonder about what kinds of things we can do that he couldn't do. So far, this question remains unanswered.

Another clue to the boy's abilities is found in his vertebrae. Remember, the vertebrae form a bony cover for the spinal cord. From the spinal cord come all the nerves that control the arms, legs, and torso. Thus, the

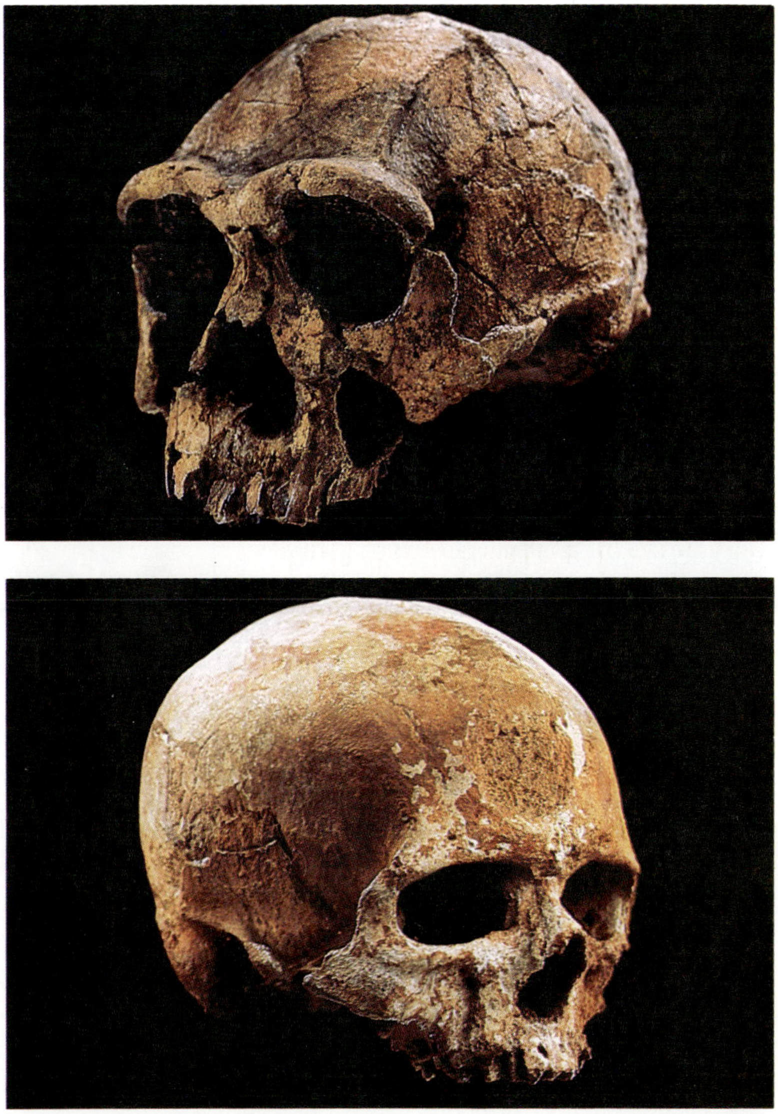

Figure 17–7 The size and shape of a fossil skull similar to the Turkana boy's (top) is not like that of a modern human's (bottom).

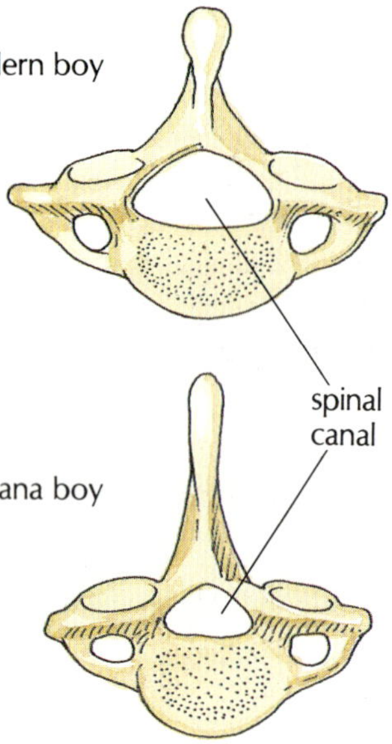

Figure 17–8 Comparison of vertebrae

size of the opening, the spinal canal, in the vertebrae tells something about the number of nerves that extend through the spine. KNM-WT 15000's spinal canal is less than half the size of a canal from a 12-year-old boy of today. (See Figure 17–8.) This fact leaves scientists with another puzzle. If fewer nerves traveled through this boy's spinal canal, how did that affect his abilities to do things? How skillful was he with his hands? Could he use tools?

More Questions

Obviously, scientists still have many questions about KNM-WT 15000. In order to find answers, they will continue to look for his missing bones. They will compare his skeleton with other ancient skeletons and they will study the way our own skeleton works to get clues as to what this boy could do.

Figure 17–9 The dig site in Kenya

Watch newspapers and magazines if you would like to keep track of the continuing story of this ancient youth.

17-4 LEARNING AND MEMORY

The brain is incredible. Together, your cerebrum, cerebellum, and medulla keep you alive and aware of your surroundings. While all parts of the brain are important, the cerebrum deserves special attention. As you learned, the cerebrum is the "brainy" part of the brain; it contains the areas for intelligence. Intelligence is not one specific characteristic. It is a combination of being able to study, memorize, and learn. Intelligence is also being able to solve problems, concentrate, and make plans. It is being both creative and analytical.

Intelligence does not just happen. Do you want to be able to solve tough math problems? Then practice! Do you want a good memory? Then work at it!

- What are five techniques that you can use to memorize something?

In order to answer this question, you will take another look at three activities you have studied before.

PROCEDURE

Work with a partner or in a small group. For this activity, two or three brains are better than one.

Part A. Memorizing Words

Sometimes learning involves "pure memory." Two memory activities that you have done were in ***The Human Skeleton***, pages 158–161, and in ***Major Muscles***, pages 186–189. In one activity you learned the names of some bones; in the other, the names of some muscles.

1. As a group, choose one of the activities to think about now.
2. Imagine that you are sitting down for the first time to learn the names of some bones or muscles. In your group, talk about how you would proceed.
3. List at least 3 different things you could do to learn the names of these bones or muscles.

Part B. Memorizing Facts

When you memorize the names of bones or muscles, you can match the words with a picture. Sometimes, however, you have to memorize something that does not go with a picture. Look at the reading **Skin Cancer** starting on page 232 for an example of this type of memory task.

1. Imagine that you want to learn the signs of skin cancer. In your group, talk about how you would proceed.
2. List at least 3 techniques that you could use to memorize the signs of skin cancer.

Part C. Finding the Facts

What you have to learn is obvious if you are given a list of words or facts; but usually you are not told exactly what you need to learn. There are many times when you have to identify the important information and then decide how you want to learn it. This time, use the reading ***Fats and Oils,*** starting on page 244, as an example activity.

1. Scan the reading to help you remember what it is about.
2. Identify 5 important things that you think the author would like you to learn from this reading. Write down these 5 things.
3. List at least 3 clues you used to decide what was important.
4. Now that you have decided what you would need to learn from this reading, list at least 3 techniques you could use to help you remember the information.
5. Imagine that you will be taking a test on this reading one week from today. Answer these questions:
   When would you start to study?
   How much time would you need to learn the material?

CONCLUSION

Write an answer to the focusing question on page 457. In your answer, list the 5 techniques that *you* are most likely to use.

EXTENSION ACTIVITY

Ask the library media specialist to help you find a book on study skills. As you read the book, make a list entitled "Good Ideas for Studying." Try to find at least one good idea in each chapter. When you finish the book, look over your list and put asterisks by the 3 ideas you think are best. Share these ideas with your friends or with the entire class.

17–5 FUTURE THINKING

A giraffe does not plan what it is going to do next week, nor do raccoons think about what they are going to be when they grow up. They couldn't even do these things if they wanted to. Humans are the only species that has a brain capable of planning for the future. Setting goals and then planning what to do to meet those goals are important parts of growing up.

- When you consider your future, what is the hardest part for you to imagine? What is the easiest? Why?

In this activity you will be thinking about your own future. You will consider such things as where you are going to live, what you are going to be, and what you will be like.

MATERIALS

worksheet: ***Future Thinking***
(Figure 17–10)

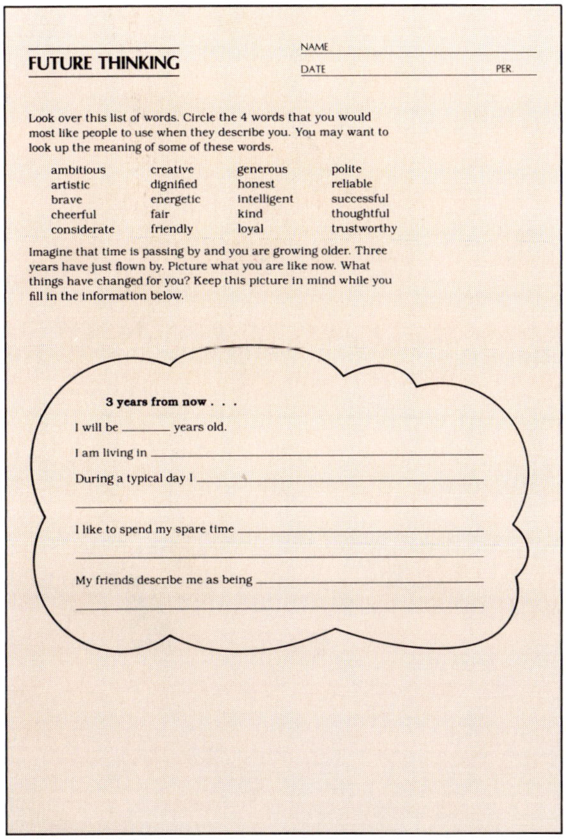

FUTURE THINKING

NAME ____________
DATE ________ PER ____

Look over this list of words. Circle the 4 words that you would most like people to use when they describe you. You may want to look up the meaning of some of these words.

ambitious	creative	generous	polite
artistic	dignified	honest	reliable
brave	energetic	intelligent	successful
cheerful	fair	kind	thoughtful
considerate	friendly	loyal	trustworthy

Imagine that time is passing by and you are growing older. Three years have just flown by. Picture what you are like now. What things have changed for you? Keep this picture in mind while you fill in the information below.

3 years from now . . .

I will be ______ years old.

I am living in ____________

During a typical day I ____________

I like to spend my spare time ____________

My friends describe me as being ____________

Figure 17–10 (Worksheet, page 1)

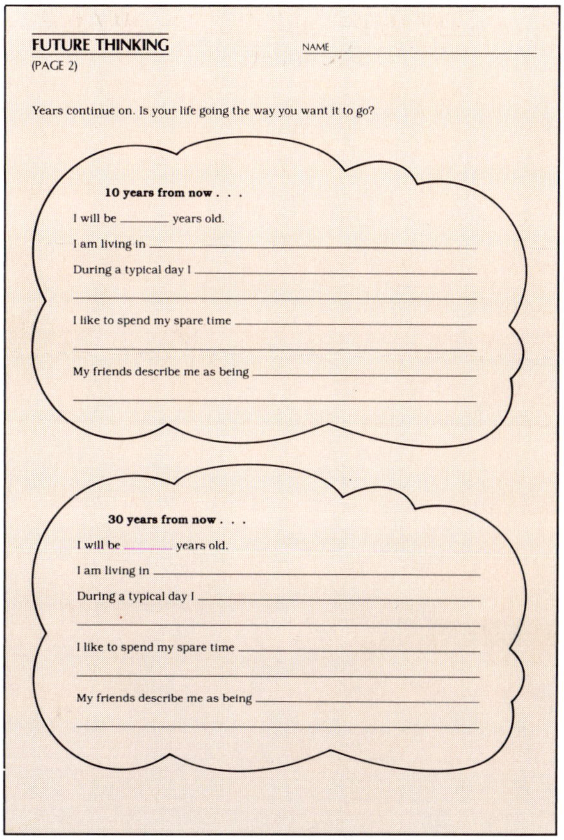

FUTURE THINKING
(PAGE 2)

NAME ____________

Years continue on. Is your life going the way you want it to go?

10 years from now . . .

I will be ______ years old.

I am living in ____________

During a typical day I ____________

I like to spend my spare time ____________

My friends describe me as being ____________

30 years from now . . .

I will be ______ years old.

I am living in ____________

During a typical day I ____________

I like to spend my spare time ____________

My friends describe me as being ____________

Figure 17–10 (Worksheet, page 2)

PROCEDURE

1. Take a minute to think about yourself.
   a. Think about what you enjoy doing, the skills you have, and the kind of person you are.
   b. If you are like most people, you have some good characteristics and some that you would like to change. You also know the kind of person you would like to be. Keep this person in mind while you do this activity.
2. Complete the worksheet. Pause to think before you fill in each of the sections.
3. Continue to use your cerebrum while you answer the analysis questions.

ANALYSIS QUESTIONS

1. Look at how you picture yourself three years from now. List at least 2 things you will need to do to make this picture come true.
2. Look at how you picture yourself ten years from now. List at least 3 things you will need to do to make this picture come true.
3. At what age do you think you will need to start to plan for your future? Explain your answer.
4. Rate the difficulty of this activity.
   very hard 5 4 3 2 1 very easy

CONCLUSION

Now that you've given some thought to your future, turn to page 459 and answer the focusing question.

17–6 DECISION MAKING

One of the most complex tasks that the human brain can do is make a decision. We all make decisions everyday, most of which are simple and routine. We usually make these decisions without giving them much thought. However, it is important to *think before* we make some decisions, especially those decisions involving major problems. What types of things do you consider when you make a decision?

- What are the basic steps in making a decision?

In this activity, you will use a decision-making model to solve a problem.

MATERIALS

worksheet: ***Decision Making*** (Figure 17–11)
Decision Card

PROCEDURE

Work in a group with one or two other students.

1. Your group will need a Decision Making worksheet and a Decision Card. Decide which person in your group will fill in the worksheet.
2. Read the situation described on the Decision Card. Discuss and decide what the problem is and record it on the worksheet.
3. Now, with your group, brainstorm a list of things that the person in this situation might be able to do.

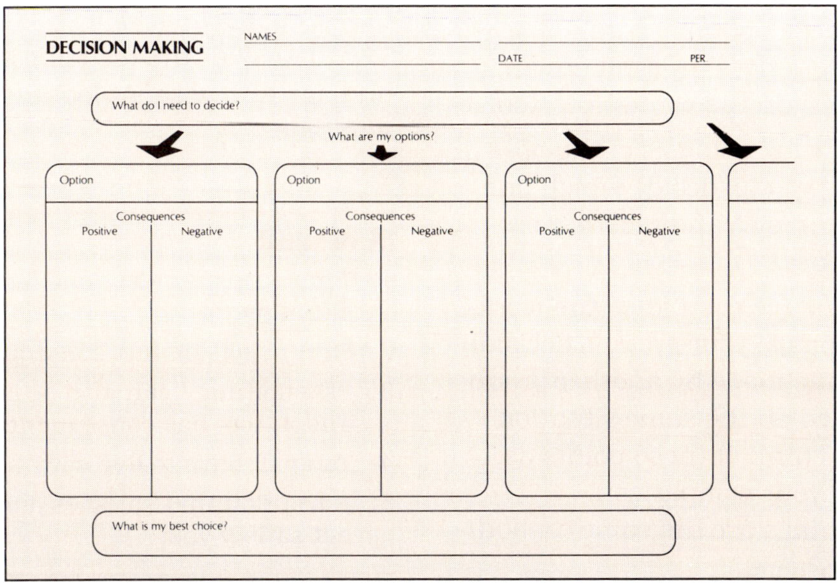
DECISION MAKING
NAMES
DATE
PER
What do I need to decide?
What are my options?
Option
Consequences
Positive
Negative
Option
Consequences
Positive
Negative
Option
Consequences
Positive
Negative
What is my best choice?

Figure 17–11

a. Have the recorder list the options on the back of the worksheet.
b. List as many options as you can. If you stop before listing at least 5, you may miss the best ones.

4. Together look over the list of options and cross off the ones that could hurt someone or have other extremely negative consequences.
5. Discuss and then choose the 3 best options from those that are left. Use the Decision Making worksheet to analyze these options. Think of several consequences for each option; record them on the worksheet, too.
6. Look over your analysis of the options. As a group, decide which one is the best choice and record it on the worksheet.
7. When you have completed the worksheet, compare your group's decision with the one made by another group that worked on the same situation as yours. Think of these points as you make your comparison.

   What things are the same on both worksheets?

   What things are different?

   Do you agree with the other group's best choice? Why or why not?

HOMEWORK

Analyze a decision of your own. It might be a decision you already made or one you are going to have to make. Use the worksheet to help you reach your best choice.

CONCLUSION

Try to answer the focusing question on page 461 without looking at the worksheet.

REMEMBER

1. There are many options for every situation.
2. Every decision has consequences.
3. "Do nothing" can also be a decision and it, too, has consequences.
4. You will have to live with your decision.
5. A solution that satisfies one person may not satisfy you.

CHAPTER 18 DECISIONS AND DRUGS

Why does everyone want you to "Just Say No"?

18–1 DRUGS AND THE NERVOUS SYSTEM

This chapter is about drugs—not about all drugs, just about the ones that act on the brain. They can change normal brain activity in a variety of ways. Think about it. The brain is responsible for everything from the most basic body functions to the wildest of dreams. It lets you talk and listen, learn and remember, plan and make decisions . . . it is even the center for your hopes, feelings, and emotions. Because of our brains, I am me and you are you. What happens then when someone takes a drug that affects the brain?

Drug Questions

There are many questions that we could consider about drugs and the nervous system. For instance, we could ask about what these drugs do to the brain. Do they affect the ability to think, learn, and make decisions? What do they do to someone's personality? How do they affect other body systems? Usually, the answers depend on which drug you are talking about because no two drugs act exactly the same. But, in order to answer some questions, all you need to know is whether or not the drug is legal.

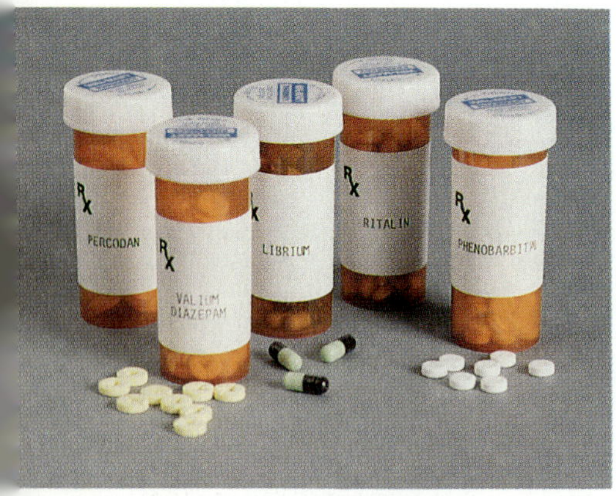

Figure 18–1 Some prescription drugs are useful for treating nervous system problems. However, these drugs can be dangerous if misused.

Legal Drugs

We tend to think of any drug that affects the mind as being "bad." Actually, some of these drugs can be very useful; they are important in the treatment of medical problems. Seizures, severe depressions, and other disorders of mood and personality can be helped by drugs. Even though these drugs are important medically, they can also be quite dangerous. They can make a mental problem worse, affect other body systems, or even cause death. For this reason, these drugs are controlled by law. They are not sold on the shelves of a drugstore or supermarket. In order to buy them, a person needs a prescription from a physician. He or she will make sure that a person gets the appropriate drug in the right quantity and will watch for side effects. (Some examples of these drugs were shown on page 464 in Figure 18–1.)

Illegal Drugs

Other substances that act on the brain have no good uses. In fact, they are just plain dangerous. For this reason, they are illegal. (See Figure 18–2.) They cannot be legally bought or sold. Cocaine. LSD. Marijuana. Heroin. PCP. These drugs are examples of illegal drugs. They are dangerous for a variety of reasons.

- All of these drugs distort reality. They make it impossible to think clearly, make wise decisions, or even sort out emotions. Vision and other senses may be distorted or dulled. Reaction time is likely to be slowed. Muscle coordination is often poor.
- Illegal drugs can make an existing mental condition worse. For instance, a person who is depressed may become suicidal under the influence of one of these

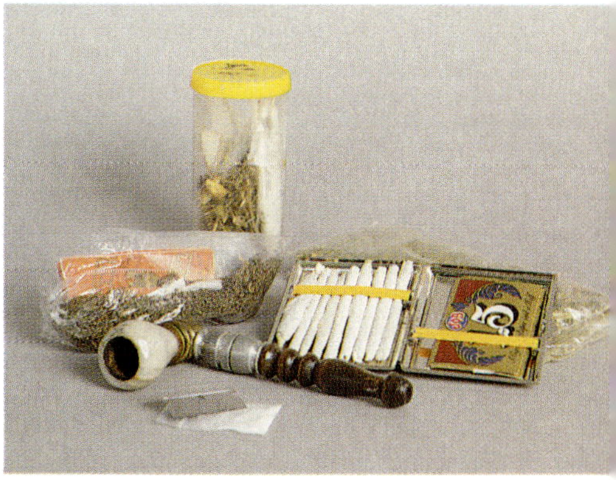

Figure 18–2 Illegal drugs come in a variety of forms. Would you recognize these as being illegal?

Legal or Illegal?

Many things are considered when deciding whether or not a drug should be legal. Here are three of the main considerations:

1. whether or not it is addicting
2. if it has any medical uses
3. how likely it is to be abused

These benefits and risks are weighed against each other.

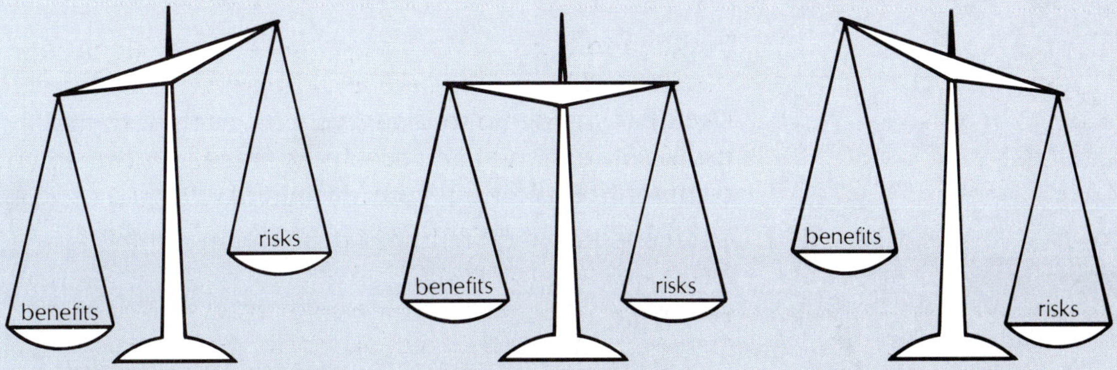

a. If there are many benefits and the risks are slight, the drug will be sold in the stores without a prescription.

b. If the drug has definite benefits, but there are some risks, the drug is sold by prescription only.

c. If the risks far outweigh the benefits, the drug is made illegal.

drugs. They can also make other medical problems worse. A person with a heart problem may go into heart failure as a result of using some of these drugs.

- Studies show that long-term use of illegal drugs may even damage brain cells. Because human nerve cells cannot divide, they must last a lifetime. When some of them are damaged, they cannot be replaced.
- Most concerning of all, these drugs are **addicting.** When a person is addicted, he or she needs a steady supply of the drug. The brain or body may not be able to function without it. Once addicted, the person becomes a slave to the drug.
- These drugs are even riskier if taken by children or teens. They interfere with the normal physical and social development that leads to adulthood. Many of these drugs affect the hormones that direct body growth and sexual development. Frequent drug use, even in small doses, may cause teens to turn away from family and friends. Teens may also fail to form career goals when drugs take over their lives.

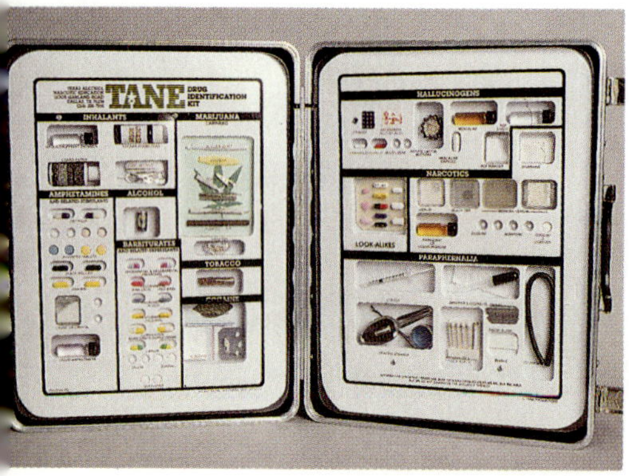

Figure 18–3 A drug identification kit used by police officers when talking to students about illegal drugs.

Going On in This Chapter

The rest of this chapter focuses on two of the most discussed and most abused drugs. One of them, alcohol, is legal, but with age limits. The other one, marijuana, is illegal. Information is provided about both of these drugs. When you read the "facts," remember that most of the data is from studies on *adults.* Very little information is available about the effects of the drugs on the developing brain and body of a young person.

TAKE A STAND

Decide if you agree or disagree with each of the following statements. You also may want to write down comments explaining your decisions.

1. Teens know more about drugs than adults do.
2. Society thinks it's worse if a teen uses drugs than if an adult does.
3. Drugs are more harmful to teens than to adults.
4. All people who use drugs, whether teens or adults, do so because they have emotional problems.
5. An addiction is the same thing as a habit.

18–2 KNOW THE FACTS ABOUT ALCOHOL

Milk, water, and alcohol are all drinks, but they are drinks with a difference. Our bodies *need* milk and water; no person's body needs alcohol. Most people can drink milk and water in fairly large quantities with little problem. In contrast, even small amounts of alcohol can have major effects on a person. For these reasons, alcohol is a drug—one that receives a lot of attention in our country. You have heard about this drug. What do you know about it?

After Swallowing . . .

Like other drinks, alcohol goes to the stomach almost immediately after it is swallowed. Some alcohol is absorbed into the bloodstream through the stomach walls, but most of it is absorbed from the small intestine. This process does not take very long. Alcohol molecules are so small that they can easily diffuse across cell membranes.

Once alcohol is in the bloodstream, it travels throughout the body—to the heart, the kidneys, the brain, the eyeballs, the liver—to every part of the body. When blood flows through the liver, the alcohol is broken down. This is the body's way of protecting itself from drugs. However, it takes time—two, four, six hours, or more—for the liver to break down all of the alcohol. The more alcohol a person drinks, the longer it takes to get rid of it all. Nothing can speed up the process. Black coffee? Cold showers? No, nothing will help a drinker sober up any faster.

Meanwhile . . .

Unlike other drinks, alcohol interferes with brain functioning. It affects the way a person thinks, feels, and acts. The more alcohol a person drinks, the more the brain is affected. At first, when the level of blood alcohol is low, a drinker may report feeling more confident, relaxed, and talkative. He may not realize that he's not

Alcohol and the Brain

Even one drink affects the **cerebrum.** That's why the drinker finds it more difficult to think, reason, focus, and talk. With more alcohol, the person starts to stagger; it is obvious that the **cerebellum** is not functioning normally. Eventually, the **medulla** is also involved. When this happens, the body's automatic functions may stop. Death may follow.

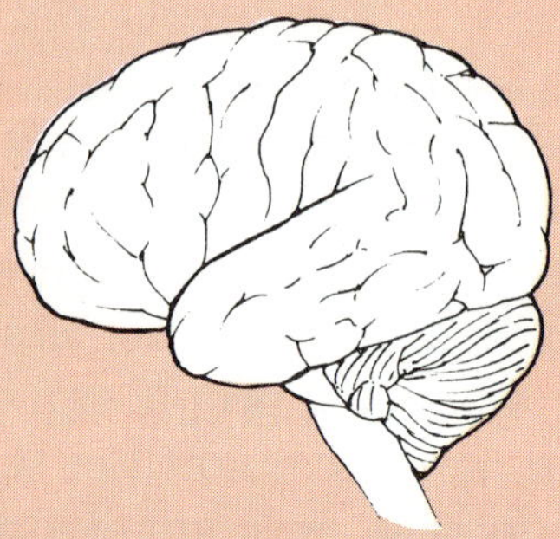

functioning as well as usual—his thinking is not as clear, his reaction time is slower, and his judgment is poorer. This can be a dangerous time. The drinker may think that he is fine but actually he's not in any shape to make wise decisions.

If the person continues to drink, other effects are noticed. His vision may blur; he may feel dizzy and his speech may slur. Other people are likely to notice that the drinker is noisier or more moody than usual.

With another drink or so, the drinker starts to stagger—he is losing his coordination. He may be feeling drowsy. These are warning signals. If he takes another drink, his blood alcohol will reach a dangerous level. He might find himself shouting, yelling, or groaning for no reason. His judgment would be seriously impaired. Yet another drink could lead to a stupor, a coma, or maybe even death.

Young People

Just one drink can result in—
- irritated stomach
- headaches
- vomiting
- flushing
- increase in blood pressure
- increase in heart rate
- disruption of cell membranes
- increased urine production

Alcohol affects young people even more than it does adults because a growing brain does not handle alcohol well. It is not unusual for a child who tries alcohol to overdose accidentally and end up in the hospital. Thousands of high school and college students have died from overdrinking or from driving while drunk. Kids who drink are more likely to have problems in school, drop out of school, and get into trouble with the law. For all of these reasons, it is illegal throughout most of the United States for people to drink until they are 21 years old.

Whether or not to drink is a serious decision. Many adults have chosen not to drink alcohol. Others drink occasionally and are careful to limit the amount they drink at any one time. But some people have real problems with alcohol; this drug has taken over their lives.

Alcoholism

Over time, alcohol can change the brain chemistry so that a person feels a need to drink. When this happens, the person is addicted. Below are some of the health problems that result:

- malnutrition
- liver disease
- ulcers
- mouth cancer
- esophagus cancer
- heart failure
- imbalance in sex hormones

Alcohol addiction can ruin the lives of not only the alcoholic, but also those of his or her family. People with alcohol problems can get help from Alcoholics Anonymous. Their families and friends can get help from Alanon; children of alcoholics can call Alateen.

ANALYSIS QUESTIONS

1. Think about what happens to alcohol once it is in the body.
   a. What happens to it in the stomach?
   b. What happens in the small intestine?
   c. What happens in the liver?
2. Even if a person drinks heavily, there will not be much alcohol in the large intestine. Why is that?
3. Think about what happens when alcohol reaches the drinker's brain. What happens to the drinker when alcohol affects the drinker's
   a. cerebrum?
   b. cerebellum?
   c. medulla?
4. What do you think are 2 of the most serious effects of alcohol on the body? Explain your answer.
5. List at least 3 things that can happen to a person's body after just one drink. Answer this without looking in your text; then, check your list with the information given in your book.

6. In most states, it is illegal for people to drink alcohol until they are twenty-one. Do you think that is fair? Explain your answer.
7. You are at a friend's house doing your homework. You decide to take a break and get something to eat. Your friend opens the refrigerator and offers you a beer. You say, "No way! But, I will take some orange juice." List at least 4 reasons for saying no in this situation.
8. Why do some teenagers drink alcohol even though it is illegal and bad for their health?
9. The law requires that a warning label be clearly visible on every package of cigarettes. Do you think warning labels should be printed on every bottle or can of alcohol? Explain.

EXTENSION ACTIVITY

Talk to the counselor or principal at your school to find answers to the following questions:

Does he or she think alcohol is a problem at your school?

What is the school policy about dealing with a student who drinks?

Are there any local programs to encourage students not to drink?

18-3 ALCOHOL QUESTIONS

It is important to *know the facts* about alcohol, but it is more important to be able to *apply* what you know. Here are six letters from people who have questions about alcohol. Write short paragraphs to answer 3 of these questions. If you do not know how to answer a question, you may want to go back and read ***Know the Facts About Alcohol*** starting on page 467 again. Underline the facts in your answers that were included in the reading.

1

I just heard about two girls, Jessica and Susan, from my school. Yesterday, after school, they went to Jessica's house. Jessica took some vodka out of the cupboard, and Susan and she decided to taste it. I guess they drank quite a bit. Susan passed out and Jessica got scared. Jessica called her mom and her mom called the ambulance. Susan was rushed to the hospital, and everyone was really worried about her. Why did she pass out? Why was everyone so worried?

2

My older brother just graduated from college. He is living at home until he can find a job. I don't know if I like having him here. He drinks beer a lot. Then he gets noisy, does dumb things, and embarrasses me. My friends say he's drunk. I don't think he is because he doesn't stagger or fall down or anything. Who's right?

3

Last night my parents gave a party. I heard Dave, our neighbor, say that he really didn't want any alcohol but, just to be friendly, he'd take one drink. Well, he had a lot more than one drink. After his first drink, he said, "I'll have just one more." I don't know how many he had but Dad had to help him home. Dave said he was going to stop after one drink. Why didn't he?

4

This morning Mom and Dad were talking about a party they went to last night. Mom said that Mrs. Hawkins shouldn't have driven home last night because she'd had too much to drink. Dad told her not to worry because Mrs. Hawkins had had two cups of strong, black coffee before she left so she wouldn't have any trouble driving. Will black coffee help sober up a person?

5

Last week, I went to the movies with two friends. When we were walking home, one of Kim's older brothers stopped and offered us a ride. He had a case of beer in the car. He handed us a six pack and said, "Here. It's about time you learned to drink. A beer or two won't hurt you." I've been thinking about that. Is that true? Can a beer or two hurt me?

6

I went to a big party last Saturday. We were talking, listening to music, eating, and drinking pop. Everything was okay until some kids showed up with a couple of cases of beer. Most of us ignored them but Joe and a couple other kids drank some beer. About a half hour later, Joe threw up. Why did he do that? My mom doesn't throw up when she drinks beer.

EXTENSION ACTIVITY

Look up Alcoholism Treatment in the yellow pages of a phone book. Call one of the treatment centers that you see listed there. Explain that you are doing a project for school. Ask about the types of services the center offers for alcoholics. Find out if children can be alcoholics and if the center has any programs for children of alcoholics.

18-4 KNOW THE FACTS ABOUT MARIJUANA

"Marijuana makes you feel really good and it doesn't harm your body like alcohol does. It's okay to use it."

"Marijuana is dangerous. It causes cell damage and ruins the brain. No one should use it."

Above are two views of marijuana. You have probably heard at least one of them, if not both. Which one do you believe? Your answer will probably depend on what you've already heard about marijuana. The trouble is that what you've heard is likely to be a little bit of truth mixed with a lot of misinformation. Here are the facts about marijuana.

Marijuana, the Drug

Marijuana is also known as "pot" or "grass." It comes from the hemp, or cannabis, plant. The leaves and flowers are usually dried and rolled into a "joint" to be smoked.

Like cigarette smoke, marijuana smoke is inhaled. However, it is usually drawn more deeply down into the lungs and held there longer than cigarette smoke. This is one reason marijuana is more harmful to lung tissue than cigarettes. Marijuana also contains more cancer-causing substances and other chemicals than do cigarettes. One of these chemicals is THC. This is the ingredient that affects the brain.

Figure 18–4 Marijuana plant

Brain Effects

The most obvious effects of THC are on the cerebrum. The exact effects differ from person to person; no two people react exactly the same to any drug. As you read on, however, you will notice that the effects involve the emotions, thinking, learning, and the senses. All of these are controlled by the cerebrum.

Most people who smoke marijuana say they like the way it makes them feel. They say they feel calmer and more relaxed, and yet are still in control. Other people may feel sleepy, "fuzzy," or uneasy. Time often seems to pass more slowly for pot smokers. They report being more sensitive to sights and sounds and they claim that their thinking is much clearer; but an observer is likely to say that marijuana users seem confused and that their conversations may not make sense. Their minds may move quickly from one thought to another and their judgment is not as good as usual.

After smoking more, the user will not be able to concentrate and will have trouble remembering even simple facts. As the amount goes up, the user becomes more confused and may lose touch with reality. Marijuana can make a person feel more and more anxious, even panicky. He or she may become fearful of going insane or dying. Strange fantasies and possibly hallucinations may follow. While even fairly high doses of marijuana do not kill a person, they can trigger mental problems.

With continued use, marijuana is psychologically addicting. It can hook a user, especially a young one. Even when they are not using it, addicts constantly think about the drug. It affects their schooling, their family lives, and their friendships. Addiction is no fun.

Effects on Other Body Systems

While marijuana's most obvious effects are on the brain, it also acts upon other body systems. It affects the circulatory system by causing an increase in heart rate and blood pressure. Anyone with heart problems is asking for trouble by smoking marijuana.

As you might expect, marijuana's effects on the respiratory system are also of serious concern. The smoke can trigger coughing fits and asthma attacks. Marijuana users are at even higher risks for developing cancer than are tobacco smokers. Think about the risks that go along with smoking cigarettes. Marijuana smokers run the same risks.

Bloodshot and dry eyes and a dry mouth are other common side effects. In addition, as the user's "high" wears off, hunger often sets in—especially a craving for sweets.

Studies have shown that marijuana can affect other parts of the body, too. There are reports of damage to white blood cells, sperm, and brain cells. In men, marijuana causes a lower sperm count and a lower level of the hormone testosterone. More studies, however, need to be done before anyone can say how dangerous these findings really are. So far, most studies have been done on cells in a petri dish or on lab animals. The real question is, what are the effects on *humans*?

Other Concerns

There is a lot of wrong information about marijuana that you may hear. For instance, people may tell you that marijuana is not at all dangerous. If so, they may be thinking about old studies that have since been proved wrong. They also may not realize that the marijuana grown today is five to ten times stronger than the marijuana used in the 1960s.

You should also realize that most of the studies have been on adults. Most people believe that the effects on young people are even more serious. It can interfere with their reasoning, their ability to learn, and their social development. No one knows the long-term consequences of early marijuana use.

And, of course, there are various legal issues. Young people who are caught using marijuana will not only be embarrassed and face disapproval, they may end up with a police record. Because it is illegal to grow and sell marijuana, there are no laws to make sure the drug being sold is clean and pure. Without knowing it, people may buy marijuana that contains insecticides or other poisons, or the marijuana may have had some other drug added to it to make it seem stronger. It might not even be marijuana; it may be some other kind of dried leaves.

ANALYSIS QUESTIONS

1. What is the active chemical in marijuana called?
2. List at least 10 effects of marijuana use that involve the cerebrum.
3. Of all the things that marijuana does to a person, which one do you think is the most concerning? Why?

4. Why do you think some teenagers use marijuana despite the risks?
5. If someone offers you a marijuana joint, list at least 4 reasons for saying no.
6. Do you think the marijuana laws are fair? Explain.

EXTENSION ACTIVITIES

1. Marijuana laws vary from state to state. Call your local police department and explain that you are studying marijuana in science class. Ask someone to explain your state's marijuana laws to you.
2. Talk to the counselor or principal at your school to find out if he or she thinks marijuana is a problem in your community. Discuss the school policy for dealing with a student who smokes pot.

18–5 MARIJUANA QUESTIONS

Facts are important, but just because you know your facts doesn't mean you know your subject. You've got to be able to *use* what you've learned. Now's your chance to test your own knowledge. Here are six letters from people who have questions about marijuana. You won't be able to find the answers spelled out in the book but, if you think about what you've learned, you will do fine.

PROCEDURE

Work with a partner or by yourself.

1. Read all 6 of the letters and their questions and choose 3 of the letters you want to answer.
2. Spend some time on your answers. It will probably take several sentences to answer each question.
3. If you want to check your facts, look back at the reading ***Know the Facts About Marijuana*** starting on page 473.
4. When you have answered the questions, read over your answers. Underline the facts in your answers that were included in the reading.

THE SIX LETTERS

1

Frank just called. He doesn't want to play basketball this afternoon. Instead, he's going over to Josh's house. I think he's going to try smoking pot. Josh's sister uses marijuana and she's offered some to Josh and Frank. Frank says it won't hurt to try it once. Is that true? Is it okay for Frank to try a joint just once?

2

I've heard that marijuana can help a person do better in school. They tell me you can concentrate better when you're high and that you do better on tests if you smoke a little pot first. It also makes it easier to get through boring classes. Is all of this true? Could marijuana help me in school?

3

My sister Meg and her friend were talking about a party they went to last night. Meg was worried because one guy smoked two joints and then drove off with a carload of people to go to another party. Her friend says it's not dangerous to drive after smoking pot; it's worse to drive after drinking alcohol. Who's right? Is it safer to drive after using marijuana than after drinking alcohol?

4

Nick's sister just had a baby boy. Nick says the baby's upper lip didn't form right so the baby has a cleft lip. He says it's his sister's fault because she used to smoke pot when she was in high school. She's 25 now. Is that true? Could marijuana have caused this birth defect?

5

Lynnette told me that she smoked pot once. She said it wasn't anything. She didn't feel any different and can't understand what all the fuss is about. Why do people use marijuana if it doesn't have any effect?

6

I went to a big party the other night. It was a lot of fun. We swam, ate, drank pop, talked, and laughed a lot. A couple of kids went behind a bush and smoked a joint of marijuana. I told my brother, and he said I was lucky that I didn't get in trouble. He said that even though I didn't use the marijuana, I would have been in trouble if the police had shown up. Is that true?

18–6 TOUGH SITUATIONS

Cocaine. LSD. Heroin. Crack. Opium. PCP. . . . The list of illegal drugs that affect the nervous system goes on and on. You could learn the facts about all of these drugs or, instead, you could remember these four basic facts.

1. They affect the nervous system.
2. They alter a person's thinking, personality, and ability to learn.
3. They are addicting.
4. They are illegal.

These drugs are even more dangerous than alcohol or marijuana. It is frightening to think that some teens try these drugs without realizing what they are getting into. Sometimes, if they would have given the situation just a little thought, they would have said, "No, thanks." Here is your chance to think about some situations that involve drugs.

MATERIALS

reading: ***The Situations*** (page 481)
worksheet: ***Decision Making*** two per student (Figure 17–11)

PROCEDURE

Work with a partner.

1. Read the 5 situations and then you and your partner should agree on one of them to analyze.
2. Next, with your partner, brainstorm a list of the options you would have if you really were in this situation.
   a. Both of you should record the options on the back of your worksheets.
   b. Try to list at least 5 options; even more would be better.
3. Together look over your list of options and cross off the ones that could hurt someone or have other very negative consequences.
4. With your partner, choose the 3 best options from the ones that are left. Use your worksheets to help you analyze the options and identify your best choice.
5. Now, imagine that you are your parents. Circle the options that you think they would choose.
6. Answer the analysis questions after you have completed the homework assignment below.

HOMEWORK

Take your worksheets and a copy of the situations home with you. Ask your parents or some other adult to work through the same situation that you did and arrive at their best choices. Then answer the analysis questions.

ANALYSIS QUESTIONS

1. Rate how well you and your parents agreed on how to handle the situation.

   We agreed completely. 5 4 3 2 1 We didn't agree at all.

2. Compare the consequences that you listed in class with the ones your parents mentioned. What did they think of that had not occurred to you?

3. What did you learn from this activity?

EXTENSION ACTIVITIES

1. Do research on one of the illegal drugs listed in the introduction on page 479. Report 5 of the most important facts you learn to your class.

2. Write a situation involving drugs that you think would be realistic for your school. Discuss your situation with at least 4 people—friend, parent, teacher, and counselor. Compare the "best choices" identified by these people.

The Situations

Baby Sitter's Dilemma

You've been baby sitting for three children while their parents were at a party. You've baby sat here before; the Haleys are friends of your parents. Mr. Haley is supposed to drive you home. (You live about three miles away.) When he gets home, you suspect that he's been drinking alcohol so you watch him closely. He doesn't seem to be walking very steadily and you think he's talking more slowly than usual. You know that a person who's been drinking shouldn't drive. What will you do now?

Hallway Conversation

You're on your way to class when you overhear someone say, "I got it from my brother. It makes you feel so good." Glancing over, you notice three kids. One of them is putting something in his pocket. You catch another part of the conversation, "It's not a drug, it's just a pill." Your school has been involved in an anti-drug program. You remember one of the slogans, "Drugs are everybody's problem." What will you do now?

Worried Friend

You're really worried about your best friend because he uses marijuana every day. You've told him that smoking pot isn't smart, but he doesn't seem to care. He says he likes the way it makes him feel and, besides, he's not hurting anybody. You can't decide if you should keep trying to talk to him yourself or tell someone about the situation. You also don't know whom you'd tell. What will you do now?

Unsupervised Party

You begged your parents to let you go to this party. Now you discover that it's not at all as you expected. Kids have been smoking, and now they've brought out some beer. You know your parents wouldn't approve. Your parents brought you and a friend to the party. Your friend's dad is supposed to pick up both of you, but not for two hours. What will you do now?

UNIT 6

BODY CHANGES

CHAPTER 19

GROWTH

Besides height, what other clues can you use to estimate the age of a person?

19-1 WHAT IS PUBERTY?

Puberty is the time between childhood and sexual maturity when a person's body changes from that of a child to that of an adult. It is also a time of changing feelings. No doubt you are experiencing some of these changes yourself. While some of these changes are obvious, you are probably unaware of many of the subtle things that are going on. However, if you take the time to think about it, you may be able to list 10, 20, or even 30 of these changes.

- What changes take place during puberty?

You will be considering this question while working with a small group of other students and while you read about puberty.

MATERIALS

reading: ***Puberty: A Time of Change***
(page 487)

PROCEDURE

Part A. Defining Puberty

Work in a group with two or three other students.

1. With your group, brainstorm a list of changes that occur during puberty.
2. Choose one person in your group to be the recorder. Have the recorder write down the changes that your group brainstorms.
3. Remember the rules for working in a group.
   - Only one person talks at a time.
   - All members of a group participate.
4. Respect each other's ideas by not laughing or judging them. (When brainstorming, some ideas may sound strange but that's all right. A suggestion that sounds funny or wrong just may be right.)
5. Work quickly and quietly. You will have 5 minutes to complete this part of the activity. At the end of that time, all of the groups' lists will be combined during a class discussion.

Part B. Learning the Facts

1. On your own paper, make a chart with three columns like the one shown below.

CHANGES THAT TAKE PLACE DURING PUBERTY

Boys Only	Girls Only	Both Boys and Girls

   a. Copy each of the changes that occur during puberty from the class list onto your chart. Put each item in the correct column.
   b. If you are not sure in which column something belongs, write it in pencil at the bottom of the page. You will be able to change your answers later.

2. Read ***Puberty: A Time of Change*** starting on page 487 and correct your chart.
   a. While you read, look for additional changes to add to your chart.
   b. Check your chart. If you find that you have written something in the wrong column, erase it and rewrite it in the correct column.
3. Answer the analysis questions.

ANALYSIS QUESTIONS

1. Decide which of the following statements you think is true and write a paragraph to defend your choice. Include at least 3 reasons for your decision.
   Puberty is a time of more change for boys than for girls.
   Puberty is a time of more change for girls than for boys.
   Both boys and girls experience about the same amount of change during puberty.
2. In many school districts, children are first taught about the reproductive system in elementary school. If you were helping the people who work on curriculum, at what grade level would you suggest they start teaching about the reproductive system? What do you think are the 3 most important topics that the students at that grade level should be taught?

EXTENSION ACTIVITY

Talk to your parents about puberty. Below are some points you might want to include in your discussion.

Find out if anyone told your parents about what changes to expect during puberty.

Find out if they remember when they went through puberty.

Ask your parents if they think it was easier to grow up when they did or if they think it is easier for you now.

Puberty: A Time of Change

Human growth is fairly predictable and continual; we all grow from infants to children, then go on through our teen years and into adulthood. Each of these life stages is different, and each can be described by its own set of body changes.

The Early Stages

We often think of birth as marking the beginning of development, but a lot of growing takes place inside the mother's body before a baby is born. This first stage of growth, which takes about nine months, is a time of rapid change. A single cell must divide again and again until there are billions of cells. These cells must specialize and organize so that eventually they become a baby.

Growth continues at an astonishing rate throughout infancy. From birth to five months of age a baby's weight usually doubles, and by one year of age its birth weight has tripled.

Between the ages of two and about six, growth and development continue, but somewhat more slowly. During this time, a child gains in both height and weight. Then growth slows down for a while and may even appear to stop during the last few years of childhood. This seems to be a resting time, allowing the body to prepare for what comes next. (See Figure 19–1 on pages 488 and 489.)

Now comes the stage of rapid growth called **puberty.** Puberty marks the beginning of the changes that lead to physical maturity. Boys and girls vary widely in the ages at which they enter puberty. Sooner or later, however, in their own time and at their own rate, boys will acquire the physical characteristics of an adult male and girls will acquire those of an adult female. Some of the changes are easy to see; some are not.

Since you are in this stage right now, you are probably aware of some of the changes taking place in you and your friends. Naturally, there are many things you want to know about these changes. You may already have picked up bits and pieces of information and,

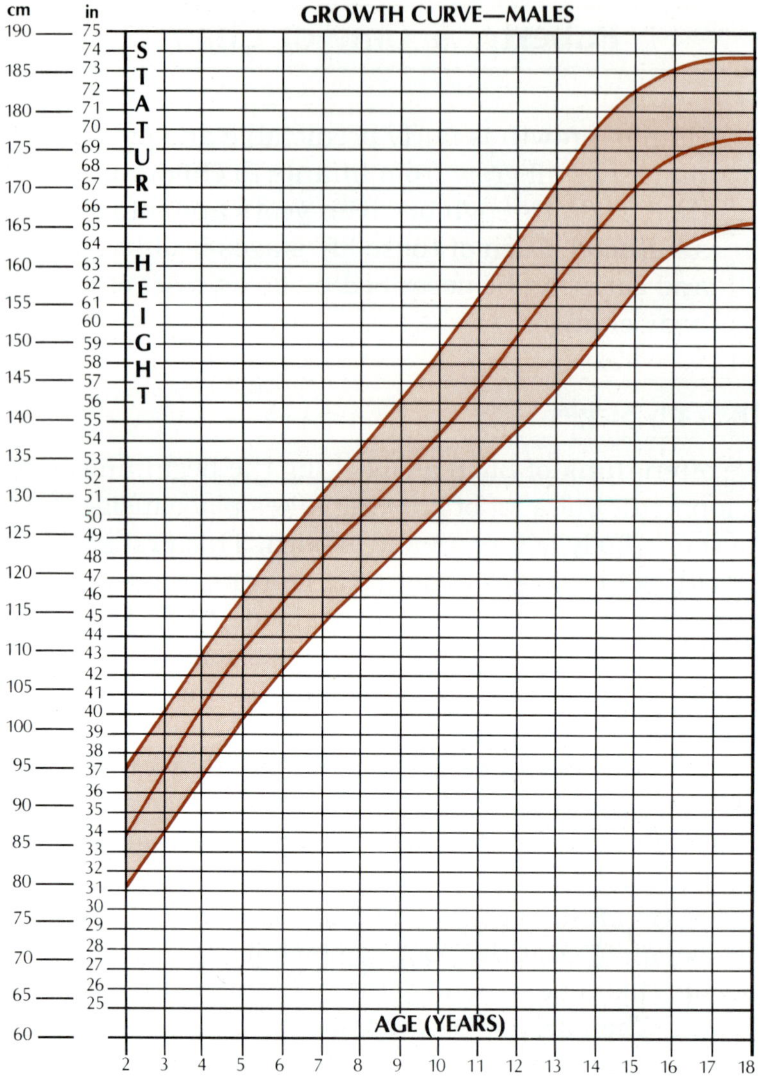

Figure 19–1 When you plot height versus age, you get growth curves such as these. What is the normal range in height for a 14-year-old girl? for a 14-year old boy? If a person is much shorter or taller than normal, he or she might want to ask a doctor or the school nurse if there is a medical reason for the height difference.

perhaps, some pieces of misinformation about the growing-up process. This unit has been written to help you get the facts straight.

Growing Taller

One of the first signs of puberty is often a growth spurt. For a girl, this usually begins between ages eight and thirteen. This is about two years earlier than for a boy. A boy's growth spurt generally begins anytime between the ages of ten and fifteen.

After this initial spurt, growth continues—but at a slower rate—until a girl is about sixteen to eighteen and a boy is about eighteen to twenty. By the time growth is complete, boys often end up being taller and heavier than girls.

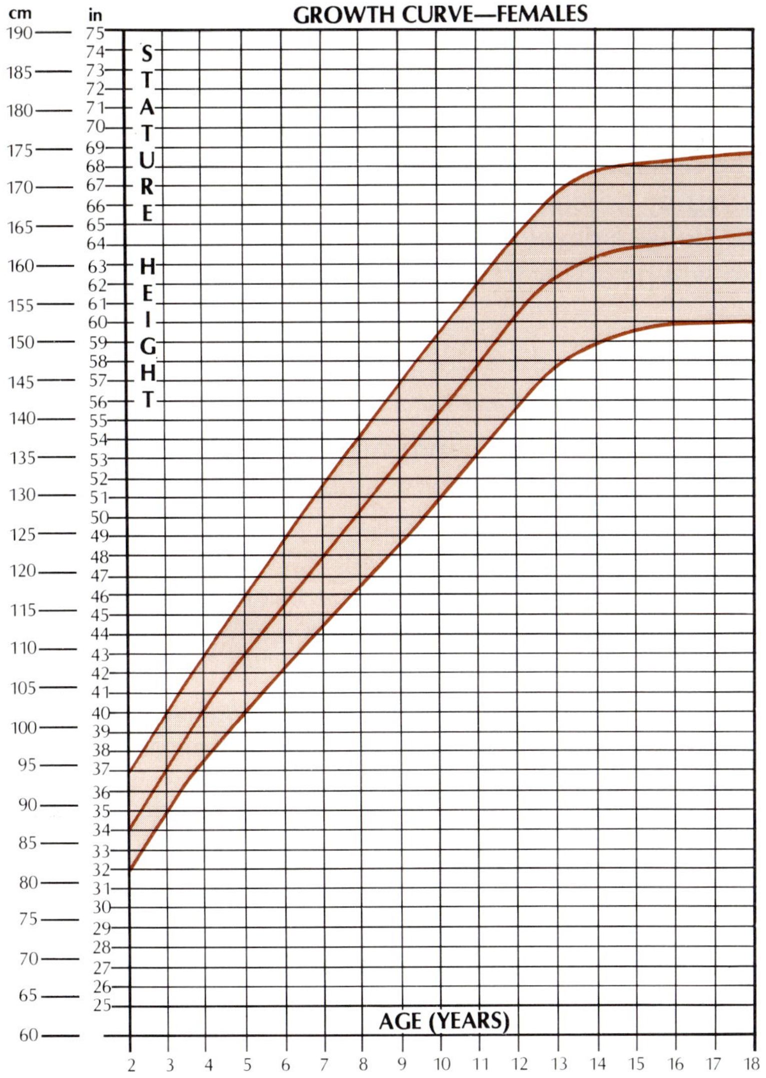

It is important to remember that there is no such thing as an "average" person. Rate of growth varies from person to person. Whenever your own growth spurt begins is right for you. You may have noticed gains in height and weight as early as age eight or nine. This indicates that physically you were maturing early. If so, you may be as tall as you are going to be by age fifteen or sixteen. Some of your friends may not start their growth spurts for several years, and then they may keep growing until they are in their early twenties.

You can get some idea of how tall you will be by looking at the heights of your parents. If shortness runs in the family, you are not likely to be very tall. Likewise, tall parents usually produce tall children. However, once again, there are no hard and fast rules. Short parents may have tall offspring, and tall parents may have one or more children who are short. It can also happen that if one parent is tall and the other short, their children may be short or tall or somewhere in between. In general, a person's growth pattern is influenced to a great extent by heredity. If your parents developed early, you are likely to develop early, too. If they developed late, you will probably follow that pattern.

Growing Tissues

An increase in height and weight is not the only change that occurs during puberty; all of the body's organs and tissues grow, too. For instance, the heart and other muscles grow larger and stronger. This means that your strength and endurance are likely to increase. Because the bones in the hands and feet grow faster than other bones, such as the vertebrae, you may feel awkward for a while. But, by the time you reach your adult height, all parts of your body will be in proportion.

All of this growth requires energy so you may notice that you need more sleep than you did earlier. There may be times when you are brimming with energy and other times when you feel just plain lazy. Luckily, this situation does not last; as your growth slows down, your energy levels will probably even out. Now while you are still growing, you may want to get to

Figure 19–2 Nutritious snacks

bed reasonably early. Then you won't find yourself struggling to wake up in the morning.

In addition to needing sleep, a growing body requires plenty of calories and nutrients—proteins, vitamins, fats, and carbohydrates. Consequently, your appetite may increase; you may always seem to be hungry—even after a meal. It is especially important to eat well and to choose nutrient-rich foods, such as milk, cheese, vegetables, fruit, poultry, fish, red meat, and eggs. As you might expect, because you are eating more food, some weight gain occurs. This is normal. However, if you find yourself gaining too much weight, try not to snack as often. High calorie foods, such as cakes, soft drinks, and candy, may satisfy your appetite but they may also result in an unwelcome amount of extra fat.

Body Shape

Some fat is important. In addition to providing insulation, fat also gives the body its shape. During puberty body fat is redistributed and some of it may be lost. Girls generally lose less body fat than do boys, giving girls a rounder appearance. Some of this fatty tissue accumulates in the developing breasts around the milk ducts.

At the start of puberty, a child's breasts are fairly flat with slightly raised nipples. Soon hormones trigger

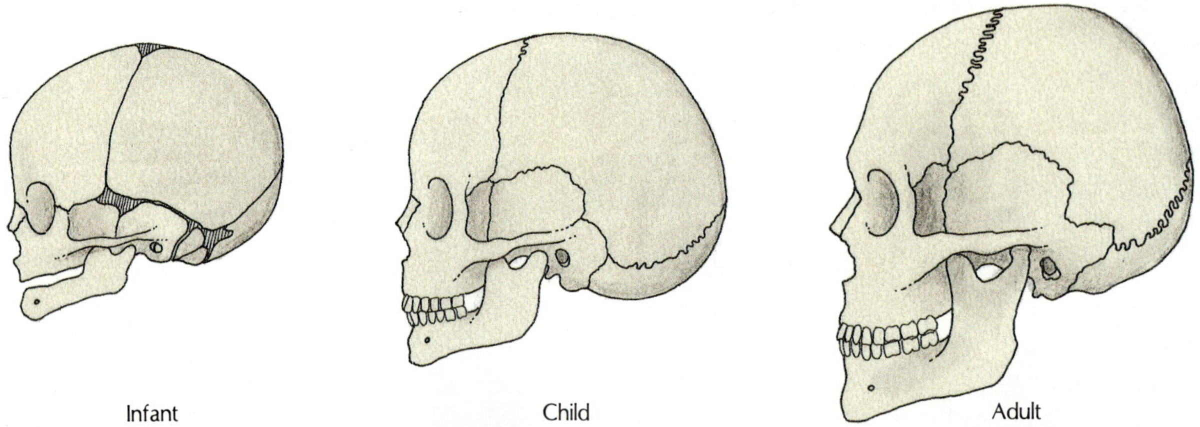

Figure 19–3 A child's face does not look like an adult's. By looking at these skulls, can you list some reasons why?

breast development, causing the nipples to stand out a little farther and the underlying tissue to swell slightly. At this time, the breasts may be tender and sensitive. This occurs in both boys and girls. However, in boys, the swelling goes down after the hormones balance out. In girls, breasts continue to develop, gradually filling out over the next three or four years. No one can say for sure when a girl's breasts will start to develop. As with puberty in general, it may begin when she is only eight or not until she is sixteen or older. The age at which breast development starts has nothing to do with final bust size. An early starter may end up with either large or small breasts; the same is true for late starters.

Changes in the skeletal system also affect body shape. Boys' shoulders widen, giving them an adult male form. By the end of puberty, girls have wider hips because their pelvises have broadened. Facial proportions also change. Gradually, the round face of a child transforms to the somewhat longer face of an adult. (See Figure 19–3.)

Hair, Voice, and Skin

The growth of body hair is another change that marks the approach of sexual maturation. One of the first areas that hair appears for both boys and girls is in the pubic area. Hair also grows under the arms; more hair appears on the arms and legs. This hair is generally not as obvious in girls as it is in boys.

Many boys anxiously await the appearance of hair on their faces—the beginnings of a beard. The first facial hairs are fine and sparse; gradually they become coarser and thicker. A little later boys may notice hair on the chest and sometimes on the back. The amount of hair varies from one male to the next. "Hairiness" has nothing to do with masculinity. Men with lots of body hair are just as masculine as men with very little body hair. Ironically, as some males approach the end of puberty and while their body hair is still developing, they may start to go bald.

Some girls may notice that they have a bit more fuzz on the upper lip than they like. A light growth softens facial features but, if the fuzz is bothersome, a doctor can suggest ways to remove it.

Both boys and girls lose the clear, high voices of children as their vocal cords lengthen and the larynx grows to a lower position in the throat. Boys may find that when they are saying something, their voices may "crack" and end up on an embarrassingly high note. This transition does not last long and when it is over, the voice is much deeper. Girls' voices also become lower and richer but the difference is not as noticeable as in boys.

Another consequence of puberty is that the sweat glands become more active. Sweat glands are located all over the body; they are especially numerous under the arms. Although the perspiration these glands produce is odorless, the bacteria that live on the skin may cause it to have an unpleasant odor. Most of this odor can be minimized by washing daily with soap and water. You may also want to experiment with deodorants and antiperspirants. Some people find that certain brands are more effective for them than others.

You may notice that your skin feels oilier. That's because the oil glands become more active during puberty. Sometimes the oil clogs the tiny duct that leads from the gland to the skin surface—causing a pimple or acne to form. (See Figure 19–4.) Pimples are especially common on the face, shoulders, chest, and back since that's where oil glands are the most numerous. Just about everyone gets at least a few pimples but many boys and girls are plagued by them.

The best thing to do for pimples is to clean your face with soap and water. You may need to wash your hair every day if it is oily. If your skin problem seems to be severe, a dermatologist—skin doctor—may be able to help you. Prompt treatment can reduce your risk for long-term scarring.

Figure 19–4 How acne develops

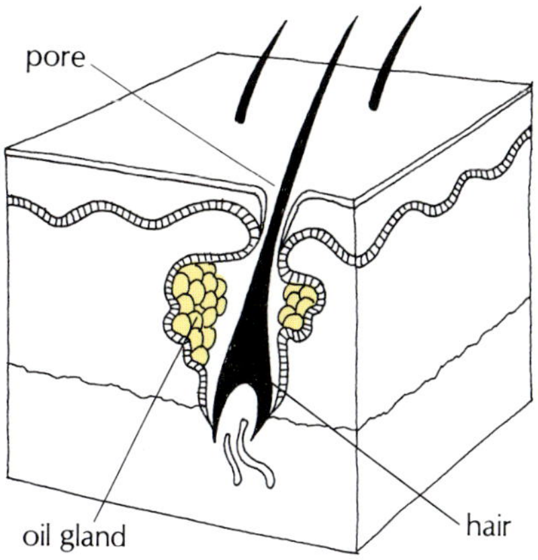

Oil is produced in glands along the hair.

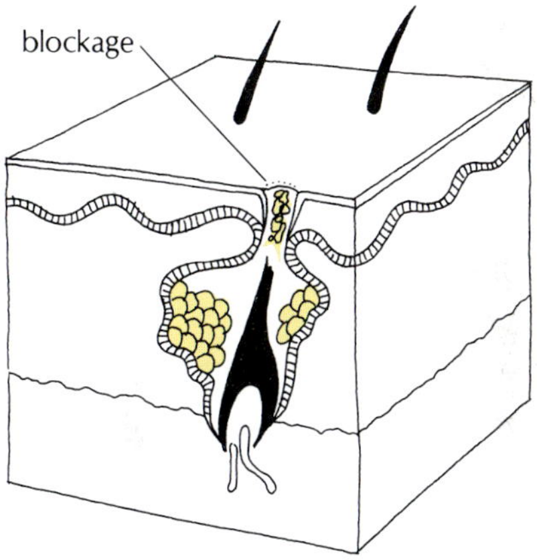

The pore may become blocked by oil and dead skin cells. The oil builds up behind the blockage.

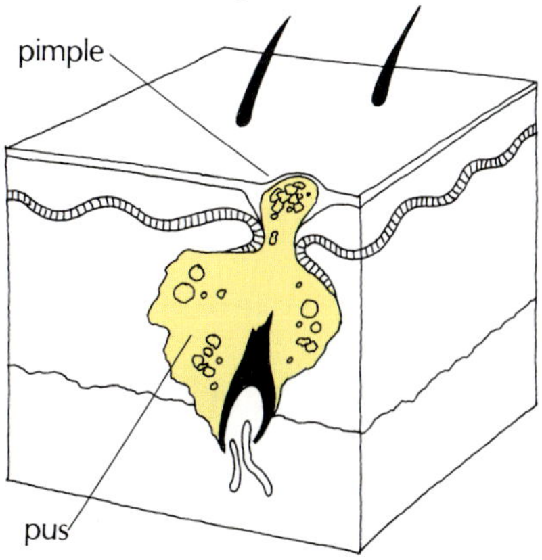

Bacteria from the skin may infect the oil. Pus and swelling result. This is now a pimple.

Did you realize that all of these things are going on in your body? It's impressive the way our bodies control and coordinate so many changes. All of us eventually experience all of these changes and, in so doing, we become physically mature men and women.

A Pink Puberty?

Flamingo chicks are gray and fluffy. They can fly when they are three months old and are almost fully grown after one year. However, they do not turn pink until they are four years old. After they have turned pink, they can reproduce. Is turning pink a sign that a flamingo is going through puberty? What do you think?

19-2 THE REPRODUCTIVE SYSTEM: EXTERNAL CHANGES

Some of the most significant changes that take place during puberty are those that involve the reproductive system—the system that includes the organs that enable humans to reproduce. This system is essential for life. If we could not reproduce, the human race would come to an end.

As children, we do not need a functioning reproductive system. When we are very young, it is important that our other body systems, such as the digestive, respiratory, and skeletal systems, develop to keep pace with our growing bodies. As we approach adulthood, the reproductive system seems to "wake-up." This chapter is about the obvious changes that result when the reproductive system begins to mature. Less obvious changes will be discussed later. Keep in mind, of course, that being able to reproduce is not all there is to being an adult.

As they go through puberty, both boys and girls grow taller, develop body hair, worry about being normal, and experience other similar changes and feelings. However, there are some changes that are unique to boys and some that are unique to girls. In this reading, you will learn about some of the unique changes that deal with the reproductive systems and involve the outside of the body. These external reproductive structures are called the **genitals** or **genitalia.**

Boy to Man—Changing Anatomy

Most boys are aware of their external genitals and so are likely to notice the changes that take place during puberty. The two most visible structures are the **penis** and the **scrotum.**

The penis functions in two body systems. As part of the urinary system, it carries urine from the bladder to the outside of the body. As part of the reproductive system, the penis transports sperm. The small opening at the end of the penis is the **urethral opening.** Both urine and semen pass through this.

During puberty, the penis grows a little longer and thicker. Just how long and how thick varies from one male to the next. At some time or another, almost every male worries about the size of his penis. It is important to remember that, because puberty starts at different times for different people, some boys' penises will be much bigger than some other boys', even though they are the same age. That's okay. In fact, the size range is normal. By the end of puberty, size differences are not nearly as great. Penis size has nothing to do with manliness, strength, or anything else; it does have to do with the genes a boy inherits.

The appearance of a penis depends on whether or not a boy has been circumcised. At birth, the end of the penis, often referred to as the glans, is enclosed in a sheath of skin called the **foreskin.** The surgical removal of the foreskin is called **circumcision.** The decision of whether or not to have a baby circumcised is usually made soon after birth. Sometimes it is done because of religious beliefs; other parents believe it is easier to keep the glans clean when the foreskin is removed. Either way is okay. Whether or not there is a foreskin does not affect the functioning of the penis.

Behind the penis is a loose sac of thin skin called the **scrotum.** It contains the **testes** or **testicles.** They are an important part of the reproductive system since sperm develop here. Muscular fibers in the scrotum automatically control the temperature of the testes. If it is cold outside, these muscles contract, bringing the testes closer to the body. In hot weather, the muscles relax so that the testes hang a little farther away from the body. When this happens, the testes may seem larger but there really is no change in their size. In this way the temperature in the scrotum can be kept slightly lower than body temperature. This lower temperature is necessary for sperm development.

Like other body parts, the size of the scrotum varies slightly from one person to another. One of the early signs of puberty is often an increase in the size of the testes and, therefore, of the scrotum, too. When this happens, the skin on the scrotum becomes darker and more wrinkled. As the scrotum grows, one side often appears lower than the other. This probably makes movements such as walking more comfortable since the testes are at slightly different levels and won't rub against each other.

During puberty, **pubic hair** appears around the base of the penis. At first, it may look like a pale fuzz. Over time, the hairs become coarse, dark, and curly. Pubic hair probably serves a protective function, keeping dirt and microorganisms away from the genitals.

All these changes will take place in every boy as he goes through puberty. The exact timing varies from one boy to another but eventually all boys will be adult men.

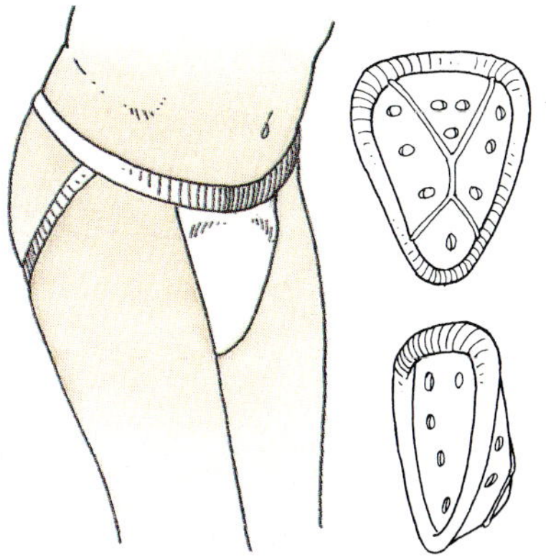

Figure 19–5 It is painful to get hit in the scrotum. For this reason, many boys and men wear athletic supporters or cup-shaped protectors when participating in active sports.

Girl to Woman—Changing Anatomy

A girl's genitals are not as obvious as a boy's because they are located between the legs. Perhaps that's also why many girls are not aware of the changes that are taking place as they go through puberty. A female's genitals are called the **vulva.** The vulva is made up of several different parts. The most obvious parts are two pairs of fleshy folds of tissue—the inner and outer labia.

The **outer labia** are two fatty folds that enclose and protect the underlying structures. In girls, the inner surface of these folds is smooth but, as puberty progresses, small, slightly raised bumps may appear. These are oil glands that keep the area moist so the skin does not get irritated.

Just inside the outer labia are two thinner folds of tissue called the **inner labia.** In childhood, they may not be very noticeable; they grow and become more obvious during puberty. They too have oil glands that produce oil. As a girl goes through puberty, the labia become fleshier and darker in color.

Where the two inner labia meet in front, there is a firm bump of tissue, the **clitoris.** In grown women, the clitoris is likely to be about the size of the head of a wooden match. Although the clitoris has no reproductive function, it contains many nerve endings and is highly sensitive to touch.

Directly behind the clitoris is the **urethral opening.** It may not be very noticeable, but it is the opening through which urine passes on its way out of the body. This opening is not part of the reproductive system; it functions only in the urinary system.

The **vaginal opening** is behind the urethral opening. The vagina itself is not visible since it is inside the body. Even though this opening is larger than the urethral opening, it is not very obvious either. That is because the sides of the vagina are usually touching each other. Sometimes there is a membranous tissue, the **hymen**, that partially covers the vaginal opening. This membrane is often so thin that it is easily broken or torn. In fact, most women are unaware that they ever had a hymen.

Early in puberty, **pubic hair** starts to grow in front of the vulva and on the outside of the outer labia but not on the inner labia. At first, the pubic hair is fine and sparse. As puberty progresses, it becomes coarser, thicker, and curlier.

Men and Women

All of these changes are gradual, taking several years to complete. Nothing happens overnight, in a week, or even in a month. The entire process is pretty amazing if you stop to think about it. During this time, the bodies of children—boys and girls—become the bodies of adults—men and women.

This reading has mentioned some of the most obvious changes, but, of course, it does not explain everything. If you notice something going on with your own body that you wonder about, or are worried about, talk to someone. Your parents, school nurse, doctor, or some other trusted adult can be good sources of information. You may even find that when they were young, they had some of the same questions that you have now.

19-3 THE REPRODUCTIVE SYSTEM: INTERNAL CHANGES

Now that you have learned about the observable changes that take place during puberty, you are ready to study the changes that are going on *inside* the body. Even though these changes are not visible, they affect the way you feel and the way your body works.

In this reading you will learn the names and locations of more structures that make up the reproductive system. You will also read a little about their functions but you will learn much more about how the reproductive system works in later readings.

Boy to Man

Although boys are often familiar with the external parts of their reproductive systems, they seldom know much about the structures *inside* their bodies. In the last reading, you learned that one of the first signs of puberty is the growth of the scrotum. It grows because the testes inside are growing. The **testes** are paired, egg-shaped structures. They are made up of a mass of microscopic coiled tubes; they do not contain any fat or muscle tissue. Testes have two main functions—to release hormones and to produce sperm. Sperm are made in the coiled tubes and the male hormone **testosterone** is made in cells between the tubes. During puberty a testis may double in size, growing from about 2 cm to a length of 4 to 5 cm. (See Figure 19–6.)

Figure 19–6 Testis size

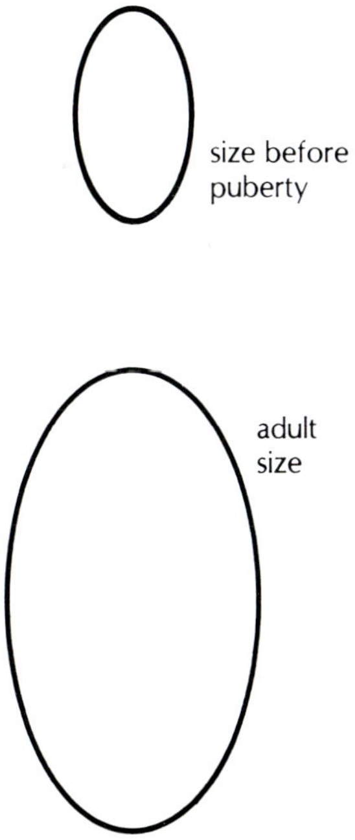

If you look at Figure 19–7 on page 500, you can see that a tube lies along the back side of each testis. This tube is the **epididymis.** Once sperm are produced in the testis, they travel into the epididymis where they finish maturing.

Leading away from the epididymis is a larger tube, the **vas deferens.** Sometimes it is referred to simply as the vas. It extends up out of the scrotum and into the abdomen. It then loops around the bladder to join the urethra. Sperm are transported through the vas and then to the urethra on their way out of the body. Follow

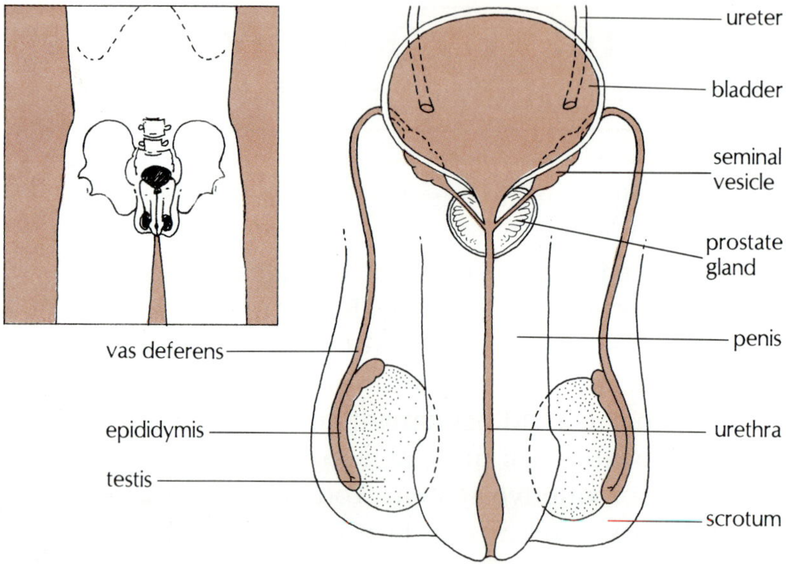

Figure 19–7 Male reproductive system—front view

the path of the vas in Figure 19–7 above and Figure 19–8 on page 501.

The **urethra** is yet another tube. (It seems like the male reproductive system is mainly tubing.) Since the urethra drains urine from the bladder, you may have discussed it when you studied the urinary system. As part of the reproductive system, it transports sperm from the vas out of the body through the penis. Even though both urine and semen pass through the urethra, they are never there at the same time. A muscle closes off the connection between the urethra and bladder when sperm are present.

Several glands produce fluid that is released into the vas. The **seminal vesicles** are a pair of such glands. (See Figure 19–7.) The largest of the glands, located right beneath the bladder, is the **prostate.** In children, it is relatively small; but, during puberty, it starts to grow and eventually reaches 2 to 3 cm in diameter. Boys and young men usually do not need to give much thought to their prostates. But, after the age of 40, men should have annual prostate checks because it is a common site of cancer. You can see these glands in Figure 19–8 on page 501.

As with any other part of the body, a physician must check the entire reproductive system, not just the prostate, for any sign of trouble. The physician will feel the testes, scrotum, and penis to check for lumps or pain. While this may seem embarrassing, it is a routine

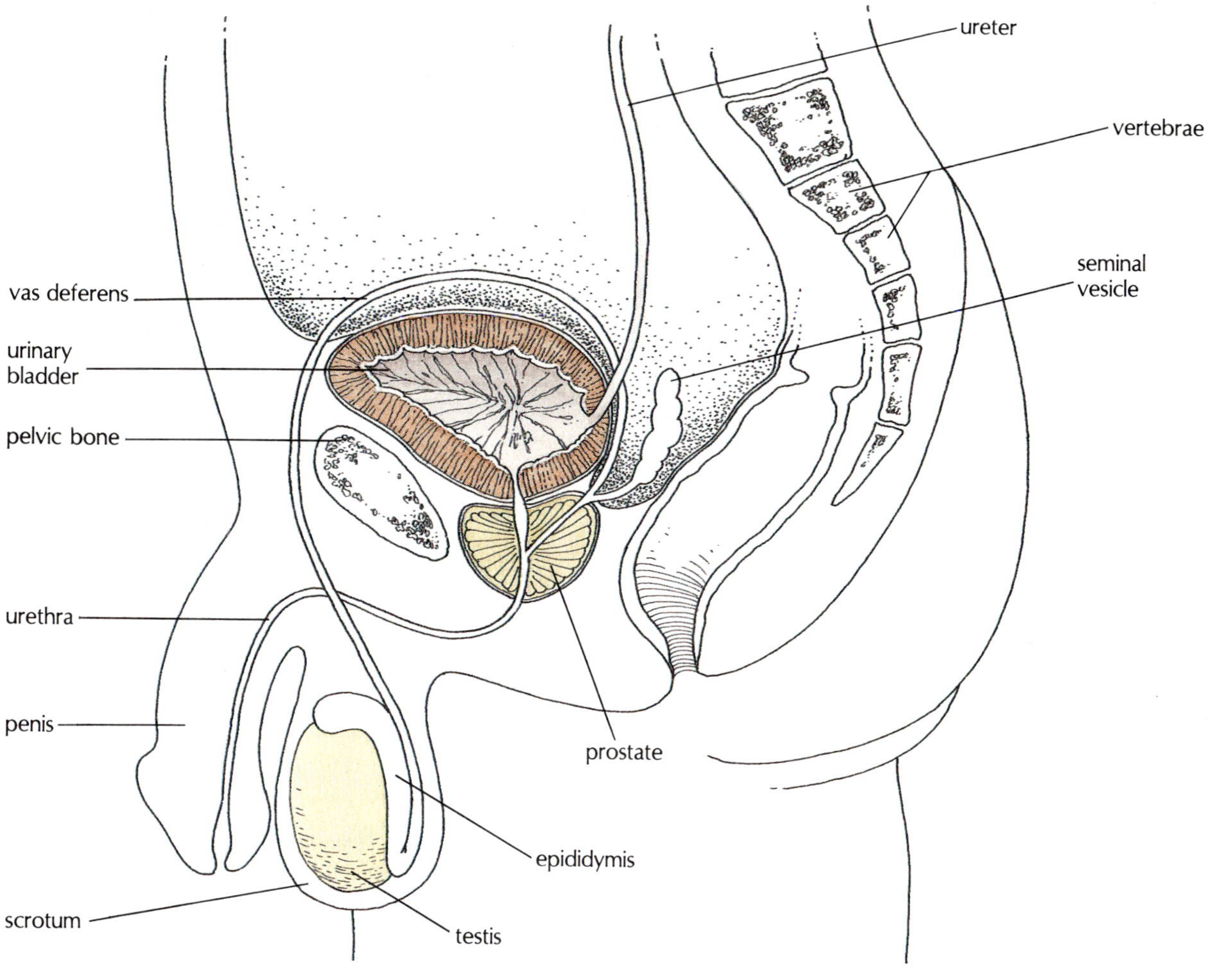

Figure 19–8 Male reproductive system — side view

part of a physical exam. Health care providers are used to seeing people without clothes so they don't think much about it. They are as comfortable examining genitals as they are checking a person's eyes or ears.

Girl to Woman

In some ways, it seems easier to learn about a girl's internal reproductive system than it does a boy's; perhaps that is because this system doesn't involve as many "tubes." The main structures of a female's reproductive system are labeled in Figure 19–9 on page 502 and Figure 19–10 on page 503.

First locate the **ovaries** in both figures. In an adult female, they are about 2 to 3 cm long. While they are fairly small, they are extremely important. In addition to releasing egg cells, they also produce sex hormones. These hormones regulate many of the characteristics

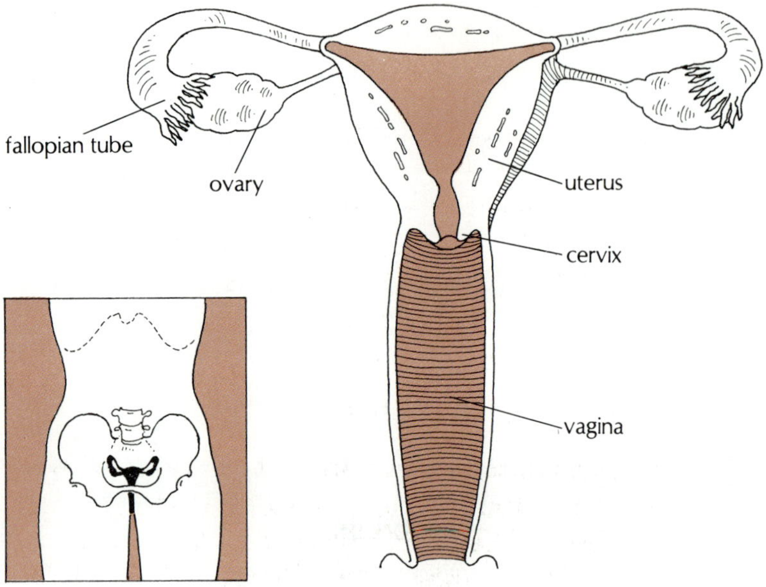

Figure 19–9 Female reproductive system — front view

that we think of as being feminine. Biologically, the ovaries are similar to the testes since they have such similar functions. But, unlike the testes, they are not on the outside of the body; they are tucked up securely in the abdomen.

When an egg is released from an ovary, it enters a **fallopian tube.** These 10 cm (4 inch) long structures extend from each ovary to the uterus. They are fairly narrow, being about the diameter of a pencil. The end of a fallopian tube flares open and has tiny fingerlike projections that help guide the egg into the opening of the tube itself. The walls of the fallopian tubes are made up of smooth muscle cells. When these muscles contract, the tiny **cilia*** that line the tubes wave back and forth. This waving motion moves the egg along the length of the tube to the uterus.

*Cilia are tiny hairlike structures that extend from the cell membrane. You may remember hearing about cilia when you studied the respiratory system since cilia line the bronchial tubes.

To most people, the **uterus** is the most familiar part of the reproductive system since it is here that a baby develops and grows. During puberty, the uterus more than doubles in size. It grows until it is about the size and shape of a pear. Like the fallopian tubes, the walls of the uterus are made up of smooth muscle cells. When a female is not pregnant, there is not much space in the uterus. But during pregnancy, the walls of the uterus thin out and stretch so much that the uterus can hold a baby—or maybe even two or three babies!

During puberty, the lining of the uterus develops new blood vessels that are capable of forming a spongy,

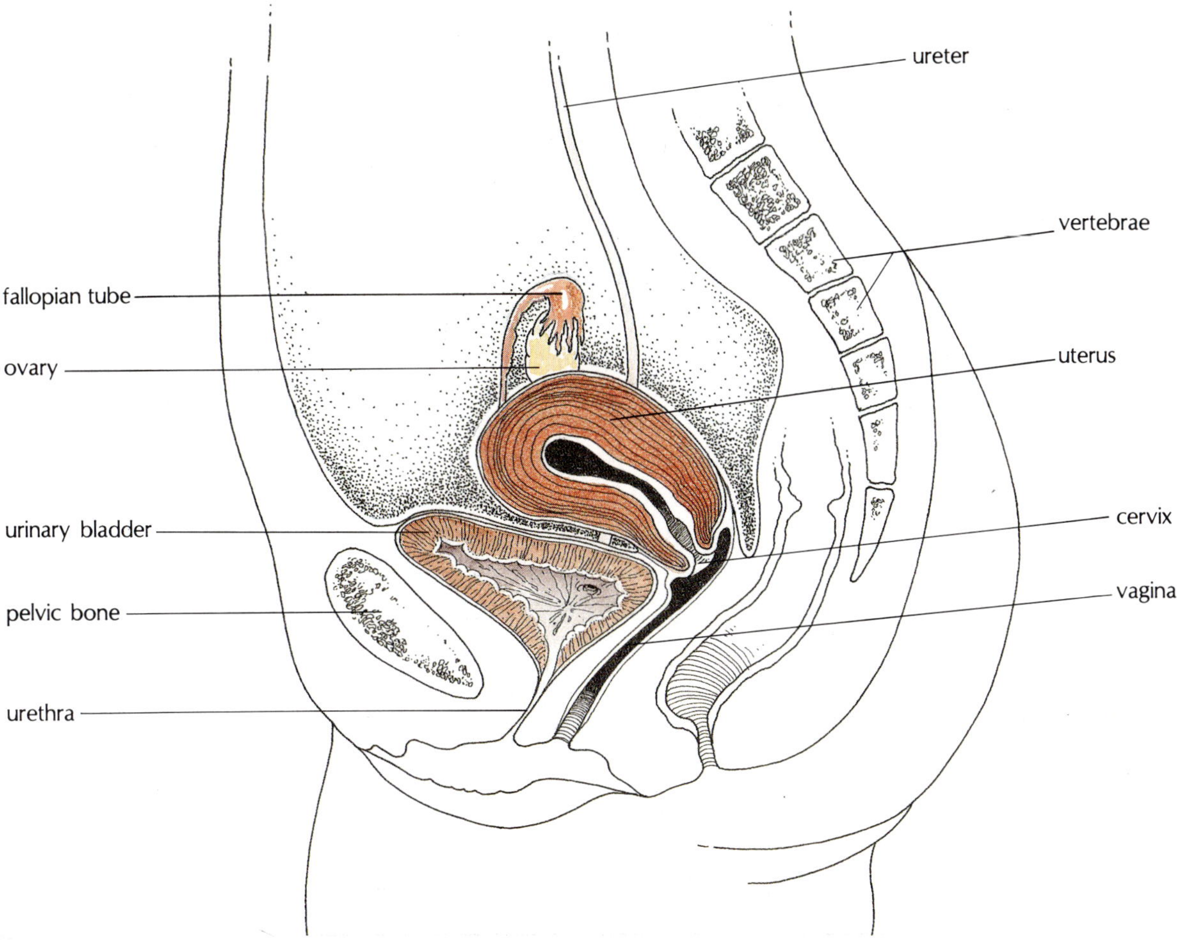

Figure 19–10 Female reproductive system — side view

cushioning tissue. This lining goes through monthly changes. Its loss is referred to as a "period" or "menstrual flow." This process is too complex to explain in this reading. So, if you have questions about it, wait a few days. You will be hearing more about menstruation in a later reading.

The lower part of the uterus is the **cervix.** It is a firm muscular ring that surrounds the opening to the uterus. The muscles in the cervix must be quite strong because they keep the uterus closed during a pregnancy. Without this muscle, a baby could "fall out" of the uterus. The cervix extends down a little way into the vagina.

The **vagina** is the passageway that extends from the uterus to the outside of the body. Like the uterus and fallopian tubes, its walls also contain smooth muscle cells. These cells can contract, relax, and stretch. Usually, the cells are relaxed so the vagina looks something like a collapsed tube. Since it is stretchable, it can increase in diameter enough for the delivery of a baby. For this reason, it is sometimes called the "birth canal."

The urinary bladder and the urethra are also labeled in Figure 19–10 on page 503. While neither one is part of the reproductive system, they are included to help you see how everything fits together. By now, you should realize that the female has three openings in the lower part of her body—the urethra, the vagina, and the anus—whereas, a male has two—the urethra and the anus.

It is harder to check a female's reproductive system for medical problems than it is to check a male's since most of the female reproductive system is inside the body. In order to check the vagina, uterus, cervix, and ovaries for changes in shape or for the presence of lumps, the physician must be able to feel them. This is easily accomplished by doing a pelvic exam. The physician gently inserts two gloved fingers into the vagina and carefully feels for any abnormalities. It may feel a little uncomfortable but it is not painful.

19-4 FEELINGS, FRIENDS, AND FAMILY

Not all questions that boys and girls ask about puberty involve physical changes. They also ask questions concerning feelings, friends, and family. You can see this for yourself in this sampling of questions.

What causes teenagers to have mood swings?
Why do so many kids at school act weird?
How can I get someone to like me?
Should we be going out at our age?
Why don't my parents let me wear what I want?

As you might expect after looking at the first part of this introduction, this reading is about emotions and relationships. While you read, keep these questions in mind:

- Do I agree with everything in this reading? If not, what would I have said differently?
- What other issues would I have included if I had been writing about this topic?

Feelings

During puberty feelings have to change. As children grow into adults, they discover that they feel differently about their lives, their interests, and the people they know. You may be experiencing these changes right now and they can be confusing. At times you may not know how you are feeling. For instance, you may still enjoy some of the games and books from your childhood, but something about them, or perhap it's you, has changed. They may not give you the same satisfaction that they did before.

As your feelings and attitudes change, you may be one of those people who experiences mood swings. One minute you may be cheerful and full of plans; then suddenly you may find yourself feeling depressed and angry with the world. No one knows for sure why feelings can change so quickly. Hormones are often blamed for the changes; but hormones probably are not the whole answer. As some researchers point out, there are lots of changes going on in your life right now and these take some getting used to.

For instance, you have probably noticed that teachers expect more from you than they did a couple of years ago. You may be expected to take better notes, read more difficult assignments, write more thoughtful answers, and solve more complex problems. Parents too expect you to take on more responsibilities—perhaps by helping out around the house a little more or by making decisions like a responsible adult. Some students are comfortable accepting these new challenges, others are not. It can be tough to take on your new roles. Sometimes you enjoy them, other times they feel like a burden. These conflicting feelings can contribute to mood swings.

Moodiness is part of the bigger issue of behavior—the way people act. Look around you. Think about some of the students in your class. What do you think of the way each one behaves? You may decide that some are childish, or stuck up, or "neat," or strange, or whatever. One reason you notice this wide range of behaviors is because everyone is at a different stage of puberty. Some people act childish because they are still children. Others act "strange" or "weird" because they think that is the way teenagers should act, because they are trying to impress someone, or perhaps because they don't know any other way to behave. You judge people by their behavior and you, in turn, are judged by your own behavior.

Friends

Recently you may have noticed that your friends are changing. The kids you hung around with in elementary school may not be as much fun as they once were. Then, too, there may be new people you would like to meet. All of this is normal. As your interests change, your friends will too. It may not feel good if you find yourself losing a friend or if you feel alone, but try to remember that lots of people feel the same way. Then, with time, the hurt feeling will go away.

One day you may realize that your best friend has a new "special" friend . . . one of the opposite sex! You may be hurt but then you may wonder if perhaps you too should find someone special or start to date. Actually, it is not at all important for you to date at this age. It is much more important for you to learn how to get along with other people and learn how to be a friend—a friend to people of both sexes.

When kids start to "go steady" or to date one special person, they limit themselves. They do not have a chance to meet lots of other people or to discover who they themselves really are. These are good years to do things with lots of friends—you can go swimming, watch a movie, play softball, or do whatever you enjoy, but do it with a group of friends. It is easier to talk to people and to meet new people when you do things in groups.

A little later, it will feel right to date. Your parents probably have some ideas about dating. Talk to them about it. It is much better to discuss this issue in advance. That way, when you do want to date, you will already understand each other's expectations.

The mention of parents brings up a whole new issue, that of family.

Family

Puberty often marks the beginning of a gradual change in your relationship with your parents.* At times they may not seem to understand you or you may think that they are being unreasonable. Actually, your parents are probably feeling as if they are going through a kind of puberty of their own right along with you. They are watching their child develop into an adult. It is not any easier for them than it is for you. They realize that you want more independence and the freedom to make your own decisions. However, you and your parents may not agree on exactly how much independence you should have or on which decisions you should be making. Even small differences in opinion can lead to major conflicts.

Many of these conflicts can be solved if you and your parents sit down and talk with each other, realizing in advance that you both may have to make compromises. It doesn't do any good to get angry and yell or to expect to always get your own way. Let your parents explain how they feel and you tell how you feel. Talk about any differences that may come up. Sometimes,

*The word *parents* is used a lot in this unit. Since not all kids live in families with two "birth parents," you should substitute the word *parents* for whatever fits your situation. Perhaps you live in a single parent family, or with foster parents, grandparents, or some other relative or friend of the family. Any of these people can be parents. Just think of the adults you live with while you are reading.

even if you try to compromise, you may not feel like you are communicating. If you find that happening to you, talk to some other adult who is not as personally involved in the situation. He or she may be able to offer some helpful suggestions.

It is also important to realize that independence does not come all at once; it takes time to achieve. As your parents watch the way you make decisions, they will see that you are starting to consider more options and that you are trying to make wise choices. This will help them acknowledge that you are growing up. Gradually you will assume more responsibilities until one day you realize that you are making all of your own decisions . . . . like an adult!

WHAT DO YOU THINK?

Now that you've completed the reading, go back and look at the two focusing questions on page 505. Write a one- or two-paragraph response to one of the questions.

19-5 PEER PROBLEMS

During puberty, many teenagers find that it is very important to have friends. Friends can provide support. A friend can be someone to talk to, to do things with, or to confide in. Unfortunately, friendships don't always go smoothly. They may be hard to get started and, even then, problems can arise.

- When there is a problem in a friendship, what options do adults suggest that teenagers may not consider?

You will be able to answer this question after you have used the decision making model to solve a problem that involves getting along with other people. You will be solving the same problem twice, once with other students and once with adults.

MATERIALS

The Problems (page 511)
worksheet: ***Decision Making*** two per student (Figure 19–11)

PROCEDURE

Part A. Small Group Decisions

Work in a group with two or three other students—your peers.

1. Get ready to solve one of the problems described on page 511.
   a. Be sure each member of your group has two worksheets.
   b. Choose one person in the group to watch the time. (This person's job is to make sure the group stays on task and is able to finish the assignment on time.)

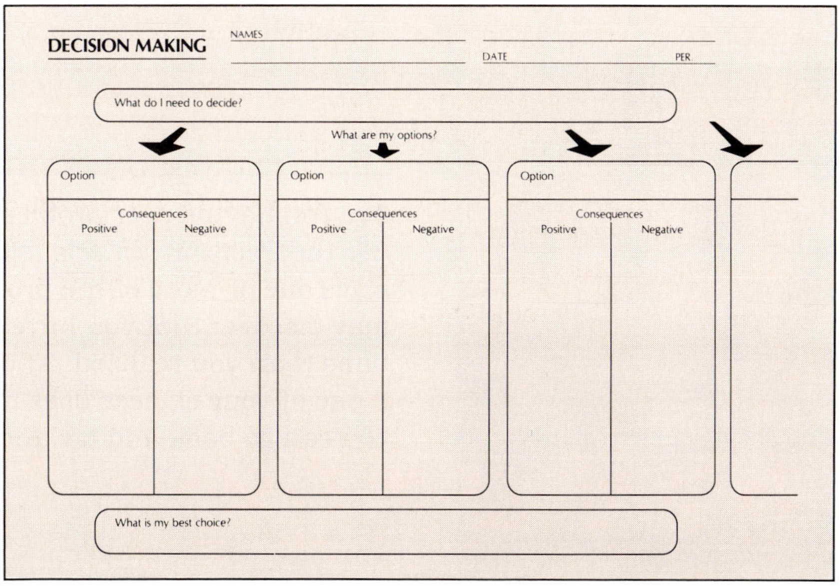

Figure 19–11

c. Help make sure that all members of your group participate.
2. Quickly read the situation that your group is supposed to analyze.
3. Next, with your group, brainstorm a list of the options that the person in the situation might have.
   a. Record these options on the back of your own worksheet.
   b. Try to think of at least 10 options. If the group stops after coming up with only three or four, you may miss some of the best ones.
4. As a group, discuss the list of options. On your own worksheets, cross off the ones that could hurt someone or have other strong negative consequences.
5. Choose the 3 best options from among the ones that are left. Use the Decision Making worksheet to analyze these options and to identify a best choice.
6. Now, imagine that you are your parents. Circle the options that you think they would choose.

7. Choose another situation to analyze. Use a situation from the ones that have been provided or make up one of your own. Follow this same procedure to analyze it.

Part B. Homework

1. Take your worksheets and a copy of the situations home with you.
2. Ask your parents to work through the same 2 situations that you did in class and to arrive at their best choices.
3. Answer the analysis questions.

ANALYSIS QUESTIONS

1. How well did you and your parents agree on the best way to handle each problem?
   We agreed completely. 5 4 3 2 1 We didn't agree at all.
2. Compare the consequences that you listed in class with the consequences that your parents brought up. What consequences did they think about that had not occurred to you?
3. What did you learn from this activity?

EXTENSION ACTIVITY

Are you having peer problems? If so, use the Decision Making model to try to solve one or more of the problems. You may discover that you have more options than you realized. And, remember, if one of your choices does not work, you can go back and try another.

The Problems

Basketball, Not Girls

Lately, my friends are more interested in girls than in basketball. We used to go to the gym every Friday. Not anymore. Most of my friends want to hang around with girls instead. My friend John wants to go to the shopping mall to meet some girls this Friday instead of playing basketball. I feel like I'm losing him as a friend. What should I do?

Steve

Getting His Attention

It seems like all my friends are talking with boys all the time. How do I get guys to notice me? Some girls flirt and giggle a lot. I don't. I try to be myself but the guys ignore me. There's this one guy in class who seems really nice, but he never looks at me. I don't even think he knows my name. How can I get him to talk to me?

Katie

Tongue-tied

I'd like to talk to Becky. In fact, I'd really like to sit with her at the game on Friday, but I don't know what to say. She doesn't seem to know I exist. Besides, she always has her friends with her and I don't want to talk to them too. I'm too tongue-tied to start a conversation with her. What should I do?

Matt

Too Young

All my girlfriends are allowed to go on dates, but my parents say I'm too young. I think they're unreasonable and old-fashioned, and I've told them so. The other day I told them I was going out with my girlfriends, and then we met some guys and went out in a group. We are supposed to do this again Saturday. I don't like lying to my parents, but I really want to go. What should I do?

Gena

CHAPTER 20 REPRODUCTION

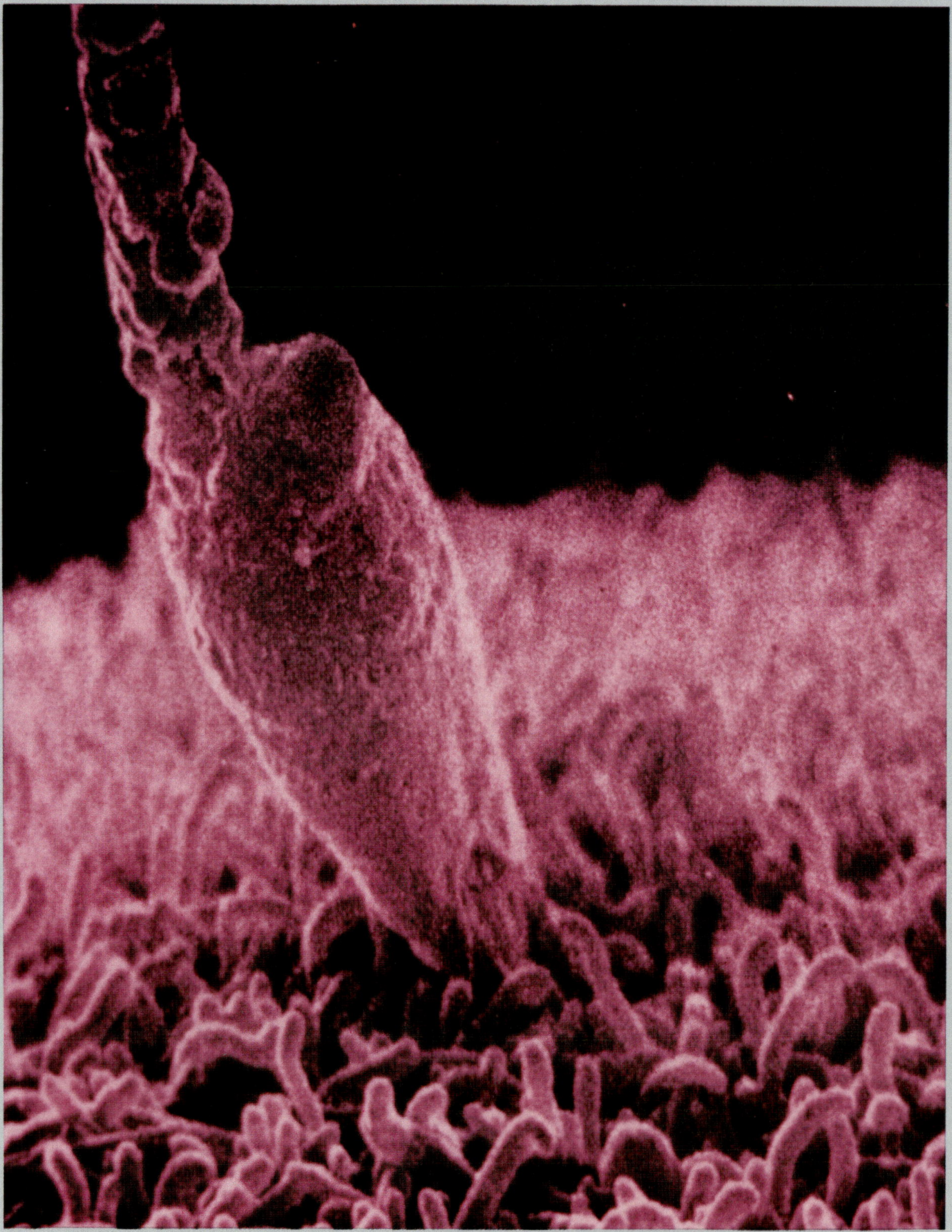

How many times larger do you think this egg is than the sperm?

20–1 UNIQUELY MALE

After resting for ten to twelve years, a boy's reproductive system seems to "wake up" during puberty. Not only do the sex organs—both external and internal—start to grow, they also take on new functions. The onset of these new functions can be startling and maybe even scary if a boy does not know what to expect. As you read the next several pages, you will learn about these functions—about what they are and why they happen.

Erections

There is one event that first occurs long before puberty. This event is called an **erection.** Males of every age, including tiny babies, experience them. In fact, a male has his first erection before he is born. During an erection, the penis stiffens, becomes larger, and stands out from the body. Because the penis becomes so much firmer than it usually is, many people mistakingly assume that it contains a bone.

Actually the penis is not made up of bone or muscle. Instead, it contains three cylinders of spongy tissue that is laced with blood vessels. This tissue runs the length of the penis. (See Figure 20–1.) During an erection, blood enters the vessels in the penis causing

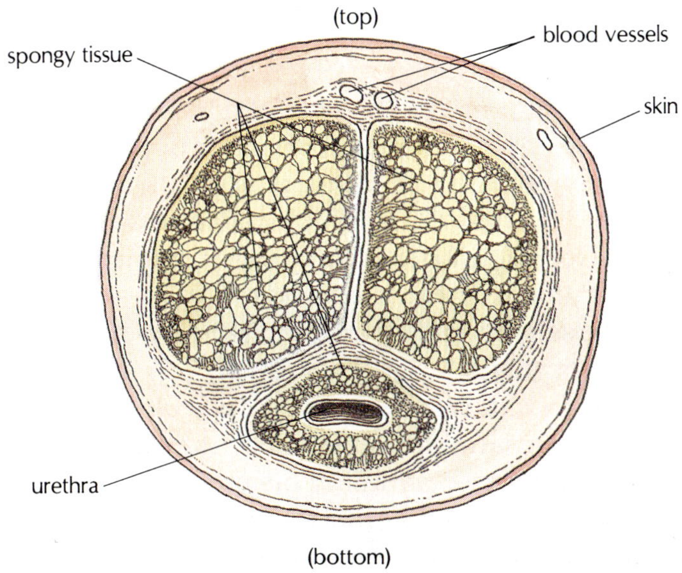

Figure 20–1 Cross section of a penis

the tissues to swell and the penis to stiffen. All of this can happen almost instantly or it may take several seconds.

Erections happen for all sorts of reasons. Getting nervous or excited can cause an erection; so can pleasant, warm thoughts. Thinking about sex or about the sex organs is the most common cause of an erection. Some erections, however, have nothing to do with thoughts or feelings. They may be caused by a full bladder or by tight clothing or they may occur for no obvious reason whatsoever. They can happen any time of the day or night. Erections are just another normal body function like any other body function, except they are unique to males.

Many boys notice that once they start puberty they have more erections than they did before; however, it is also normal for boys not to notice any change in their frequency. Some males experience more erections than do others. They may have them once a month or many times a day. Whatever number of erections a male experiences will be normal for him.

Most erections simply "go away." That is, the excess blood flows back out of the penis so it becomes limp again. This may take a few seconds, a few minutes, or even a half hour or so. Other erections end following an ejaculation.

Ejaculation

Unlike erections, ejaculations do not happen in babies or young boys. A male does not ejaculate until a year or two after he has started puberty. For a boy, his first ejaculation is a sign that his body is developing into that of a man. During an **ejaculation,** about 3–6 ml of fluid is propelled from the end of the erect penis. (5 ml is about a teaspoonful.) This whitish fluid, called **semen,** is a mixture of sperm, nutrients, and water. Immediately after ejaculation, the erection starts to go away. The pleasurable sensation that accompanies an ejaculation is called an **orgasm.**

Many boys experience their first ejaculation as a result of masturbating. That is, a boy may be touching, stroking, or otherwise handling his erect penis when suddenly the fluid buildup inside his body is released. It is normal for boys to masturbate; it is also normal not to. For years, people thought masturbation could be dangerous. Now researchers have found that it is not harmful to the body. Even so, some people object to masturbation. You probably know how your family feels about the subject. If you don't, ask them. They may be startled by your question but they usually appreciate it when you take the time to discuss topics with them that can be controversial.

While erections can occur just about any time, ejaculations don't. Semen will not appear unexpectedly following an erection except occasionally when a boy is sleeping. An ejaculation that takes place when a boy is sleeping is called a **wet dream.**

Wet Dreams

A boy's first ejaculation is often a wet dream. Even if he has heard about wet dreams, the first time it happens is usually startling. He is likely to wake up and find that his skin, pajamas, or sheets feel wet and sticky. If he does not know what to expect, he may think he has wet the bed. However, the fluid does not look or smell like urine and there is not nearly as much of it. Even so, many boys are embarrassed although there is no need to be; wet dreams are another normal part of growing up. Some boys have them frequently. Other boys may never have a wet dream and that, too, is normal.

Another term that you may hear for wet dreams is nocturnal emission. Nocturnal means "during the night" and emission means "sending something on." So nocturnal emission refers to semen that is sent forth, or ejaculated, at night.

Semen

Many people ask, "Where does semen come from?" The answer to the question is easy if you remember that semen is a mixture of sperm and fluid. Sperm are produced in the testes. They then spend several weeks traveling slowly through the epididymis before moving into the vas deferens. When they reach the end of the vas deferens, the sperm can be stored until they are released from the body.

When a male ejaculates, smooth muscles around the prostate and other parts of the reproductive system

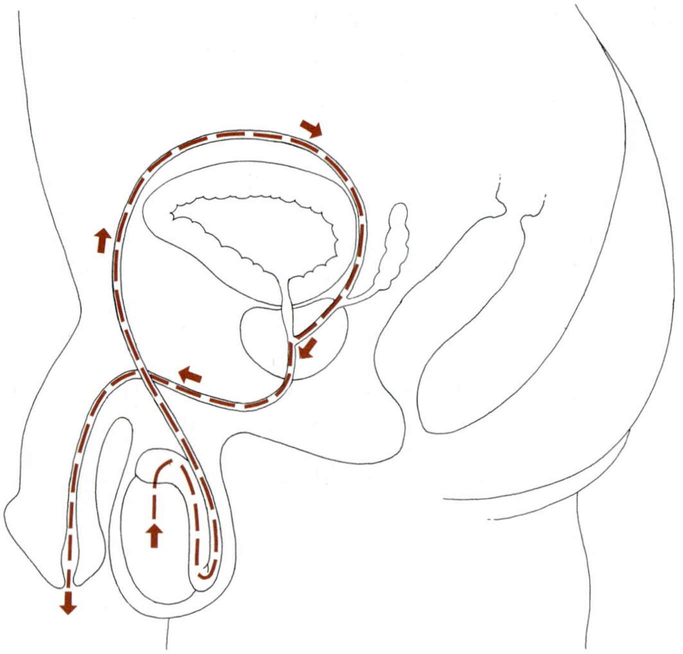

Figure 20–2 This drawing shows the path sperm follow on their way out of the body. Make sure you can name each of the structures that the sperm passes through.

contract, forcing the sperm out of the vas and through the urethra. These contractions also squeeze fluid from the prostate and other glands. This fluid mixes with the sperm to form semen. As soon as the fluid mixes with the sperm, the sperm become much more active and start to wiggle and "swim" along with the fluid as it is propelled through the urethra. The sperm pathway is illustrated in Figure 20–2.

A Continuing Process

Although the production of sperm and the onset of ejaculations are a sign that a boy is maturing, they do not mean that he is a man yet. His body must continue to grow for several more years until he reaches physical maturity. It also takes many years of living and learning before he will be ready to take on the responsibilities of an adult.

ANALYSIS QUESTIONS

1. What is the difference between an erection and an ejaculation?
2. What is **semen**?
3. List these structures in the order that sperm pass through them on their way out of the body: vas deferens, urethra, testes, epididymis.

Decide if each of these statements is true or false. If it is false, rewrite the statement so it is correct.

4. T F An erection results when the penis muscle contracts.
5. T F Erections are always caused by sexual thoughts.
6. T F Ejaculation always follows an erection.
7. T F An erection always comes before an ejaculation.
8. T F Once wet dreams have started, they occur about once every month.
9. T F An erection is a sign of sexual maturity.

EXTENSION ACTIVITY

Look up semen in a biology book, medical book, or encyclopedia. Find out what each gland contributes to the semen. Show where these glands are located on a diagram of male internal anatomy.

Animal Sperm

Many animals reproduce sexually. That is, the females of the species produce eggs and the males produce sperm.

All sperm look similar but the heads of the sperm are often quite distinctive. Within the head of each sperm is the nucleus. This nucleus, along with that of the egg, determines what kind of offspring is produced.

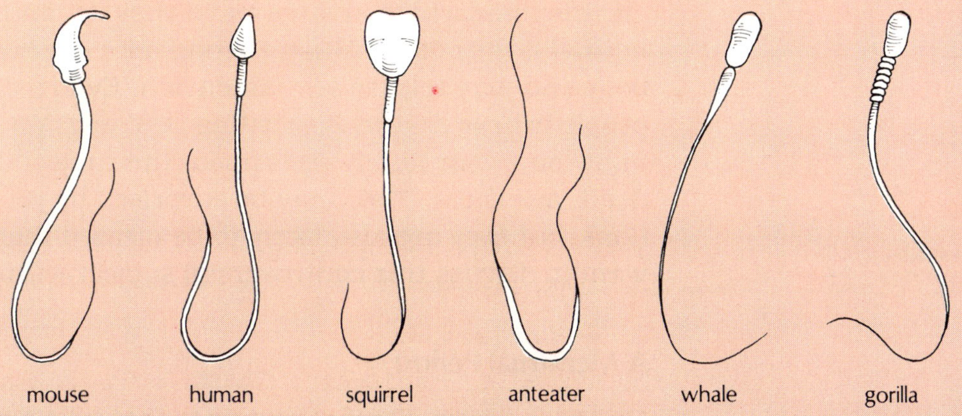

20-2 UNIQUELY FEMALE

Once puberty starts, a girl becomes aware that her body is changing. She can see the changes that involve body shape and size, those that involve functions are less noticeable. Both girls and boys have many questions about these female functions. This reading will answer some of the questions, but probably not all of them. If you think of other questions while you are reading, jot them down so you can ask them later.

Vaginal Responses

Early in puberty, the cells in the vagina become more active. They start to produce small quantities of fluid that keeps the lining of the vagina clean and healthy. This fluid is clear, with perhaps a whitish or a pale yellow tint.

Later on in puberty, a girl may notice that there are some times when she produces more of these fluids than usual. Some of this variation may be due to her awakening sexuality. Watching a romantic movie or thinking about sex may cause more fluid production. Sexual thoughts can also trigger stronger reactions such as a pleasurable tingling or a throbbing sensation in the genitals.

These feelings may also be a result of masturbation. That is, a girl may be touching or stroking her labia and clitoris when she notices these feelings. Not all girls masturbate. It's normal if they do, and also normal if they don't. Masturbation does not harm the body and may release sexual tension. Even so, some people believe that masturbating is wrong. You may know how your family feels about the subject. If you don't, ask them. They may be surprised by your questions, but they are also likely to be pleased that you want to discuss this controversial subject with them.

A Menstrual Period

One of the most significant events that tells a girl she is becoming a woman is her first menstrual period. This usually occurs sometime between the ages of 11 and 14. However, some girls may begin as early as 8 while others may not start until they are 17. Many girls start to menstruate at about the same age that their mothers started. If a girl's mother started her periods when she was 15, the girl, too, is likely to start around age 15.

When people talk about a **menstrual period** or **menstruation,** they are referring to the loss of a few tablespoons of blood and mucus from the vagina. This discharge does not happen all at once; instead, it is lost a little at a time, over several days. There is then a span of 3 to 4 weeks before a girl has another period. This pattern repeats itself over and over again.

3–6 days menstrual flow	21–28 days no flow	3–6 days menstrual flow	21–28 days no flow

This pattern reflects events that are going on inside the body. These internal changes are known as the **menstrual cycle.** The next reading explains some of the biology of this cycle.

Menstruation is a normal body process. Most of the time it does not affect a girl's activities or her daily life. During her period, she can go ahead and do all of the things that she normally does. Sometimes, however, girls have cramps just before or during their periods; some get cramps more often than others. Cramps may be very mild or they can be fairly uncomfortable. No one knows for sure what causes the cramping. The discomfort usually can be relieved by a little heat such as

Figure 20–3 Swimming and other exercises can be continued during menstruation.

from a hot water bottle or by a mild pain reliever such as aspirin. Exercise helps, too. The movement probably stretches and relaxes the muscles in the lower abdomen, and this somehow eases the cramping. Girls who have painful cramps should talk to their physicians; sometimes a stronger medication is recommended.

There has been a lot said about the way periods make a person feel. The feelings are varied. Some girls and women feel more energetic or happier during their periods. Others feel quieter, more restless, or even "out of sorts with the world." A period, however, may be thought of as a kind of message or signal that all is well with the female body.

A Continuing Process

The whole process of growing up is pretty amazing. While the start of the menstrual cycle is a sign that a girl is maturing, it does not mean that she is a woman yet. Her body must continue to grow for several more years until she reaches physical maturity. It will also take a lot more living and learning before she will be ready to take on the responsibilities of an adult.

ANALYSIS QUESTIONS

1. What is the difference between a menstrual period and a menstrual cycle?
2. Why does the vagina produce fluid?

Decide if each of the statements below is true or false. If a statement is false, rewrite it so it is correct.

3. T F A menstrual period usually lasts 3 to 6 days.
4. T F A menstrual cycle usually lasts 3 to 6 days.
5. T F A girl can go swimming during her period.
6. T F A girl can run, play tennis, or do other active exercises during her periods.
7. T F Periods are caused by sexual thoughts.
8. T F A girl may feel weak and shaky from loss of blood during her period.
9. T F A girl knows that something is wrong if she gets cramps during her period.

EXTENSION ACTIVITY

Find out more about menstrual periods. The book *Period*, by Loulan, Lopez, and Quackenbush, is short and interesting reading. It includes basic information as well as answers to many of the questions that people ask about menstruation.

Sanitary Pads and Tampons

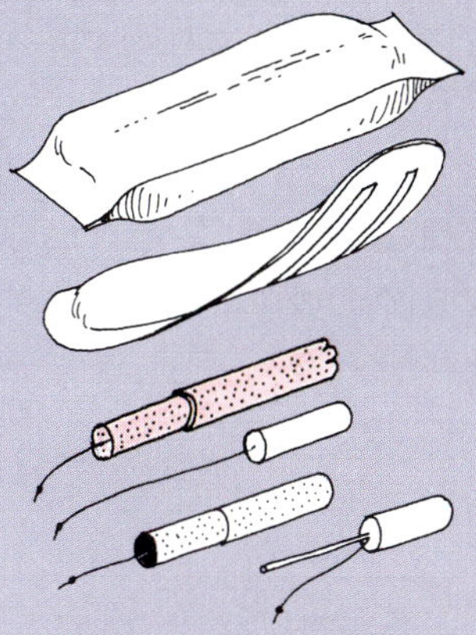

Sanitary pads or tampons are used to absorb menstrual flow. Both products come in a variety of sizes and absorbencies. When the flow is heaviest, a girl may choose a thick pad or a highly absorbent tampon. When the flow is lighter, a thin pad or slender tampon is more appropriate. There are advantages to both products so it is a matter of individual choice which one to use.

Tampons are inserted into the vagina where they fit snugly and securely. They can't slip out nor can they work their way into the uterus because they are blocked by the cervix. They are easily removed by pulling on the string that is attached to the end of the tampon. The first few times a girl uses tampons she may feel awkward but eventually using tampons becomes routine.

Some people are concerned about a rare disease that has been associated with tampons. This disease, called toxic shock syndrome (TSS), is caused by bacteria. Some of its symptoms are like those of a flu—fever, vomiting, diarrhea, headaches, and aching muscles, but, unlike a flu, TSS is accompanied by a rash. Somehow, tampons may make it easier for the bacteria to grow. However, if the tampons are changed at least every 4 to 6 hours, the risk for TSS is very low. TSS also is not unique to menstruating women; anyone can get it.

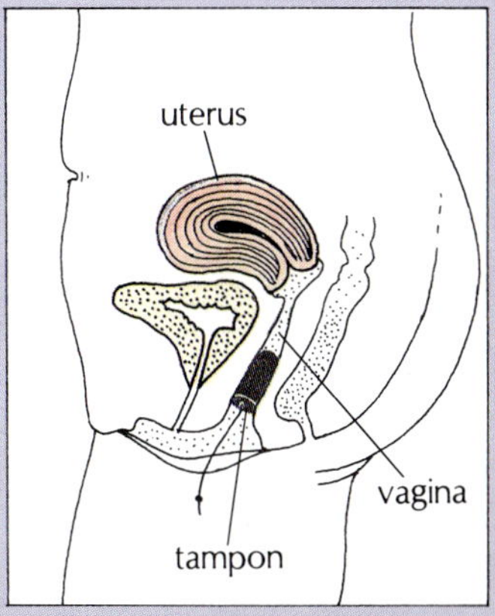

20-3 THE MENSTRUAL CYCLE

Menstruation is the part of the menstrual cycle that we usually hear about. However, it is relatively short, lasting only a few days. Do you know what is going on during the rest of the menstrual cycle?

Changes in the Uterus and Ovaries

Even though the menstrual cycle involves both the uterus and the ovaries, the changes in the uterus are the most obvious—especially the changes that involve its lining. The uterine lining is unique because it changes constantly; it is always building up or breaking down.

The lining is thinnest immediately after a menstrual period. Over the next several weeks it gradually becomes spongier and develops a network of new blood vessels. This thicker tissue is capable of nourishing a growing embryo. However, if it is not needed, it breaks down and passes out of the body through the vagina. The blood and mucus that is lost during menstruation comes from this lining. By the end of a menstrual period, the uterine lining is thin once again and ready to start a new cycle. (See Figure 20–4.)

While the uterine lining is developing, an egg is maturing in one of the ovaries. As the egg matures, it forms a tiny bulge on the surface of the ovary. This bulge continues to swell until it bursts, releasing the egg. The rupturing and release of an egg is called **ovulation.** Most females do not feel anything when this happens. However, some women do feel a slight twinge or cramp that could indicate ovulation. (See Figure 20–5 on page 523.)

Figure 20–4 During a menstrual cycle, the lining of the uterus gradually thickens and then is lost.

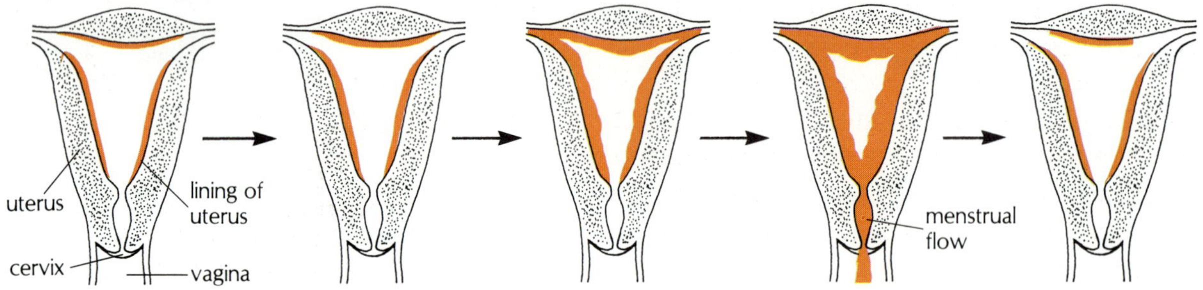

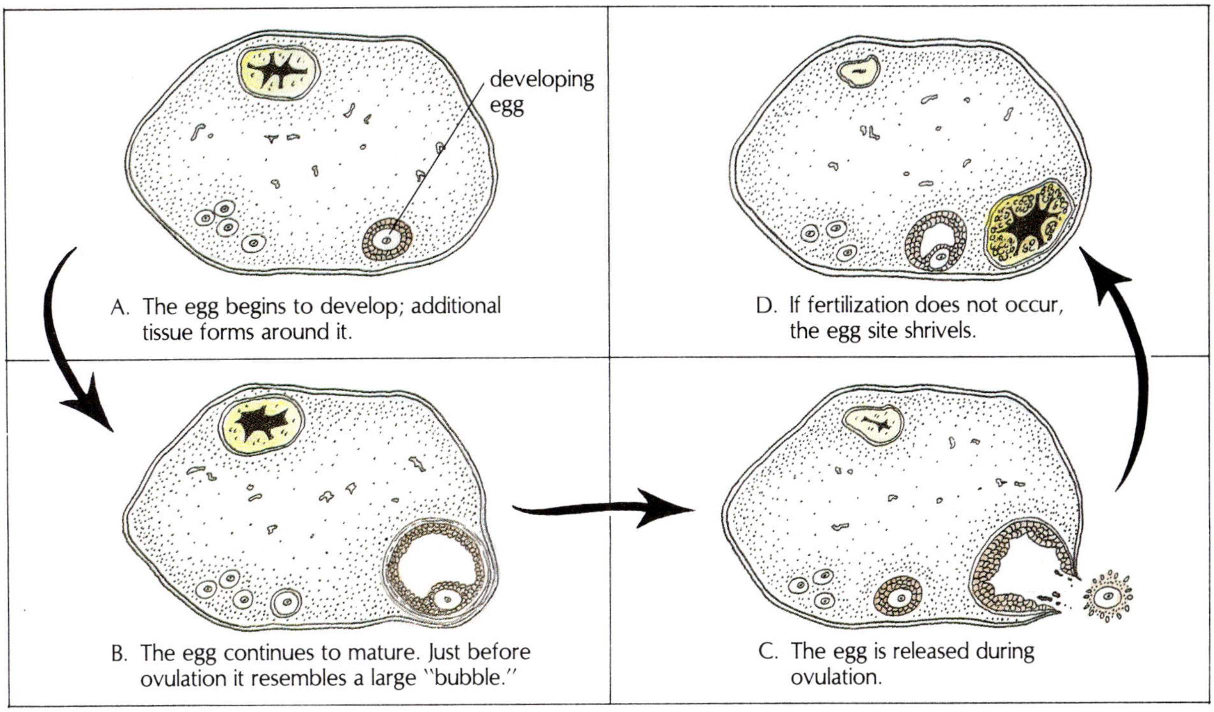

Figure 20–5 During every menstrual cycle, an egg develops in one of the ovaries.

After ovulation, a small pit is left on the surface of the ovary. It will gradually heal until only some scar tissue remains. Meanwhile, the egg is drawn into the fallopian tube and moves along to the uterus. This journey takes several days. Since the egg cannot live for much more than a day, it usually dies somewhere along the way. Within a week or two, another egg will start to develop.

Coordination

The changes in the lining of the uterus and the development of an egg are closely coordinated. This guarantees that if a fertilized egg reaches the uterus, the lining will be rich with blood. If the egg dies, the lining must break down. Then, as another egg matures, it must build back up again.

These changes are coordinated by hormones from the ovaries and the pituitary gland. (The pituitary gland is located on the underside of the brain. See Figure 20–6.) The pituitary hormones cause the egg to mature. As the egg matures, the tissues around it produce another hormone—estrogen—that causes the uterine lining to develop. In this way the lining is thickest when the egg arrives.

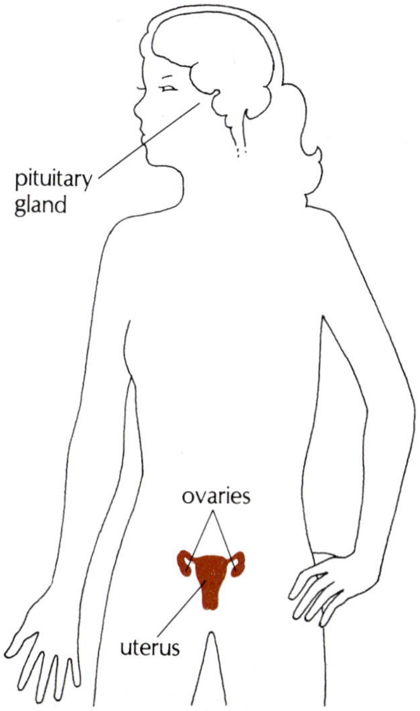

Figure 20–6 Hormones from the pituitary gland and ovaries coordinate the menstrual cycle.

Variation

The length of the menstrual cycle varies from one individual to another. While some women may have cycles that last about 21 days, the cycles of other women are longer, lasting up to 40 days. The average length of a cycle is around 28 days or about one month.

Some women are very regular; that is, their cycles are always the same length, such as every 26 days or every 30 days. Many women, however, have cycles that vary from month to month; they may have a cycle length of 27 days one time, 31 days the next, and then 29 days for the next three cycles. When a girl first starts to menstruate, her cycles are likely to be very irregular. It is not unusual for a girl to skip a cycle or two completely. It may take months or even years before a regular pattern develops.

Days in a Cycle

If you understand the menstrual cycle, you can predict what is happening in the uterus and ovaries on any particular day. Figure 20–7 summarizes the daily changes that go on during a 30-day cycle. You can see that, on day 1, the uterine lining starts to break down

Figure 20–7 For any given day of the menstrual cycle, you can predict what is happening in the uterus and ovaries.

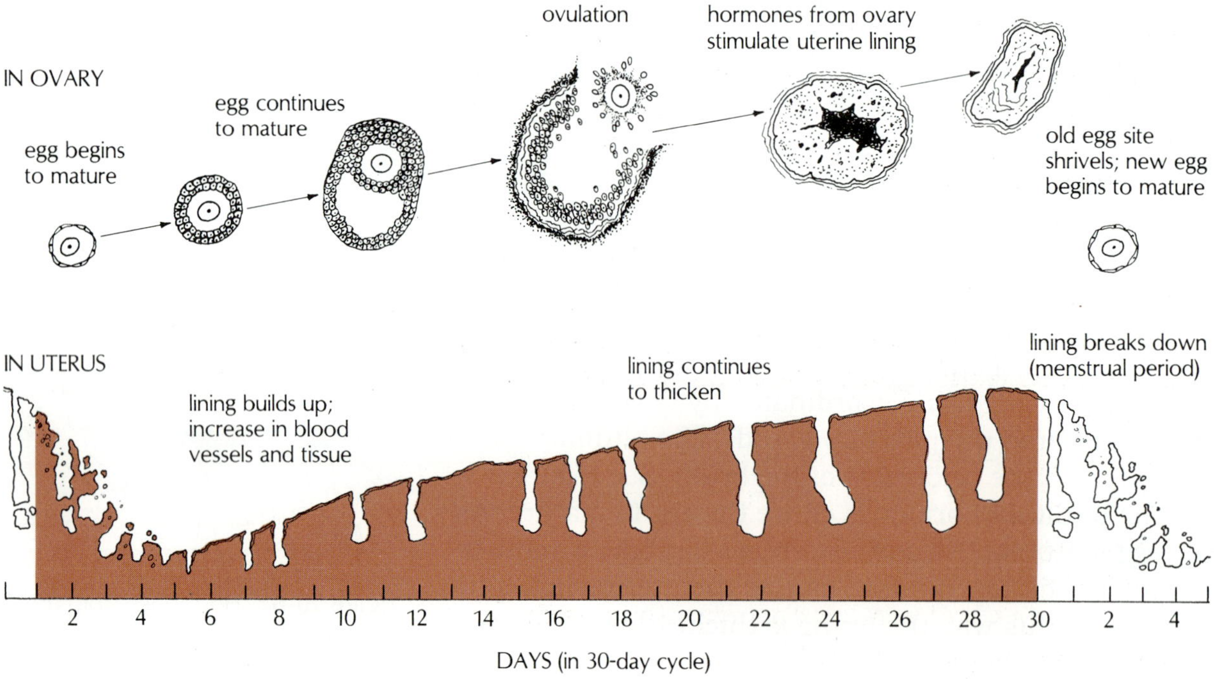

and an egg starts to develop in the ovary. When the egg is mature, ovulation occurs. Two weeks later, the lining of the uterus is lost. Study the rest of the figure. See if you can figure out what is happening on any given day of the menstrual cycle.

When the Cycle Stops

The cycle continues month after month, year after year, unless a woman becomes pregnant. A pregnant woman does not have menstrual periods since the uterine lining is needed to nourish the growing embryo. After the baby is born, the mother's menstrual cycle starts again.

Eventually, between the ages of 40 and 55, a woman's menstrual cycle will come to an end. It may stop all at once or her cycle may become irregular for a while before it stops. This time in a woman's life is called **menopause.**

ANALYSIS QUESTIONS

1. Summarize the changes that take place during one menstrual cycle in
   a. the ovary.
   b. the uterus.
2. Define the word **ovulation.**
3. Explain what is happening in the uterus and ovary on each of these days of a 30-day cycle. (Use Figure 20–7 on page 524 to help you answer this question.)
   a. day 9
   b. day 18
   c. day 27
4. If a woman menstruates every 27 days, on what day of her cycle is an egg most likely to be released from an ovary?
5. During which days of a menstrual cycle is there most likely to be an egg in a fallopian tube?
6. Write a complete answer to this girl's question, "I'm really worried. I got my first period 2½ months ago and I haven't had another one since then. Is there something wrong with me?"

Estrous Versus Menstrual Cycles

Most animals will mate only when the female is ovulating or in "heat." This fertile period is known as estrus. The time from one estrus to another is called an estrous cycle.

While a menstrual cycle and an estrous cycle are similar, they are not exactly the same. During a menstrual cycle, the uterine lining is shed if the egg is not fertilized. In an estrous cycle, the uterine lining simply shrinks back to its original size; there is no discharge. The bleeding that occurs in some animals right before they are in heat is caused by red blood cells that pass through the wall of the vagina.

The length of an estrous cycle varies from species to species. Dogs go into estrus twice a year, cows and horses once a month, and mice, every few days! Wild animals usually come into heat so the young will be born when environmental conditions are most favorable. For instance, in wolves and deer, estrus occurs only in the fall so the young will be born in the spring.

20–4 FERTILIZATION: A NEW LIFE BEGINS

In many ways, sperm and eggs are like every other body cell that you've studied. For instance, they too are enclosed by cell membranes and they contain nuclei. They also have specialized functions and characteristic shapes that you can recognize.

There is, however, one major difference between eggs and sperm and other body cells. When an egg and sperm combine, a baby will develop. This does not happen with other types of cells. One or two skin cells cannot grow into a new individual; nor can bone cells, muscle cells, or any other cells. Only eggs and sperm have this amazing ability.

In this reading you will learn about these two kinds of cells and about the process that brings them together to form a new life.

Sperm

A sperm is one of the smallest human cells. As you can see in Figure 20–8, it looks somewhat like a tadpole with an oval head and a thin tail. The "head" contains the nucleus of the sperm cell. The tail can whip rapidly back and forth to move the sperm forward.

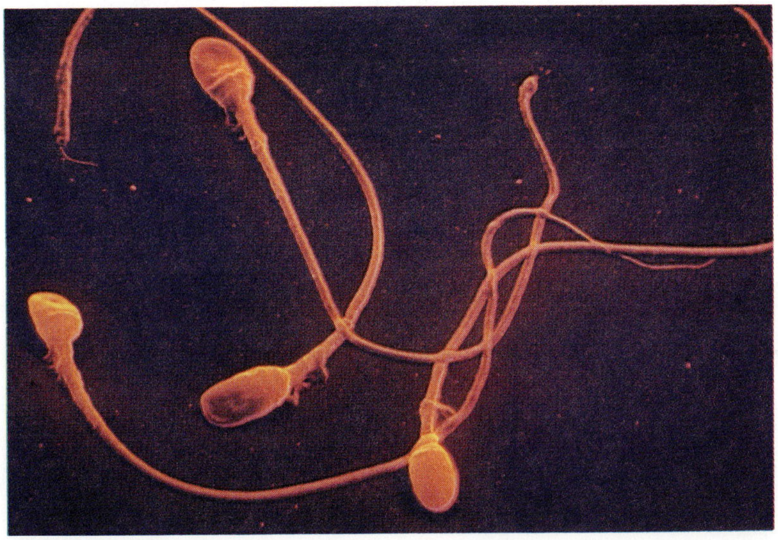

Figure 20–8 This photograph of a human sperm was taken through a microscope. Compare it with the drawing of the sperm.

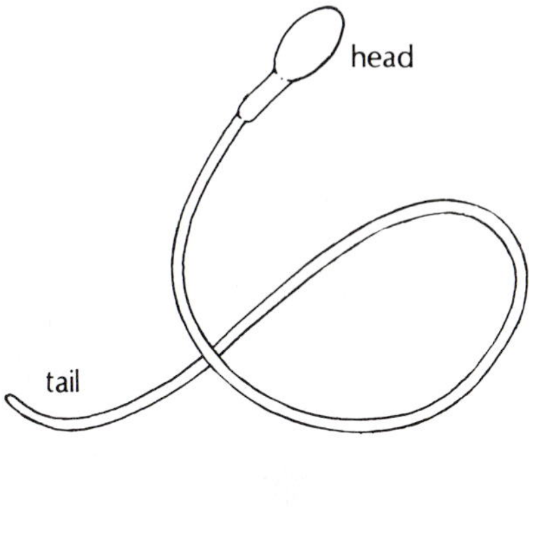

Sperm are produced in the coiled bundles of microscopic tubules that make up a testis. If you could uncoil the tubules, they would extend for about 500 meters. This provides plenty of room for making sperm. Sometime during puberty, the testes will produce enough testosterone to trigger sperm production. Eventually, each testis produces about 75 million sperm every day.

When sperm leave the testis, they are not fully formed. They spend the next 50 or so days in the epididymis where they complete their development. The mature sperm then move on to the vas deferens where they are stored until they either disintegrate or are released from the body.

Eggs

A female is born with all of the eggs she will ever have. At birth, there are about a million eggs in her ovaries. Over the next several years, many of these egg cells die so that, by puberty, fewer than half—perhaps about 400,000 of the original eggs—still remain. Most of these will eventually disintegrate, too.

Figure 20-9 This human egg was photographed through a microscope. Compare it with the drawing of an egg. What can you learn about an egg cell from looking at the photo that you cannot see in the drawing?

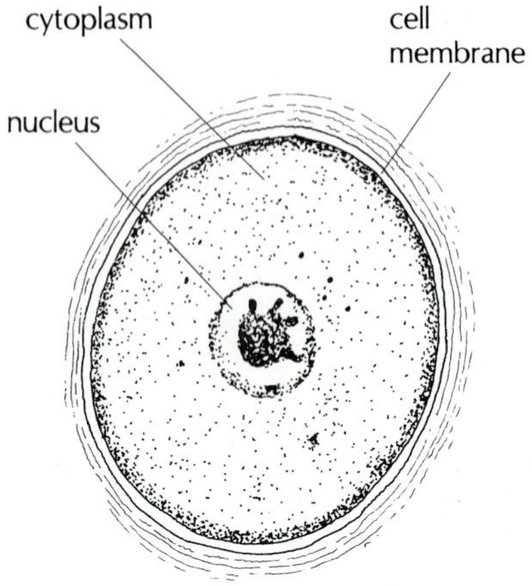

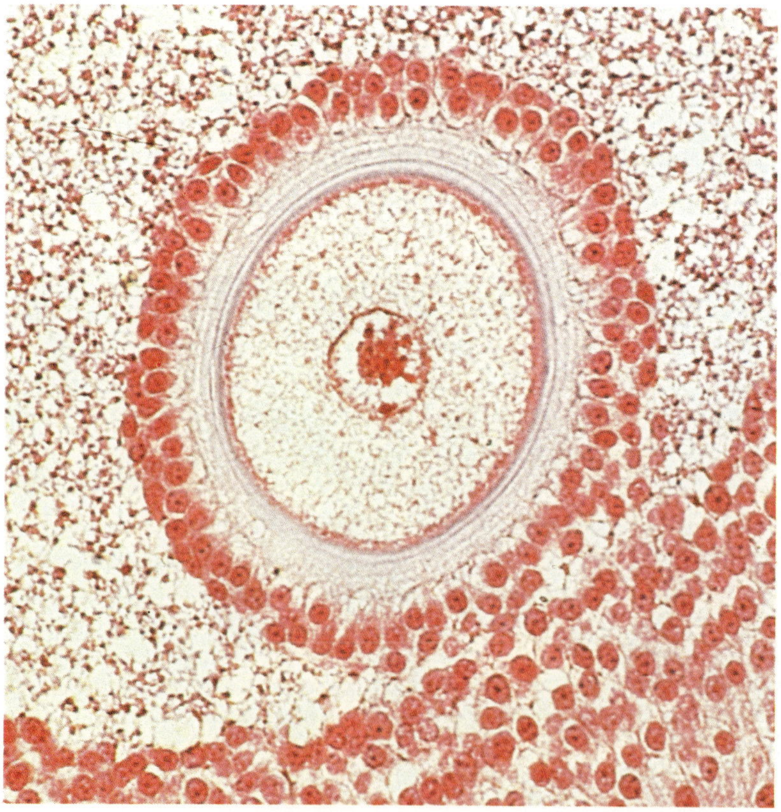

Once a girl enters puberty, one egg cell matures and is released from the ovary about every 28 days. During a woman's lifetime, only about 400 eggs ever mature. This is significantly different from the male system where millions of sperm are produced daily.

Even though an egg is thousands of times larger than a sperm, it is barely visible without a microscope. (See Figure 20–9 on page 530.) Compared with other cells, the cytoplasm of an egg contains a fairly large supply of nutrients. If the egg is fertilized, these nutrients will be used to nourish the embryo. When an egg is released from the ovary, it is drawn into a fallopian tube and moves toward the uterus. It usually dies and starts to break down somewhere along the way, since this "trip" takes several days but an egg lives only about 12 to 24 hours.

Fertilization

Sometimes the egg does not die; instead, it is joined by a sperm and a new life begins. This joining of egg and sperm is called **fertilization.** Fertilization usually results from sexual intercourse. During sexual intercourse, the man's penis becomes erect and is inserted into the woman's vagina. Semen, containing about three hundred million sperm, is then ejaculated close to the cervix. Many of the sperm die almost instantly. The survivors, their tails whipping rapidly, are propelled into the uterus. They move at a speed which would be about the same as that of a swimmer covering twelve meters in a second. Many more sperm die in the uterus; only a few hundred ever reach the fallopian tubes. The dead sperm cells break down and are cleaned up by white blood cells as are all other dead cells in the body.

If there is an egg in one of the fallopian tubes, hundreds of sperm surround it. One sperm may cross the egg's membrane. If so, the surface of the egg changes immediately, making it impossible for other sperm to enter. In this way, only one sperm can fertilize an egg.

The dot-sized fertilized egg divides as it continues through the fallopian tube. (See Figure 20–10.) It also produces a chemical that causes the uterine lining to become even richer with blood. If it weren't for this chemical, the lining would be lost during menstruation.

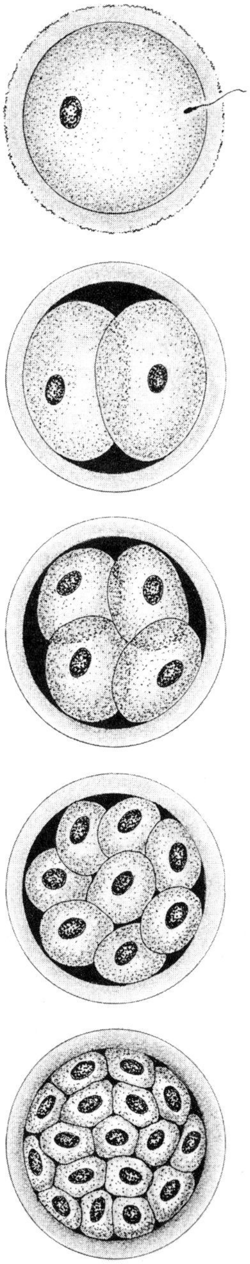

Figure 20–10 Soon after the egg and sperm join, the fertilized egg divides into two cells, then four, and so on. After several days, it is a "ball" of cells.

It's almost as if the fertilized egg is sending a signal that the lining is needed because an egg is on the way. Once it reaches the uterus, the fertilized egg burrows into the spongy lining where it continues to develop into a baby.

More Than "Sex"

Sexual intercourse was mentioned very briefly a few paragraphs ago. It is important to spend a little longer discussing it because no other behavior causes the same emotions or the same strong feelings. Ideally, sexual intercourse is an expression of strong love and respect between a husband and wife. Their shared love is one of the deepest there is. It involves a sharing of more than just a physical attraction. In a marriage, many things are shared—a home, money, good and bad times, interests, and a deep commitment to each other. Under these circumstances, sexual intercourse is an expression of this deep love.

Under other circumstances, however, sexual intercourse can have some serious consequences. Some teenagers get carried away by their emotions and have intercourse without stopping to think about the consequences. Before you read on, close your eyes and think about the consequences. How many can you name?

Five of the most serious consequences are pregnancy, sexually transmitted diseases, emotional turmoil, depression, and parental disapproval. (How many of these did you think of?) Some of these have life-long impact. All of them are stressful and none will improve a relationship or make a person feel good about himself or herself. These are some of the reasons that teens your age should not be involved in sexual intercourse. **Abstinence**—not having sexual intercourse—is your best choice. There are many ways to enjoy a male-female relationship without "having sex."

Some students wonder when they are "supposed to have sex." There is not a magical age when the time is suddenly "right" for sexual intercourse. The early teen years are definitely too young. There are lots of things in addition to age that should be considered. If you have not already talked with your parents about "the right time to have sex," do so soon. They will have their opinions and reasons and will probably be happy to share them with you.

ANAYLSIS QUESTIONS

1. How many sperm does a male produce each day?
2. About how many sperm does a male produce in a month?
3. How many egg cells mature every month?
4. How many sperm are released in an ejaculation?
5. Define the word **fertilization.**
6. What happens to the sperm that don't fertilize an egg?
7. List 5 serious consequences that can result from sexual intercourse.
   a. Circle the consequence that you think is most serious.
   b. Put a box around the consequence that your parents would think is the most serious.
8. Do you think it would be possible for a female to get pregnant if she had intercourse only once? Explain your answer.
9. List at least 3 reasons why you think some teens have sexual intercourse despite the possible serious consequences that could result.

20-5 EGGS AND SPERM

Humans, as well as most other animals and many plants, reproduce sexually. Sexual reproduction requires two kinds of cells, usually from two parents. One parent usually produces fairly large cells that provide nourishment for the future embryo. These cells are called **eggs.** The other parent often produces smaller cells that can move about. These cells are the **sperm.** Since eggs and sperm are such special cells, they deserve a closer look.

- How can you recognize an egg cell on a microscope slide of an ovary?
- What do guinea pig sperm look like?

Look at microscope slides of an ovary and sperm to learn the answers to these questions. This activity will challenge your skill with a microscope since eggs and sperm are not easy to observe.

MATERIALS

worksheet: ***Eggs and Sperm*** (Figure 20–11)
microscope
microscope slides: guinea pig sperm
mammalian ovary

PROCEDURE

Work with a partner.

1. With your partner, read the two pages of the worksheet.
   a. Put asterisks by the questions that you think will be hard to answer.
   b. If you think you already know the answer to one or more questions, write it lightly with a pencil. (Change the answer later if you find that your answer is wrong.)
2. Look at the microscope slides. Fill in the appropriate part of the worksheet for each slide. You may want to reread parts of the reading ***Fertilization: A New Life Begins*** that starts on page 527 to help you answer some of the questions.

CONCLUSION

Based on what you have learned, write answers to each of the focusing questions. Be sure you give complete answers for each question.

EXTENSION ACTIVITY

Some plants and animals do not reproduce sexually. Instead, they reproduce asexually. Look up asexual reproduction in an encyclopedia or biology textbook. List several examples of organisms that reproduce asexually. Explain how they do reproduce since they do not produce eggs or sperm.

EGGS AND SPERM

NAMES ______________________

____

DATE ______________ PER. ______

Observe the microscope slide of guinea pig sperm while you answer these questions.

1. Make a sketch that shows exactly what you see through the microscope. Check if your sketch

   has a title. ______

   includes the magnification. ______

   is the size as you see it in the field. ______

   is neat. ______

2. Which of the following is closest to the size of a sperm? (Check one.)

   ______ virus ______ muscle cell ______ red blood cell

   ______ bacterium ______ egg cell ______ fat cell

CHALLENGE

Look at a slide of human sperm. Make a sketch that shows exactly what you see through the microscope.

Observe the microscope slide of a cross section through an ovary to answer these questions.

3. What magnification are you using to observe the ovary? __________

4. Look at this sketch to help you locate the egg. Can you find an egg on the microscope slide? __________

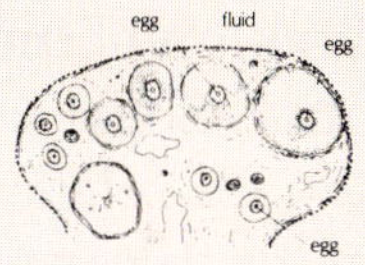

5. Why do you think they don't make microscope slides that show hundreds of eggs?

Figure 20–11 (Worksheet, page 1)

EGGS AND SPERM
(PAGE 2)

NAMES ______________________

____

Based on what you have learned about sperm and eggs, answer this last question.

6. Fill in the beginning of each sentence to show whether it applies to an egg, a sperm, both, or neither. You may want to change the first word in the sentence ending.

   ______________ is a single cell.

   ______________ contains lots of nutrients.

   ______________ has a tail.

   ______________ has a cell membrane.

   ______________ is barely visible without a microscope.

   ______________ is produced in tubules.

   ______________ contains many tiny capillaries.

   ______________ has a nucleus.

Figure 20–11 (Worksheet, page 2)

20-6 BIRTH CONTROL

Since students your age should not be involved in sexual intercourse, you don't need to know how to use birth control at this time in your life. But because living in our society you hear so much about birth control, you should understand what it involves.

As we explained earlier, intercourse is a caring and pleasurable way for couples to express their continued love for each other. However, a couple does not want a pregnancy to result each time they have intercourse. For this reason, most couples use some form of birth control.

The topic of birth control can cause some heated debates because people have very different ideas about whether or not it is right to use birth control. Some people strongly object to its use, often for religious reasons. Others feel that using birth control is necessary to limit their family size. You may already know how your family feels about this issue. If you don't, talk to your parents. They will probably be very willing to share their thoughts and beliefs with you. The following information may help you focus that discussion.

Family Planning

The desire to control family size is not new to the twentieth century. Over hundreds of years, people have tried a variety of methods to control pregnancies. One of the safest things that people tried was avoiding intercourse whenever they thought that the chance for pregnancy was high. Eventually, this evolved into what we now call **natural family planning** or **rhythm.** It involves not having intercourse when the woman is ovulating. This method is inexpensive and medically safe but its effectiveness varies.

Problems arise because the human body is not 100% predictable. As you learned, a woman's cycle can vary from month to month. This makes it hard to predict when ovulation will occur. It is also hard to predict how long an egg will live. About one out of every four couples who use rhythm will be expecting a baby by the end of one year. For this reason, some people say that it is not a good method of birth control.

Contraceptives

Other methods of birth control involve some kind of chemical or a device. These methods are sometimes called **contraceptives.** They work in a variety of ways. One contraceptive, the birth control pill, is often discussed. It prevents egg cells from maturing so ovulation does not occur. Since birth control pills contain hormones, they are available only by prescription.

Other contraceptives prevent the sperm and egg from joining. They work by blocking or killing the sperm before they reach the egg. Condoms, diaphragms, sponges, and spermicides are examples of this type of contraceptive.

Sterilization is a more permanent way to prevent sperm and egg from joining. When a male is sterilized, the *vas* deferens are cut. This procedure is called a *vas*ectomy. In females, the fallopian tubes are cut.

Intrauterine devices (IUDs) are another method of contraception. These devices must be placed in the uterus by a health care professional. They prevent a fertilized egg from implanting in the uterus.

The three most talked about and most frequently used contraceptives are condoms, spermicides, and birth control pills. They are explained briefly on page 536.

Another "method" that some people try is called **withdrawal.** It doesn't require any special devices and it doesn't cost anything, but it is very risky. In fact, it is so risky that people who use it are likely to become parents.

CONTRACEPTIVES

condom
spermicide
diaphragm
sponge
vasectomy (male sterilization)
tubal ligation (female sterilization)
IUD (intrauterine device)
birth control pill

Effectiveness

It is important to realize that none of these methods works 100% of the time. From the information about condoms, you can see that they are 90% effective. That means they fail 10% of the time; or, in other words, 10 out of every 100 couples who use this method for one year will end up expecting a baby. Condoms are more

Birth Control Methods

NATURAL FAMILY PLANNING

- involves timing of intercourse to avoid the "fertile days"
- requires careful planning and control
- 60–80% effective
- requires help to learn woman's cycle from doctor or trained professional

BIRTH CONTROL PILLS

- contain hormones that prevent release of an egg from the ovary
- must be taken regularly by the female
- are more than 99% effective
- require a prescription

CONDOMS

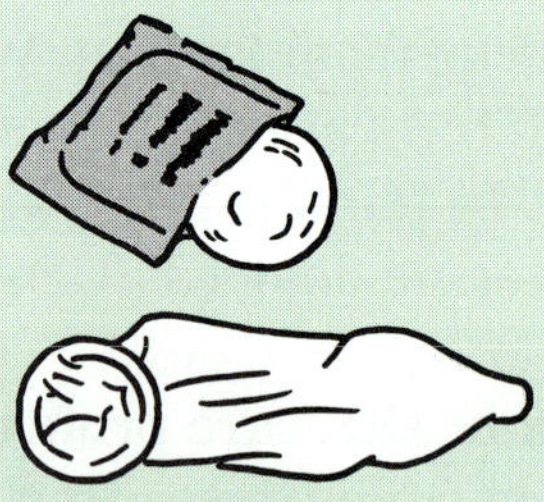

- covering for penis that holds sperm so they don't enter the vagina
- are 98% effective if used with spermicide; about 90% effective if used alone
- provides some protection from sexually transmitted diseases
- does not require a prescription

SPERMICIDES

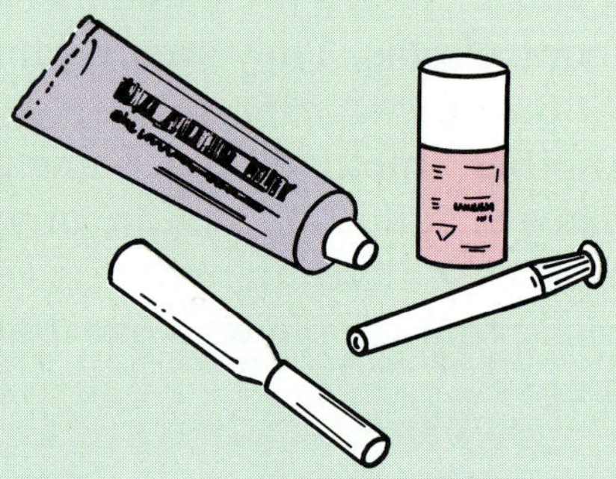

- foams, creams, and jellies containing chemicals that kill sperm
- are put into the vagina shortly before intercourse
- 98% effective if used with a condom; 75%-80% if used alone
- does not require a prescription

WITHDRAWAL

- penis is removed from vagina before ejaculation
- not very effective (about 60%)
- some sperm are usually released before ejaculation
- timing is very difficult because a man does not know when sperm will be released

effective if used with a spermicide. Of course, the effectiveness of any method will go down if it is not used correctly. Keep in mind that the only 100% effective way to prevent a pregnancy is not to have intercourse.

That's it. You should now understand the basics of birth control. For instance, you should know that the number of methods is limited. The methods that are mentioned here have been declared safe and effective by physicians and our government.* Any other method you may hear people talking about is probably based on their own misconceptions. Such methods are likely to be unsafe or ineffective.

*The Food and Drug Administration (FDA) is the governmental agency that monitors the safety of food and drugs and other health care products.

ANALYSIS QUESTIONS

1. Make a table like the one shown below on your own paper. It should have six blank lines for you to fill in.

BIRTH CONTROL

Method	Requires a Prescription?	One Important Fact	Rating

   a. Use the information from the reading to fill in the first three columns of the table.
   b. Use stars to fill in the rating column.
      *** = very effective
      ** = effective
      * = better than nothing
2. List at least 3 places someone could go to get more information about one of these methods.
3. What are 3 reasons people may not want to have children at a certain time in their lives and so may want to use birth control?
4. What are 3 reasons people may decide not to use any method of birth control?
5. What is the only 100% certain way not to get pregnant?

20-7 SEXUALLY TRANSMITTED DISEASES

STDs. Sexually **t**ransmitted **d**iseases. These diseases are passed from one person to another during sexual activity. Millions of Americans have caught them. They are even a major problem among teenagers. If untreated, they can cause serious health problems or even death.

- What are four examples of STDs?
- What are the symptoms of STDs?
- What treatments are available for STDs?

You will learn the answers to these questions by working in a small group to learn the information printed in an STD pamphlet.

MATERIALS

4 STD pamphlets
12 blank "question cards"
envelope

PROCEDURE

Work with a team of three other people. Each of you will be responsible for reading and then explaining one section of an STD pamphlet to the rest of your team.

1. Join the others in your team and look over the pamphlet. Decide who will read which part of the pamphlet. (If one section is longer than the others, ask the fastest reader to choose that section.)
2. Spend 5 minutes reading your part of the pamphlet. As you read, jot down any words you do not understand. List the main ideas from your section.
3. Join your "expert group"—all of you will have read the same part of the pamphlet.
   a. Choose someone to be the discussion leader and another person to be the noise monitor. (The noise monitor should make sure that the noise level stays down and everyone can hear what is being said.)
   b. Have each person take notes during the discussion.
   c. If you read something you did not understand in your part of the pamphlet, ask about it now. Someone in the group will probably be able to explain it.
   d. As a group, write a sentence that summarizes the main idea in your part of the pamphlet.
   e. Agree on 3 or 4 other important facts in your section that you think all students should know. Write down these facts.
4. Return to your team.
   a. Explain your section of the pamphlet to the rest of your team.

b. Listen and take notes while the other members of your team explain their sections.
c. If you do not understand something, ask questions! It is important that you learn the facts about STDs.

5. Spend the next 10 minutes writing quiz questions about STDs. Work as a team to think of these questions.
   a. For each section of the pamphlet, write 3 questions and their answers.
   b. When your teacher gives your team question cards, copy one question and its answer on each card.

EXAMPLE QUESTION CARD

Q: What do the letters S, T, D stand for?
A: Sexually Transmitted Disease

6. If you have a few minutes left, try out your team's questions with your team. Make sure that each person knows the facts about STDs. It will take teamwork and practice to win the quiz game you will be participating in later!
7. Have one member of the team put your team's questions in your team envelope and write the names of all the members of the team on the envelope. Give the envelope to your teacher to save until it is needed again.

EXTENSION ACTIVITY

Call your local or county health department. Ask to talk with someone who can tell you which STDs are the biggest problem in your city or county. Try to find out also which ones are the biggest problem in the state and in the nation.

20-8 AIDS

AIDS is a sexually transmitted disease that is currently receiving a lot of attention. Unlike other STDs, AIDS is an STD that cannot be cured and is almost always fatal. You should know about AIDS so that you can protect yourself from getting this deadly disease.

- How is AIDS passed from one person to another?
- How can a person protect himself or herself from getting AIDS?

Watch a film about AIDS and then work with your team to write AIDS questions.

MATERIALS

your team's envelope of STD questions
(See pages 538 and 539.)
6 blank question cards

PROCEDURE

Work with the same group with whom you read and discussed STDs.

1. Listen carefully while you watch the movie about AIDS. Take notes to help you remember the most important information.
2. Still working in your team, decide which facts are most important. Then, as a team, write 4 or 6 questions about AIDS for the STD quiz game.
3. Practice answering your questions with each other if you have any extra time!

EXTENSION ACTIVITY

For one week, cut out every article, cartoon, or editorial that has to do with AIDS. Read all of them and write a one- or two-paragraph summary of the most important AIDS news for the week.

CHAPTER 21

BIRTH

What do you think a baby's life is like a week before it is born?

21-1 FETAL DEVELOPMENT

A fertilized egg soon divides to form two cells; the two cells divide to form four cells, the four form eight cells, and so on until after nine months there are billions of cells that form a baby. Obviously, there is more going on than just cell division or else we would all be nothing more than huge clusters of cells. The cells also have to specialize and organize to form various body tissues and organs.

- When is the most critical period of development?
- What's the difference between an embryo, a fetus, and a baby?
- Where can you find information about fetal development?

Work in a group to research one stage of fetal development and summarize your findings on a poster which you will explain to the rest of the class.

MATERIALS

resource books
poster board or large sheet of paper
colored markers
glue, tape, scissors, and other miscellaneous materials for your poster

PROCEDURE

Work with a small group of students.

Part A. Poster Preparation

1. Record the stage of fetal development that your teacher assigns to your group, then read through these directions so you know what you will be doing.
2. Your group will need to research one stage of fetal development so that you can include the following information on your poster:
   - a title that indicates the age of the fetus
   - the actual size of the fetus at this time of development
   - important developmental happenings
   - a status report on the heart, brain, fingers, and toes
   - an illustration
3. Each member of your group should help with the research. Plan to do part of your research as homework.
4. Keep track of the resources you use. Record the title, author, and publication date for each of your resources on the back of your poster.

Part B. Poster Reports

1. One or two members of your group should be prepared to explain your poster to the rest of the class.

2. Take notes while you listen to the presentations by other students. You may want to make a table like the one shown below to help you organize your notes.

Age of Fetus	Size	Developmental Happenings

CONCLUSION

Write a short paragraph to answer each of the three focusing questions on page 542.

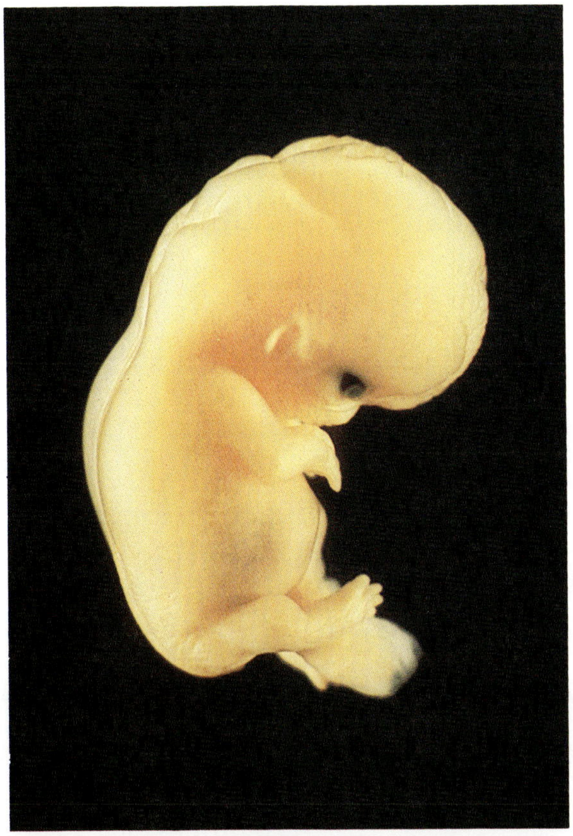

Figure 21-1 How old do you think this fetus is? What clues are you using to help determine its age?

EXTENSION ACTIVITIES

1. Call the newborn or intensive care nursery at a nearby hospital. If possible, arrange to visit the newborn or intensive care nursery or ask to talk with someone who could answer some questions for you. Here is a list of questions you might want to ask.
   - How early in its development can a baby be born and still survive?
   - Is it possible to tell by looking at a baby how many weeks into its development it was when it was born? (For instance, can you tell if a newborn is a 36-week baby or a 38-week one?)
   - What kinds of problems do you check for when examining a newborn?
   - Why is it hard for a baby who is very premature to survive?
   - What are some of the most common birth defects?
   - Why do some babies need to be placed in incubators?
2. Be a "mother" yourself—a mother to a chicken. Check with your science teacher to see if you can get fertile chicken eggs. Find out what you would need to do to incubate the eggs. If you think you can do it, try to hatch two or three eggs. Record what you are doing and note any changes in the appearance or weight of the eggs. Try to answer these questions:
   - How many days do the eggs have to incubate before they hatch?
   - How long does it take the chicken to get out of the egg?
   - What does a newly hatched chicken eat?

Which Is Which?

During the first few weeks of development, all embryos look very similar. Look at the embryos shown below. One is a human, one is a lizard, and one is a rabbit. Which is which?

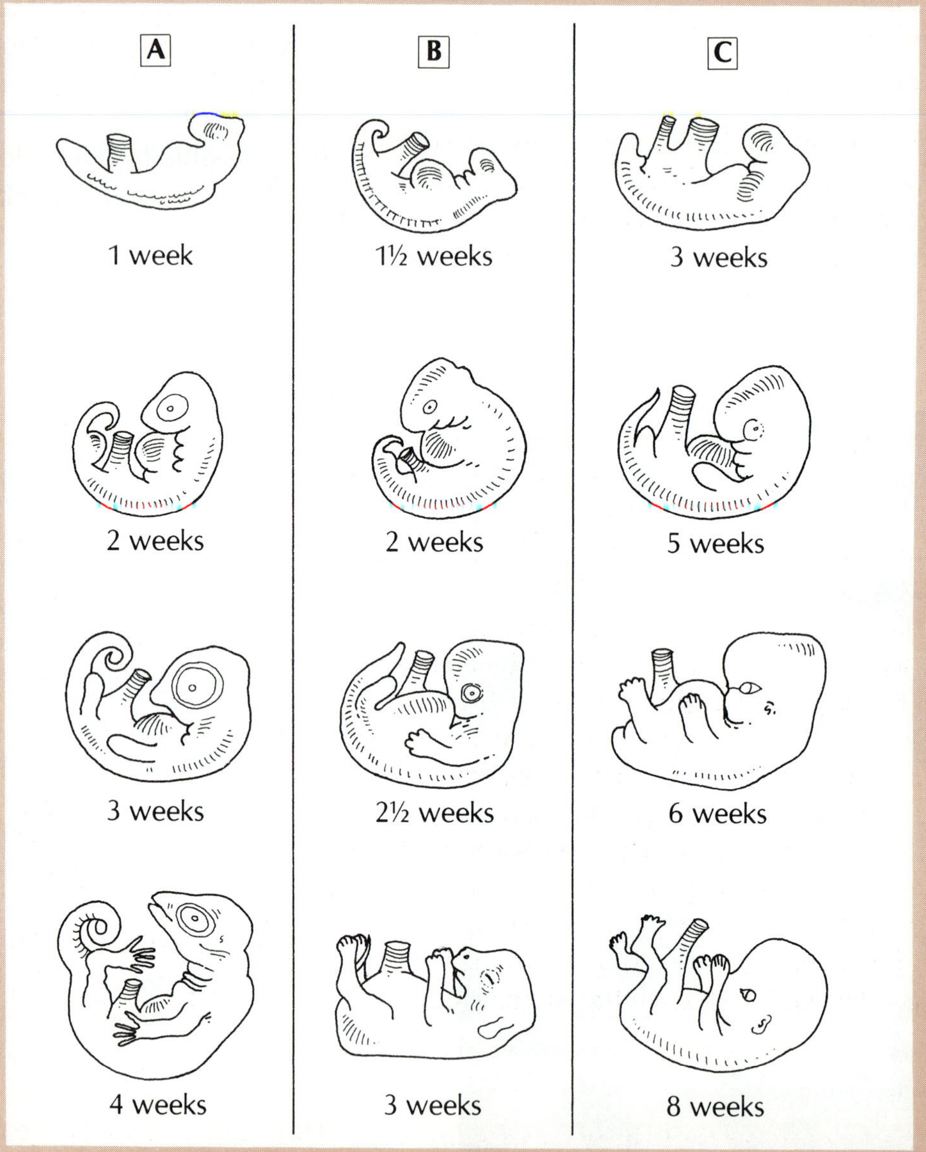

Answers: A—lizard, B—rabbit, C—human

21–2 PREGNANCY AND BIRTH

Pregnancy is a time of change for the mother, father, and the rest of the family, as well as for the developing baby. This reading will focus on a few of the hundreds of physical changes that the mother experiences during the nine months it takes the baby to develop.

Signs of Pregnancy

A woman cannot tell the moment she becomes pregnant; she has to wait for some "sign." This sign is often a missed menstrual period. A doctor then should confirm the pregnancy since there are other reasons for missing a period. By using a simple laboratory test and doing a physical exam, the doctor will be able to tell for sure if she is pregnant.

Frequent urination is another early sign of pregnancy. As the uterus grows, it presses against the bladder, producing the sensation that the bladder is full. This usually goes away but will return again during the last few weeks of pregnancy when the uterus is much larger.

In the early stages of a pregnancy, some women are affected by morning sickness. They may be nauseated and feel like vomiting. Even though it is called "morning" sickness, this queasy feeling may last throughout the day and into the night. It seldom continues much past the sixteenth week of pregnancy.

As time passes, there are other signs of pregnancy. A woman's breasts will grow and become more sensitive because the milk ducts are developing. And, by the fourth or fifth month, she has usually gained several extra pounds and noticed an increase in her waist size.

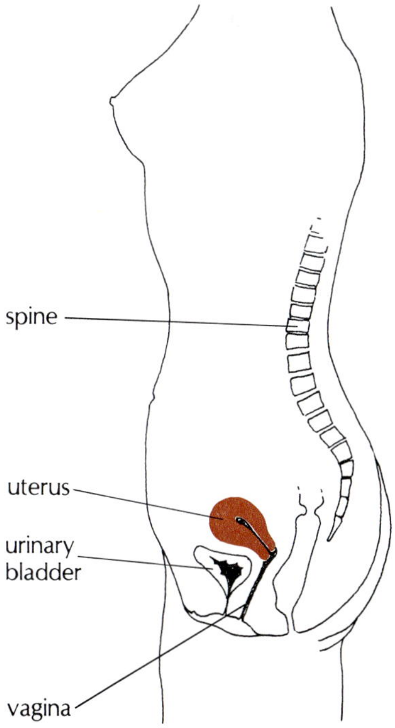

Figure 21–2 This drawing shows the size and position of the uterus in a woman who is not pregnant.

Early Care

One of the most important things a woman can do during her pregnancy is to see a health professional regularly. She will learn what she needs to do in order to take care of herself and the developing baby. Some suggestions for a pregnant woman to follow are listed on page 546.

- See a health professional regularly.
- Eat a well-balanced diet.
- Drink plenty of water.
- Get plenty of sleep.
- Exercise daily.
- Keep from smoking.
- Avoid alcohol.
- Check with a health professional before taking any medication.

You can see that a pregnant woman should think before she puts anything into her body. That is because everything she eats, drinks, or smokes gets into her system and could cross the placenta and get into the fetus' system, too.

The Placenta

The **placenta** nourishes the developing baby. It starts to form as soon as the embryo settles into the uterine lining. Cells from both the uterus and the fertilized egg divide and multiply to form the placenta. The entire placenta soon becomes laced with blood vessels.

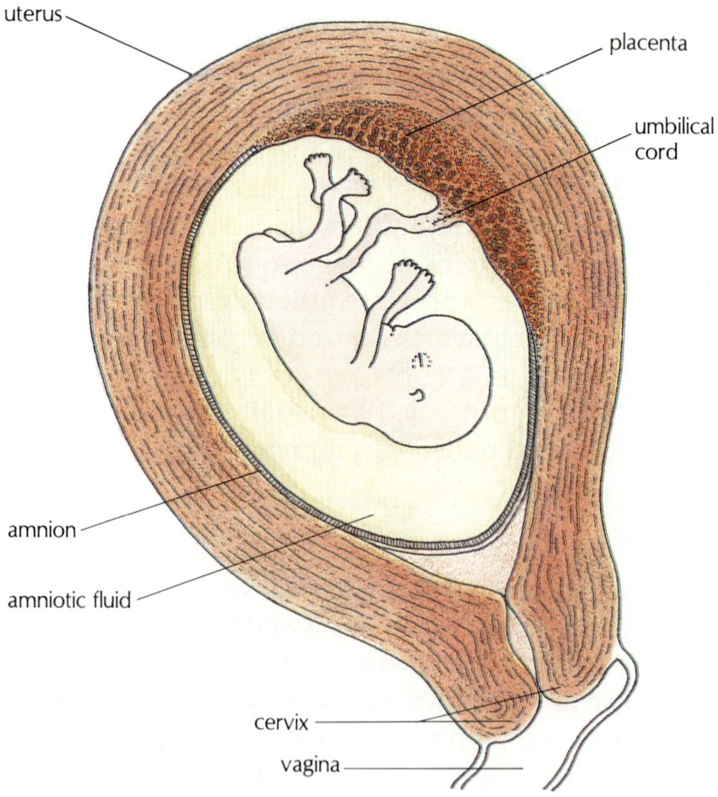

Figure 21-3 The placenta keeps the growing fetus alive.

Blood vessels from the placenta enter the embryo at the site of the baby's future belly button. These vessels form the **umbilical cord.** Oxygen, glucose, calcium, and other nutrients from the mother's blood diffuse into the blood vessels that nourish the embryo. Carbon dioxide and other wastes from the embryo travel back through the umbilical cord and diffuse into the mother's blood. In this way, the placenta serves as the lungs, kidneys, intestines, and liver for the developing baby.

The embryo does not "hang" from its umbilical cord in the uterus. Instead it floats in a fluid-filled membranous sac called the **amnion.** The **amniotic fluid** cushions the baby from bumps and jars. The amnion and placenta continue to develop throughout pregnancy right along with the fetus. (See Figure 21–3 on page 546.)

Labor and Delivery

The average pregnancy lasts about 265 days (38 weeks or 9 months). Its actual length usually varies by 5, 10, or even 20 days.

As the time for birth approaches, the baby usually turns upside down and its head drops into the mother's pelvis. If the mother notices this "settling," she knows that the baby will probably be born within the next several weeks. (See Figure 21–4.)

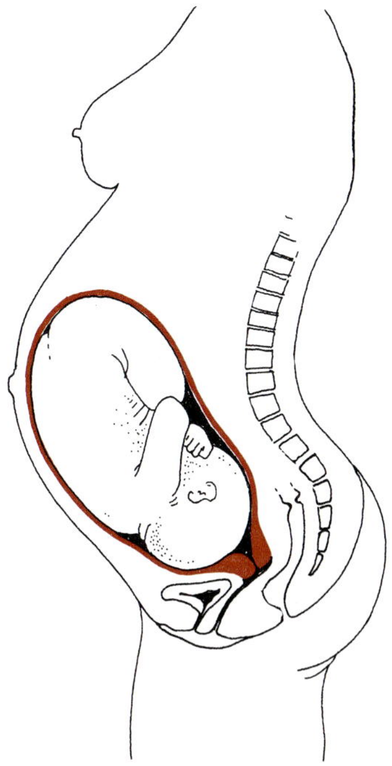

Figure 21–4 By the end of nine months, the baby takes up quite a bit of room in the mother's abdomen.

The first sign that the baby is on its way is the beginning of **labor.** During labor, the smooth muscles that make up the walls of the uterus contract and relax. These contractions begin the process of birth. Right before or during labor the amnion may break, releasing the amniotic fluid through the vagina. As labor continues, the contractions get stronger and come closer and closer together. When they begin, the mother or father will usually contact their doctor. They will be asked about the timing and strength of the contractions in order to determine whether or not it is time for the mother and father to go to the hospital.

Meanwhile, the muscles that make up the cervix have to relax. They usually keep the uterus tightly closed; now they gradually allow the uterus to open. Eventually, the baby's head can be seen and, after what may seem like forever to the parents, the contractions push the baby out into the vagina. The muscles

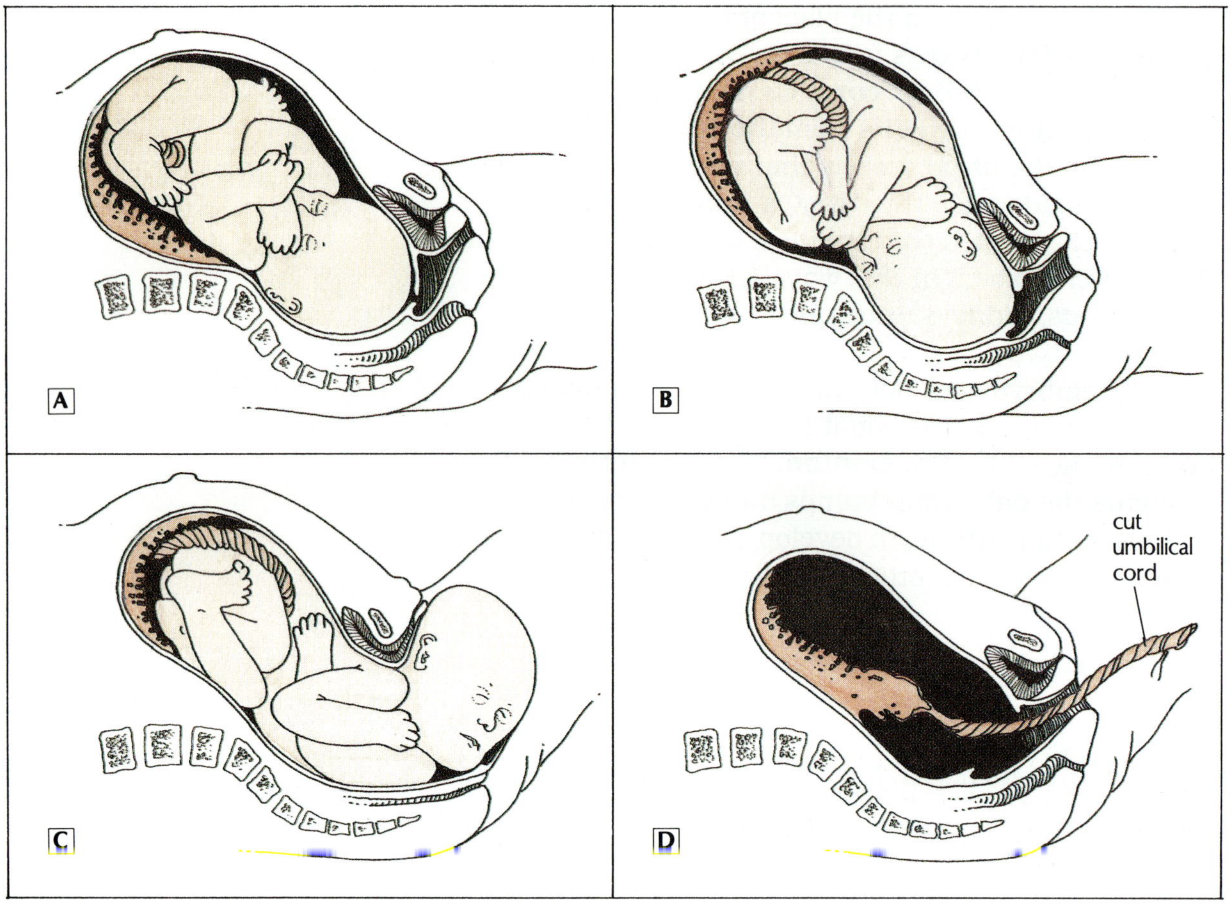

Figure 21–5 How would you describe what is happening in each of these drawings? Which of these stages do you think would be the most uncomfortable for the baby? for the mother? for the father?

that make up the walls of the vagina stretch to let the baby pass through. Once the baby's head is delivered, the rest of the body follows quickly. (See Figure 21–5.)

Even though the baby has been delivered, labor is not quite over. The uterus continues to contract until the placenta is also expelled.

The length of time from the beginning of labor to the delivery of the baby varies from woman to woman and from pregnancy to pregnancy. With a first child, it averages about 14 hours, although it may be much shorter (3 hours or less) or longer (24 hours or more). Labor is usually shorter for mothers who have given birth before.

Attention to the Baby

As soon as the baby is delivered, it needs attention. Since the baby's lungs are now working, the umbilical cord is no longer needed. It can be clamped and cut fairly close to the baby's body. This is not painful since

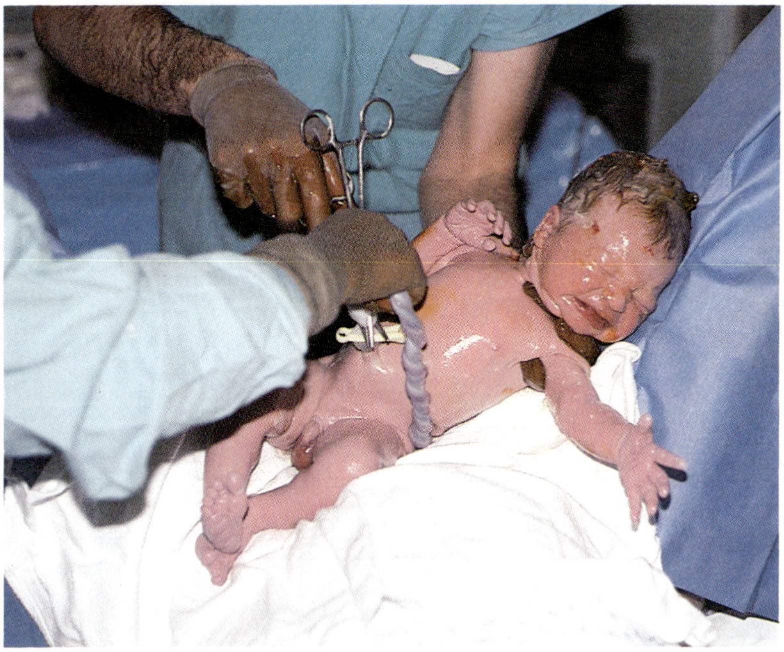

Figure 21-6 How would you describe the appearance of the umbilical cord?

there are not any nerves in the umbilical cord. Over the next several days, the stump of the cord dries and withers. When it finally falls off, it leaves the navel or "belly button" as evidence of a prebirth existence. (See Figure 21–6.)

Mucus is gently suctioned from the baby's mouth and nose. The baby's weight and the time of birth are recorded. The parents may then want to spend a few minutes with their baby before its check-up is finished. A few drops of antibiotic ointment or silver nitrate will be placed in the baby's eyes to prevent blindness in case the mother has an STD. A quick prick on the heel provides a few drops of blood for some laboratory tests and then the baby is ready for a warm sponge bath. Welcome to the world!

Small Babies

The average weight of a newborn is about 3200 grams (7 pounds). Newborns who weigh less than 2500 grams (5.5 pounds) are said to be **low birth weight** infants. They are more likely to have problems during the first few weeks of life than are other babies. They may not be able to control their body temperature or they may have respiratory or heart problems. All of these problems are serious. This is why small babies have higher death rates during the first few weeks of life than do

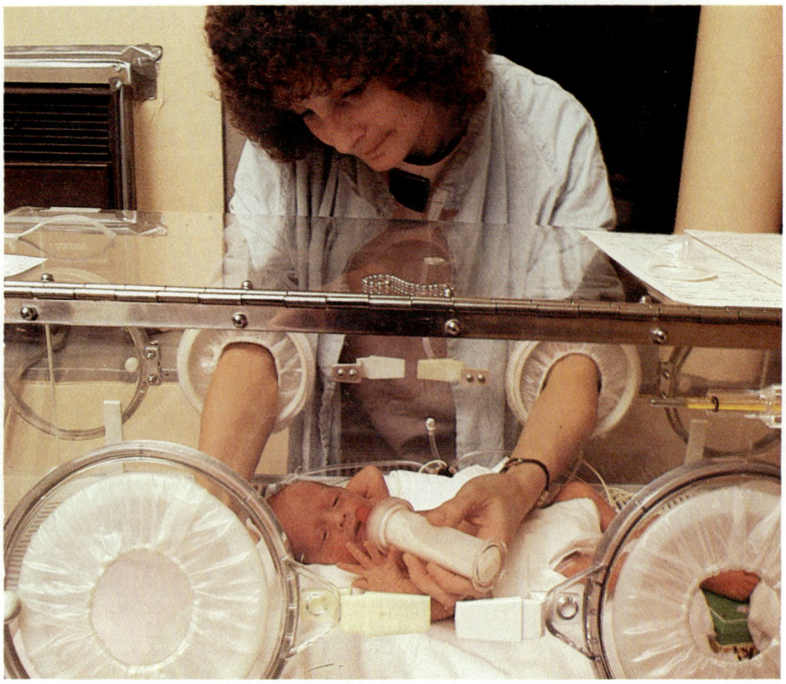

Figure 21-7 An incubator can be used to help a low birth weight baby maintain its body temperature.

slightly heavier babies. Of course, the smaller the baby, the more likely it is to have problems. (See Figure 21–7.)

There are many reasons for low birth weight. One common reason is that the baby may have been born early; these babies are often referred to as being **premature.** Mothers who smoke or drink during pregnancy also have smaller babies; so do very young mothers or mothers who do not eat well during pregnancy. Of course, a baby may also be small at birth because its parents are small. These babies usually are not at much risk. (See Figures 21–7 and 21–8.)

Figure 21-8 These babies are newborns. The one on the left is a normal birth weight baby, weighing around seven pounds. The baby on the right is a low birth weight baby. Compare the size of the arms of the two babies.

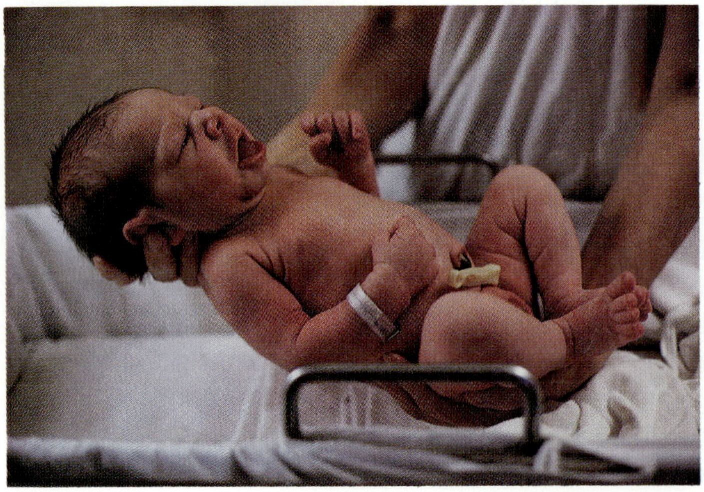

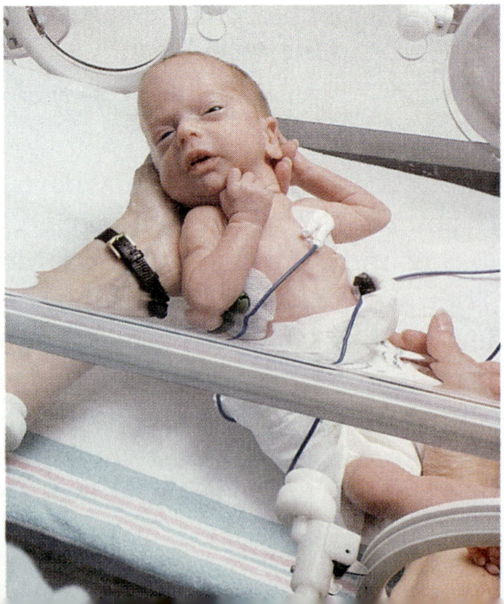

Cesarian Section

While most newborns are healthy and most pregnancies go smoothly, problems can and do occur anywhere along the way. Sometimes problems develop either before or during labor that make a vaginal delivery risky for the mother or the baby. When this happens, the baby may be delivered through an incision in the mother's abdomen and uterus. This procedure is called a **cesarian section.** Once a woman has had a cesarian section, she usually must have cesarian sections for any future deliveries. However, it may be possible for her to give birth vaginally.

EXTENSION ACTIVITIES

1. Call an obstetrician or your local health department. Ask about the newborn screening program in your state. Find out what is screened for, how many babies who have problems are detected each year, and how much the screening costs. Find out what would happen to the affected babies if they were not detected by this screening program.
2. The goal of the March of Dimes is to help all babies be born healthy or free from birth defects. Call the nearest office of the March of Dimes and ask someone to send you pamphlets about having healthy babies. Read the pamphlets and make a list of things that can cause birth defects or other problems for a newborn. Make another list of things that parents can do to improve their chances of having a healthy baby.

21-3 PREGNANCY: ANOTHER POINT OF VIEW

You have read about an "average" pregnancy and delivery, but who is completely average? In actuality, every pregnancy is unique. A pregnancy differs from woman to woman and from pregnancy to pregnancy for the same woman.

- In what ways do real pregnancies differ from the process described in the book?
- What kinds of things do people recall when they describe a pregnancy?

Interview two people about pregnancies they have experienced. Then compare what they say with what you have learned in your text. By doing so, you will find that these two focusing questions are easy to answer.

MATERIALS

interview form: ***Pregnancy: Another Point of View*** (Figure 21–9)
scissors

PROCEDURE

Part A. Getting Ready

1. Look over the interview form so you will be familiar with the questions you will be asking.
   a. Think about the questions as you read them. Are there any that don't make sense? If so, ask your teacher about them.
   b. Answer analysis questions 1–3.
2. Decide which two adults you would like to interview.
   a. Choose people who will be comfortable talking with you about their pregnancies.
   b. Consider what you are going to say. Remember to tell the adults that this is a class assignment. Also, explain that it is all right if they choose not to answer a question.

Part B. Summarizing the Interviews

1. Cut the interview questions apart. Keep the questions in order when you do so.
2. Place your questions and responses at the appropriate question station. When all students have done this task, everybody's responses for F-1 will be in one spot, those for F-2 will be in another spot, and so on.
3. Summarize the responses for the question that your teacher assigns to you. As you read the responses, watch for anything that surprises you.
4. Be prepared to present your summary to the class.
5. After the class discussion, answer the rest of the analysis questions.

PREGNANCY: ANOTHER POINT OF VIEW

NAME ______________________

DATE ______________ PER. ______

Ask two adults you know, one male and one female, who have experienced a pregnancy if you could ask them a few questions. Explain that you are studying about pregnancy in one of your classes. Have a pencil ready to record their answers.

Remember to thank the people for their time and for sharing this information with you.

QUESTIONS FOR FATHERS

F-1. How old were you when your first child was born? ______

F-2. Did anything about the pregnancy surprise you? ______
If yes, please explain.

F-3. What, if anything, did you do to prepare for the birth of the baby?

F-4. Were you present at the delivery? ______

F-5. How did you feel when you first held your baby?

F-6. What was the biggest change in your life after the baby was born?

F-7. What new duties and responsibilities did you have after the baby came home?

F-8. What do you think is a good age for someone to have their first child? Why?

Figure 21-9 (Interview form, page 1)

PREGNANCY: ANOTHER POINT OF VIEW
(PAGE 2)

NAME ______________________

QUESTIONS FOR MOTHERS

M-1. How old were you when your first child was born?

M-2. How did you feel while you were pregnant?

M-3. Did anything about the pregnancy surprise you? ______
If yes, please explain.

M-4. How long was the labor? ______

Did you have an anesthetic or pain killer? ______

M-5. How did you feel when you first held your baby?

M-6. What was the biggest change in your life after the baby was born?

M-7. What new duties and responsibilities did you have after the baby came home?

M-8. What do you think is a good age for someone to have their first child? Why?

Figure 21-9 (Interview form, page 2)

ANALYSIS QUESTIONS

1. Which question on the interview form do you think looks the most interesting? Explain your answer.
2. How do you predict that adults will answer
   a. question 6?
   b. question 8?
3. Do you think that people will be uncomfortable with any of these questions? If so, why?
4. How did the adults you interviewed feel about answering these questions?
5. How did the adults answer
   a. question 6?
   b. question 8?
6. What was the most interesting thing you learned from this activity?

CONCLUSION

Think about the interviews you conducted and the class discussion of the interview questions. Then write a paragraph to answer each of the focusing questions on page 552.

UNIT 7

CELLS AND GENETICS

CHAPTER 22

CELLS AND CHROMOSOMES

How could a powerful microscope like this help you learn more about cells?

22–1 IDENTIFYING CELLS

Bone or blood, lung or leaf, they all come from living organisms —so they're all made up of cells. Throughout the year, you've been looking at cells—all kinds of cells. Some of them have been easier to see than others. If you close your eyes, how many kinds of cells can you picture?

- Which cells can you correctly identify?
- What clues are most helpful when trying to identify cells?

Come back and answer these questions after you've looked at eight microscope slides. Their labels will be covered so you'll have to recognize them from memory; but that shouldn't be a problem since you've seen them all before.

MATERIALS

microscopes
microscope slides with covered labels

PROCEDURE

Work with a partner.

1. Make a data table like Figure 22–1. Leave plenty of room for each cell type so you will be able to record all of your observations.

Letter	Cell Sketch and Magnification	Clues (at least 2)	Cell Type
A	○ ____		
B	○ ____		
C	○ ____		

Figure 22–1

2. Read the following steps so you will know what you will be doing.
   a. Go to a station and observe the microscope slide.
   b. With your partner, discuss what to record on the data table.
   c. Take turns filling in the data table.
   d. When you need to make a sketch, draw only *one cell.*
   e. When your teacher calls time, move on to the next station.

3. When you have been to all of the stations, return to your seat.
   a. Look over your data table. You may find that you identified two different slides as being the same cell type. Make any changes that you need to.
   b. If you'd like to go back and look at a slide, do so.
4. After the class discussion, answer the analysis questions and write your conclusion.

ANALYSIS QUESTIONS

1. Look at your sketches of cells and the magnification.
   a. Which microscope slide had the smallest cells?
   b. Which had the largest?
2. If someone wanted to see cell nuclei, which microscope slide would you suggest they observe?
3. Why do so many of the cells look pink or purple? Why don't they look green or brown instead?
4. List at least 3 things that all of the cells you observed had in common.

CONCLUSION

Remember to answer the focusing questions on page 557.

22-2 MORE HUMAN CELLS

We cannot count all of the cells in a human body, but, if we could, the results would look something like the list below. As you can see, it is estimated that there are about one hundred trillion cells in one body. That's hard to imagine, isn't it?

You have looked at some of the evidence showing that our bodies are made up of this immense number of cells. You looked at muscle tissue through a microscope and saw that it is made of muscle cells. You looked at a drop of blood and saw that it contained red and white blood cells. If you inspected a body microscopically, you would find cells just about everywhere you looked. That's because almost every part of the body is made up of cells.* The cell is the basic unit of life.

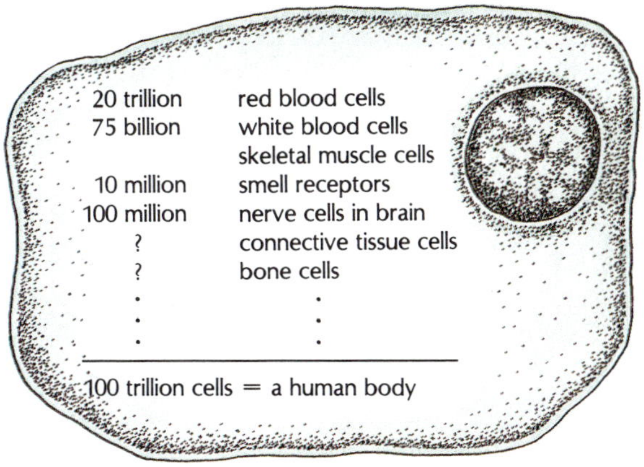

- How do you make a microscope slide of living cells?
- How do fat cells and cheek cells compare with other cells you have seen?

After you have looked at two more cell types—your own "cheek cells" and fat cells on a prepared microscope slide—you will be able to answer these questions.

MATERIALS

2 microscope slides
2 cover slips
2 flat toothpicks
prepared microscope slide: fat
dropper bottle with stain
microscope
eyedropper
bucket with 10% bleach solution

*Some structures, such as hair and finger nails, are not made up of cells but they are produced by cells.

PROCEDURE

Work with a partner.

Part A. Observing Fat Cells

1. Observe fat cells through your microscope.
   a. Move the slide until you find what looks like some good, clear cells.
   b. Sketch what you see.
   c. Label the cell membrane, nucleus, and cytoplasm.

Remember

Check your sketch. Remember to:

- fill the circle the way the cells fit in the field
- include enough detail for someone else to recognize it
- record the magnification
- label everything you can recognize in the sketch
- title the sketch

---

2. Compare your sketch with your partner's. Do they look the same? Did you label the same things? If there are any differences, look at the microscope slide again and fix your sketch if necessary.
3. Answer analysis question 1.

Part B. Preparing Cheek Cells

Both you and your partner should make microscope slides of your own cheek cells. Then compare the cells on both microscope slides. Follow these steps.

1. Clean your microscope slide and cover slip if they are dirty.
2. Place a drop of water on your microscope slide.
3. Using the flat end of a clean toothpick, *gently* scrape the inside of your cheek. Mix the scrapings in the drop of water and cover with a cover slip. (See Figure 22–2.)
4. Examine the slide under low power. You may need to dim the light to see the cells. Switch to high power and look again.
5. Remove the slide from the microscope.
   a. Place a drop of stain along one edge of the cover slip. Hold a piece of paper towel next to the opposite edge of the cover slip. This helps "pull" the stain under the cover slip. (See Figure 22–3.)
   b. Observe the slide again under both low and high power.
   c. Remember to move the slide until you find a clear view of cells.
   d. Look at your partner's slide, too.

Figure 22–2 Making a slide

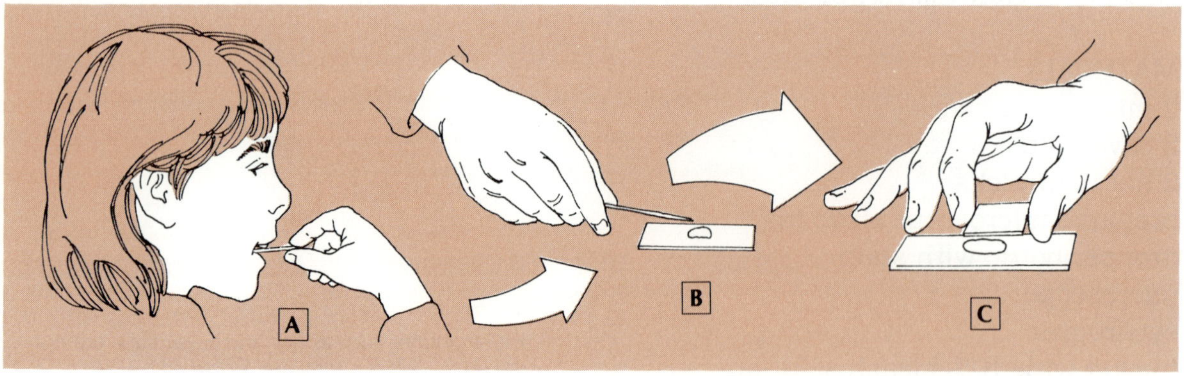

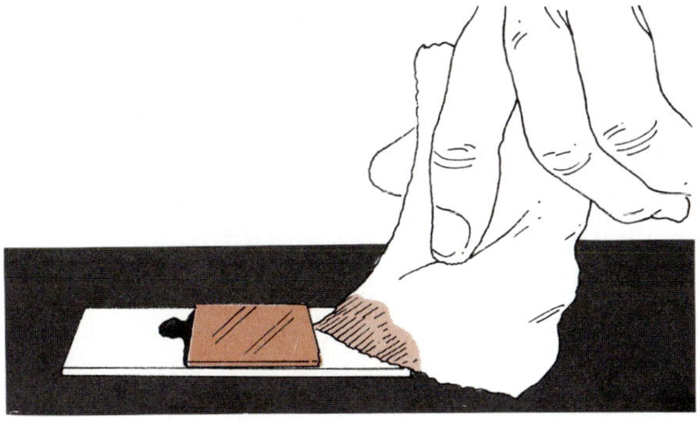

Figure 22–3 Adding stain

6. Sketch your cheek cells. Label the nucleus, cytoplasm, and cell membrane. Check your sketch. (See the Remember section in Part A.)
7. Answer analysis questions 2–5.

Part C. Finishing Up

1. Clean up your equipment. After you have washed your microscope slide, rinse it in the 10% bleach solution. This is to kill any bacteria or viruses that may still be on the slide.
2. Answer the rest of the analysis questions.

ANALYSIS QUESTIONS

1. Describe the appearance of a fat cell.
2. Why did the directions tell you to put a drop of water on the microscope slide?
3. How did the stain affect the appearance of your cheek cells?
4. What did you see besides cheek cells on the slide?
5. Compare your cheek cells with your partner's.
   a. How are they alike?
   b. How are they different?
6. Think about the shapes of the cells that you just observed. The shape of the cell may give you a clue about its function.
   a. What do you think is the function of a fat cell? Explain why you think this is the function.
   b. What do you think is the function of a cheek cell? Again, explain your answer.
7. The introductory paragraphs mentioned that hair and fingernails are not made up of cells. What other parts of the body do you think are not made up of cells?
8. Why are cells called "the basic unit of life"?

CONCLUSION

Reread the two focusing questions on page 559. Write two sentences that explain something you think is important to remember when you make a slide of living cells. Answer the second question by comparing one of the types of cells you saw today with a type you saw in a previous activity.

22-3 CELLS: THE BASIC UNIT OF LIFE

So far, you have looked at several types of cells. Most of these have been human cells because you've been studying about the human body. If you had been studying mice, you would have looked at mouse cells. Likewise, if you had been interested in flies, you would have studied fly cells. No matter which organism you study, you would want to look at its cells since all organisms are made up of cells. Therefore, as a part of Life Science, it is important that you understand a few things about the basic unit of life—the cell.

Organization

A jumble of trillions of cells would not make a human being. The cells must be assembled in specific combinations, in the correct number, and in the right place. In other words, the cells must be organized in some way.

A group of similar cells that have a similar function is called a **tissue.** You have already studied several kinds of tissues. For instance, you saw muscle cells organized to form muscle tissue and you saw blood cells forming blood tissue. However, when you think about parts of your body, you are more likely to think about organs than about tissues. An **organ** is a group of tissues that functions together. For example, the heart is an organ; it is made up of blood tissue, muscle tissue, nerve tissue, and connective tissue. An eye, a kidney, or a stomach—each is an organ, too; and so, some say, is every separate bone in the body.

Organs, too, are organized. A group of organs working together is called a **system.** To continue using the heart as an example, we would say that the heart—an organ—is part of the circulatory system. (See Figure 22–4.) The digestive system, as you learned, is made up of many organs. Some of these are the stomach, tongue, and esophagus. In all, there are 10 systems. The systems cannot work by themselves; they work together as parts of a team to make an **organism.**

Working backwards, we would say that an organism is made up of systems; systems are made up of

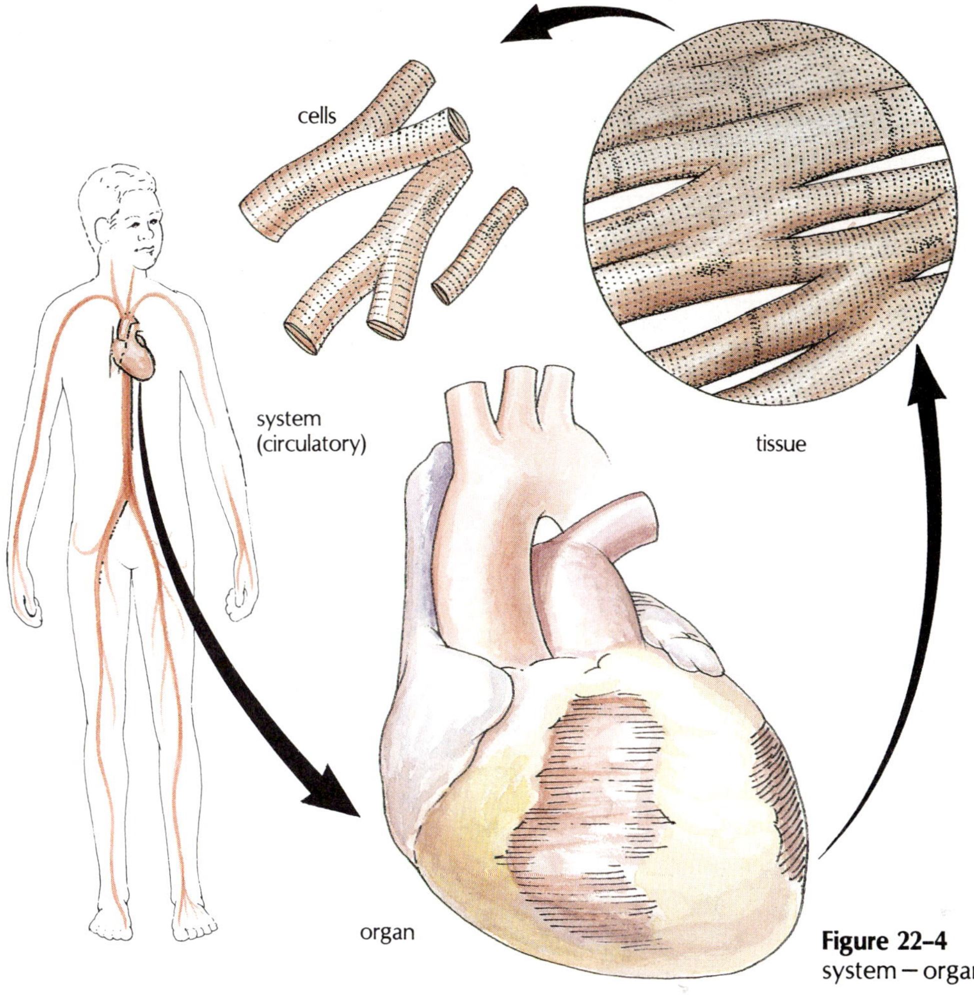

Figure 22-4 The circulatory system—organ to tissues to cells

organs; organs are made up of tissues; and tissues are made up of cells. Thus, the cell is the smallest living level of organization. For this reason, it is called the basic unit of life.

Cell Size and Shape

Most cells are extremely small. Sperm and red blood cells are so tiny that they are hard to see with the light microscopes you have at school. On the other end of the size range are the large cells. The largest cells of all are egg cells, especially those produced by birds. The largest human cells are also eggs but even they are hard to see without some magnification. Most human cells are neither extremely large nor exceptionally small. Most of them can be viewed when they are magnified 100 to 400 times.

Ten Systems

Usually, the organs that make up a system are related in an obvious way and you can figure out which ones go together. However, some organs function in more than one system. For example, the liver functions in the circulatory system since it breaks down old blood cells; but it also functions in the digestive system since it is important for digesting fats. The ten human body systems are shown below.

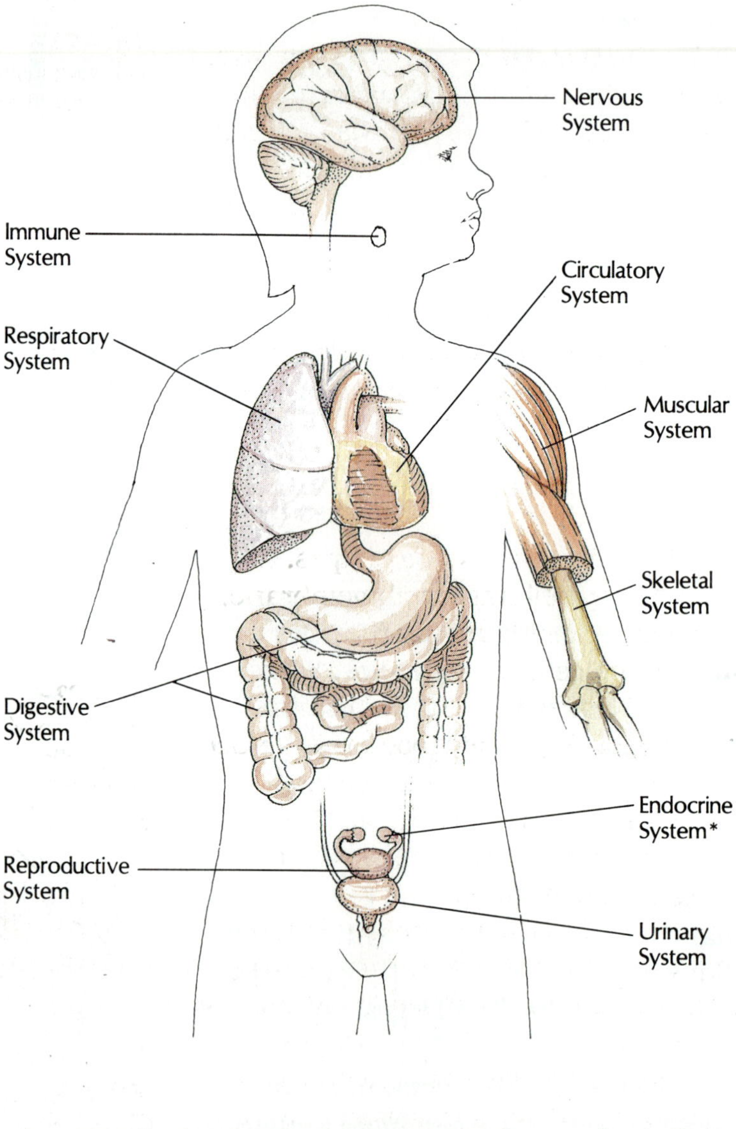

*The endocrine system, like the nervous system, is important for controlling body functions. This system is made up of glands – such as the pituitary gland, ovary, and testis – that produce hormones such as growth hormone, estrogen, and testosterone.

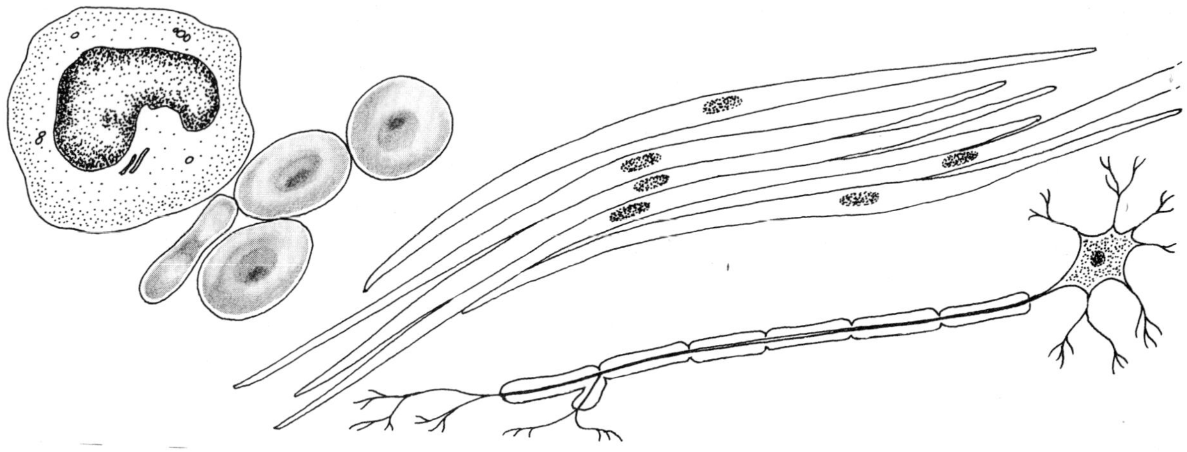

Figure 22–5 Each cell type has a characteristic shape. Do you recognize these cells?

The shape of cells varies even more than their size. It is often helpful to look closely at the shape of a cell since its shape may give you a clue about what the cell does. Think about the cells you have studied. Nerve cells are long and skinny; they extend from one part of the body to another. This shape is ideal for carrying messages but it would be very impractical for carrying oxygen throughout the body. The small, round shape of a red blood cell is a much better design for flowing through blood vessels. Each cell is specialized for the job it does. (See Figure 22–5.)

Cell Structure

Even though cells differ in their sizes and shapes, they all have similar parts. Every cell has a cell membrane, cytoplasm, and a nucleus. (See Figure 22–6.)

The **cell membrane** holds a cell together. It also controls what enters and leaves the cell. For example, since a cell is alive, it needs food. This food must diffuse through the membrane to enter the cell. Once inside the cell, the food molecules are broken down. The waste molecules that are formed will then diffuse out of the cell by crossing the membrane.

For being so important, this structure is extraordinarily thin. It would take 100,000 membranes to make a stack 1mm thick. If a cell membrane is damaged in any way, the cell may die.

Inside the cell membrane is the **cytoplasm.** Cytoplasm is a watery gel that contains many very tiny structures. Most of these structures are too small to see with a light microscope but they do the work of the cell. Some serve as a support system to give the cell its

Figure 22–6 Like other animal cells, this white blood cell has a membrane, a nucleus, and cytoplasm.

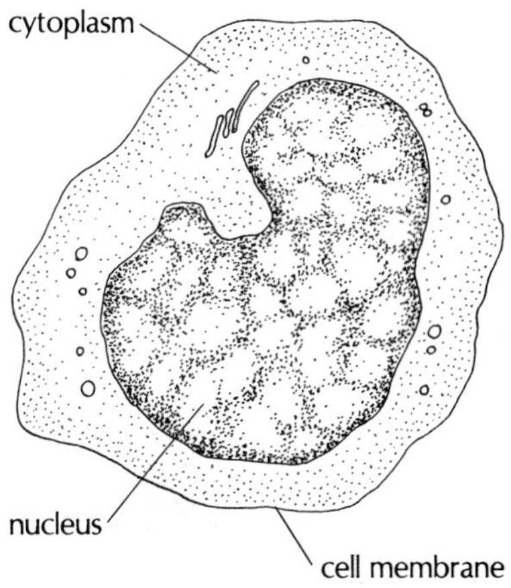

shape. Others contain enzymes and are sites for important chemical reactions such as the production of energy or the breakdown of waste molecules. Still others function during cell division. Each structure has its own function. As you can see, cells are quite complex—so complex that some scientists spend their entire careers studying these small but important structures.

The **nucleus** is the largest structure inside the cell. You can usually see it with a light microscope but sometimes it's necessary to stain the slide before the nucleus will be visible. The nucleus is the "control center" that directs the cell's activities. It controls all of the chemical reactions that go on in the cell. It can do so because it contains the **chromosomes.** A set of chromosomes is like an instruction manual on how to run a cell. You will learn more about chromosomes later in this chapter.

The cytoplasm, nucleus, and cell membrane are the three basic parts of every animal cell. Plant cells have these same three basic parts as well as a few additional structures. For instance, plant cells have sturdy **cell walls.** A cell wall gives a plant cell its shape and provides support for the entire plant. Cell walls are very important since plants do not have skeletons. Some plant cells also contain **chloroplasts.** Photosynthesis takes place within the chloroplasts. You should have seen both of these structures if you did the ***Plant Cells*** activity on pages 51 and 52 when you studied ecosystems and ecology. (See Figure 22–7.)

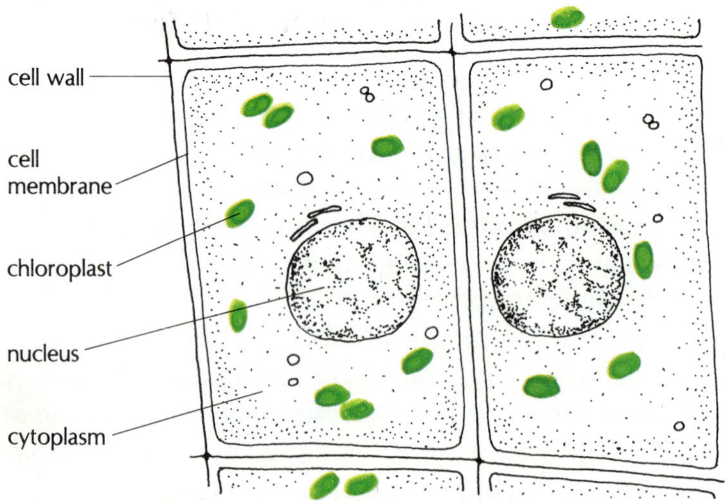

Figure 22–7 Unlike animal cells, a plant cell has chloroplasts and a cell wall.

Alive and Active

A single living cell can do many of the same things that a living organism can do. A cell can move, respond to its environment, grow, and reproduce. These activities all require energy—in order to get this energy, a cell needs food and will produce waste.

Cells are complex and fascinating. The more you learn about them, the more you will want to learn. In this course, you are just being introduced to cells. If you take any Biology courses, you will learn much more about them. Of course, if you want to learn more about cells now, you can do so. Several suggestions for further study are suggested in the Extension Activities.

ANALYSIS QUESTIONS

1. List the following structures in order of size—from smallest to largest. (Try to answer from memory before you look up the answer.)
   cell   organ   organism   system   tissue
2. List at least 2 tissues that make up each of the following organs.
   a. small intestine
   b. scapula
   c. eye
3. To which *system(s)* does each of the following organs belong?
   a. trachea
   b. gluteus maximus
   c. urethra
4. Look at the cell types shown below. Explain why each cell is well-shaped for the job it does.
   a. skeletal muscle cell 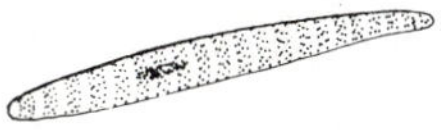
   b. sperm cell 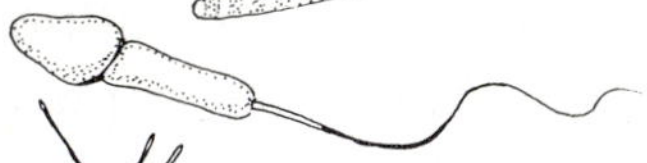
   c. bone cell 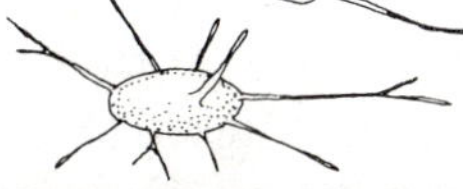
5. List the 3 basic parts of an animal cell and explain one important function of each one.
6. What do you think would happen to a cell if its membrane were damaged?
7. Read about an amoeba experiment on page 568 and then answer the questions that follow.

An Amoeba Experiment

Amoebas are tiny one-celled animals. They are often used in experiments to study cells. In one experiment, a scientist cut 100 amoebas in half. One half of each amoeba had the nucleus; the other half did not. The scientist watched these half-amoebas and recorded the following results.

Day of Experiment	Number of Half-amoebas with Nuclei	Number of Half-amoebas Without Nuclei
1	100	100
2	74	78
3	66	60
5	59	48
10	52	14
20	47	3
30	61	0
40	88	0

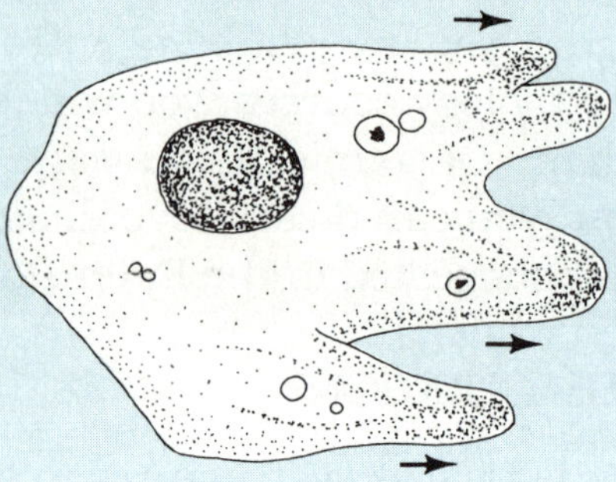

a. How many days did the scientist work on this experiment?
b. On the second day of the experiment there were 74 half-amoebas with nuclei. What happened to the other 26?
c. Describe what had happened to the half-amoebas with nuclei by the fortieth day.
d. Describe what happened to the half-amoebas without nuclei.
e. What do you think the scientist concluded?

8. Imagine that you are being questioned by inhabitants of another planet. They've been studying life on earth. They find it very unusual that every form of earth life seems to be made up of very small things. You tell them these small things are called cells.
   a. They ask you many questions about these tiny structures. Answer one question below and explain your response as completely as you can.
      Are nerve cells smarter than other cells?
      Both red blood cells and egg cells are round. Why are eggs larger than red blood cells?
      Humans all seem to have different skin colors. Is that because their cell membranes are colored differently?
   b. They would like to take some cells back to their planet for further study. They want to make sure

that the cells will be alive when they reach home so they ask you, "What do these cells need to stay alive?" Provide at least 2 suggestions as to how they could safely transport the cells.

EXTENSION ACTIVITIES

1. Look through a book that has pictures of cells taken through an electron microscope. (It magnifies much more than a light microscope.) In that book, find pictures of several cell types that you have already studied. Compare these pictures with what you saw through your light microscope.
2. Search for one-celled animals. Look for them in drops of water from a pond or puddle. Use a microscope on your hunt. Sketch the animals you find.
3. When Charles Wallace becomes seriously ill, we learn that it is due to a problem with his mitochondria. Mitochondria are tiny structures within every cell that provide the energy a cell needs to stay alive. Read *A Wind in the Door* by Madeleine L'Engle to find out what happens to Charles Wallace. Then look up *mitochondria* in a biology book. Write a report that compares Ms. L'Engle's view of mitochondria with the scientific explanation.

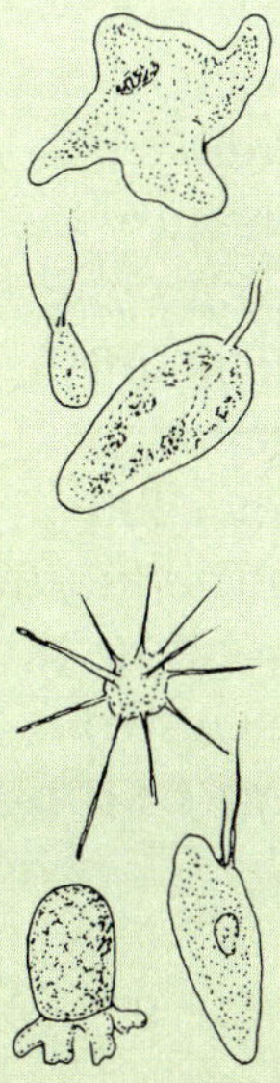

Single-Celled Organisms

The plants and animals that you see every day are made up of millions of cells. In addition to the large organisms, there are thousands of microscopic animals that are made up of just one cell. A one-celled animal can move, eat, and breathe. It also reproduces, senses what is going on around it, and gets rid of waste.

Like other cells, a single-celled organism has a cell membrane, cytoplasm, a nucleus, and other structures. Some of these organisms look like blobs while others have very specific shapes. Look at the single-celled organisms pictured here. If you were a single-celled organism, which one would you like to be? Why? What do you think your life would be like as this little critter?

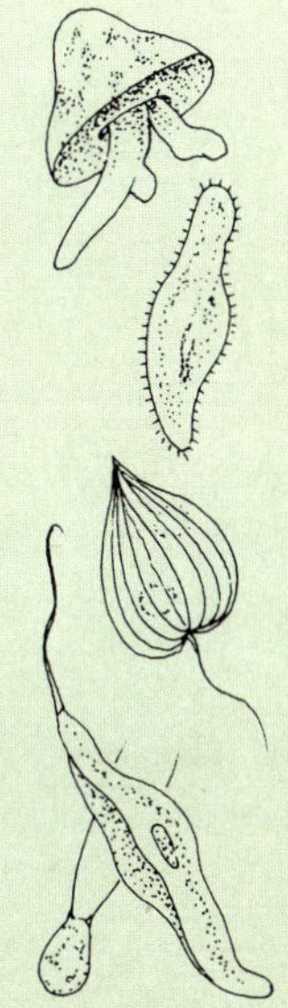

22-4 GROW AND DIVIDE

From a microscopic fertilized egg . . . to a 4-inch fetus . . . to a sleepy newborn . . . to an active three-year-old . . . to an eager student . . . to a young adult. For about 15 years the human body is constantly growing—sometimes very slowly, sometimes in sudden spurts. Think about what is going on with cells during growth. For instance, you might picture an infant as being made up of extremely tiny cells. If so, you might think the baby grows because its cells grow larger. Does this sound reasonable to you?

Growth

Actually, the body does not grow because its cells are growing; the body grows because the number of cells increases. Cells do grow somewhat but they never get very large. When a cell reaches a certain size, it splits in half. The process sounds easy but the cell has a lot to do before it can divide. It has to make another nucleus and make copies of all of the structures in the cytoplasm. After doing so, the cell will begin to pucker around the middle and pinch in until it has separated into two approximately even halves. Where there was once one cell, there will now be two, almost identical, new cells. This process is called **cell division.**

When a cell divides, the new cells are usually just like the original one. In this way, skin cells only give rise to new skin cells and, likewise, blood cells only form new blood cells. (Imagine what a hodgepodge would result if this were not the case; kidney cells might form in the eyeball and brain cells in the kidney.) The information in each chromosome controls the division process. Remember, the nucleus is the control center of the cell and chromosomes are found in the nucleus. (See Figure 22–8.)

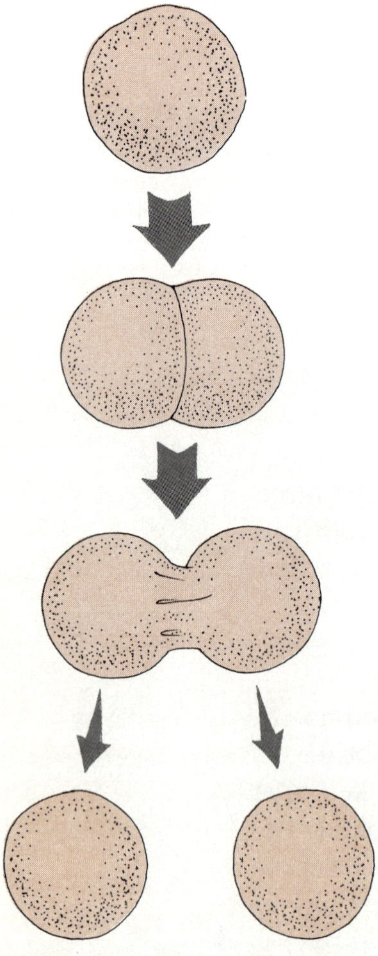

Figure 22-8 What size would the new cells be if the original cell did not grow before it divided?

Replacement of Dead Cells

In most tissues, cell division continues even after a person reaches adulthood. Whether or not particular cells divide depends on the cell type. Epidermal cells must divide fairly often to replace themselves since some skin cells live only about 8 hours. If these cells were not replaced, we would soon be left without an epidermis. You may also remember that the cells lining the small intestine must also be constantly replaced. These cells live an average of a day and a half. Since white blood cells live only about 13 days and a red blood cell lives an average of 100 days, they too need to be replaced.

Altogether millions, perhaps even *billions*, of your cells die *every minute*. What happens to these dead cells? They self-destruct. They produce enzymes which break themselves down from the inside out. All of the molecules that are released can be recycled; that is, the molecules are used again to build new cells.

Not All Cells Divide

Cells divide for three reasons: so growth can occur, to replace dead cells, or to replace damaged cells. For instance, if the surface of the skin is cut, the area will soon heal since skin cells are constantly being formed anyway. The liver, too, if injured, can produce new liver cells to repair itself. However, other cells lose the ability to divide once the body reaches maturity. If lung or kidney cells are lost from disease or an accident, they cannot be replaced. The reason why some cells can divide and others cannot is not known, but the answer lies somewhere within the chromosomes. Perhaps someday researchers will learn how the chromosomes direct cell division.

TEST YOURSELF

- If 100,000,000 (one hundred million) of your cells are dying every *minute*, how many new cells are *forming* in your body every *hour?*
- Imagine that you are studying a type of cell that divides every 60 minutes. At noon, you have one cell. How many cells will you have at noon the next day?

EXTENSION ACTIVITIES

1. Adopt a few amoeba or paramecia. As a "parent," find out what your pets need to stay alive. See if you

can keep them alive for a month. How many did you start with? How many did you have at the end of a month?

2. With your teacher's approval, invite someone who studies cells to talk to your class. For instance, a cytologist does research on cells and many lab technicians either observe or grow cells in their jobs. Find out which cells your speaker studies and why he or she does so. Have the person also explain what can be learned by looking at these cells. Ask the speaker to discuss what he or she likes about the job.

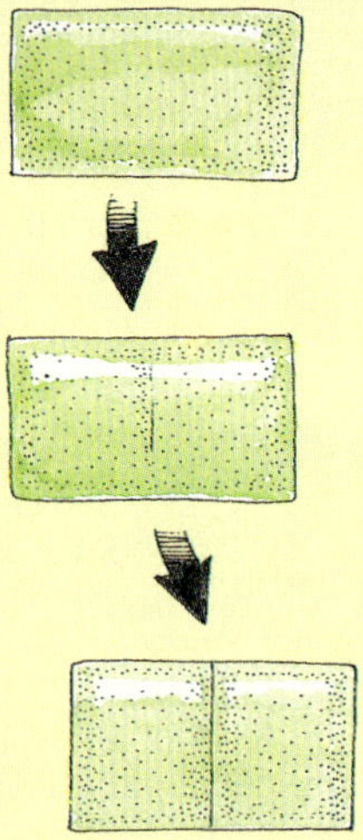

Plant Cells

Plant cells must be able to divide, too. But, because of their rigid cell walls, these cells cannot pinch together in the middle. Instead, a new cell wall must form down the middle of the plant cell. This cell wall divides the original cell into two cells.

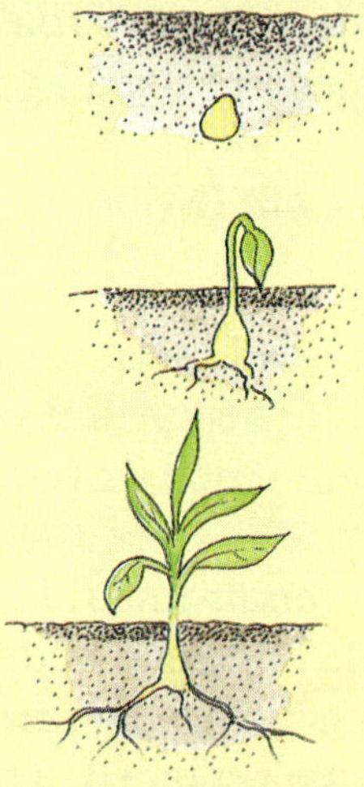

22-5 FINDING CANCER FACTS

Normally, cell division is an orderly, controlled process; but sometimes it goes out of control. When this happens, the new cells may not look like the original ones, they may invade surrounding tissues, and they may not function properly. Uncontrolled cell division is the cause of cancer.

Many people fear cancer. Part of this fear is fear of the unknown. One way to combat fear of cancer is through education and by learning some facts about cancer.

- What are some risk factors that can increase a person's chances of getting cancer?
- What can you do to lower your chances of getting cancer?

Work with two other students to find the answers to these questions in pamphlets about cancer.

MATERIALS

pamphlets about cancer
worksheet: ***Finding Cancer Facts*** (Figure 22–9)

PROCEDURE

Part A. Finding the Facts

1. Join your group and get your materials. As a group, read through this procedure so you know what you will be doing.
2. Each person will be responsible for two pamphlets. Decide who will read which ones. This activity works best if everyone finishes at the same time so you may want to have the person who reads the fastest take the longest pamphlets.

PLANNING YOUR TIME

3 minutes	– get in groups – look over pamphlets
1 minute	– decide who will read which pamphlet
12 minutes	– read pamphlets and fill in worksheet
6 minutes	– share information
4 minutes	– answer focusing questions

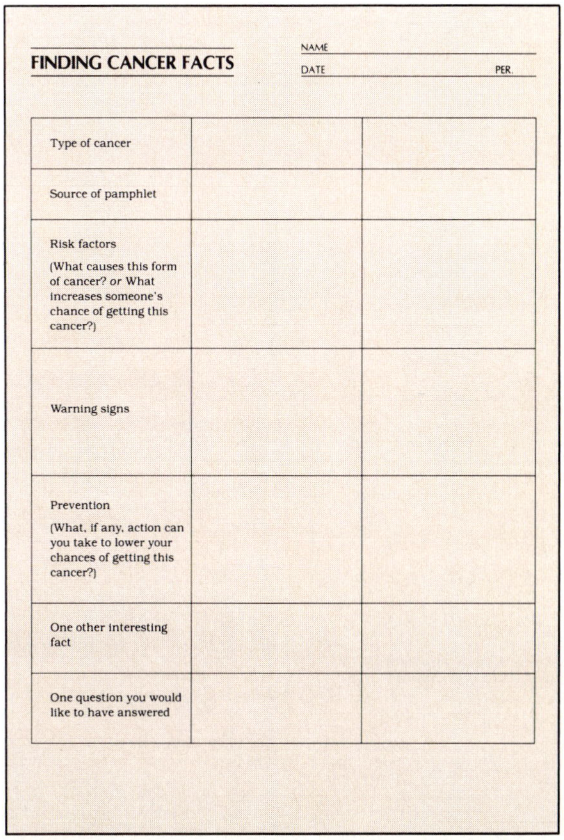
FINDING CANCER FACTS

NAME

DATE PER.

Type of cancer		
Source of pamphlet		
Risk factors (What causes this form of cancer? *or* What increases someone's chance of getting this cancer?)		
Warning signs		
Prevention (What, if any, action can you take to lower your chances of getting this cancer?)		
One other interesting fact		
One question you would like to have answered		

Figure 22–9

3. Quietly, read one pamphlet and fill in the first column on your worksheet.
4. Read your second pamphlet and fill in the second column of the worksheet.

Part B. Sharing Information

1. Take turns sharing what you learned with the other members of your group.
   a. Briefly explain the causes, risk factors, and ways to prevent a particular type of cancer.
   b. Answer questions that other members of your group may have.
2. Look back at the two focusing questions. As a group, write your answers to these questions. Use what you learned from all of the pamphlets when you write your group answers.
3. All three of you should sign your name to the paper indicating that you helped answer the questions and that you agree with what was written.

CHALLENGE

If your group finishes before the other groups have finished, go on a Scavenger Hunt! Quickly hunt through your pamphlets in search of definitions. See if you can find the definitions for these words.

benign	carcinogen
malignant	radiotherapy
biopsy	chemotherapy
metastasis	tumor

22-6 PUBLIC AWARENESS

Most types of cancer can be cured if they are detected early—before the cancerous cells have invaded nearby tissues or moved to other parts of the body. If everyone knew the seven warning signs for cancer, maybe more cancers could be caught early. If everyone knew *why* it is important to catch cancer early, maybe people would seek help more quickly. If everyone knew. . . .

How can we "get the word out"? Health care providers are always trying to find ways to reach people, to teach people the facts. They have tried a variety of approaches; they write pamphlets, give talks, design posters, buy commercial time, and advertise in magazines and newspapers—anything that will catch people's attention.

In this activity, you are going to help them out. You will help spread the facts about cancer.

PROCEDURE

1. Design a poster. Write and put on a commercial. Design a pamphlet for 11–14 year olds. Select one of these or another method that you think would be effective in reaching and teaching other people.
2. Choose a title for your presentation. Here are some examples. You can use one of them or think of your own.
   Cancer Prevention
   Early Detection
   What Is Cancer?
   Know the Facts
3. Plan how to promote your idea.
   - If you make posters, you might want arrange for an exhibit in the lobby of your school.
   - Television commercials could be performed for fifth or sixth grade classes at a nearby elementary school.
   - Think of an original way to present your project.

EXTENSION ACTIVITY

Find examples of advertisements that provide information about cancer. Choose a magazine for adults such as *Better Homes and Gardens*, *Good Housekeeping*, or *Time*, and read the advertisements. Cut out all of the ads that mention cancer. What is their message? What techniques are used to get people to read them? Who sponsors these ads?

22-7 HUMAN CHROMOSOMES

Since the information in chromosomes controls everything about a cell, chromosomes are obviously very important. So far you have been reading about chromosomes but you have not seen what they look like. That's because chromosomes are very small—so small that you cannot even see them when you look at cells through your microscope. Sometimes you can barely see the nucleus, let alone see the chromosomes inside it!

A Karyotype

It would be interesting to look at your own chromosomes, but that requires more laboratory equipment than you are likely to have at school and very careful laboratory work. So, instead of looking at your own chromosomes, imagine that you are looking at someone else's through a microscope with a 100 power objective. First, using the low power objective, you peer through the eyepiece. Carefully, you focus and slowly move the slide until you find a cell. Switching to medium power, you focus on the nucleus. Now, you are ready to use the high power objective. As you focus, the contents of the nucleus become visible; you can see the chromosomes. They look like the ones pictured in Figure 22–10.

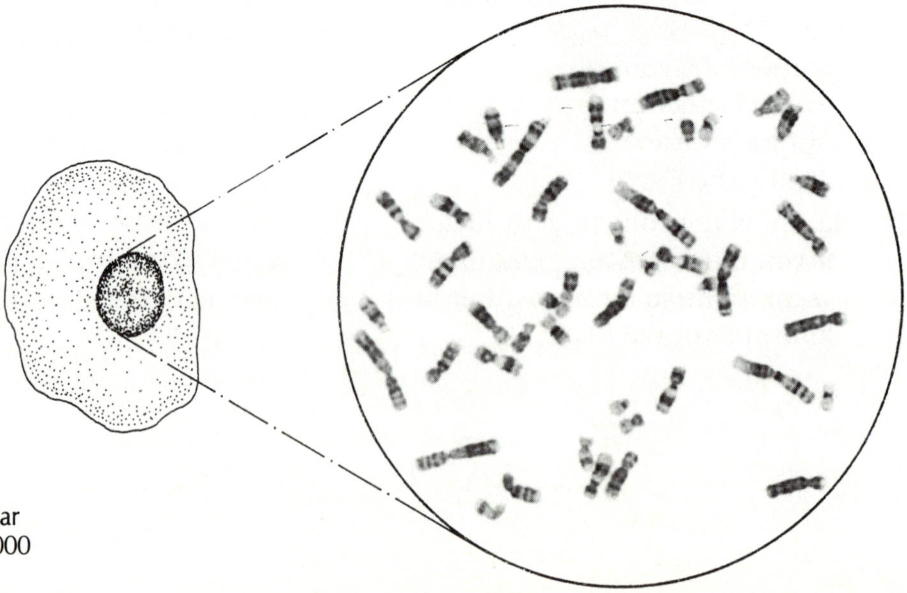

Figure 22–10 Human chromosomes as they appear when they are magnified 1000 times.

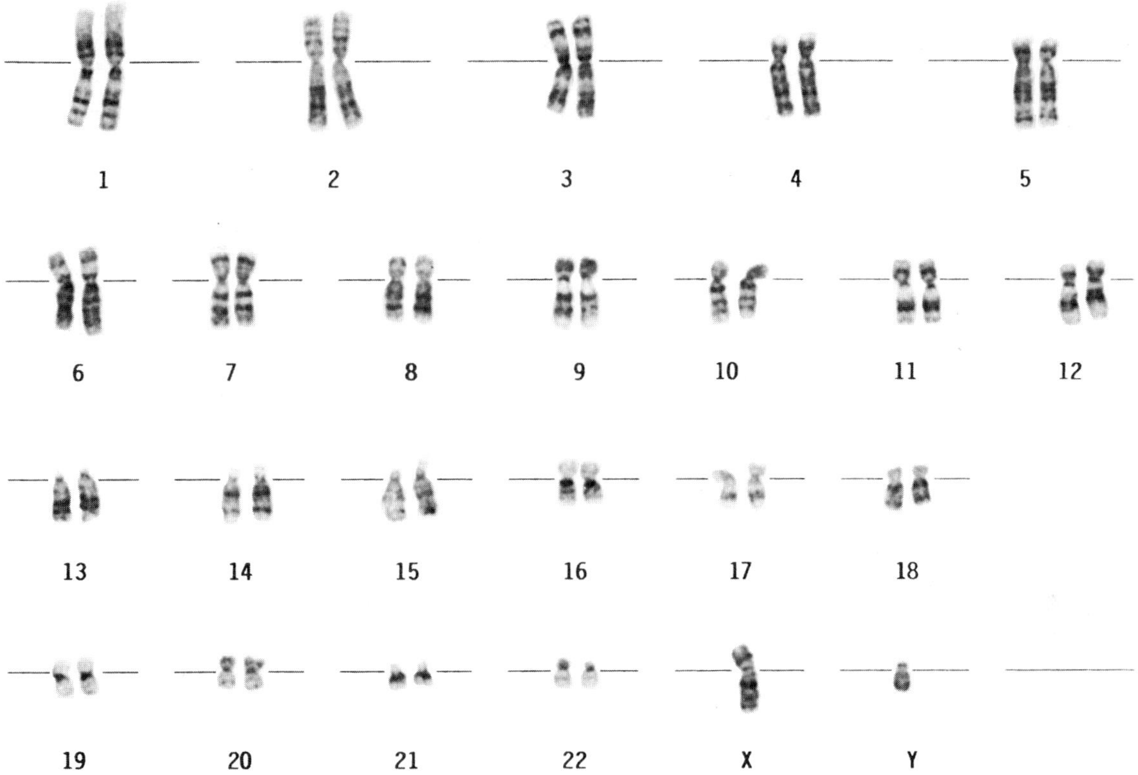

Figure 22–11 A male karyotype

Figure 22–12 A female karyotype

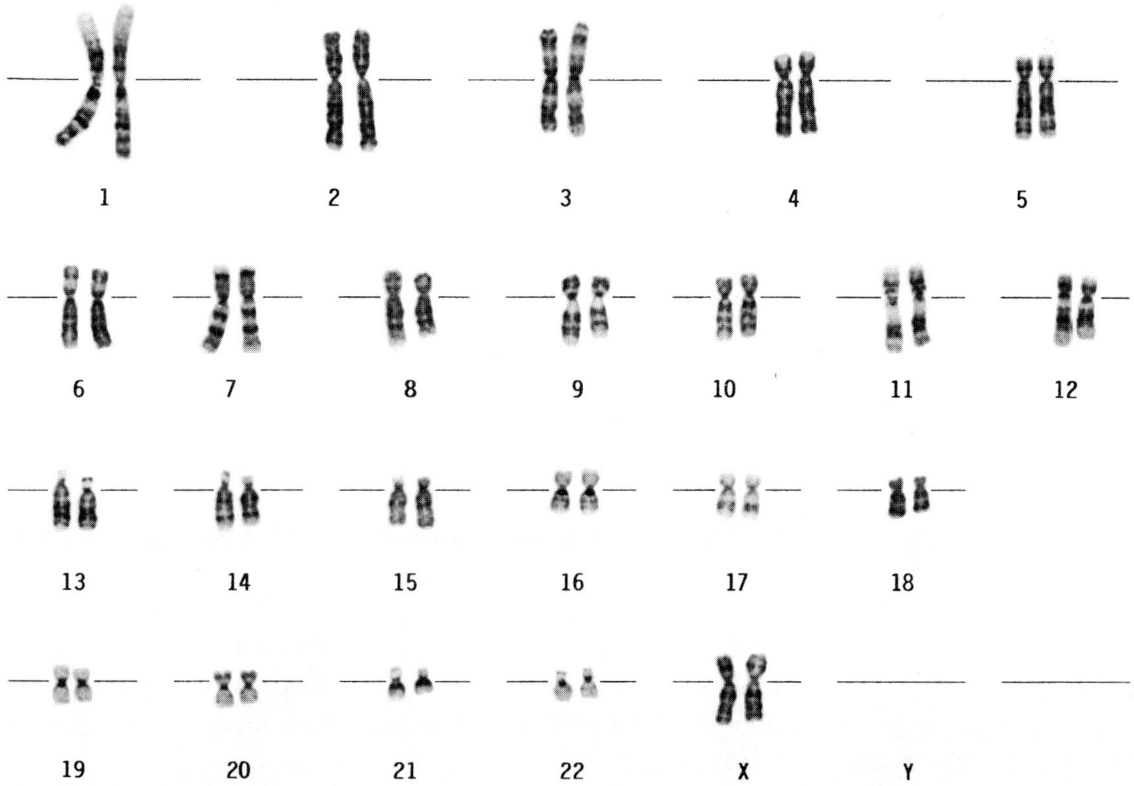

Do the chromosomes in Figure 22–10 on page 576 look the way you imagined that they would look? Remember, these structures are incredibly small. These pictures were photographed through a microscope while being magnified 1000 times. If they are to be studied, the picture can be cut up and the chromosomes arranged and glued down as shown in Figure 22–11 on page 577. Compare Figure 22–10 on page 576 and Figure 22–11. For instance, look at the two chromosome 1s in Figure 22–11. Can you find these same two chromosomes in Figure 22–10? A set of chromosomes that has been arranged as in Figure 22–11 is called a **karyotype.**

The cells for the karyotype in Figure 22–11 came from a boy. Cells from a girl were used to prepare the karyotype for Figure 22–12 at the bottom of page 577. Look carefully. You will be able to see the difference between a male karyotype and a female karyotype.

The boys in your classroom have karyotypes similar to Figure 22–11. The girls have karyotypes like Figure 22–12. That's because all humans have the same basic set of human chromosomes. Even though chromosomes from all humans look very similar, they contain slightly different information. How do you know this must be true?

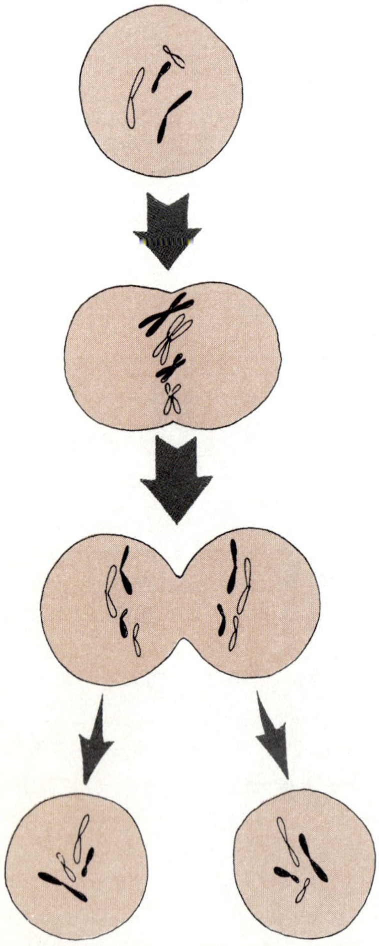

Figure 22–13 Cell division. How many chromosomes were in the original cell shown here? How many are in each of the new cells?

Chromosomes During Division

Remember, every cell in the body has a set of chromosomes. Lung cells, liver cells, bone cells, and brain cells; old cells and new cells; all of them have chromosomes. Where do all of these chromosomes come from? They are produced inside the cells. Just before a cell divides, it copies every one of its chromosomes. Then, when the cell divides, each new cell gets a copy of each chromosome. In this way, the two new cells are identical to the original cell. (See Figure 22–13.)

The whole process is amazing if you stop to think about it. Normally, a cell has 46 chromosomes. Right before the cell divides, it contains 92 chromosomes. All of these chromosomes have to be equally shared so that each new cell gets 46 of them—two chromosome 1s, two number 2s . . . just any combination of 46 chromosomes won't do. Aren't you glad that your cells do all of this automatically so you don't have to tell them what to do?

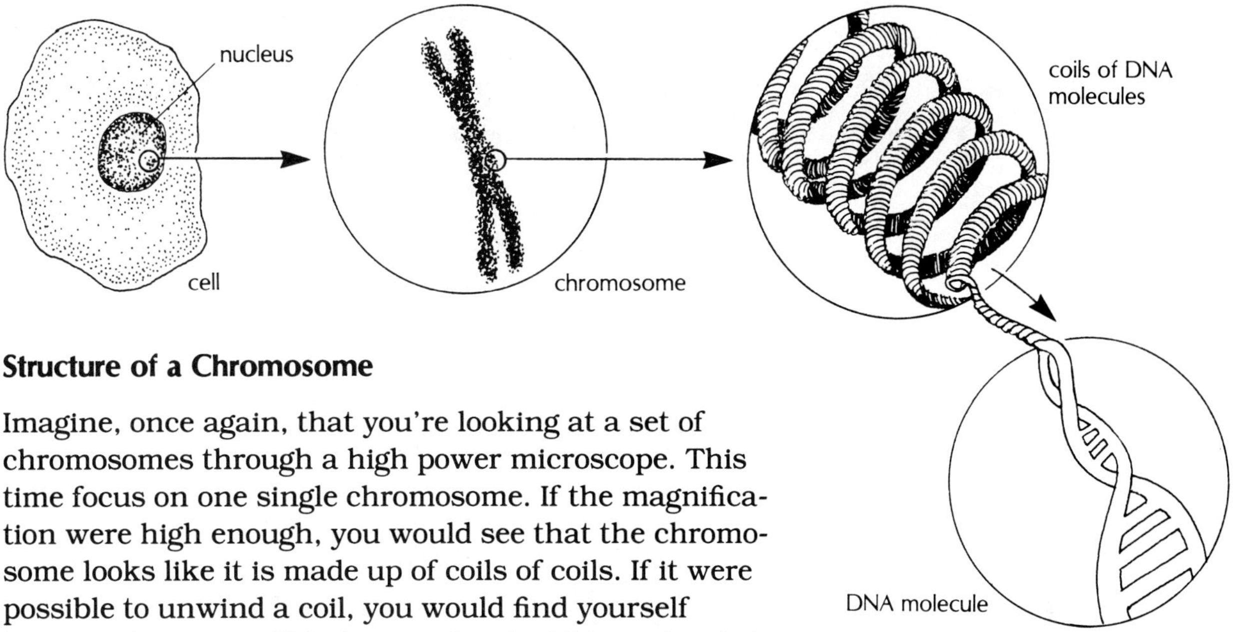

Figure 22–14 Chromosomes are made up of DNA.

Structure of a Chromosome

Imagine, once again, that you're looking at a set of chromosomes through a high power microscope. This time focus on one single chromosome. If the magnification were high enough, you would see that the chromosome looks like it is made up of coils of coils. If it were possible to unwind a coil, you would find yourself looking at an incredibly long molecule. This molecule is DNA, or *d*eoxyribo*n*ucleic *a*cid. DNA is the molecule of life. It contains all of the information needed to operate a cell, organize tissues, and control the action of organ systems to make a body. (See Figure 22–14.)

Thousands of scientists from all over the world are studying the way DNA works. Keep your ears open because you will be hearing more about this molecule.

ANALYSIS QUESTIONS

1. Close your eyes and picture a typical chromosome.
   a. List 3 words that describe a chromosome.
   b. Sketch what a chromosome looks like.
2. How many chromosomes are in Figure 22–10 on page 576? (Be careful not to mark on the picture when you count.)
3. The chromosomes in the reading have been magnified 1,000 times. This line, – , is 2 mm long.
   a. How long would it look if it were magnified 10 times?
   b. How long would it look if it were magnified 1,000 times?
4. How many chromosomes are in the karyotype shown in Figure 22–11 on page 577?
5. Why do you think researchers and doctors would rather look at a karyotype than at the chromosomes as they appear through a microscope?

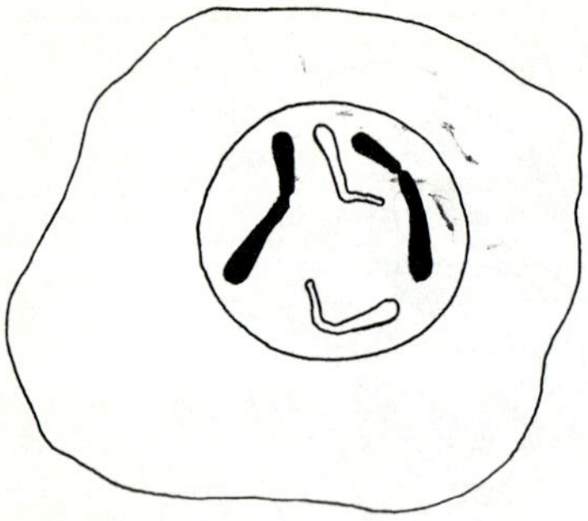

Figure 22–15

6. List at least 3 chromosome characteristics that are used to make a karyotype. Look at Figure 22–10 on page 576 and Figure 22–11 on page 577 to help you.
7. Compare the karyotypes shown in Figures 22–11 and 22–12.
   a. How are they alike?
   b. How are they different?
   c. How can you tell a female karyotype from a male karyotype?
8. Can you tell the height of the boy whose karyotype is shown in Figure 22–11 on page 577? Explain.
9. Look at the cell in Figure 22–15.
   a. How many chromosomes does it contain?
   b. How many pairs of chromosomes?
   c. Sketch the cells that will result after this one divides.
10. How many chromosomes do you think a goldfish has in one of its cells? Explain.

CHALLENGE

The pictures show chromosomes that are magnified thousands of times. Actually, cells and chromosomes are very small. Eight thousand white blood cells can fit into a box that is 1 millimeter on each side. How many chromosomes would be in this box of cells?

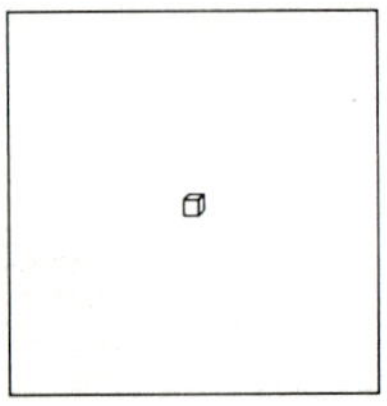

EXTENSION ACTIVITIES

1. Call a local hospital or a physician's office. Explain that you are studying genetics and would like to talk with someone about chromosome analyses. These are some questions you might want to ask:
   Why do people have their chromosomes studied?
   What kinds of cells are used?
   How long does it take to get results?
   How much does it cost?
   Where do you send the cells to have the chromosomes analyzed?
2. There are about 100 trillion cells in the human body. Figure out what that means in terms of chromosomes.
   a. Write out 100 trillion in numbers. How many zeros did you use?
   b. Since each cell has a set of chromosomes, how many *sets* of chromosomes are there in the human body? How many individual chromosomes?
   c. If a chromosome were the size of a pencil, how much room would it take to store all of the chromosomes from one person?

22-8 SEX CELL FORMATION

Each of us started life as a single cell. This single cell formed when a sperm from Dad joined an egg from Mom. It then divided again and again and again until it became a baby made up of millions of different cells. In each of the baby's cells there are 46 chromosomes. Think about this for a minute.

- How can a baby end up with 46 chromosomes in each of its cells when each parent has 46 chromosomes in each of his or her cells?

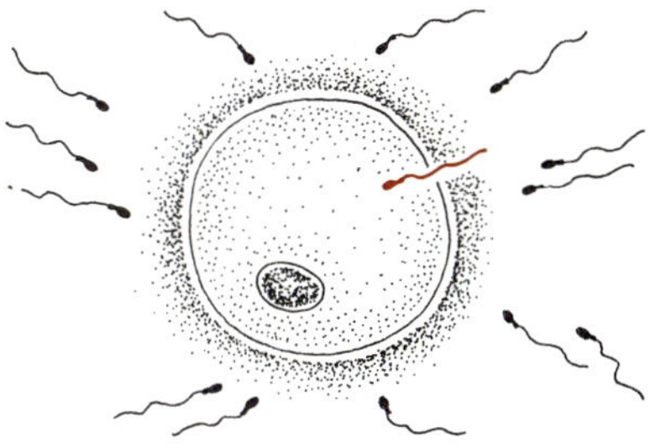

Work with model chromosomes and read about sex cell formation to learn the answer to this question.

MATERIALS

set of Chromosome Models
2 pairs of scissors
3 sheets of paper
tape or glue
reading: ***Sex Cell Division*** (page 584)
stapler

PROCEDURE

Work with a partner.

Part A. Using Chromosome Models

1. Get your set of Chromosome Models and look them over. Notice that ten of the chromosomes are from the "father's cells" and the other ten are from the "mother's cells." (Actually there would be 46 chromosomes in a real human cell but it would get very confusing for you to handle that many so only ten are provided for each "parent.")
2. Make sure your understand your goal. You are to use chromosomes from both the "mother" and the "father" to form a new "cell" that will contain 10 chromosomes. This new cell must, of course, be able to grow into a "baby."

Figure 22–16

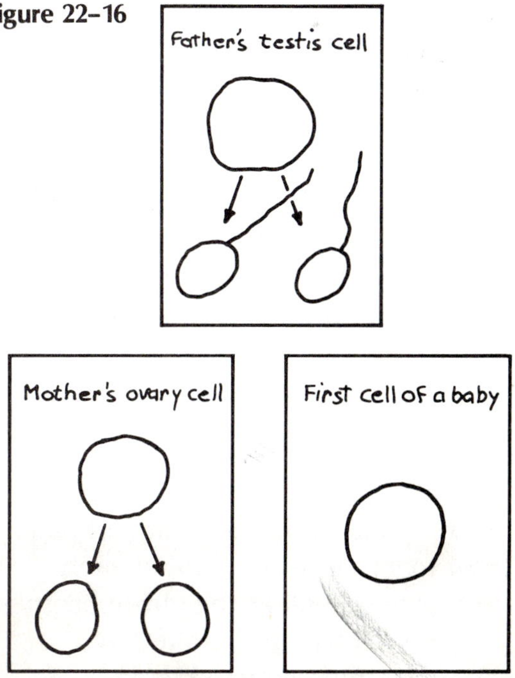

3. Mark your 3 pieces of papers as shown in Figure 22–16.

4. Cut apart the 10 chromosomes from the father and put them in his "testis cell." Cut apart the other 10 chromosomes and put them in the mother's ovary cell.
5. Throw away the scraps of paper on your desk, but make sure you don't throw away any chromosomes!
6. Now imagine that these cells are going to divide—you are ready to form sperm cells and egg cells. Think about what will happen to the chromosomes.
   a. The bottom half of the father's page shows two newly formed sperm. Move the chromosomes from the testis cell into these two new sperm cells.
   b. The bottom half of the mother's page shows newly formed eggs. Move the chromosomes from the ovary cell into these two new eggs.
7. Take the chromosomes from one egg and from one sperm and put them into the "baby cell."
8. Look at your baby cell. Does it contain ten chromosomes? If you want to go back and redo the chromosomes for the egg and sperm, do so. If you are satisfied with the baby's chromosomes, go to Part B. **Do not tape or glue the chromosomes yet.**

Part B. Doing the Reading

1. You and your partner should read ***Sex Cell Division*** starting on page 584 together. At the end of each paragraph, look at the chromosomes in your "cells" and make any changes that you need to do.
2. When you have finished the reading, tape or glue down the chromosomes, sign your names to each of the papers, and staple the three pages together.
3. Answer the analysis questions.

ANALYSIS QUESTIONS

1. Imagine that the cell in Figure 22–17 is from an ovary or a testis.

Figure 22–17

   a. How many chromosomes are in this cell?
   b. How many chromosome pairs?
   c. Sketch the new cells that will form when this cell divides.
   d. How many chromosomes are in each of the new cells?
   e. How many matched pairs of chromosomes are in each new cell?
2. A body cell from a frog has 26 chromosomes.
   a. How many chromosomes are in a frog egg?
   b. How many are in a frog sperm?
   c. How many are in a nerve cell from a a male frog?
3. A sperm cell from a dog contains 39 chromosomes.
   a. How many chromosomes are in an egg from a dog?
   b. How many are in a dog's body cell?
4. What do you think is the most important thing for you to remember about sex cell division?

5. Think about one of your skin cells.
   a. How many chromosomes does it contain?
   b. How many did you inherit from your mother?
   c. How many did you inherit from your father?
6. Imagine that the cell in Figure 22–18 is about to divide.
   a. Draw the cells that will form if this cell divides to form body cells.
   b. Draw the cells that will form if this cell divides to form sex cells.

Figure 22–18

7. People with Down syndrome may have problems with just about every system in their bodies. Why do you think one small, extra chromosome can cause so many problems?
8. Why don't they just remove the extra chromosome 21 so people with Down syndrome will have 46 chromosomes?

THINK ABOUT IT

How many chromosomes are in a body cell of a "test tube" baby?

CONCLUSION

Look back at the focusing question on page 581. With your partner, write one paragraph that provides a clear answer to this question.

Sex Cell Division

Humans have 23 pairs of chromosomes for a total of 46 chromosomes. Not just any combination of 46 will do. In order to develop and grow correctly, a person needs 46 of the right chromosomes. If a cell is missing just one chromosome, that cell is likely to die. If a cell has one too many, it too will probably die. An exact balance of chromosomes is important for growth and survival.

Two Kinds of Cell Division

If sperm and eggs formed from the same kind of cell division as do all other body cells, they would all contain 46 chromosomes. Think about what would happen then—46 chromosomes in an egg, 46 in a sperm. After fertilization, the new cell would contain 92 chromosomes. That would never work.

A second kind of cell division, **sex cell division**, is needed. It occurs only in specialized organs. In females, sex cell division occurs in the ovaries. In males, it occurs in the testes. Sex cell division results in the formation of eggs and sperm (gametes).

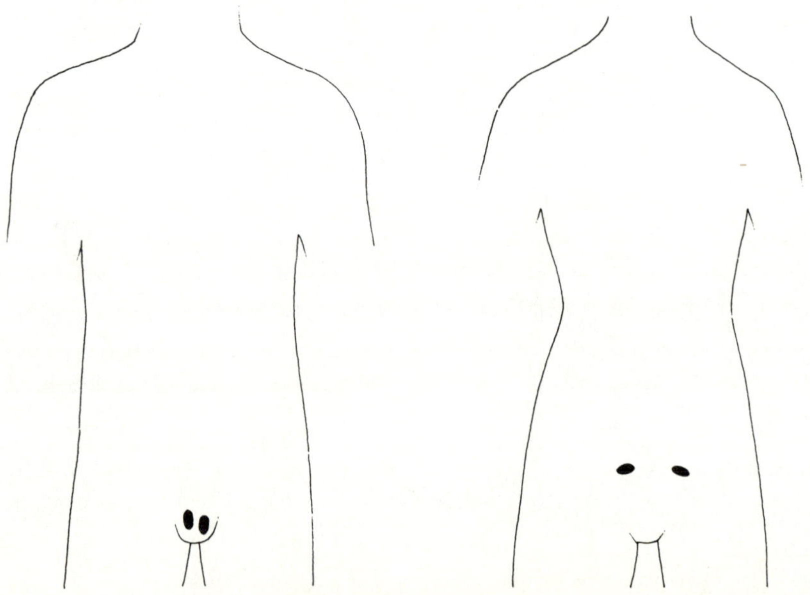

Figure 22–19 Sex cell division takes place only in ovaries and testes.

The Process

The important result of sex cell division is that the new cells have half as many chromosomes as did the original cell. If the original testis cell had 46 chromosomes, the new sperm will have 23 chromosomes. Likewise, an ovary cell with 46 chromosomes will give rise to egg cells with 23 chromosomes. When the original cell divides, the chromosome pairs separate. This way, each new cell contains one chromosome from each pair. Sex cell division must be very exact. Only one of each of the 23 pairs of chromosomes must end up in a gamete.

An Occasional Mistake

Sometimes, there is a mistake in sex cell division. As a result, a gamete may get too many or too few chromosomes. Usually this gamete is not very healthy and soon dies. If it does survive and ends up in a ferilized egg, that egg may die or it may divide a few times and then die. When this happens, the woman probably will probably never know that she was pregnant.

Figure 22–20 This is a karyotype from a person with Down syndrome. Is this person a boy or a girl?

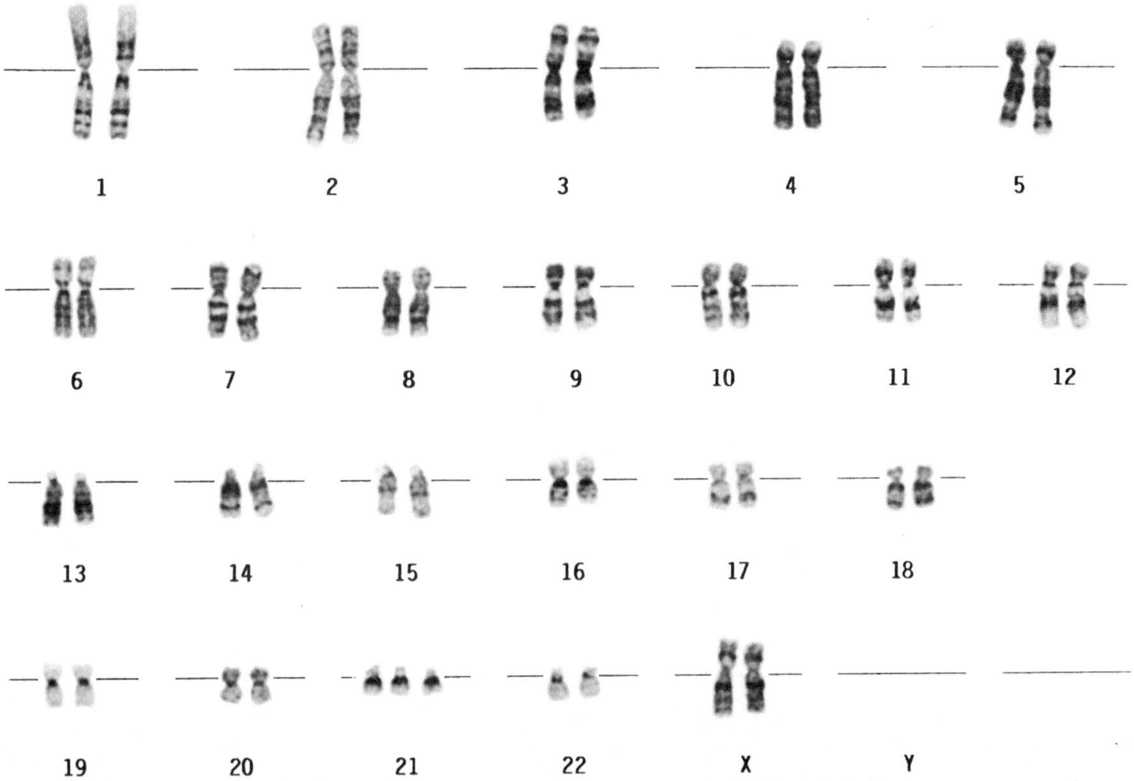

Sometimes the fertilized egg not only divides, but may survive for several weeks before it dies. In this case, the woman experiences a miscarriage. (An abnormal chromosome number is a common cause of miscarriages.) Occasionally, however, a baby manages to survive all of the problems that go along with an unbalanced chromosome number. For example, children who have an extra chromosome 21 have a total of 47 chromosomes. (See Figure 22–20 on page 585.) They have a condition known as Down syndrome. These children usually have some degree of mental retardation and may have heart problems, ear problems, intestinal problems—problems with just about every system in their bodies. All because of one extra chromosome.

The reason why some cells with an extra chromosome 21 survive is probably because that chromosome is so small. You would never find a cell, let alone a person, with an extra chromosome 1. That chromosome is just too big. If a cell had an extra number 1, too much information would be off balance and the cell could not survive.

(Go back and read Part B of the Procedure on page 582 to make sure you have remembered to do what you were supposed to do.)

22-9 MALE OR FEMALE?

When you studied sex cell division, you may have wondered about the sex chromosomes. What happens to the X and Y chromosomes when eggs and sperm are formed? This is an important question to consider because these two chromosomes determine if a baby will be a boy or a girl.

First, think about what happens during sex cell division in a female. Look at Figure 22–21 (left) while you read this. Find the circle that represents an ovary cell. As you can see, it contains 46 chromosomes—2 X chromosomes and 44 others. When ovary cells divide, they produce egg cells that contain 23 chromosomes—one X chromosome and 22 others. Every egg cell contains one X chromosome.

Next, think about what happens during sex cell division in a male. This time, find the circle in Figure 22–21 (right) that represents a testis cell. It contains 46 chromosomes—an X, a Y, and 44 others. When testes cells divide, they produce sperm. Each sperm contains 22 chromosomes plus *either* an X *or* a Y chromosome.

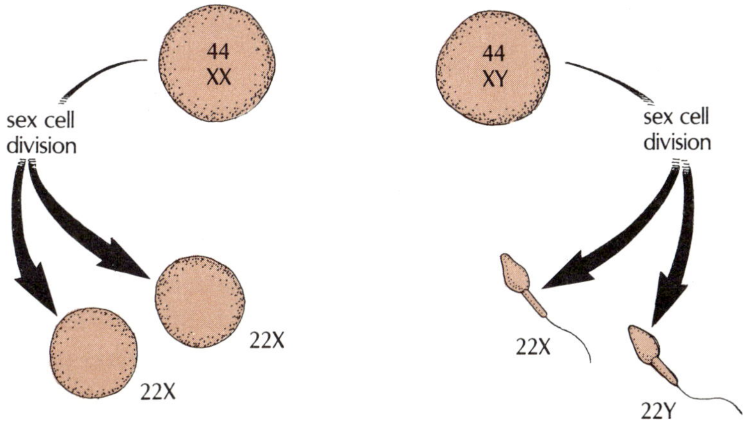

Figure 22–21 Sex cell division in females (left) and in males (right)

When one of these sperm fertilizes an egg, a new baby will result. Depending on the sex chromosome contained in the sperm, the baby will be either a boy or a girl.

- What are the chances that a baby will be a girl? What are the chances that it will be a boy?
- If a couple already has three sons, what are the chances that their next child will be a girl?

In this activity, after making a data page, you will play a game to find out what these chances are.

MATERIALS

2 sheets of paper

PROCEDURE

Work with a partner. You and your partner need to prepare the materials for this activity.

1. One of you should make a data page while the other makes the gametes.
   a. To make the data page, copy the data table shown below onto your paper. On the same sheet, set up up a place to record your predictions and actual results.

Number	Child's Chromosomes	Sex
1		
2		
• • • 19		
20		
Prediction: ______ boys and ______ girls Actual: ______ boys and ______ girls		

   b. To make the gametes, neatly fold and tear (or cut) a sheet of paper in fourths.
      1) Use two of the pieces of paper to represent egg cells. Label each piece "Egg—X + 22 other chromosomes."
      2) Use the other two pieces of paper to represent sperm. Label one "Sperm—X + 22 other chromosomes." Label the other "Sperm—Y + 22 other chromosomes."
      3) Fold all four of the gametes and give your partner either the two sperms or the two eggs. You keep the other two.
2. Now, with your partner, put together a "family" of 20 children.
   a. Make some predictions. How many of the 20 children do you think will be boys? How many girls? Record your predictions on your data page.
   b. Determine the sex of the first child in the "family."
      1) Without looking at the pieces of paper, pick one "egg" and one "sperm." Open up the two pieces of paper and record this child's chromosome make-up on your data table—for example: 44 + X + X.
      2) Decide if the child is a girl or a boy and record the child's sex on your data table.
   c. Refold the pieces of paper that represent the gametes. Mix the two eggs. Mix the sperm. Again without looking, pick one egg and one sperm. Record the data for child number 2.
   d. Repeat this process until you have data for all 20 children.
3. Consider your results.
   a. How many boys and girls are actually in this family? Record these figures on your data page and on the class data table.
   b. Answer the analysis questions.

ANALYSIS QUESTIONS

1. From which parent does a boy inherit
   a. his X chromosome?
   b. his Y chromosome?

2. From which parent does a girl inherit
   a. her X chromosome?
   b. her other X chromosome?
3. *In this game*, what were the chances that each child would be a girl? a boy? Explain.
4. If you played this game again and had another 20 children, how many boys do you predict you would have? How many girls?
5. *In real life*, what are the chances that a baby will be a girl? a boy? Explain.
6. Mr. and Mrs. Jones have three daughters. They are expecting another child. What are the chances that the baby will be a boy? Explain.
7. King John wants a son. After his sixth daughter was born, he decided to divorce Queen Jane and marry someone else. He thought maybe then he would have a son. Queen Jane tells him that a new wife won't solve his "problem." Who is right? Why?
8. Can a son inherit the information on his father's X chromosome? Explain. (You may want to use a diagram to illustrate your answer.)

CHALLENGE

Can a grandson inherit the information on his grandfather's X chromosome? Explain.

EXTENSION ACTIVITIES

1. Find out how many babies were born in your state last year. You can get this information from your state health department or from an almanac. How many of them were girls? How many were boys? How do these figures compare with what you expected?
2. Did you enjoy figuring out the problems that involve chances? If so, here are a few more problems for you to try.

A

What are the chances that a two-child family will have

1) a boy and a girl—in any order?
2) first a girl and then a boy?
3) two girls?

B

In a three-child family, what are the chances that

1) that the fourth child will be a boy?
2) there will be three girls?
3) there will be two boys and one girl in any order?

C

What are the possible combinations of boys and girls that can be born to a four-child family? Which combination is most likely to occur?

CHAPTER 23

GENETICS

Which traits do you think all four generations of this family share?

23–1 FAMILIES AND PEDIGREES

When studying genetics, it is important to know about families. The word *family* means different things to different people. To some people, it means brothers, sisters, parents, and grandparents. Others think of a family as being made up of two people—perhaps a parent and a child. Families differ not only in size but also in the way they function and in the way the people are related to one another.

Relationships

Family relationships are especially important to understand when you study genetics. You need to know if people are biologically related or not. If people are biologically related, they share at least one ancestor. For example, two brothers who have the same parents are biologically related; so are a mother and the child who was born to her.

Figure 23–1 Which of these families is most like yours? In the upper left is a family with three generations living together. In the lower left is a single-parent family. On the right is a family in which each parent had children from a previous marriage.

There are many relationships that are not biologic ones. A husband and wife are not biologically related, nor are brothers and sisters who have different parents.

Think about your own family for a minute. How many people does it include? How are the members of your family related to each other? Which relationships are biologic ones? Which ones are not biologic relationships?

Pedigrees

A person may want to make a family tree to show the way everyone in the family is related to one another. When a geneticist makes a family tree, it is called a **pedigree.** In pedigrees, symbols are used to stand for people. A circle means that the person is female; a square indicates a male.

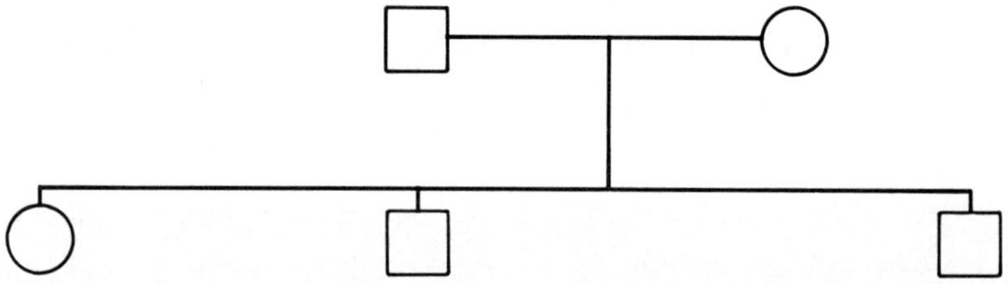

Figure 23–2 The Lincoln family

Figure 23–2 shows a simplified pedigree of the Lincoln family. You can see that this family is made up of a father, a mother, and three children—one girl and two boys.

Figure 23–3 is a more complete pedigree of the Lincoln family. It includes the name and birth date of each family member. You can see that the person who drew this pedigree did not know Mrs. Lincoln's or Thomas' birthdays.

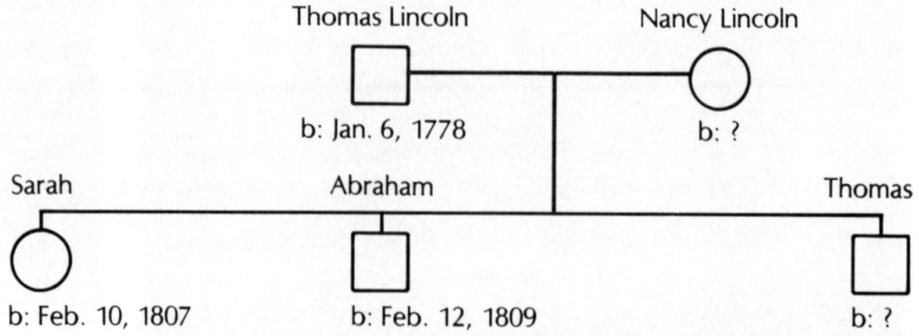

Figure 23–3 The Lincoln family

The more information included in a pedigree, the more useful it is. Figure 23–4 on page 594 is even more detailed than the others. By now you may have realized that we've been talking about Abraham Lincoln's family. This larger pedigree provides a lot of information about his parents, brother and sister, and children. The chart below will help you understand how to read the pedigree on page 594.

Make sure that you understand the rules and symbols for drawing pedigrees that are explained in the chart. You will be using this information throughout

Symbols and Rules for Drawing a Pedigree

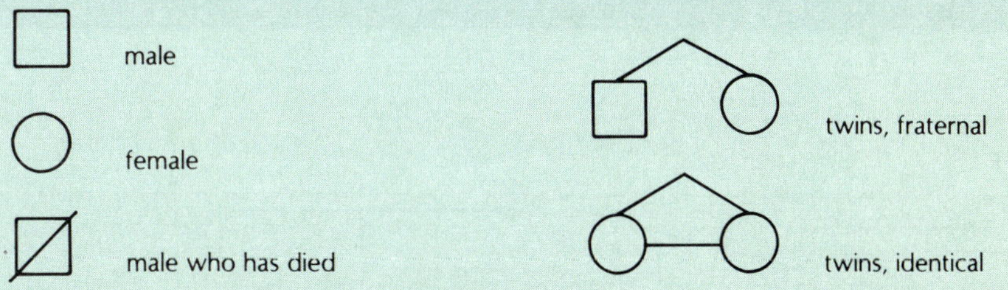

1. A line between two symbols means a marriage.

2. If a person has been married more than once, the marriage lines are numbered.

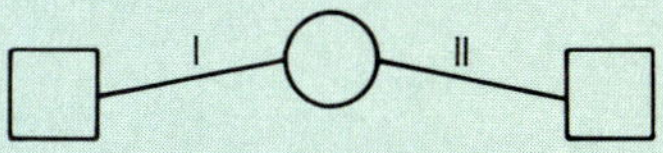

3. A line above the symbols means brothers and sisters.

4. Each person's name is printed neatly by his or her symbol.

5. Shaded symbols show what trait is being traced. The key to the shading is usually in the lower corner of a pedigree.

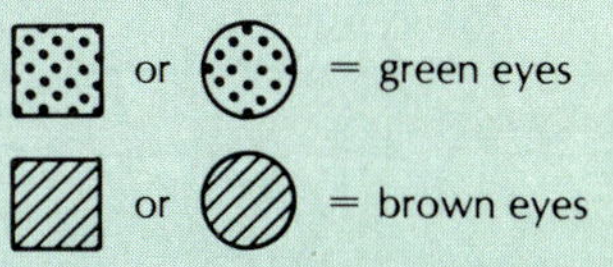

6. The generations may be numbered with Roman numerals on the left side of the pedigree. (A generation is all of the people in a family who are born at about the same time. For example, your grandparents are one generation, your parents are another generation, and you and your brothers and sisters make up a third generation.)

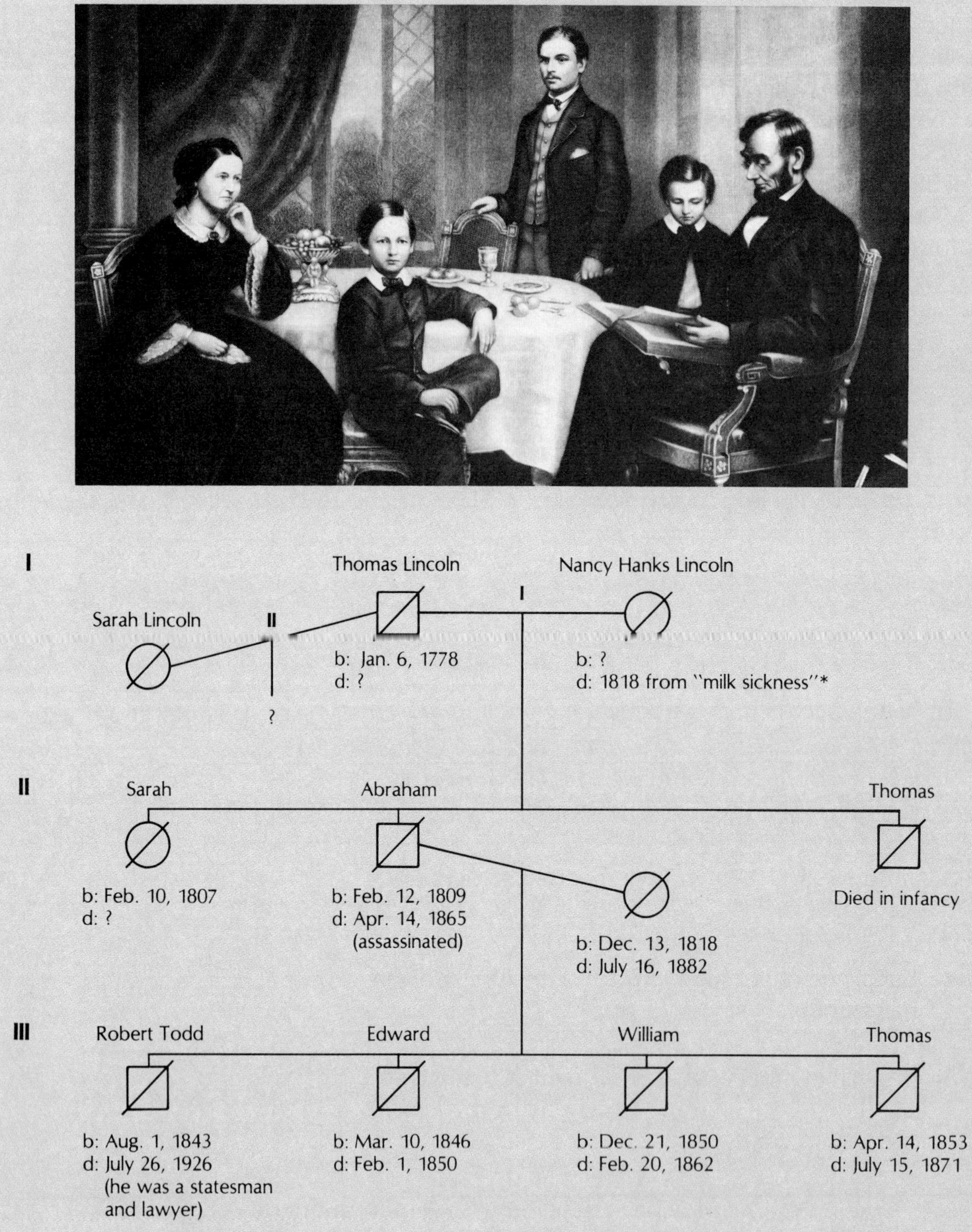

*Nancy Hanks Lincoln is said to have died of "milk sickness." She might have drunk milk that was poisoned as a result of plants that the cows had eaten.

Figure 23–4 The Lincoln family

the chapter. The more you look at pedigrees, the easier they will be for you to understand. Learning to read a pedigree just takes practice.

ANALYSIS QUESTIONS

1. What is meant by the term *biologically related?*
2. Define the word **pedigree.**
3. Refer to Figure 23–4 on page 594 to answer these questions about the Lincoln family.
   a. How many brothers did Lincoln have?
   b. What was Abraham Lincoln's mother's name?
   c. How many daughters did Lincoln have?
   d. Which people in this pedigree have birthdays in December?
   e. How old was Abraham Lincoln when he died?
   f. Which people make up the first generation?
   g. Which people in this pedigree are *not* Abraham Lincoln's *biological* relatives?

Use the information in the following paragraph to answer questions 4 and 5.

> Mr. and Mrs. Smith had three children—Susie, Sandra, and Steven. When the children were still young, Mr. Smith was killed in an automobile accident. Three years after the accident, Mrs. Smith married Mr. Jones. Mr. Jones had one daughter, Joy. Mr. Jones and Joy's mother had been divorced when Joy was six years old.

4. Draw a pedigree for each of the following families.
   a. The Smith family before the automobile accident.
   b. The Jones family before the divorce.
   c. The Smith-Jones family after Mrs. Smith and Mr. Jones' marriage.
5. Indicate with *yes* or *no* if each of the following pairs of people are biologically related or not.
   a. Mrs. Smith and Mr. Jones
   b. Susie and Sandra
   c. Steven and Mrs. Smith
   d. Sandra and Joy
   e. Mr. and Mrs. Jones
6. Some doctors include pedigrees in patients' medical charts. List at least 3 types of things that a doctor might include in a pedigree that most people would probably not include in a family tree.

Figure 23–5 Abraham Lincoln with his stepmother, Sarah Bush Johnston Lincoln

EXTENSION ACTIVITIES

1. Ask the librarian to help you choose a biography about Lincoln. Combine the information from Figure 23–4 on page 594 with the information in the book to make a more complete pedigree for the Lincoln family.
2. Many cities have genealogical societies. A genealogical society is a group of people who are interested in tracing their family histories. Some people find enough information to produce books about their families. Contact someone from your local genealogical society. Your library media specialist will be able to help you locate an address or phone number. Find out what kinds of things they include in their family trees, where they get their information, and what interested them in researching their family histories.

23-2 DRAWING PEDIGREES

Some people make pedigrees of their families as a way to record their families' histories. Animal breeders use pedigrees to keep track of their animals' ancestries. Medical professionals use pedigrees to help them understand a patient's medical background. A pedigree helps geneticists determine how traits are inherited. Pedigrees are useful tools to many people.

- What are three problems that people may have when trying to draw a pedigree?

You will be able to answer this question after you have drawn a pedigree for either your own family or for someone else's.

MATERIALS

pencil and eraser
chart: **Symbols and Rules for Drawing a Pedigree** (page 593)
information source about a family (relative, biography, encyclopedia—depends on which family you use to make a pedigree)
ruler
penny (to make circles for females)
cardboard square—1.5 cm on each side (to make squares for males)

PROCEDURE

1. Decide whose pedigree you will draw. You may do one for your own family, for a friend, for a neighbor, or for someone famous.
2. Make a first copy of your pedigree.
   a. Using a pencil, sketch a pedigree that includes as much as you already know about the family.
   b. Try to include at least three generations in your pedigree.
   c. Make a list of the other things you need to know in order to make your pedigree complete. Talk with an adult who can provide you with some or all of the missing information.
3. Recopy your pedigree. Look at your sketch to help you decide where to place the symbols on the paper.
4. When you think you have finished the pedigree, check to make sure that you have written down each person's name by the correct symbol and that you have used Roman numerals to number the generations.
5. Depending on whose pedigree you are drawing, do either step *a* or *b*.
   a. If you are drawing a pedigree for your family or for a friend's family, pick a trait to record on the pedigree. On page 598 are some examples of traits you could use.

nearsightedness
allergies
lefthandedness
height more than 6 feet

Shade the symbols for all of the people who have the trait. On your pedigree, remember to record which trait you are considering.

b. If you are making a pedigree for someone famous, record the birthdates of as many people on the pedigree as you can.

6. Answer the analysis questions.

ANALYSIS QUESTIONS

1. Look at your pedigree and tell how many of each of the following you included.
   a. people
   b. males
   c. females
   d. generations
2. What information could you add to your pedigree to make it more interesting?
3. Imagine that you are making a very complete pedigree on a family—your own or someone else's. List at least 3 people you could talk to or places you could go for more information.

CONCLUSION

Think about what you learned from doing this activity. What problems did you have? Do you think that this activity was worth doing? Do you think you will remember anything about this activity three years from now? Briefly write down your thoughts in answer to these questions.

EXTENSION ACTIVITIES

1. Call an animal breeder or a zoo veterinarian. Explain that you are a student studying genetics. Ask if they have pedigrees or "blood line" information on any of the animals. If they do, find out if you could have a copy of one. Compare it with the pedigree you drew. In what ways are they alike? different?
2. Ask the librarian to help you find a biography about someone who interests you—possibly a scientist, famous American, or athlete. After you have read the book, make a pedigree for this person.

Figure 23–6 A breeder of guinea pigs

23-3 HUMAN TRAITS

Imagine what the world would be like if everyone were *exactly* the same. No matter where you looked, you'd see the same person and that person would be you. Life would be very different. For instance, it wouldn't take very long to shop for clothes since you wouldn't have to sort through a variety of sizes. After all, everyone would wear the same size. Then, too, there would be fewer choices of color and style since everyone would like the same thing. Just think how the food and entertainment businesses would be affected. And how would this "sameness" affect doctors and health care?

Luckily, people are not all identical. We don't look alike. We don't act alike. We don't think alike. When we look at all of the ways in which people differ, we are looking at human **variability.**

- What are some ways in which people differ?
- What causes so much variability?

Think about how you would answer these two questions. Then, in this activity, talk about variability with other students. Afterwards, consider these two questions again.

MATERIALS

large piece of newsprint or chart paper
3 colored markers—red, blue, black

PROCEDURE

Work in a group with three to five other students.

Part A. Looking for Variability

1. Spend a minute organizing your group.
   a. Decide who will be the group recorder.
   b. Each person in the group should help the others stay on task. Since this is a fairly large group, it may be tempting to talk about other things.
2. Title your chart paper "Human Characteristics."
3. Have the recorder use the black marker to list the traits each group member recorded on his or her pedigree. (This should take about 2 minutes.)
4. Think of other ways in which people differ. Try to think of *at least 40* characteristics for the recorder to add to your list. You might want to start by considering all of the ways the people in your group differ from each other.

Part B. Analyzing Characteristics

1. Look over your list of characteristics. As a group, decide which ones might

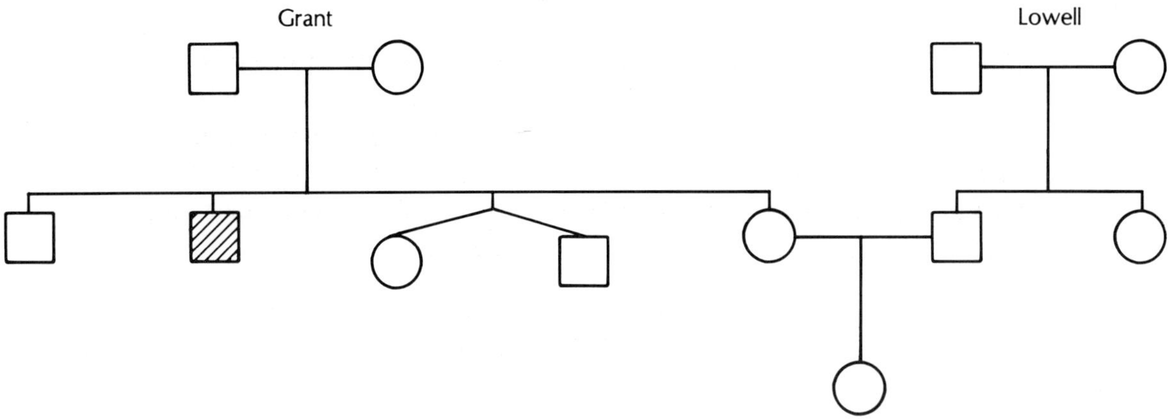

Figure 23–7

have a genetic basis—that is, which ones do you think tend "to run in the family"? Have someone other than the recorder circle each of these with a red marker.

2. Using a blue marker, someone else should circle the characteristics that have nothing to do with genetics. These will be the characteristics that are most influenced by the environment.
3. Hang your list where your teacher tell you to.
4. If you finish before the other groups are done, start to talk over the analysis questions.

ANALYSIS QUESTIONS

1. Give one good example of a trait that is totally controlled by
   a. what you inherit. (It doesn't matter what you do, you can't change this trait.)
   b. the environment. (It doesn't matter who you are; you can choose how you want to be for this trait.)
2. Define the two words below.
   a. genetic
   b. environmental
3. Do you think that more traits are genetic or environmental?
4. Imagine that you are studying a trait called "missing incisor tooth" in the Grant-Lowell family. Figure 23–7 shows their pedigree.
   a. Give several reasons why this trait may be environmental.
   b. Give several reasons why it might be genetic.
   c. Can you think of any traits in real life that might appear like this in a pedigree?
5. Think of at least 5 traits for which there is no variability. For these 5 traits, all people should be the same.
6. What do you think are some advantages to having lots of variability in a population?

23-4 TYPING BLOOD

Blood type is an example of a trait that you could show on a pedigree. Everyone has a blood type but they may not know what it is. The pedigree below shows the blood types for people in one family. As you can see, everyone is either type O or type A. (If you looked at the blood types of thousands of families, you would find only four blood types—A, B, AB, and O.)

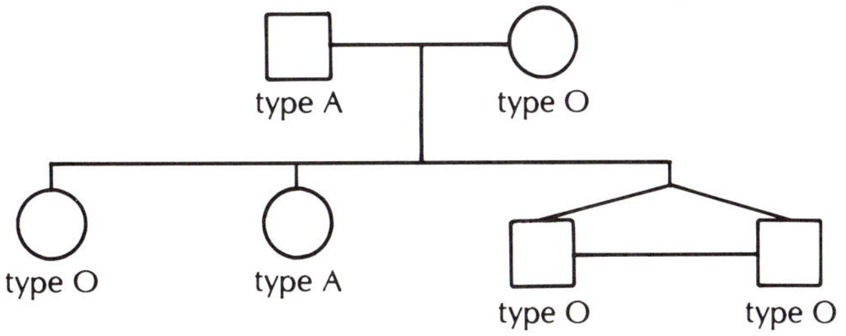

- How do you type a person's blood?
- What is your blood type?

In order to learn the answers to these questions, you will type your own blood.

MATERIALS

marking pen
clean microscope slide
cotton ball
alcohol
sterile, disposable lancet
2 toothpicks
antiserum—anti-A and anti-B
bleach solution for clean-up

PROCEDURE

1. Wash and dry your glass slide if it is dirty. Mark it with the marking pen as shown in Figure 23–8.
2. Your hands must be very clean. Wash them with soap and water.
3. Put rubbing alcohol on a cotton ball. Use it to clean the finger you are going to prick. After you have cleaned your finger, do not let anything touch it.

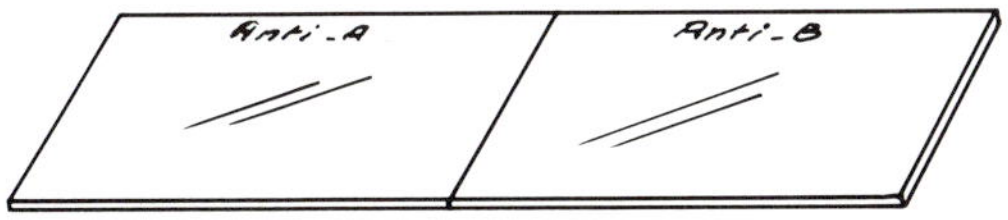

Figure 23–8

CAUTION
Blood can carry infectious diseases. This activity describes safe procedures for handling blood. Follow it carefully.

4. Take your lancet out of the wrapper. Remember! It is sterile. Do not let the sharp end of the lancet touch anything.
5. Sit down and relax. Then quickly prick the tip of your clean finger with the lancet. Do not set down your lancet. Put it into the beaker labeled "Used Lancets." **It must *not* be reused.**
6. Squeeze one drop of blood onto each half of your slide.
7. Immediately, place one drop of anti-A onto the drop of blood on the anti-A half of the slide. Add one drop of anti-B to the other drop of blood.
8. Use a different toothpick to mix the blood and antiserum gently on each half of the slide.
9. Wait about one minute, then use the chart below to determine your blood type.

If Your Blood Clumps With	Your Blood Type Is
anti-A	A
anti-B	B
anti-A and anti-B	AB
does not clump	O

10. Add your tally to the class data chart.
11. Clean up! Wash your microscope slides and your desk top with the bleach solution. (Bleach should kill any infectious microorganisms that are present.)

ANALYSIS QUESTIONS

1. Based on this lab activity, what is your blood type?
2. Give at least 2 reasons why your blood type results may be wrong.
3. Look at the class data chart.
   a. Which blood type is most common in your class?
   b. Which is the rarest?
4. If the science department had enough money to buy only one lancet, would it be okay to do this activity? Explain.
5. List at least 2 reasons why it was important to have a special place for putting the used lancets.
6. Use the pedigree in Figure 23–9 to answer the questions below it.

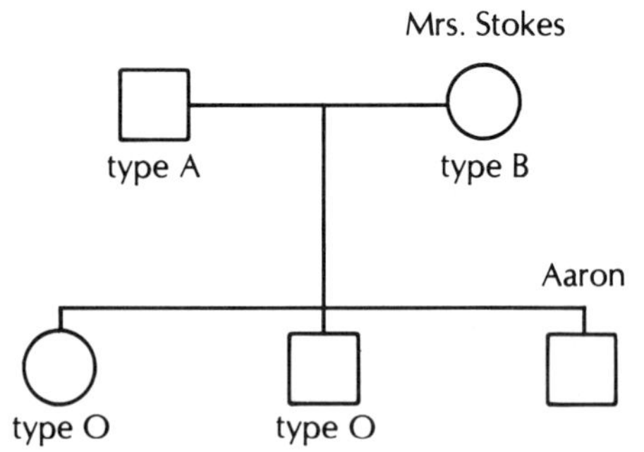

Figure 23–9

   a. What blood type does Mrs. Stokes have?
   b. Do you think it is possible for Aaron to have an AB blood type since his brother and sister both have type O?
   c. Do you think blood type is an inherited trait? Explain.
7. Matt, a student, has type AB blood. When he had surgery, he got a transfusion of type O blood. What type blood does he have now? Explain.

CONCLUSION

Look back at the focusing questions on page 601. Make sure you can answer

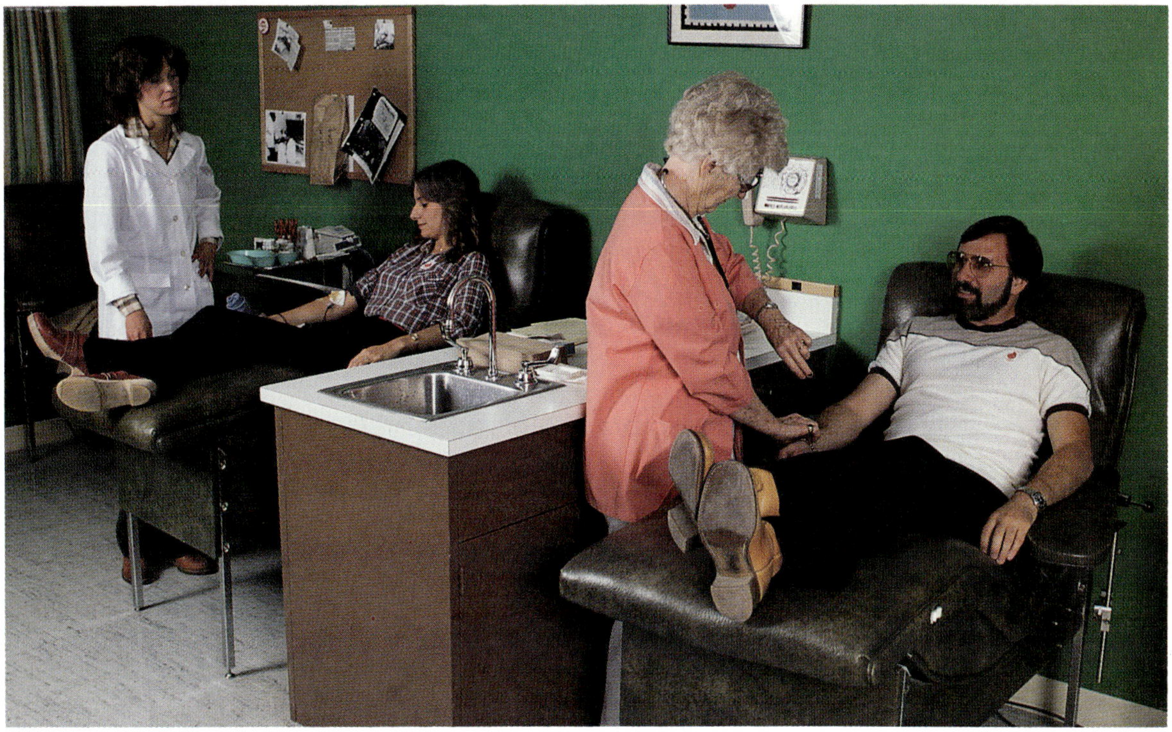

Figure 23–10 When people donate blood, their blood is typed.

them. Record at least 2 questions you still have about blood types.

EXTENSION ACTIVITIES

1. Contact a local hospital or blood bank. Explain that you are studying blood types in science class and would like to talk to someone who does blood typing. Find out what techniques are used. Also find out which blood type is the most common, which is the rarest, and why it is important to know a person's blood type. If possible, visit the lab.
2. Humans are not the only animals who have blood. Call a local veterinarian and ask about blood types of domestic animals. Find out if vets check an animal's blood type before doing surgery. (They may be able to tell you the blood types for animals such as horses, parakeets, and cats. If so, find out what is the most common blood type for each of these species.)

23-5 THE GENETICS OF BLOOD TYPE

Everyone has a blood type—it can be either A, B, AB, or O. It's fairly easy to type someone's blood. All you need is a microscope slide, some antiserum, and two drops of blood. Now you may be wondering what is in a drop of blood that determines a person's blood type.

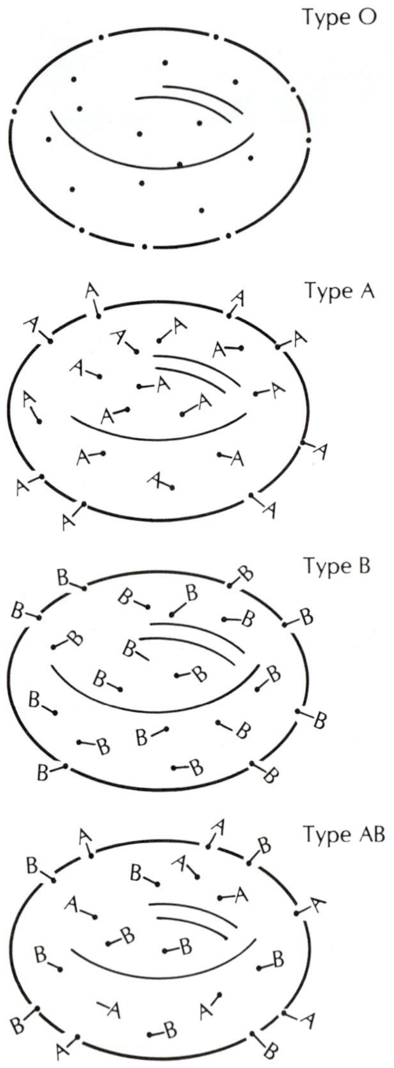

Figure 23-11 Every person has molecules sticking out of their red blood cell membranes. The kind of molecules that are present determine the person's blood type.

Cell Membrane Molecules

A drop of blood is mostly water and red blood cells. If you could look closely at one of the red blood cells, you would see that it is shaped like a doughnut without a hole in its center. Instead, it has a smooth, bowl-like depression on both sides. You would also notice hundreds of molecules sticking up out of the cell's membrane. If you were actually to count these molecules, you would find that there are about a million of them on every cell. These membrane molecules determine a person's blood type.

There are three different forms of membrane molecules. A person's blood type depends on which forms of the membrane molecules are on his or her cells. The simplest form is found in people who have blood type O. A person with blood type A has "A attachments" on each of his or her membrane molecules. "B attachments" make a person type B and a type AB person has both A and B attachments. Cells representing the four blood types are shown in Figure 23–11. In this figure, letters are used to represent the attachments instead of the actual chemical structures.

Reacting with Antiserum

When you type someone's blood, you use antiserum to find out which attachments are present. Antiserum is used because it contains antibodies; these antibodies bind with the A and B attachments. Antibodies work because they are very specific; they only recognize one thing. For instance, A antiserum contains antibodies that bind only A membrane molecules; they do not bind

Bs. The B antibodies are in B antiserum; they bind only Bs, not As. There aren't any O antibodies.

Now imagine what happens when a drop of A antiserum is added to a drop of blood. If the blood comes from a person who is either type A or type AB, the red blood cells are covered with A attachments. The antibodies in the antiserum will bind to these attachments and clumps will form. However, if the person is type B or type O, no clumps will form. That's because there are no A groups on the cells. There is nothing to which the anti-A can bind. Likewise, anti-B forms clumps when mixed with type B or type AB blood, but not with types A or O.

A Genetic Trait

Blood type is one example of a **genetic trait.** This means you inherit it from your parents. Your blood type was determined from the moment the sperm fertilized the egg that became you. Inside every sperm and egg are chromosomes, and those chromosomes contain all of the information it takes to make a person.

Your blood type is determined by information near the end of your chromosome 9s. This area is where the blood type gene is located. The blood type gene contains information for putting together membrane molecules. (See Figure 23–12.)

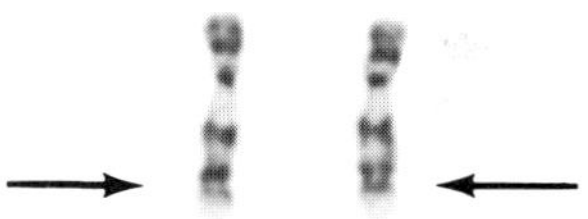

Figure 23–12 These chromosome 9s are from a karyotype. The arrows show the area where the ABO blood type gene is located.

There are several forms of the blood type genes and each form contains slightly different information. The A gene form directs the cell to attach As to the membrane molecules; the B gene form directs the cell to attach Bs; and, with the O form, neither As nor Bs are attached.

Remember, however, that you inherit *two* chromosome 9s, one from each of your parents. This means you can have two forms of the blood type gene. If both of these gene forms say "add As to the membrane molecules," you have an AA gene pair and you will have type A blood. People who have the BB gene pair have type B blood. Their genes give directions for attaching B groups to their membranes. If someone has an OB gene pair, what do you think is his or her blood type? If you said type B, you're right. Even though the O gene form doesn't direct the cell to add anything to its membrane, the B gene form causes "Bs" to be added. Just by "looking" at a blood cell, you could not tell if the person has an OB gene pair or a BB pair. In either case, their blood cells would be covered with "Bs."

Remember that while there are three forms of the blood type gene—A, B, and O—a person can inherit only two of them. One person cannot have all three forms of the gene.

ANALYSIS QUESTIONS

1. Lyle, a student in your class, has type B blood. He does not understand why his blood clumps with anti-B but not with anti-A. How would you explain this to him?
2. What are the gene forms for the three blood types?
3. How many different pairs of the three gene forms are possible? List them. (Example = AO)
4. Copy the table below and then fill in the possible gene pairs for each blood type. The first one is done for you.

Blood Type	Possible Gene Pairs
AB	AB
A	
B	
O	

5. Using your own words, define each of these terms.
   a. gene
   b. gene pair
   c. genetic trait
6. Here is a list of six cell types. Which ones do you think contain genes for blood type?
   bone cells — fat cells
   connective tissue cells — nerve cells
   skeletal muscle cells — white blood cells
7. Use the pedigree in Figure 23–13 to answer these questions.
   a. What gene pairs could Mrs. Ward have for blood type?
   b. Mike has an AB gene pair. From whom did he get his B gene?
   c. From whom did he get his A gene?

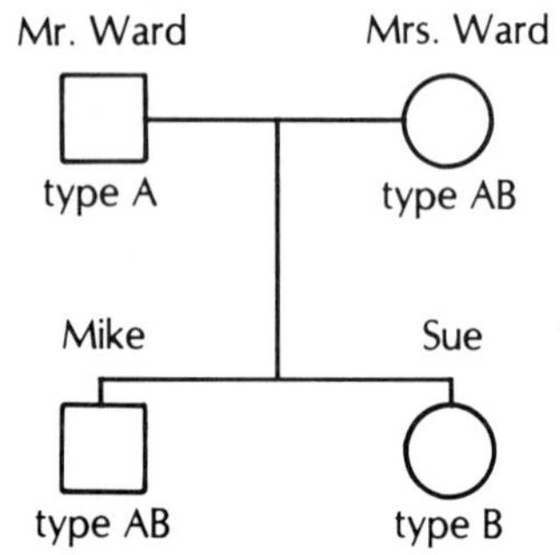

Figure 23–13

d. What gene pair does Sue have? Explain.
e. What gene pair could Mr. Ward have?

8. Blood type is one example of a genetic trait. List at least 3 other traits that you think are genetic.

EXTENSION ACTIVITIES

1. A surgeon must know a patient's blood type before doing a transfusion. If the patient is accidentally transfused with the wrong type of blood, he or she could die. Look up the ABO blood group and blood transfusions in a biology book or an encyclopedia. Find out which transfusions are safe and which could result in death. Report your findings to your class. Be prepared to explain why some transfusions are fatal.
2. As you learned, there are about one million ABO blood type molecules on the surface of every single red blood cell. Do you have any idea how many a million is? Use a 3″ × 5″ notecard to help you with this idea. Predict how many dots you can fit on one side of the notecard and how long it would take you to make a million dots. Then time yourself while you count the number of dots you can fit on the notecard. In conclusion, decide what size card you would need to hold a million dots and how long it would take you to make that many dots.

23-6 A TASTE TEST

Do you like the taste of cauliflower? licorice? How about sausage? Most people have definite likes and dislikes when it comes to food. Sometimes their attitudes depend on how the food was prepared the first time they tasted it or whether or not they ate the food when they were growing up. Another reason for our different opinions about foods lies in our genes. To some degree, we inherit our tastes.

The Taste of PTC

Foods may not taste the same to you as they do to someone else. The way something tastes depends on such things as the structure of your taste buds and the enzymes in your saliva. PTC is an example of a substance that may taste different to one person than it does to another. Before you read on, take the PTC taste test described in the box at the bottom of the page.

Well, how does PTC taste to you? Whether or not you can taste this chemical depends on your PTC genes. If you have inherited at least one "taster gene," you will be a PTC taster.

So far, no one has figured out how the PTC gene works. However, we can come up with a theory about

The PTC Taste Test

1. Place the piece of PTC paper that your teacher gives you on your tongue.
2. Leave it there for about 10 seconds. Don't chew it but you may want to move it around with your tongue.
3. Throw away the piece of paper where your teacher tells you.
4. Think of one or two words to describe the way the paper tasted.

its action. For instance, we know that PTC is a chemical. Perhaps this chemical is tasteless but, if it is cut in half, the small molecules that result may have a very distinctive taste. If this is the case, people who taste PTC would have the enzyme that breaks down PTC molecules. People who do not have the enzyme would not be able to split PTC molecules so they would not taste anything when they put PTC in their mouths. This would explain the two forms of the gene. The tasting form would produce an enzyme to split PTC molecules; the nontasting form would not.

Genetic Traits

You have now studied two genetic traits—PTC tasting and blood type. Keep these traits in mind while you read the following four "genetic rules." These rules apply to all genes and to all genetic traits.

- *Every gene has more than one form.* There are three forms of the blood type gene (A, B, and O) and two forms of the PTC gene (taste and nontaste). The various forms of the genes are represented by letters. In the case of the PTC gene, T stands for the tasting form and t stands for the nontasting form. The letter chosen to represent a gene is usually obvious, such as T for taste.
- *For any trait, a person only has two forms of the gene.* These two forms are shown by writing two letters. This "rule" means a person can have an AO gene pair but not an OAO. Likewise, a PTC taster could be TT, but not TTT or TTTT.
- You inherit your genes from your parents. *For every trait, you inherit one gene form from your mother and one from your father.*
- *A genetic trait is unchangeable.* Since you cannot change your genes, you cannot change the traits. A type O person cannot change to type AB nor can a nontaster become a taster.

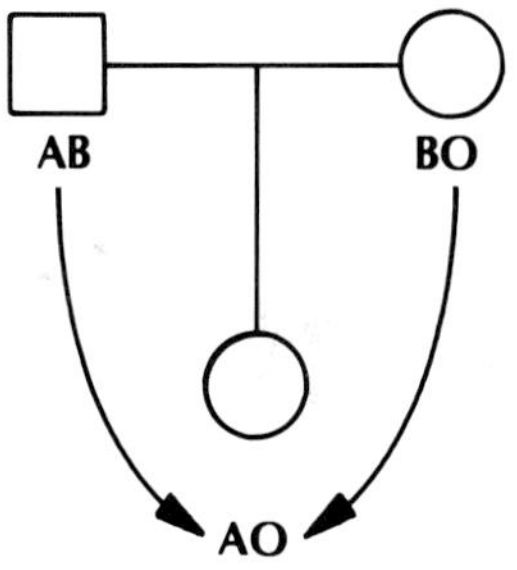

Figure 23–14 Which genetic rules are illustrated in this pedigree?

Close your eyes and try to repeat these four rules. If you can't remember some of them, go back and reread this section. If you understand these rules, the rest of this reading will be easier for you.

Tt—Taster or Nontaster?

It is obvious that someone who has the TT gene pair will be a PTC taster and that a tt person will be a nontaster. But what about people who have the Tt gene pair—are they tasters or nontasters? Think about what you have learned before you try to answer this question. The T gene form makes it possible for a person to taste PTC. Perhaps it does so by producing a "PTC enzyme." All it would take is one T gene for a person to taste PTC; so people who have either the Tt gene pair or the TT pair are tasters.

The T gene form is said to be **dominant** since a person only needs one T gene in order to be a taster. To say it more generally, you always see the effects of a dominant gene form. By comparison, the t gene form is said to be **recessive.** You see the effects of recessive gene forms only when the dominant form is not present.

Of the three blood type gene forms, A and B are both dominant. That's because people who have either the AA or the AO gene pair are type A. Likewise, the BB and BO gene pairs result in type B blood and the AB gene pair results in type AB blood. The O gene form is recessive. It must be paired before you know it is there. The OO gene pair is the only one that gives type O blood.

Figure 23–15 This scientist is studying the DNA molecule.

Thousands of Genes

The PTC tasting gene and the ABO blood type gene are two examples of human genes. Humans may actually have as many as 100,000 different genes. Each of these genes has a specific role to play in either the structure or function of the body.

While it would be interesting to study the effects of some of these 100,000 genes, it isn't easy to do so because most genes are responsible for the production of enzymes. The enzymes then do the work of the cell—whether it is to transport oxygen, digest a particular amino acid, or regulate calcium deposits in bones. Thousands of scientists in research labs around the world are working to figure out *what* all of the genes do, exactly *how* they work, and *where* each gene is located on which chromosome.

23-7 KEVIN, TOO?

Most of the time our genes do their jobs so we take them for granted. However, when a gene does not function normally, a person may have a genetic disorder. Just as there are thousands of genes, there are thousands of genetic disorders. This story is about a little girl, Laurie Ramsey, who has a genetic disorder called cystic fibrosis. The Ramseys have just had another baby and everyone is wondering if he too will be affected. After you have read this story, you decide. What do you think are Kevin's chances of having cystic fibrosis?

A Baby Brother

"Mom! When will Kevin wake up?" Laurie Ramsey was excited about having a baby brother. Today was his first day home from the hospital nursery and Laurie was asking all kinds of questions. She wanted to know if Kevin could play on the swings and if he liked pancakes. She also wanted to know if he would have to take pills. Obviously, Laurie doesn't realize what a baby can and can't do; but that's understandable since she's only four years old.

Mrs. Ramsey smiled at her daughter. She and her husband had been asking themselves some of the same questions. They too were wondering if Kevin would have to take pills every day. They also wanted to know if he would have a cough like Laurie's and be required to go through the same daily lung exercises. You see, the Ramseys remember all too clearly those first few months following Laurie's birth.

Figure 23–16 Laurie, at 4 years

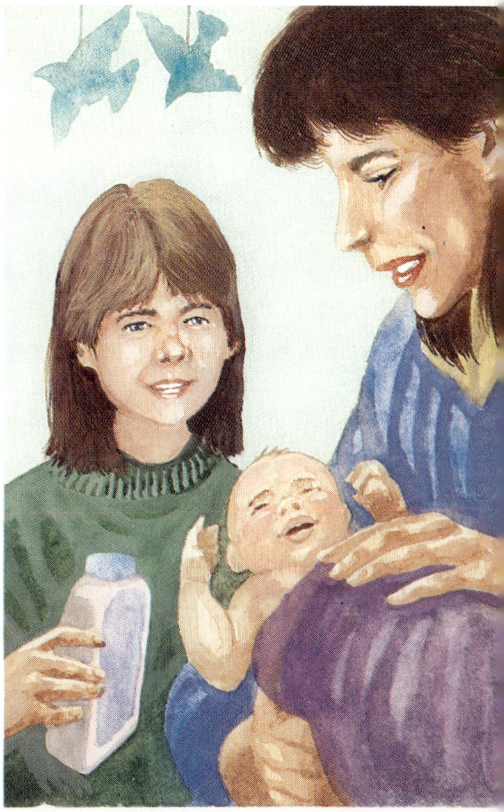

When Laurie Was a Baby

Laurie had seemed healthy when she was born. She had been a small, good-natured baby who seldom cried and loved to be held. Then, when she was about three months old, she developed a cough. It wasn't much of a cough; but since it wouldn't go away, the Ramseys had asked the doctor about it.

Figure 23–17

At first Dr. Fisk had been more worried about Laurie's size than about her cough. The fact that she was smaller than other babies her age even though she ate well was puzzling. How could a baby who ate all of the time gain so little weight?

Unknowingly, Mr. Ramsey provided an important clue when he mentioned that he and his wife noticed a salty taste whenever they kissed Laurie. "Oh," Dr. Fisk had said, "I'm glad you mentioned that; it might help us understand the problem. Laurie's symptoms sound like cystic fibrosis. We'll need to do some laboratory tests to check that possibility."

The Ramseys had returned to Dr. Fisk's office when the results of the tests were back. Dr. Fisk's suspicion had been right; Laurie had cystic fibrosis. The Ramseys had heard of cystic fibrosis but they didn't know what it was; so the doctor explained:

> "Cystic fibrosis is the most common genetic condition in white people. Many people have the recessive gene for this condition but do not realize it. That was the situation in your case; both of you have a gene for cystic fibrosis but you've been unaware of it. We say that you are *carriers* for cystic fibrosis. Laurie inherited a cystic fibrosis gene from each of you and so, with a double dose of the gene, she is affected.
>
> "The main problem with cystic fibrosis seems to involve the body's ducts and glands. For example, the sweat glands usually are affected. For some unknown reason, the body loses more salt than it should. That's why Laurie's skin seems salty when you kiss her. Also, the glands in the digestive system are easily clogged. When this happens, the enzymes needed for digestion can't get to the food. That's why Laurie isn't gaining weight even though she eats a lot."
>
> "Another serious problem involves the respiratory system. The small air passages in the lungs become filled with a thick mucus. The mucus makes it hard to breathe. Laurie coughs because she's trying to clear this mucus out of her lungs."

As you can imagine, the Ramseys had been very concerned. Laurie had breathing problems, she couldn't digest her food properly, and her body was

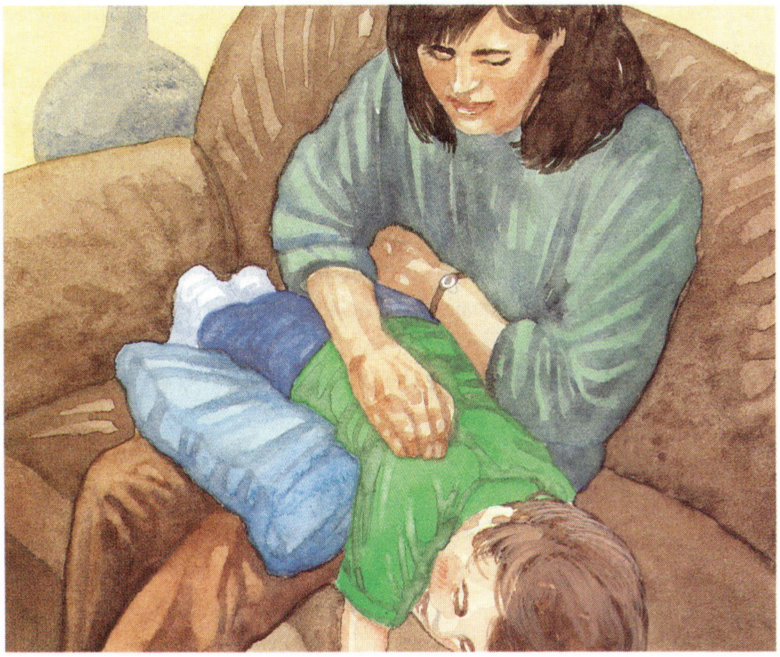

Figure 23–18 By gently "thumping" Laurie's back over her lungs, her parents can loosen the thick fluids so they will drain from her lungs.

losing too much salt. They had looked worriedly at Laurie but all they saw was a healthy, happy baby. It was hard to believe that she had so many medical problems. "What," they had asked, "can we do to help her?"

Dr. Fisk explained, "At this time cystic fibrosis cannot be cured. All we can do is try to manage her condition. It will be important for you to keep Laurie's lungs free of mucus. By gently thumping her chest and back, you can help her cough up the mucus. Think of it as a type of lung exercise. However, because of this mucus, she'll always have the cough so you'll have to learn to live with it. The mucus provides the right environment for bacterial growth; that will put Laurie at risk for developing lung infections. Consequently, we'll be using antibiotics to help her fight these infections.

"Because of her digestive problems, Laurie will need to take several other kinds of pills. I'll prescribe enzyme tablets to help her digest her food so that she can gain weight and grow. She'll also need to take vitamins to supply what her body can't absorb from her food."

Kevin's Chances

This all happened more than four years ago. By now the Ramseys know all too well how cystic fibrosis affects a person. Laurie, too, understands some things about cystic fibrosis. For instance, she knows that her breathing problems may prevent her from playing rough games or running as much as the other children. However, in most ways she's like other four year olds—she loves peanut butter sandwiches, chocolate milk, and cookies and, like other kids, she's fascinated by the new baby.

Right now she wants Kevin to wake up. Finally, she can't wait any longer, "Daddy, can I go see Kevin?"

Mr. Ramsey nodded, "Just don't make any noise." The Ramseys watch Laurie run to the baby's room. Then they continue their conversation. They are talking about Kevin's future. They know that cystic fibrosis is genetic. Now they're wondering—does Kevin have cystic fibrosis, too?

Today the Ramseys probably would not need to wonder whether or not Kevin has cystic fibrosis. As a result of recent DNA research, a fetus can now be tested for numerous genetic conditions, one of which is cystic fibrosis.

Well, what do you think are Kevin's chances of having cystic fibrosis? Consider what you have learned about cystic fibrosis while you answer the analysis questions.

ANALYSIS QUESTIONS

1. List at least 4 symptoms of cystic fibrosis.
2. List at least 4 ways that Laurie is being treated for cystic fibrosis.
3. Do you think other children can catch cystic fibrosis from Laurie? Explain.
4. Could Laurie outgrow cystic fibrosis? Explain.
5. Describe how you would feel if you found out that you had cystic fibrosis.
6. Now consider the Ramsey family.
   a. Draw the family's pedigree. Use these symbols to indicate carriers. 
   b. Record the gene pair for each person. Let **f** stand for the gene form for cystic fibrosis and **F** stand for the normal form of the gene.
7. Consider Kevin's chances of having cystic fibrosis.
   a. What are the possible gene pairs that he could have inherited from his parents?
   b. What do you think are the chances that he got the gene pair for cystic fibrosis?
8. Would it be possible for Mr. or Mrs. Ramsey to come down with cystic fibrosis as they grow older? Explain.

9. Imagine that Laurie is now grown up and has married a man named Larry. He is a carrier of cystic fibrosis. They have a daughter named Lisa.
   a. Draw a pedigree that includes Lisa, Laurie, and Larry.
   b. Show which gene pair each person in the family has.
   c. What are the chances that Lisa has cystic fibrosis? Explain your answer.
10. Why is cystic fibrosis said to be a recessive genetic condition and not a dominant one?
11. Do you think that you could be carrying the gene for cystic fibrosis? Explain.

EXTENSION ACTIVITIES

1. Find out more about cystic fibrosis. Hundreds of doctors and laboratory researchers are trying to figure out its cause. They know that the gene for cystic fibrosis is on chromosome 7 but they do not yet understand exactly what the gene does. Contact the Cystic Fibrosis Foundation for pamphlets about the condition or ask a library media specialist to help you find a recent article about it.
2. Cystic fibrosis is the most common genetic disorder among American whites; sickle cell anemia is the most common genetic disorder among blacks; and Tay-Sachs is the most common among people of Jewish descent. Learn the facts about one of these other two conditions and write a story about a day or a month in the life of a person who has the condition.

23-8 TWIN STUDIES

Many researchers study twins. They are especially interested in identical twins since these people have identical genes. You'd expect identical twins to be alike in most ways. However, it is important to ask whether these twins are alike because they have the same genes or because they live in the same environment. For instance, if a pair of identical twins is very shy, you might ask if they are shy because they both have "shyness genes" or if it could be because their parents are very quiet and seldom do things socially.

When trying to decide whether genes or environment control a trait, it's especially helpful to study identical twins who were raised in different homes. These twins have the same genes but were exposed to different environments.

- How can identical twins who have been raised in different environments help us understand the way traits are inherited?

Learn about one such set of twins and then answer this question.

MATERIALS

worksheet: ***Twin Studies*** (Figure 23–19)

PROCEDURE

Work in a group with two other people.

1. Imagine that your group is a team of researchers who will be studying a set of twins who volunteered for this study. The twins will be coming to your laboratory where for one week you will have them answer questions, take psychological tests, and undergo a series of medical tests. Before you meet the twins, discuss analysis questions 1 and 2 within your group and have one person record the group's answers.
2. Prepare to meet the twins by reading the information on the note card at the right.
3. Answer analysis question 3.

Names of twins: Peter and Jules

Early data: They were born in Kentucky in 1954. Their parents were killed in an automobile accident soon after they were born. There were no other children in the family.

Later data: Peter was raised by one set of grandparents—his mother's parents—in a small farming village in France. He lived and worked there.

Jules was raised by an aunt and uncle—his father's brother—in the southern part of the United States. He has worked in that family's textile business since he was 10.

The brothers have never talked or written to each other.

TWIN STUDIES

NAME ______________________

DATE ______________ PER ______

Trait/Characteristic	1 Prediction—Alike (A) or Different (D)?	2 Actual—Alike (A) or Different (D)?	3 Genetic Environmental, or Both?
a. shirt style			
b. occupation			
c. location of hair part			
d. dimples			
e. eye color			
f. eyesight			
g. height			
h. facial hair			
i. language spoken			
j. favorite subjects			
k. activities/sports			
l. food allergies			
m. taste of PTC			
n. blood pressure			
o. blood type			
p. number of cavities			
q. aggressiveness			

Figure 23–19

4. Look over the traits listed on the worksheet.
   a. For each trait, discuss whether you think Peter and Jules will be alike or different.
   b. Record your predictions in column 1 on the worksheet.
5. Listen while your teacher reads the results of the interviews and tests in which Jules and Peter participated. Fill in the second column on your worksheet while you do so.
6. In your group, discuss and answer analysis questions 4 through 6.
7. Look at the data on your worksheet and discuss the results in your group. Based on this study and what you know about these traits, decide if you think each is controlled by genes, by the environment, or by both. Record your decisions in the third column on the worksheet.
8. As a group, discuss and answer the rest of the analysis questions.

ANALYSIS QUESTIONS

1. Define each of the following terms:
   a. identical twins
   b. fraternal twins
   c. identical triplets
2. Think about identical twins.
   a. List at least 3 ways in which any pair must be the same.
   b. List at least 3 ways in which a pair of identical twins is likely to differ.
3. Sketch a pedigree for Peter and Jules. Use question marks to indicate where you don't have enough information.
4. If, for a given trait, Peter and Jules appear to be alike, does that mean the trait is genetic? Explain.

Figure 23–20 Identical twins

5. If, for a given trait, Peter and Jules are different, does this mean that trait is *not* genetic but that it is determined by the environment? Explain.
6. Imagine what would have happened if Peter and Jules had not been separated. If they both had been raised by their French grandparents, do you think they would have been more alike or more different? Explain.
7. Would studies of this type work using fraternal twins instead of identical twins? Explain.

CONCLUSION

Discuss the focusing question in your small group until you decide upon an answer that satisfies everyone.

EXTENSION ACTIVITY

In Minnesota some researchers have been studying pairs of identical twins who were raised apart. You may want to learn about their findings. Ask a library media specialist to help you find more information about these twin studies. Tell your family or one of your friends about what you learn.

Figure 23–21 These thirteen pairs of twins attend the same school in Sulphur, Louisiana. Which ones do you think must be fraternal twins?

Identical or Fraternal?

"Guess what! We're going to have twins!" While people are usually surprised by the announcement that twins are expected, twins really aren't that rare. In the United States, about 1 in 80 pregnancies results in twins.

Twins can be either identical or fraternal. Which kind they are depends on how they were formed.

Identical twins are formed from the same egg and sperm. At first, the fertilized egg starts to develop as it would for a single child. It divides repeatedly to form a tiny ball of cells. However, for some unknown reason, this ball of cells may separate into two groups of cells. Each group of cells then goes on to form a baby. When this happens, identical twins result.

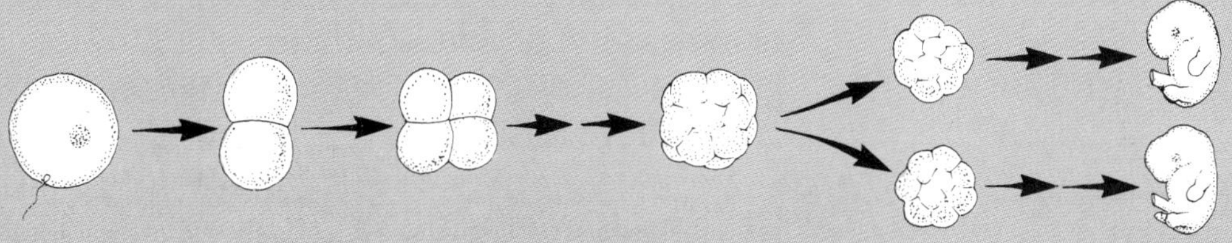

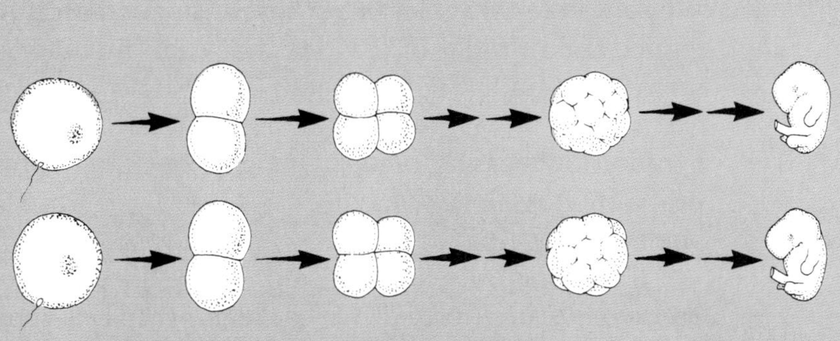

Fraternal twins result when two eggs are fertilized. Each twin is formed from its own egg and sperm.

Genetically, identical twins are exactly alike since they develop from the same egg and sperm. Since fraternal twins are formed from different eggs and sperms, they do not have exactly the same chromosomes. As a result, fraternal twins are not alike genetically. They are no more alike genetically than are ordinary brothers and sisters.

23–9 IT INHERITED?

"My brother has 8 cavities. Mom and Dad have lots of cavities, too. Are bad teeth inherited?"

Nicole

"Our family is very musical; each person in the family either sings well or plays a musical instrument. Is musical ability inherited?"

Scott

Students asked these questions after they had drawn their family pedigrees. When you drew your pedigree after reading pages 597 and 598, you may have used it to trace a characteristic that is found in your family and you, too, may have wondered if it were inherited. It usually is not easy to decide whether or not a trait is genetic. Take a closer look at the two examples given above to see why.

Dental Caries (Cavities)

When Nicole Sininsky drew her family pedigree, she recorded the number of cavities that each member of her family had had. (See Figure 23–22.) She concluded that getting lots of cavities is an inherited trait. Based on her own family's cavity record, she thinks that she too will probably have lots of cavities. What do you think? Is she right?

If we look only at her pedigree, we may decide that bad teeth are inherited. We can even think of genetic

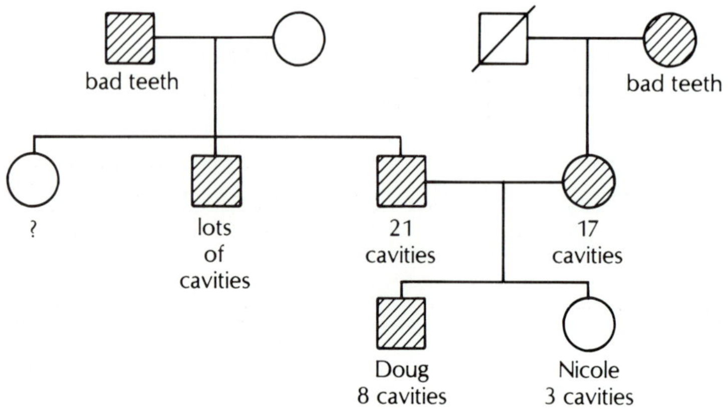

Figure 23–22 The Sininsky family and dental caries

explanations for tooth decay. For instance, people may inherit a gene that produces weak tooth enamel or the environment in their mouths may be good for the growth of the bacteria that cause decay. However, there also may be nongenetic reasons for the dental problems. The family may live in an area where fluoride is not added to the water or they may not be thorough about brushing and flossing their teeth. So, as you can see, a tendency for tooth decay may be inherited—but other causes are also possible.

Musical Ability

Figure 23–23 shows the Morris family pedigree. Scott claims that he plays the trombone so well because he comes from a musical family. His father plays in a band and his mother plays the piano and sings. Again, you could think of reasons why musical ability could be inherited. Genes affect the structure of the vocal cords, the anatomy of the ear, and the development of specific parts of the brain.

You could also think of nongenetic reasons for having so many musical people in one family. Perhaps these people enjoy listening to music. They may constantly listen to the radio or tape player. Symphonies and concerts could be their main form of entertainment and they may belong to musical groups and practice their instruments for hours each day. In other words, could Scott be good in music because he was raised in a musical environment and not because of an inherited musical talent? While we can't say for sure, it's likely that both heredity and environment are involved in musical ability. Scott probably inherited some "musical genes" and he was also raised in a musical environment.

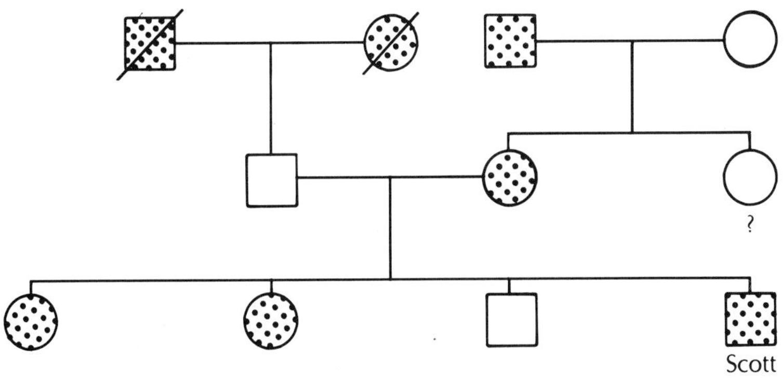

Figure 23–23 The Morris family and musical ability

Other Traits

Musical ability and dental problems are only two examples of traits that may be traced through families. There are thousands of other traits that can be considered. Nose shape and height are examples of traits that involve the way a person looks. Other traits such as shyness and artistic talent involve personality or ability. Body function traits such as blood pressure and lung volume are important traits for medical reasons. These and other examples of traits are listed below.

HUMAN TRAITS

Appearance	Personality/Ability	Body Functions/Health
nose shape	musical talent	allergies
eye color	intelligence	asthma
hair color	shyness	blood pressure
toe length	artistic talent	body temperature
freckles	fear of heights	color vision
height	creativity	distance vision
weight	math ability	lung volume

Which of these traits can "run in the family"? Or, in other words, which traits are inherited? Sometimes the answer seems obvious; but usually it is difficult to decide exactly what we inherit from our parents. Usually, you can think of ways that both genes and the environment can affect a trait. For this reason, these traits are said to be **multifactorial.** That is, they are controlled by many factors.

ANALYSIS QUESTIONS

1. In your own words, explain what is meant by the term **multifactorial trait.**
2. Todd drew a pedigree for his neighbor's family. It's shown in Figure 23–24. The symbols he shaded show those people who are "very strong."
   a. How many people in this family are "very strong"?
   b. What do you think Todd means when he says someone is "very strong"?
   c. Give one argument that supports the idea that strength is genetic.
   d. Give one argument that supports the idea that strength is not genetic.

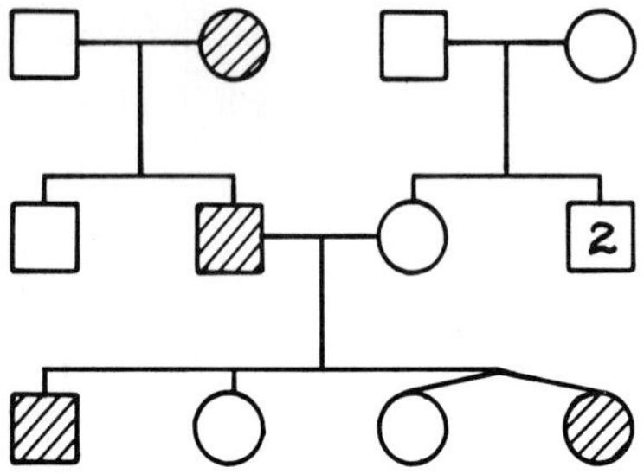

Figure 23–24

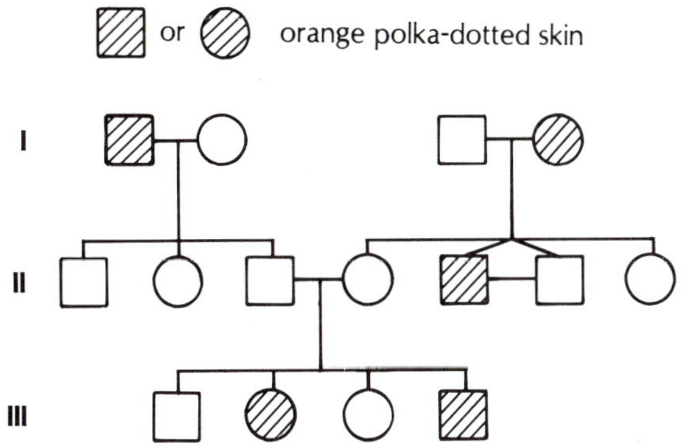

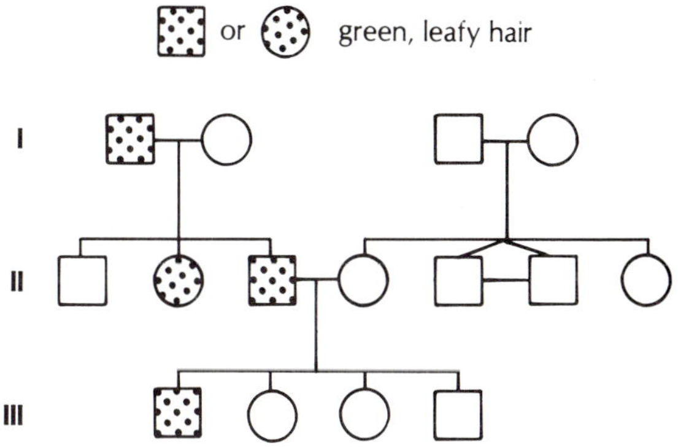

Figure 23–25

3. Several examples of human traits are listed in the table on page 622. List at least 2 more examples of traits for each of the following categories:
   a. appearance
   b. personality or ability
   c. body functions or health
4. Pick 3 traits from the table on page 622 that you think are definitely inherited. Explain why you chose those traits.
5. Figure 23–25 shows two copies of the same pedigree. Each pedigree has been used to show an imaginary trait. Decide which trait you think is most likely to be inherited. Explain your decision using the pedigree as evidence.

EXTENSION ACTIVITY

Talk to one of your parents about traits that seem to run in your family. Consider obvious physical traits but also ask about medical problems that you might be at risk for developing. Examples of multifactorial conditions are high blood pressure, diabetes, and allergies. Try to identify at least 5 traits that people in your family may inherit.

A–1 INTRODUCTION TO MICROSCOPES

pen: writer	scalpel: surgeon
camera: photographer	wrench: mechanic
telescope: astronomer	microscope: biologist

Do you recognize what these word pairs have in common? The first word in each pair names a tool; the second word is the name of a person who uses that tool.

The microscope is the tool a biologist uses to see structures and organisms that are normally too small to see. Dissections and x-rays allow us to look at and into large objects; we need microscopes for close-up views of very tiny structures.

When studying Life Science, you will need to know how to use a microscope. It will take time and practice for you to learn this skill.

- What is the correct way to carry a microscope?
- How do you set up a microscope?
- How do you focus on an object?

Learn the parts of a microscope and the basic rules for its use in order to answer these questions.

MATERIALS

microfilm #1
microscope
worksheet: ***Microscope Parts*** (Figure A–1)
checklist: ***Using a Microscope*** (Figure A–3)

PROCEDURE

Part A. Microscope Parts

Label the diagram of the microscope on the worksheet Figure A–1 while your teacher discusses each of the parts.

Part B. Using a Microscope

Work with a partner.

1. One of you should bring a microscope to your table. When you do so, carry the microscope with two hands. Grasp the arm with one hand while the other hand supports the

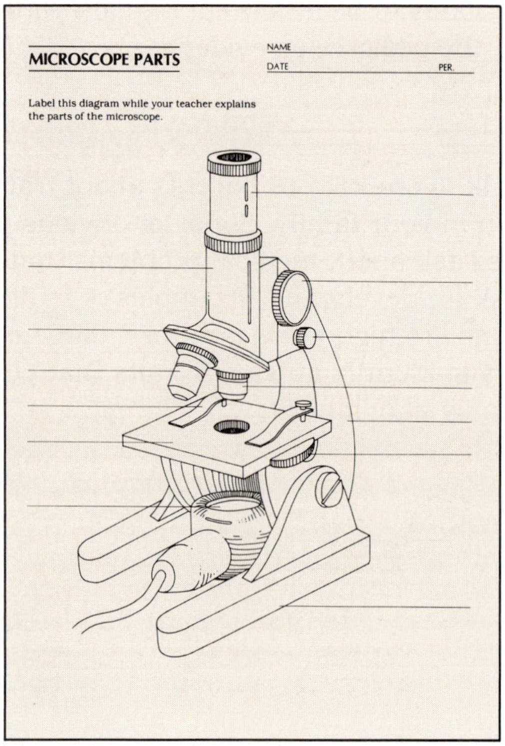

Figure A–1

Figure A-2

base. (See Figure A–2.) Place the microscope near the center of the table so it cannot be bumped accidently.

2. Each of you will have a task. Decide who will do what.
   checker—reads directions, makes sure learner understands procedure
   learner—listens to directions and does as instructed, asks questions if something is unclear

   You will switch roles later so you will get to work in both roles.

3. Take turns completing the first part of the checklist.
   a. The *checker* should read from the *learner's* checklist.
   b. While the *checker* reads, the *learner* follows the directions.
   c. When the *learner* has successfully completed one of the steps, the *checker* initials that step and reads the next one.
   d. When the *learner* has completed the first section, the *checker* should sign the *learner's* paper.
   e. Trade roles and repeat this procedure.

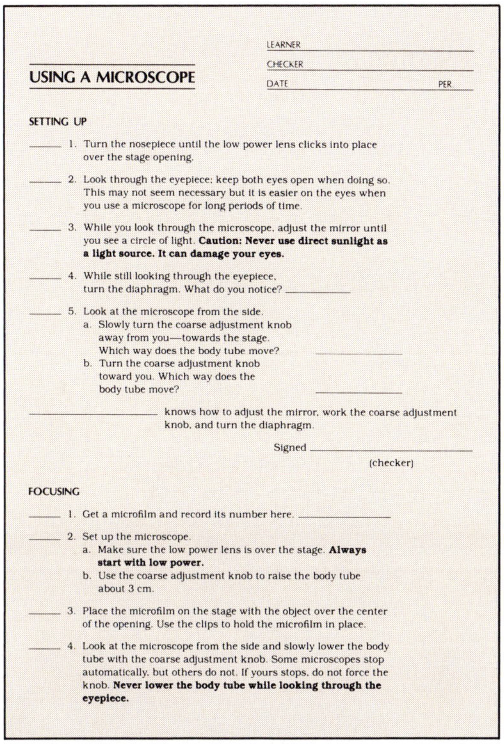

USING A MICROSCOPE

LEARNER ______
CHECKER ______
DATE ______ PER. ______

SETTING UP

____ 1. Turn the nosepiece until the low power lens clicks into place over the stage opening.

____ 2. Look through the eyepiece; keep both eyes open when doing so. This may not seem necessary but it is easier on the eyes when you use a microscope for long periods of time.

____ 3. While you look through the microscope, adjust the mirror until you see a circle of light. **Caution: Never use direct sunlight as a light source. It can damage your eyes.**

____ 4. While still looking through the eyepiece, turn the diaphragm. What do you notice? ______

____ 5. Look at the microscope from the side.
a. Slowly turn the coarse adjustment knob away from you—towards the stage. Which way does the body tube move? ______
b. Turn the coarse adjustment knob toward you. Which way does the body tube move? ______

______ knows how to adjust the mirror, work the coarse adjustment knob, and turn the diaphragm.

Signed ______ (checker)

FOCUSING

____ 1. Get a microfilm and record its number here. ______

____ 2. Set up the microscope.
a. Make sure the low power lens is over the stage. **Always start with low power.**
b. Use the coarse adjustment knob to raise the body tube about 3 cm.

____ 3. Place the microfilm on the stage with the object over the center of the opening. Use the clips to hold the microfilm in place.

____ 4. Look at the microscope from the side and slowly lower the body tube with the coarse adjustment knob. Some microscopes stop automatically, but others do not. If yours stops, do not force the knob. **Never lower the body tube while looking through the eyepiece.**

Figure A–3 (Checklist, page 1)

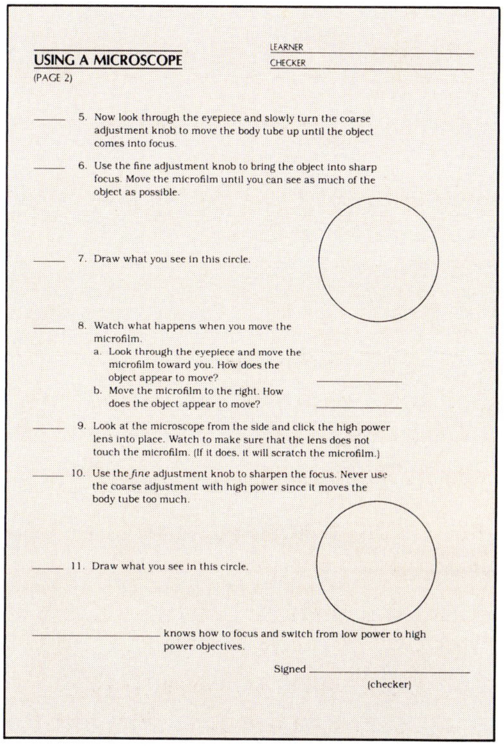

USING A MICROSCOPE
(PAGE 2)

LEARNER ______
CHECKER ______

____ 5. Now look through the eyepiece and slowly turn the coarse adjustment knob to move the body tube up until the object comes into focus.

____ 6. Use the fine adjustment knob to bring the object into sharp focus. Move the microfilm until you can see as much of the object as possible.

____ 7. Draw what you see in this circle.

____ 8. Watch what happens when you move the microfilm.
a. Look through the eyepiece and move the microfilm toward you. How does the object appear to move? ______
b. Move the microfilm to the right. How does the object appear to move? ______

____ 9. Look at the microscope from the side and click the high power lens into place. Watch to make sure that the lens does not touch the microfilm. (If it does, it will scratch the microfilm.)

____ 10. Use the *fine* adjustment knob to sharpen the focus. Never use the coarse adjustment with high power since it moves the body tube too much.

____ 11. Draw what you see in this circle.

______ knows how to focus and switch from low power to high power objectives.

Signed ______ (checker)

Figure A–3 (Checklist, page 2)

4. When you both know how to set up a microscope, switch roles again and complete the focusing section of the checklist.
5. When both you and your partner have completed the checklist, quiz each other with the following questions:

   Why is it important to look at the microscope from the side while lowering the body tube with the coarse adjustment knob?

   When can you see more of an object—under low power or high power?

   Why are there two adjustment knobs on the microscope?

   Why are you supposed to start with low power?

   Can you answer the 3 questions asked at the beginning of the activity?

Microscope Rules—Summary

Always

1. Carry a microscope with two hands.
2. Place the microscope near the center of the table.
3. Watch from the side when you lower the body tube with the coarse adjustment knob.
4. Focus with low power, then switch to high power.
5. Keep both eyes open when looking through the microscope.
6. Use *lens* paper to clean microscope lenses.

Never

Use direct sunlight for a light source.

A–2 MICROSCOPE SKILLS

Now that you know how to set up your microscope, you are ready to develop your microscope skills. There are two new terms that you will want to know when you use your microscope. These are **field** and **total magnification.**

The **field** of view is the area you see when you look through a microscope. A circle such as the one on the right is used to represent the field.

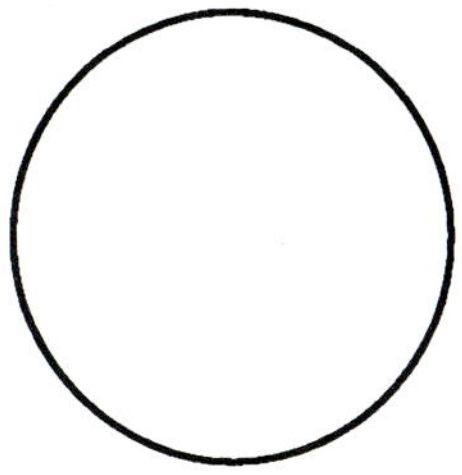

Total magnification is the number of times an object is enlarged when you look at it through the microscope. When you look through your microscope, the lens in the eyepiece magnifies the object and so does the lens on the nosepiece. To find total magnification, you multiply the magnification of the two lenses.

$$\text{eyepiece magnification} \times \text{low power lens magnification} = \text{total magnification at low power}$$

The number of times a lens can magnify is printed on the lens itself. Suppose the eyepiece magnifies 5 times and the low power lens also magnifies 5 times. (See the drawing on the right.) The total magnification would be 25 times.

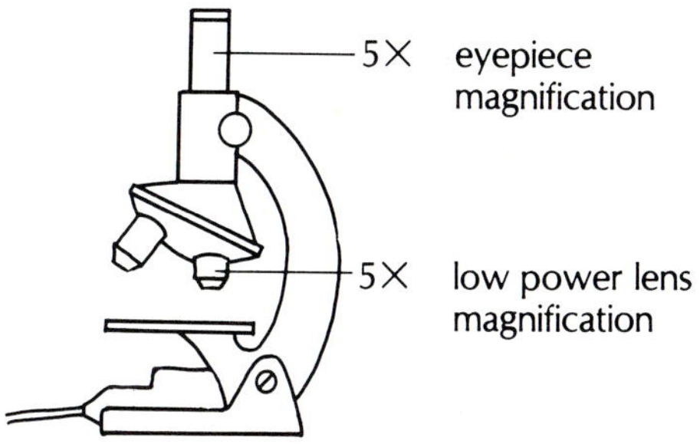

- What is the magnification of your microscope?
- What should you include in every microscope drawing?

Work through this activity to find the answers to these two focusing questions.

MATERIALS

microscope
microfilms #2 and #3
clear plastic metric ruler
worksheet: ***Microscope Skills***
(Figure A–6)

MICROSCOPE SKILLS

NAME ______________________

DATE ______________ PER ______

PART A. SCOPE WORK

1. Use microfilm #2 to check your focusing skills. Five terms are listed on the microfilm. You will observe each of these things with your microscope sometime during the year. List the five terms here. ______ ______ ______ ______ ______

2. Look at microfilm #3 under low power.
   a. Sketch what you see in the circle to the right.
   b. Make sure your drawing includes everything it should.

3. Find out how many times an object is magnified under low power with your microscope.

   ________ × ________ = ________
   low power magnification × eyepiece magnification = total magnification

4. Still using microfilm #3, switch to high power. Sketch what you see in the circle at the right.

5. Figure out the magnification under high power.

   ________ × ________ = ________
   high power magnification × eyepiece magnification = total magnification

Figure A–6 (Worksheet, page 1)

MICROSCOPE SKILLS
(PAGE 2)

NAME ______________________

6. Compare your two drawings.
   a. What is the main difference between the two? ______________________
   b. How many times larger is the object under high power compared with low power? ________

7. Get a clear plastic ruler that is divided into centimeters (cm) and millimeters (mm).
   a. Look at the ruler. How many mm are in a cm? ________
   b. List at least 2 objects that might be about 1 mm in length, width, or height.
      ________ ________
   c. Sketch what the ruler looks like under low power.
   d. Observe the measurement markings on the ruler under low power. How many mm can you see at one time? ________
   e. Switch to high power and sketch what you see.
   f. How many mm can you see at one time under high power? ________

8. Look back at your sketches of microfilm #3. (These are the answers to questions 2 and 4.) How long do you think that object is? Refer to question 7 to help you answer this question. ________

PART B. SEATWORK

1. How well did you learn the microscope rules? Fill in the word that is missing from each of these rules.
   a. Always carry a microscope with ________ ________.
   b. Place a microscope near the ________ of the table.

Figure A–6 (Worksheet, page 2)

MICROSCOPE SKILLS
(PAGE 3)

NAME ______________________

c. Watch from the ________ when you lower the body tube.

d. Always focus with the ________ ________ lens first.

e. Keep ________ eyes ________ when you look through the microscope.

f. Use ________ ________ to clean microscope lenses.

g. Never use direct ________ for a light source.

2. The chart below will help you review the parts of a microscope. Write the name of each of the following parts by its function.

arm	diaphragm	nosepiece	coarse adjustment knob
stage	eyepiece	low power lens	fine adjustment knob
base	mirror or light	high power lens	body tube
clips			

PART	FUNCTION
________	a. use for high magnification (40×, 45×, 65×)
________	b. holds slide in place on stage
________	c. controls amount of light that enters body tube
________	d. you look through it; it magnifies the object
________	e. use to carry the microscope and support body tube
________	f. supports the microscope
________	g. holds eyepiece and nosepiece
________	h. moves body tube quickly when focusing
________	i. holds slide; allows light to pass through slide
________	j. moves body tube slightly; use for fine focusing
________	k. holds low and high power lenses
________	l. directs light through diaphragm
________	m. use for low magnification (5×, 10×)

3. This square is 1 mm × 1 mm. Draw what it would look like if it were magnified 10 times.

Figure A–6 (Worksheet, page 3)

MICROSCOPE SKILLS
(PAGE 4)

NAME ______________________

4. Calculate the magnification for these three microscopes.

   a. eyepiece magnifies 10×
      lower power lens magnifies 15×
      total magnification = ________

   b. eyepiece magnifies 10×
      high power lens magnifies 30×
      total magnification = ________

   c. If the eyepiece of a microscope magnifies 10 times and the high power lens magnifies 50 times, what is the total magnification of this microscope when using high power? ________

5. Here is a picture of a 5 cm ruler. How long is the line underneath the ruler?

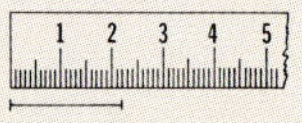

   ________ cm

   or ________ mm

6. Here is a sketch of the mm marks on a ruler as seen under the low power lens of a microscope.

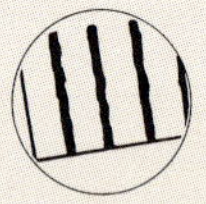

   a. How many mm wide is the field of view? ________
   b. What is the longest an object could be and still allow you to see the whole thing? ________
   c. Imagine that you are using the same microscope to observe cells. This is what you see. About how wide is one of these cells? ________

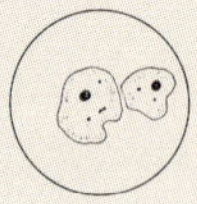

CHALLENGE

If you finish this worksheet early, ask your teacher for a microscope slide of cells. Try to figure out the size of one cell. (Look at your answer for ***Part A, #7*** to help you answer this question.)

Figure A–6 (Worksheet, page 4)

PROCEDURE

There are two parts to this activity. You need a microscope to do Part A; all you need to do Part B is a pencil. It does not matter which part you do first.

As you work, you will need to make microscope drawings. Memorize these five steps and follow them each time you make a microscope drawing.

A Microscope Drawing

1. Neatly sketch the object so it takes the same amount of space in the circle as the actual object takes in your field of view.
2. Include enough detail so someone else can recognize the exact object.
3. Title the drawing so the viewer will know what you were looking at when you made it.
4. Label everything you recognize on your drawing.
5. Record the total magnification by the drawing.

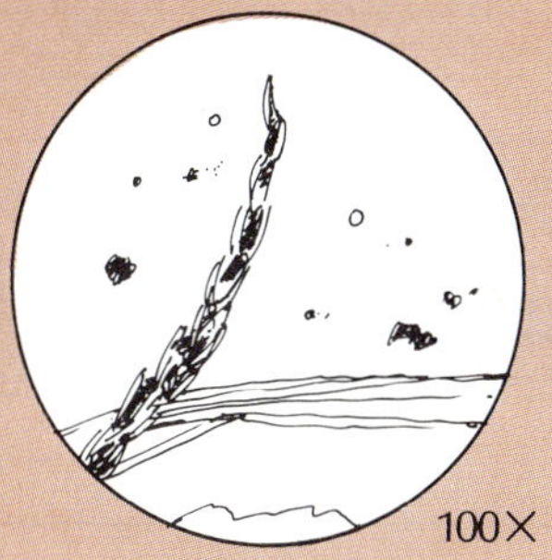

water from puddle

A-3 NUTRIENTS IN FOODS CHART*

MILK GROUP

MILK GROUP	Calories	Protein g	Carbohydrate g	Fat g	Vitamin A IU	Vitamin C mg	Thiamin (B_1) mg	Riboflavin (B_2) mg	Niacin mg	Calcium mg	Iron mg
Buttermilk, 1 cup	99	8	12	2	81	2	0.08	0.38	0.1	285	0.1
Cheese, American, 1 oz, 1 slice	106	6	–	9	343	0	0.01	0.10	Trace	174	0.1
Cheese, Cheddar, 1 oz, 1 slice	114	7	–	9	300	0	0.01	0.11	Trace	204	0.2
Cheese, Cottage, ½ cup	109	13	3	5	171	Trace	0.02	0.17	0.1	63	0.2
Cheese, Swiss, 1 oz, 1 slice	107	8	1	8	240	0	0.01	0.10	Trace	272	0.1
Cocoa, ¾ cup	164	7	19	7	239	2	0.08	0.33	0.3	224	0.6
Ice Cream, Vanilla, ½ cup, ¼ pint	135	2	16	7	272	–	0.03	0.17	0.1	88	0.1
Milk, 1 cup	150	8	11	8	307	2	0.09	0.40	0.2	291	0.1
Milk Chocolate, 1 cup	208	8	26	8	302	2	0.09	0.41	0.3	280	0.6
Milk, Lowfat (2%), 1 cup fortified with vitamin A	121	8	12	5	500	2	0.10	0.40	0.2	297	0.1
Milk, Skim, 1 cup fortified with vitamin A	86	8	12	–	500	2	0.09	0.34	0.2	302	0.1
Milkshake, Chocolate, 10.6 oz container	356	9	63	8	258	0	0.14	0.67	0.4	396	0.9
Pudding, Chocolate, ½ cup	161	4	30	4	169	Trace	0.03	0.20	0.1	133	0.4
Yogurt, Strawberry, 1 cup	225	9	42	3	111	1	0.08	0.37	0.2	314	0.1

MEAT GROUP

MEAT GROUP	Calories	Protein g	Carbohydrate g	Fat g	Vitamin A IU	Vitamin C mg	Thiamin (B_1) mg	Riboflavin (B_2) mg	Niacin mg	Calcium mg	Iron mg
Bacon, 2 slices	92	5	1	8	0	–	0.08	0.05	0.8	2	0.5
Beans, Refried, ½ cup	142	9	26	1	–	–	–	–	–	5	–
Beef, Roast, 3 oz	182	26	0	8	17	–	0.04	0.20	3.9	11	3.2
Beef Liver, 3 oz	195	23	5	9	45,417	23	0.22	3.56	14.0	9	7.5
Bologna, 1 slice	86	3	–	8	0	–	0.05	0.06	0.7	2	0.5
Chicken, Fried, 3 oz	201	26	2	9	148	–	0.06	0.38	5.9	12	2.0
Egg, Fried, large	83	5	1	6	286	0	0.03	0.13	Trace	26	0.9
Egg, Hard-cooked, large	79	6	1	6	260	0	0.04	0.14	Trace	28	1.0
Egg, Scrambled, large	95	6	1	7	311	.13	0.04	0.16	Trace	47	0.9
Frankfurter, 2 oz	172	7	1	15	–	–	0.09	0.11	1.4	3	0.9
Ham, Baked, 3 oz	179	26	0	8	0	–	0.56	0.26	4.9	11	3.2
Meat Loaf, 3 oz	230	15	13	12	106	Trace	0.27	0.24	3.4	68	2.4
Meat Patty, 3 oz	186	23	0	10	17	–	0.08	0.20	5.1	10	3.0
Peanut Butter, 2 tbsp	186	9	6	16	–	0	0.04	0.04	5.0	20	0.6
Peanuts, Salted, ¼ cup	211	9	7	18	–	0	0.12	0.05	6.2	27	0.8
Peas, Blackeye (immature), ½ cup	134	10	22	1	434	21	0.37	0.14	1.7	30	2.6
Peas, Blackeye (mature), ½ cup	94	6	17	–	12	–	0.20	0.05	0.5	21	1.6
Perch, Fried, Breaded, 3 oz	193	16	6	11	–	–	0.09	0.09	1.5	28	1.1
Pork Chop, 3 oz	308	21	0	24	0	–	0.82	0.24	4.9	9	2.7
Sausage, 2 links	135	5	–	13	0	–	0.22	0.10	1.1	2	0.7
T-Bone Steak, 3⅓ oz	212	29	0	10	19	–	0.08	0.22	5.6	11	3.5
Tuna, 3 oz	168	25	0	7	68	–	0.04	0.10	10.1	7	1.6

*REPRINTED WITH PERMISSION FROM "FOOD POWER – A COACH'S GUIDE TO IMPROVING PERFORMANCE," 1983, NATIONAL DAIRY COUNCIL, ROSEMONT, IL. 60018

FRUIT-VEGETABLE GROUP

	Calories	Protein g	Carbohydrate g	Fat g	Vitamin A IU	Vitamin C mg	Thiamin (B_1) mg	Riboflavin (B_2) mg	Niacin mg	Calcium mg	Iron mg
Apple, medium	80	–	20	1	124	6	0.04	0.03	0.1	10	0.4
Applesauce, ½ cup	116	–	30	–	51	1	0.03	0.01	Trace	5	0.6
Apricots, Dried, 4 halves	39	1	10	–	1635	2	Trace	0.02	0.5	10	0.8
Asparagus, 4 spears, ½ cup	12	1	2	–	540	16	0.10	0.11	0.8	13	0.4
Banana, medium	101	1	26	–	226	12	0.06	0.07	0.8	10	0.8
Beans, Green, ½ cup	16	1	3	–	338	8	0.04	0.06	0.3	31	0.4
Beans, Lima, ½ cup	94	7	17	–	238	14	0.15	0.09	1.1	40	2.1
Broccoli, stalk, ½ cup	20	2	4	–	1938	70	0.07	0.16	0.6	68	0.6
Cabbage, ⅙ head, ½ cup	13	1	3	–	87	17	0.01	0.01	0.1	30	0.2
Cantaloupe, ¼ medium	29	1	7	–	3273	32	0.04	0.03	0.6	13	0.4
Carrots, ½ cup	22	1	5	–	7613	4	0.04	0.04	0.4	24	0.4
Carrot Sticks, 5″ carrot	21	1	5	–	5500	4	0.03	0.03	0.3	19	0.4
Cauliflower, ½ cup	13	1	3	–	36	33	0.05	0.05	0.4	13	0.4
Celery Sticks, 8″ stalk	10	1	2	–	136	5	0.02	0.02	0.2	22	0.2
Coleslaw, ½ cup	82	1	3	8	91	16	0.03	0.03	0.2	25	0.2
Corn, ½ cup	70	2	16	1	291	3	0.02	0.04	0.8	4	0.4
Corn, 5″ ear	114	4	26	1	500	11	0.15	0.13	1.8	4	0.8
Fruit Salad, ½ cup apple, orange, banana, lettuce	99	2	25	1	530	44	0.11	0.08	0.7	45	0.9
Grapefruit, Pink, ½ medium	48	1	13	–	94	45	0.05	0.02	0.2	19	0.5
Grapes, ½ cup	48	–	12	–	71	3	0.04	0.02	0.2	9	0.3
Greens, ½ cup mustard greens, kale, turnip greens	17	2	3	–	5306	36	0.08	0.13	0.4	104	1.4
Lettuce, ⅙ head, ½ cup	10	1	2	–	250	5	0.05	0.05	0.2	15	0.4
Okra, 4 pods, ½ cup	12	1	3	–	208	9	0.06	0.08	0.4	39	0.2
Orange, medium	65	1	16	–	263	66	0.13	0.05	0.5	54	0.5
Orange Juice, ½ cup	56	1	13	–	249	56	0.11	0.01	0.4	11	0.1
Peaches, ½ cup	100	1	26	–	550	4	0.01	0.03	0.8	5	0.4
Pear, medium	101	1	25	1	33	7	0.03	0.07	0.2	13	0.5
Peas, Green, ½ cup	54	4	9	–	480	10	0.22	0.07	1.4	15	1.5
Pineapple, large slice	90	–	24	–	61	9	0.10	0.02	0.2	13	0.4
Potato, Baked, large	132	4	30	–	Trace	28	0.14	0.06	2.4	13	1.0
Potatoes, Boiled, 2 small	79	2	18	–	Trace	20	0.11	0.04	1.5	7	0.6
Potatoes, French-Fried, 20 pieces	233	4	31	11	Trace	18	0.11	0.07	2.6	13	1.1
Potatoes, Mashed, ½ cup	63	2	13	1	20	10	0.08	0.05	1.0	23	0.4
Potato, Sweet, ½ medium	78	1	18	–	4455	12	0.05	0.04	0.4	22	0.5
Raisins, 4½ tbsp	123	1	33	–	9	Trace	0.05	0.03	0.2	26	1.5
Squash, Summer, ½ cup	16	1	3	–	462	12	0.05	0.08	0.8	26	0.4
Squash, Winter, ½ medium, ½ cup	56	2	14	–	1435	13	0.05	0.13	0.7	40	1.1
Strawberries, ½ cup	28	1	6	–	45	44	0.02	0.05	0.4	16	0.7
Tomato, ½ medium	22	1	5	–	900	23	0.06	0.04	0.7	13	0.5
Tomato Juice, ½ cup	26	1	5	–	972	19	0.06	0.04	1.0	9	1.1
Tossed Salad, ¾ cup lettuce, green pepper, radish, carrot	13	1	3	–	1380	26	0.03	0.04	0.3	26	0.6
Watermelon, 1 cup	52	1	13	–	1180	14	0.06	0.06	0.4	14	1.0

GRAIN GROUP

GRAIN GROUP	Calories	Protein g	Carbohydrate g	Fat g	Vitamin A IU	Vitamin C mg	Thiamin (B_1) mg	Riboflavin (B_2) mg	Niacin mg	Calcium mg	Iron mg
Bagel	165	6	28	2	30	0	0.14	0.10	1.2	9	1.2
Biscuit, Baking Powder, enriched	103	2	13	5	Trace	Trace	0.08	0.08	0.8	34	0.4
Bread, White, slice, enriched	61	2	12	1	Trace	Trace	0.09	0.06	0.8	19	0.6
Bread, Whole Wheat, slice	55	2	11	1	Trace	Trace	0.06	0.03	0.6	22	0.5
Cornbread, 2½" × 3", enriched	191	6	30	5	264	1	0.14	0.20	0.9	93	1.2
Cornflakes, ¾ cup	72	2	16	–	0	0	0.08	0.02	0.4	3	0.3
Crackers, Graham, 2	54	1	10	1	0	0	0.01	0.03	0.2	6	0.2
Crackers, Saltines, 5	60	1	10	2	0	0	Trace	0.01	0.1	3	0.2
Hominy Grits, ½ cup, enriched	62	2	14	–	74	0	0.05	0.04	0.5	1	0.4
Noodles, Egg, ½ cup, enriched	100	3	19	1	56	0	0.11	0.06	1.0	8	0.7
Oatmeal, ½ cup	66	2	12	1	0	0	0.10	0.02	0.1	11	0.7
Pancake, 4" diameter, enriched	61	2	9	2	68	Trace	0.06	0.08	0.3	58	0.3
Rice, ½ cup	112	2	25	–	0	0	0.11	0.07	1.0	10	0.9
Roll, Frankfurter, enriched	119	3	21	2	Trace	Trace	0.16	0.10	1.3	30	0.8
Roll, Hamburger, enriched	119	3	21	2	Trace	Trace	0.16	0.10	1.3	30	0.8
Roll, Hard, enriched	156	5	30	2	Trace	Trace	0.20	0.12	1.7	24	1.2
Toast, White, slice	61	2	12	1	Trace	Trace	0.09	0.06	0.8	19	0.6
Tortilla, Corn, 6" diameter, enriched	63	2	14	1	6	0	0.04	0.02	0.3	60	0.9
Waffles, 2, 3½" × 5½", enriched	130	4	17	5	109	Trace	0.09	0.14	0.6	113	0.6

COMBINATION FOODS (foods made with ingredients from more than one food group)

COMBINATION FOODS	Calories	Protein g	Carbohydrate g	Fat g	Vitamin A IU	Vitamin C mg	Thiamin (B_1) mg	Riboflavin (B_2) mg	Niacin mg	Calcium mg	Iron mg
Beans, Baked, Pork and Tomato Sauce, ½ cup	156	8	24	3	166	3	0.10	0.04	0.8	69	2.3
Beef and Vegetable Stew, 1 cup	209	15	15	10	2303	16	0.14	0.16	4.5	28	2.8
Chili Con Carne with Beans, 1 cup	333	19	31	15	150	–	0.08	0.18	3.3	80	4.3
Custard, Baked, ½ cup	152	7	15	7	464	Trace	0.05	0.25	0.1	148	0.5
Macaroni and Cheese, ½ cup	215	8	20	11	430	Trace	0.10	0.20	0.9	181	0.9
Pizza, Cheese, ¼ of 14" pie, enriched	354	18	43	13	945	12	0.38	0.49	3.8	332	2.7
Soup, Chicken Noodle, 1 cup	59	3	8	2	45	Trace	0.02	0.02	0.7	9	0.5
Soup, Cream of Tomato, 1 cup	173	7	23	7	1200	15	0.10	0.25	1.3	168	0.8
Spaghetti, Meat Balls and Tomato Sauce, 1 cup	332	19	39	12	1587	22	0.25	0.30	4.0	124	3.7
Taco, Beef	216	17	15	10	352	4	0.10	0.19	3.0	174	2.6

"OTHERS" Category (fats, sweets, alcohol)

"OTHERS" Category	Calories	Protein g	Carbohydrate g	Fat g	Vitamin A IU	Vitamin C mg	Thiamin (B_1) mg	Riboflavin (B_2) mg	Niacin mg	Calcium mg	Iron mg
Bar, Milk Chocolate, 1 oz	147	2	16	9	77	Trace	0.02	0.10	0.1	65	0.3
Beer 1½ cups	151	1	14	0	–	–	Trace	0.11	2.2	18	Trace
Butter, 1 tsp	36	–	–	4	153	0	Trace	Trace	Trace	1	Trace
Cake, Devil's Food, 1/16 of 9" cake	234	3	40	9	104	Trace	0.02	0.06	0.2	41	0.6
Cake, Sponge, 1/12 of 10" cake	196	5	36	4	297	Trace	0.03	0.09	0.1	20	0.8
Chocolate Syrup, 2 tbsp	93	1	24	1	Trace	0	0.01	0.03	0.2	6	0.6
Coffee, Black, ¾ cup	2	–	–	–	0	0	0	Trace	0.5	3	0.2

"OTHERS" Category (continued)	Calories	Protein g	Carbohydrate g	Fat g	Vitamin A IU	Vitamin C mg	Thiamin (B_1) mg	Riboflavin (B_2) mg	Niacin mg	Calcium mg	Iron mg
Cookie, Sugar, 3" diameter, enriched	89	1	14	3	22	Trace	0.04	0.04	0.4	16	0.3
Doughnut, Cake Type, enriched	125	2	17	6	26	Trace	0.07	0.07	0.5	13	0.4
French Dressing, 1 tbsp	66	–	3	6	–	–	–	–	–	2	0.1
Gelatin Dessert, ½ cup	71	2	17	0	–	–	–	–	–	–	–
Highball, 1½ oz whiskey with soda	97	0	–	0	–	–	–	–	–	–	–
Jelly, Currant, 1 tbsp	49	0	13	0	2	1	Trace	0.01	0	4	0.3
Mayonnaise, 1 tbsp	101	–	–	11	39	–	Trace	0.01	Trace	3	0.1
Pie, Apple, ⅙ of 9" pie, enriched	403	4	60	18	47	2	0.15	0.13	1.7	13	1.1
Popcorn, Plain, 1 cup	23	1	5	–	–	0	–	0.01	0.1	1	0.2
Potato Chips, 10 chips	114	1	10	8	Trace	3	0.04	0.01	1.0	8	0.4
Roll, Danish Pastry	274	5	30	15	202	Trace	0.05	0.10	0.5	33	0.6
Sherbet, Orange, ½ cup	135	1	29	2	93	2	0.02	0.04	0.1	52	0.2
Soft Drink, Cola, 1 cup	96	0	25	0	0	0	0	0	0	–	–
Sugar, 1 tsp	14	0	4	0	0	0	0	0	0	0	Trace
Wine, Rosé, 3½ oz	87	–	4	0	–	–	Trace	0.01	0.1	9	0.4

CREDITS

Many thanks to all the students, faculty, and others who volunteered to be in some of the photos used in this text. Special thanks to Hendrick Middle School, Plano, Texas, for making their facilities available to us.

Photographs

Unit Openers

1: Thomas Kitchin/Tom Stack & Associates; **2:** Globus Brothers, Inc./The Stock Market; **3:** Robert Huntzinger, Inc./The Stock Market; **4:** Comstock, Inc.; **5:** Peter Beck/ The Stock Market; **6:** Frank Cezus/ FPG, International; **7:** Dan McCoy/ The Stock Market.

Chapter 1

Opener: Carolina Biological Supply; **1-1(top left):** Alese & Mort Pechter/ The Stock Market, **(bottom left):** Breck P. Kent, **(right):** E. Nagele/ FPG, International; **1-4:** Stephen P. Parker/Photo Researchers, Inc.

Chapter 2

Opener: J. Zimmerman/FPG, International; **p. 20:** V. Mc Millan/ Visuals Unlimited; **2-2(top):** Allen Green/Photo Researchers, Inc., **(bottom):** Don Rutledge/Tom Stack & Associates; **p. 23:** Michael S. Thompson/Comstock; **2-3:** R. Stottlemeyer, Michigan Tech. Univ./ BPS; **2-4:** Judy Capra; **2-5:** Walt Anderson/Visuals Unlimited; **p. 32:** William Harlow, National Audobon Society/Photo Researchers, Inc.; **p. 33:** Ron Thomas/FPG International; **p. 42:** Photoworld/ FPG International; **2-8(top):** Superstock, Inc., **(bottom):** Harvey Lloyd/Peter Arnold, Inc.; **2-9:** Stan Osolinski/FPG International; **2-10:** The Bettmann Archive; **2-11:** © Bob Evans/Peter Arnold, Inc.; **2-12:** D. J. Wrobel, Monterey Bay Acquarium/BPS; **2-13:** John D. Cunningham/Visuals Unlimited; **2-14:** Photo courtesy of FMC Corporation; **2-16:** David O. Norris, EPO Biology, University of Colorado.

Chapter 3

Opener: Clyde H. Smith/Peter Arnold, Inc.; **p. 62:** Lisa Means; **p.66:** Calvin Larsen/Photo Researchers, Inc.; **3-2(left):** Leonard Lee Rue III, **(right):** Lynn M. Stone; **3-3(left):** C.A. Morgan/ Peter Arnold, Inc., **(right):** Ed Robinson/Tom Stack & Associates.; **3-4(left):** Tim Christie, **(right):** B. Evans/Peter Arnold, Inc.; **3-8(left):** Jany Sauvanet/Photo Researchers, Inc., **(top right):** Manfred Danegger/ Peter Arnold, Inc., **(bottom right):** Stephen Dalton/Photo Researchers, Inc.; **3-9:** Rod Planck/Tom Stack & Associates; **3-10:** Breck P. Kent; **3-11:** Leonard Lee Rue, III; **p. 82:** Jack Fields/Photo Researchers, Inc.; **p. 83:** Junebug Clark/Photo Researchers, Inc.; **p. 84:** Stephen J. Lang/Visuals Unlimited; **p. 85:** Ty and Julie Hotchkiss/Photo Researchers, Inc.; **3-12:** Lynn M. Stone; **3-13:** Farrell Grehan/Photo Researchers Inc..; **3-14:** John D. Cunningham/Visuals Unlimited; **3-15:** Robert Winslow/Tom Stack & Associates; **3-18:** Pat and Tom Leeson/Photo Researchers, Inc.; **3-19:** Breck P. Kent; **3-20:** George D. Lepp/Comstock.

Chapter 4

Opener: John D. Cunningham/ Visuals Unlimited; **4-8:** © C. Allan Morgan/Peter Arnold, Inc.; **4-9:** Jack K. Clark/Comstock; **4-10:** John D. Cunningham/Visuals Unlimited; **4-11(top)**: David M. Dennis/Tom Stack & Associates, **(bottom):** Marilyn Shaw; **4-13:** Richard P. Smith/Tom Stack & Associates; **4-18:** Bruce F. Molnia/ Terraphotographics/PBS.

Chapter 5

Opener: Frank M. Hanna/Visuals Unlimited; **5-11:** Joe McDonald/ Visuals Unlimited.

Chapter 6

Opener: Tom McHugh/Photo Researchers, Inc.; **6-3(left):** Steve McCutcheon/Visuals Unlimited, **(right):** Barbara J. Miller/Biological Photo Service.

Chapter 7

Opener: Howard J. Sochurek, Inc./ The Stock Market; **7-2:** Peggy and Yoram Kahana/Peter Arnold, Inc.; **7-8:** Don Valenti/Tom Stack & Associates; **p. 164:** James H. Carmichael, Jr.; **7-26:** The Bettmann Archive; **7-27:** Lisa Means photo with permission of Borden, Inc.

Chapter 8

Opener: James L. Shaffer; **p. 184, p. 190, 8-4, 8-5, 8-6, p. 194:** Lisa Means; **8-7:** (1) © 1988 Time Inc. Reprinted by permission. (2) From NEWSWEEK, Oct. 10, 1988. © 1988, Newsweek, Inc. All rights reserved. Reprinted by permission. (3) Reprinted with permission from The Toronto Star Syndicate. (4) *American Health Magazine* © 1987, David Groves; **8-8:** NASA; **8-11:** A. Copley/Visuals Unlimited; **8-12:** Arthur M. Siegelman/Visuals Unlimited; **8-15, 8-18, p. 211:** Lisa Means.

Chapter 9

Opener: Comstock, Inc.; **9-4:** William Patterson/Tom Stack & Associates; **9-5a:** Joel Arrington/ Visuals Unlimited, **b (top):** David M. Phillips/Visuals Unlimited, **b (bottom):** Gary A. Robinson/ Visuals Unlimited, **c:** John D. Cunningham/Visuals Unlimited, **d:** David M. Phillips/Visuals Unlimited, **e:** K.G. Murti/Visuals Unlimited; **p. 221, 9-7:** Lisa Means; **p. 222:** Lester V. Bergman, NY; **p. 228:** Lisa Means; **9-14:** Lester V. Bergman, NY; **9-15:** Lisa Means.

Chapter 10

Opener: Superstock, Inc.; **10-2:** Lisa Means photo. KRAFT is a registered trademark of Kraft, Inc. Package reproduction used with permission by Kraft, Inc., Glenview, Illinois; **10-3:** Lisa Means photo. Used with the permission of General Mills, Inc.; **10-8, 10-9, 10-10:** Lisa Means; **10-15:** Courtesy of Maryknoll Missioners; **10-17, 10-18:** Lisa Means; **10-20:** Lisa Means photo. Pepperidge Farm is a registered trademark. Used with the permission of Pepperidge Farm, Inc.; **10-21, 10-26, 10-29:** Lisa Means; **10-30:** Lisa Means photo. Courtesy of Keebler Co.; **10-31:** Lisa Means photo. JELL-O is a registered trademark of General Foods Corporation; **10-33:** Lisa Means photo. Courtesy of Kentucky Fried Chicken, Wendy's International Inc., Pizza Hut, "Have It Your Way®" Nutrition Guide, Burger King Corporation; **p. 279:** Robert Lee/ Photo Researchers, Inc.

Chapter 11

Opener: Manfred Kage/Peter Arnold, Inc.; **11-4:** D.C. Lowe/ MediChrome/The Stock Shop; **11-10:** Photoworld/FPG International; **11-11:** The Bettmann Archive; **11-30:** Lisa Means photo. POST BRAN FLAKES is a registered trademark of General Foods Corporation.

Chapter 12

Opener: Museum of Science and Industry, Chicago, IL; **12-3:** Lisa Means; **12-4:** Lynn M. Stone; **12-5,**

12-6: Lisa Means; **12-9:** C. Yokochi and J.W. Rohen *Photographic Anatomy of the Human Body*, 2nd edition, Igaku-Shoin Ltd. 1978; **p. 329:** The Bettmann Archive; **p. 342:** Lynn M. Stone; **12-21:** Chuck Brown/Photo Researchers, Inc.; **12-22:** Lennart Nilsson, *Behold Man*, Little, Brown and Company; **12-24:** J.F. Gennaro and L.R. Grillone/Photo Researchers, Inc.; **12-27:** Lisa Means; **12-33(left):** Chuck Brown/Photo Researchers, Inc., **(right):** Sloop-Ober/Visuals Unlimited; **12-38:** Richard Humbert/Biological Photo Service; **12-40:** Photo Researchers, Inc.; **12-41:** Lee Moskowitz/ The Stock Shop; **p. 365:** The Bettmann Archive.

Chapter 13

Opener: Dan McCoy/Rainbow; **13-2:** Lisa Means; **13-4:** Vernon C. Bailey/Photo Finders Int.; **13-5:** Vernon C. Bailey, II/Photo Finders International; **13-10:** Visuals Unlimited.

Chapter 14

Opener: Zeva Delbaum/Peter Arnold, Inc.; **14-8:** Lisa Means; **14-9:** Baylor College of Medicine.

Chapter 15

Opener: Jeffry Myers/FPG, International; **15-8:** Lisa Means; **15-18:** Photo Researchers, Inc.; **15-20:** Jeff Reed/MediChrome/The Stock Shop; **p. 415:** Frans Lanting/ Minden Pictures; **p. 416:** Lisa Means; **p. 417:** Yoram Lehmann/ Peter Arnold, Inc.

Chapter 16

Opener: Martin Dohrn/Photo Researchers, Inc.; **16-1:** James H. Carmichael, Jr.; **16-3:** John Cunningham/Visuals Unlimited; **16-4:** Lisa Means; **16-5:** Steve Raymer © 1978 National Geographic Society; **p. 434:** James H. Carmichael, Jr.; **p. 436:** Will McIntyre/Photo Researchers, Inc.; **p. 437:** UPI/Bettmann Newsphotos; **16-14:** Richard P. Smith/Tom Stack & Associates.

Chapter 17

Opener: Science Photo Library/ Photo Researchers, Inc.; **p. 450:** Lisa Means; **17-4:** Will and Deni McIntyre/Photo Researchers, Inc.; **17-5:** Photo by David L. Brill, © National Geographic Society. Housed, National Musesums of Kenya, Nairobi; **17-7(top):** Photo © 1985 David L. Brill, Housed, National Museums of Kenya, Nairobi, **(bottom):** Photo © 1985 David L. Brill, Housed, Musee de l'Homme, Paris; **17-9:** Photo by David L. Brill, © National Geographic Society.

Chapter 18

Opener: Ed Gallucci/The Stock Market; **18-1, 18-2, 18-3:** Lisa Means; **p. 470:** Jim Olive/Peter Arnold, Inc.; **18-4:** Kent Wood/Peter Arnold, Inc.

Chapter 19

Opener: Comstock; **p. 490, 19-2:** Lisa Means; **p. 494:** Peter A. Simon/The Stock Market; **p. 507, 508, 510:** Lisa Means.

Chapter 20

Opener: David M. Phillips/Visuals Unlimited; **20-3:** Lisa Means; **20-8:** David M. Phillips/Visuals Unlimited; **20-9:** © Manfred Kage/Peter Arnold, Inc.

Chapter 21

Opener: James R. Levin/FPG, International; **21-1:** © Garbis Kerimian/Peter Arnold, Inc.; **21-6:** Douglas M. Munnecke/Biological Photo Service; **21-7:** Tom Tracy/ MediChrome/The Stock Shop; **21-8(right):** Henley & Savage, Inc./ The Stock Market, **(left):** Ed Lettau/ Peter Arnold, Inc.

Chapter 22

Opener: Brian Parker/Tom Stack & Associates; **p. 557:** Lisa Means; **22-10, 22-11, 22-12, 22-20:** Cytogenetics Laboratory, University of Colorado Health Sciences Center, Denver, Colorado.

Chapter 23

Opener: James L. Shaffer; **23-1(top left):** Comstock, **(bottom left):** Jeffry Myers/FPG, International, **(right):** Sybil Shelton/Peter Arnold, Inc.; **23-4:** The Bettmann Archive; **23-5:** Photoworld/FPG International; **23-6:** William J. Weber/Visuals Unlimited; **23-10:** Larry Mulvehill/Photo Researchers Inc.; **23-12:** Cytogenetics Laboratory, University of Colorado Health Sciences Center, Denver, Colorado; **23-15:** James A. Prince/ Photo Researchers, Inc.; **23-20:** Dennie Cody/FPG International; **23-21:** Rob Nelson/Picture Group.

Art

Key: Top (t); Center (c); Bottom (b); Left (l); Right (r).

Scott Bodell: 432, 433, 438, 439, 441, 442, 444, 446, 447, 449, 454, 456, 468, 563, 564, 567, 569, 570, 572, 578, 584, 587, 619.

Lisa Key: 172 (r), 210, 212.

Lyle Miller: 12, 13, 15, 34, 57, 63, 64, 72, 75, 78, 82, 83, 84, 85, 94, 102, 103, 104, 106, 108, 113, 114, 115, 116, 120, 121, 122, 129, 139, 141, 145, 152, 153, 154, 156(b), 161(t), 166, 173, 177, 181, 189, 196, 224, 226, 244, 271, 307, 308, 309(b), 314, 321, 327, 348, 349, 351, 352(l), 354, 360, 368, 380(l), 381, 382, 395, 404, 405, 419, 443, 462, 471, 487, 511, 560, 561, 608, 611, 612, 613.

Mary Nelson: 105, 111, 130, 149, 155, 156(t), 159, 165, 167, 168, 169, 172(l), 174, 187, 188, 197, 200, 201, 208, 209, 225, 227, 230, 231, 266, 282, 283, 285, 290, 297, 298, 301, 302, 303, 311, 312, 332, 334, 335, 337, 339, 340, 341, 344, 345, 346, 347, 373, 375, 376, 385, 396, 397, 398, 401, 402, 403, 411, 492, 493, 497, 500, 501, 502, 503, 513, 516, 517, 521, 522, 523, 524, 526, 527, 528, 529, 544, 545, 546, 547, 548, 559, 565, 566, 568, 576, 579, 580, 581(t), 582, 583, 625, 627, 629.

Seda Sookias-Maurer: 8, 9, 16, 21, 25, 35, 36, 38, 39, 40, 41, 42, 51, 53, 71, 73, 74, 92, 109, 125, 126, 127, 161(r), 163, 206, 207, 214, 215, 242, 252, 257, 261, 262, 286, 289, 295, 304, 305, 306, 309(t), 310, 333, 352(r), 355, 380(b), 393, 400, 409, 413, 426, 453, 465, 488, 489, 536, 581(b), 592, 593, 594, 600, 601, 602, 604, 606, 609, 614, 620, 621, 622, 623.

INDEX

Boldface number indicates page on which the word is defined.

D

H

I

J

K

L

M

N

O

P

R

S

W

X

Y